NATO Advanced Science Institutes Series

A series of edited volumes comprising multifaceted studies of contemporary scientific issues by some of the best scientific minds in the world, assembled in cooperation with NATO Scientific Affairs Division.

This series is published by an international board of publishers in conjunction with NATO Scientific Affairs Division

A Life Sciences **B Physics**	Plenum Publishing Corporation New York and London
C Mathematical and Physical Sciences	D. Reidel Publishing Company Dordrecht, Boston, and London
D Behavioral and Social Sciences **E Applied Sciences**	Martinus Nijhoff Publishers The Hague, Boston, and London
F Computer and Systems Sciences **G Ecological Sciences**	Springer Verlag Heidelberg, Berlin, and New York

Recent Volumes in Series B: Physics

Volume 89a—Electrical Breakdown and Discharges in Gases: Fundamental Processes and Breakdown
edited by Erich E. Kunhardt and Lawrence H. Luesen

Volume 89b—Electrical Breakdown and Discharges in Gases: Macroscopic Processes and Discharges
edited by Erich E. Kunhardt and Lawrence H. Luessen

Volume 90 —Molecular Ions: Geometric and Electronic Structures
edited by Joseph Berkowitz and Karl-Ontjes Groeneveld

Volume 91 —Integrated Optics: Physics and Applications
edited by S. Martellucci and A. N. Chester

Volume 92 —The Physics of Superionic Conductors and Electrode Materials
edited by John W. Perram

Volume 93 —Symmetries in Nuclear Structure
edited by K. Abrahams, K. Allaart, and A. E. L. Dieperink

Volume 94 —Quantum Optics, Experimental Gravitation, and Measurement Theory
edited by Pierre Meystre and Marlan O. Scully

Volume 95 —Advances in Laser Spectroscopy
edited by F. T. Arecchi, F. Strumia, and H. Walther

Advances in Laser Spectroscopy

Edited by

F. T. Arecchi

University of Florence and
National Institute of Optics
Florence, Italy

F. Strumia

University of Pisa
Pisa, Italy

and

H. Walther

University of Munich and
Max Planck Institute for Quantum Optics
Garching, Federal Republic of Germany

Plenum Press
New York and London
Published in cooperation with NATO Scientific Affairs Division

Proceedings of a NATO Advanced Study Institute on
Advances in Laser Spectroscopy,
which was the tenth Course of the Europhysics School of Quantum Electronics,
held July 26–August 7, 1981,
in San Miniato, Italy

Library of Congress Cataloging in Publication Data

NATO Advanced Study Institute on Advances in Laser Spectroscopy (1981: San Miniato, Italy)
Advances in laser spectroscopy.

(NATO advanced science institutes series. Series B, Physics; v. 95)
"Published in cooperation with NATO Scientific Affairs Division."
"Proceedings of a NATO Advanced Study Institute on Advances in Laser Spectroscopy, which was the tenth Course of the Europhysics School of Quantum Electronics, held July 26–August 7, 1981, in San Miniato, Italy"—Verso t.p.
Includes bibliographical references and index.
1. Laser spectroscopy—Congresses. I. Arecchi, F. T. II. Strumia, F. III. Walther, H. (Herbert), 1935– . IV. North Atlantic Treaty Organization. Scientific Affairs Division. V. Title. VI. Series.
QC454.L3N37 1981 535.5′8 83-4196
ISBN 0-306-41355-8

A Division of Plenum Publishing Corporation
233 Spring Street, New York, N.Y. 10013

Printed in the United States of America

PREFACE

This volume contains the lectures and seminars presented at the NATO Advanced Study Institute on "Advances in Laser Spectroscopy" the tenth course of the Europhysics School of Quantum Electronics, held under the supervision of the Quantum Electronics Division of the European Physical Society. The Institute was held at Centro "I Cappuccini" San Miniato, Tuscany, July 26, - August 7, 1981.

The Europhysics School of Quantum Electronics was started in 1970 with the aim of providing instruction for young researchers and advanced students already engaged in the area of quantum electronics or wishing to switch to this area from a different background. From the onset, the School has been under the direction of Prof. F. T. Arecchi, then at the University of Pavia, now at the University of Florence, and Dr. D. Roess of Siemens, Munich. In 1981, Prof. H. Walther, University of Munich and Max-Planck-Institut für Quantenoptik joined as director. Each year the Directors choose a subject of particular interest, alternating fundamental topics with technological ones, and ask colleagues specifically competent in a given area to take the scientific responsibility for that course.

The past courses were devoted to the following themes:

1) 1971: "Physical and Technical Measurements with Lasers"
2) 1972: "Nonlinear Optics and Short Pulses"
3) 1973: "Laser Frontiers: Short Wavelength and High Powers"
4) 1974: "Cooperative Phenomena in Multicomponent Systems"
5) 1975: "Molecular Spectroscopy and Photochemistry with Lasers"
6) 1976: "Coherent Optical Engineering"
7) 1977: "Coherence in Spectroscopy and Modern Physics"
8) 1979: "Lasers in Biology and Medicine"
9) 1980: "Physical Processes in Laser Material Interactions"

The first five courses were held in Erice, Sicily, at the Centre for Scientific Culture "Ettore Majorana", the next four were held at Villa Le Pianore, Camaiore, Tuscany. Starting this year, thanks to the generosity of Cassa di Risparmio di San Miniato, we have held the school in a conference center resulting from the restoration of an ancient convent in the countryside, very close to the small city of San Miniato, that is, just in the heart of Tuscany (equidistant from Florence, Pisa and Siena).

Prof. F. Strumia, University of Pisa, and Prof. H. Walther, University of Munich, undertook the scientific direction of the present course, selecting the specific topics and lectures.

During the ASI, we had the visit of a scientist from Moscow, Prof. V. S. Letokhov, who was in Tuscany as a joint guest of the University of Florence and Pisa. His lectures are also included in the proceedings.

We wish to express our appreciation to the NATO Scientific Affairs Division, whose financial support made this Institute possible. We also acknowledge the contributions of the following institutions:

Cassa di Risparmio di San Miniato
Comune di San Miniato
CISE
IBM Italia
Coherent GmbH
Spectraphysics GmbH

The Deutscher Akademischer Austauschdienst contributed with three fellowships for German participants. We finally thank Mrs. G. Ravini of CISE for her valuable assistance in the organization of the Institute, Miss A. Camnasio of Servizio Documentazione, CISE for their assistance during the course itself, and the staff of Cassa di Risparmio die San Miniato. Mrs. I. Arecchi for the organization of the social events, and finally Mrs. S. Böhm for help in the preparation of these proceedings.

We would like to thank Murray Sargent for preparing the subject index of this book. He convinced us that his "little" home computer and the PS Technical Word Processing System are of considerable help in doing this work.

F.T. Arecchi
F. Strumia
H. Walther

CONTENTS

DENSITY MATRIX EQUATIONS AND DIAGRAMS FOR HIGH RESOLUTION NON-LINEAR LASER SPECTROSCOPY: APPLICATION TO RAMSEY FRINGES IN THE OPTICAL DOMAIN

Christian J. Bordé

Laboratoire de Physique des Lasers, Associé au CNRS n°282
Université Paris-Nord
Avenue J.-B. Clément - 93430 - Villetaneuse, France

ABSTRACT

We present a semiclassical approach to the line shape problem in high resolution laser spectroscopy which includes, in a unified density matrix formalism, the influences of the beam geometry, of the molecular recoil and of the second-order Doppler effect. A fully covariant second-quantized extension of the formalism is outlined. We give also a general derivation of intensities for all non-linear processes using Racah algebra in Liouville space. We describe diagrammatic representations of the perturbative solutions of the equations suitable for all laser spectroscopy techniques. Simple topological rules characterize Doppler-free processes. The diagrams and their associated diagrammatic rules are illustrated by their applications to the computation of Ramsey fringes in the case of single photon, Doppler-free two-photon and saturation spectroscopies. Finally we show how light-shifts can be calculated by this method and find, as expected, that the shifts for Ramsey fringes are reduced by the ratio of the laser beam radius to the zone separation when compared to the shifts for the single zone signal.

INTRODUCTION

This paper presents basic equations and diagrammatic techniques adapted to the computation of the line shape in high resolution laser spectroscopy. High precision sub-Doppler techniques now require that the effects of the beam geometry (transit broadening,[1] curvature shift[1], Ramsey fringes), of the atomic recoil and of the second-order Doppler shift should be simultaneously taken into account. While writing this paper we mostly had in mind saturation methods but most considerations and equations are applicable to Doppler-free two-photon spectroscopy, CARS, beam spectroscopy and related sub-Doppler techniques. In a series of papers[1-3] the influence of the Gaussian laser beam geometry was investigated in saturation spectroscopy, two-photon spectroscopy and Ramsey interference fringes techniques. In two other papers[4,5] we studied the influence of a strong saturation field on the recoil splitting or shift in the case of plane monochromatic waves. In fact, in conditions where recoil effects can be observed, the transverse geometry of the beam plays a major role on the line shape. In the same conditions the transverse Doppler effect is also usually very important. For example, in the case of CH_4 at 3.39 μm, the natural width is of the order of 3 Hz, the recoil splitting is 2.16 kHz, the half-width contribution of the geometry can be reduced to 1 kHz and the second-order Doppler shift is 150 Hz[6]. For ^{40}Ca at 6573 Å, these numbers are 408 Hz, 23.1 kHz, 1 kHz and 1.1 kHz respectively[7,8]. Any realistic calculation of the line shape has therefore to include all these effects and will generally be feasible only with a computer. In the first part of this paper, we shall only discuss the equations which are the starting point for such a calculation. We recall the equations for a unimolecular density operator, both in Hilbert space and in Liouville space, and we review the various expressions for the induced macroscopic polarization, transition rate and absorbed laser power. The treatment of the molecular recoil requires the quantization of the translational motion of the center of mass of the molecules. Usually this implies the choice of a representation among the position $|\vec{r}\rangle$ representation, the momentum $|\vec{p}\rangle$ representation, the Wigner[9,10] or the Shirley-Stenholm[11] representations. For sub-Doppler techniques, equations are easier to understand and simpler to solve in the momentum representation[12,13]. Unfortunately it is much easier to deal with the spatial dependence of the electromagnetic field in the position representation. We show how to combine the advantages of both representations with a wave-packet approach that keeps, in the position representation, an image of the discrete comb structure existing in the momentum space. The equations are reduced to a set of coupled ordinary differential equations that can be solved by standard computer techniques or perturbation expansions. Then we examine the covariance of the density matrix equations in Lorentz transformations of the coordinates, to obtain the second-order Doppler shift for any choice of reference frame and this leads us to suggest tentatively a fully covariant formalism using second-quantization. The second part of the paper is devoted to diagrammatic representations of the integral form of the density matrix equations. First we introduce space-time domain diagrams to describe a sequence of density matrix elements at consecutive perturbation orders along the time axis, then we show that these sequences can also be displayed in an energy-momentum frame. Both kinds of diagrams help to classify and compare the various non-linear processes of laser spectroscopy, and simple diagrammatic rules can be used to write down automatically the corresponding susceptibilities. Furthermore a simple inspection of the energy-momentum domain diagrams tells whether the process is Doppler-free or not. We illustrate this point with the examples of the last part of the paper.

I - THE DENSITY MATRIX EQUATIONS

1 - The physical system, its density operator and the boundary conditions

We shall consider N identical molecules (or atoms) in a finite volume V illuminated by one or several laser beams. For the sake of simplicity we shall restrict our considerations to the case of a single laser frequency ω in the laboratory frame. The molecular system may actually be either gas in a cell or a molecular beam. A unimolecular density operator ρ can be defined by taking a trace over the N-1 other molecules and over the rest of the universe. We shall choose the normalization condition :

$$\mathrm{Tr}\rho = N$$

The average value of any additive quantity for the N molecules will be simply obtained by $\mathrm{Tr}(\rho A)$ where A is the corresponding operator for a single molecule. We shall furthermore restrict our attention to the few energy levels and to the part of the volume V directly affected by the electromagnetic field, and assume that the rest of the system is unperturbed by the field. To account for the coupling with this "reservoir", it is convenient to introduce an elementary pure case density operator $\rho(\psi,t_o,t)$ corresponding to molecules created at some initial time t_o in the state $|\psi\rangle$, that is, such that :

$$\rho(\psi,t_o,t_o) = |\psi\rangle\langle\psi| \qquad (1)$$

and we shall associate the formation rate $\lambda(\psi,t_o)$ with this particular boundary condition.

We shall usually assume that $|\psi\rangle$ is an eigenstate $|\alpha_o,\vec{p}_o\rangle$ of the Hamiltonian H_o and of the momentum operator $\vec{p}_{op}$ of the isolated molecule. In a more general theory, one could introduce an initial density operator which would not be diagonal in the $|\vec{p}_o\rangle$ representation to account for a non-uniform initial spatial distribution (this is the case of a molecular beam for example). This generalization offers no special difficulty but, to avoid cumbersome notations, we shall limit ourselves to the case of diagonal pumping. It is always possible to introduce later a slow dependence upon the spatial coordinates.

By summation over all possible initial times t_o we define:

$$\rho(\psi,t) = \int_{-\infty}^{t} dt_o\ \lambda(\psi,t_o)\rho(\psi,t_o,t) \qquad (2)$$

and :

$$\rho(t) = \sum_{\psi} \rho(\psi,t) \qquad (3)$$

2 - Basic equations for the density operator and the Liouville space formalism

The elementary operator $\rho(\psi,t_o,t)$ satisfies the equation :

$$i\hbar \frac{\partial\rho}{\partial t} = [H,\rho] + i\hbar R\rho \qquad (4)$$

where H is the total Hamiltonian for one molecule (H is the sum of an internal Hamiltonian H_o, of a kinetic energy term $\vec{p}^{\,2}_{op}/2M$ and of an interaction Hamiltonian V_{int}).

The previous equation differs from the usual Liouville - Von Neumann equation by a relaxation term. A "non-hermitian Hamiltonian" $H_o - i\hbar\Gamma/2$ would result in a simple formal expression of $R\rho$ in Hilbert space :

$$R\rho = - 1/2\,[\Gamma\rho + \rho\Gamma] \qquad (5)$$

but it has the consequence that $\gamma_{\alpha\beta} = 1/2(\gamma_\alpha + \gamma_\beta)$ which is usually not the case owing to correlations in the dephasing effect of collisions. We shall use the following more general relaxation model(*) :

$$\langle \alpha,\vec{p}\,|\,R\rho\,|\,\beta,\vec{p}\,'\rangle = -(\gamma_{\alpha\beta} + i\,\delta_{\alpha\beta})\langle\alpha,\vec{p}\,|\,\rho\,|\,\beta,\vec{p}\,'\rangle \qquad (6)$$

and the shifts $\delta_{\alpha\beta}$ will be included in the eigen-frequencies $\omega_{\alpha\beta} = (E_\alpha - E_\beta)/\hbar$. The relaxation operator cannot anymore be expressed as a Hilbert space operator ; R is called a superoperator and acts on ρ in a space where ρ is itself a vector, the so-called Liouville space[15] (**).

Equations (4) can therefore be written more simply with Liouville space notations :

$$i\,\frac{\partial\,|\rho\gg}{\partial t} = (L + i\,R)\,|\rho\gg \qquad (7)$$

(*) This model ignores elastic collisions which could be taken into account by a transfer term between $\vec{p}_o$ classes and described by a collision kernel $W_{\alpha\beta}(\vec{p}\,'_o \to \vec{p}_o)$[14,13]

(**) The Liouville space vector corresponding to the Hilbert space projector $|\,\alpha\rangle\langle\beta\,|$ is written $|\alpha\rangle|\beta\rangle^\dagger = |\alpha\beta^\dagger\gg$.

where L is the Liouvillian of the molecules(*).

An equation for $\rho(\psi,t)$ is easily derived from the previous ones :

$$i\hbar \frac{\partial \rho}{\partial t} = [H,\rho] + i\hbar R\rho + i\hbar\Lambda \tag{8}$$

where the pumping operator Λ is given by :

$$\Lambda(\psi,t) = \lambda(\psi,t)\,|\,\psi\rangle\langle\psi\,| \tag{9}$$

and in Liouville space notation :

$$i\,\frac{\partial\,|\,\rho\gg}{\partial t} = (L + iR)\,|\,\rho\gg + i\,|\,\Lambda\gg \tag{10}$$

with :

$$|\,\Lambda(\psi,t)\gg = \lambda(\psi,t)\,|\,\psi,\psi^{\dagger}\gg \tag{11}$$

and equivalent equations for $\rho(t)$ and $|\,\rho(t)\gg$ with partial or complete summation over statistical parameters of formation :

$$\Lambda(t) = \sum_{\psi}\Lambda(\psi,t) \quad \text{and} \quad |\,\Lambda(t)\gg = \sum_{\psi}|\,\Lambda(\psi,t)\gg$$

3 - The interaction representation

Very often it is simpler and formally more elegant to use the interaction representation :

$$\tilde{\rho} = \exp(iH_1 t/\hbar)\,\rho\,\exp(-iH_1 t/\hbar) \tag{12}$$

$$\tilde{V}_{int} = \exp(iH_1 t/\hbar)\,V_{int}\,\exp(-iH_1 t/\hbar) \tag{13}$$

$$\text{with } H_1 = H_o + \vec{p}^{\,2}_{op}/2M$$

This representation will be especially useful for the comparison with the covariant theory based on the Tomonaga-Schwinger equation.

The density operator equations become :

$$i\hbar\frac{\partial\tilde{\rho}}{\partial t} = \left[\tilde{V}_{int}, \tilde{\rho}\right] + i\hbar\tilde{R}\tilde{\rho} + i\hbar\tilde{\Lambda} \tag{14}$$

and in Liouville space notation :

$$\frac{\partial\,|\,\tilde{\rho}\gg}{\partial t} = |\,\tilde{\Lambda}\gg - i(\tilde{L}_{int} + i\tilde{R})\,|\,\tilde{\rho}\gg \tag{15}$$

(*) The Liouvillian L can be defined by $\hbar L|\,\rho\gg = |\,[H,\rho]\gg$ or as the following combination of tensor products $\hbar L = H\otimes 1^{\dagger} - 1\otimes H^{\dagger}$.

4 - Formal solution and integral form of the density operator equation

The previous equation can be formally integrated if we make use of the time-ordering operator T (Dyson's chronological operator)(*) :

$$|\,\tilde{\rho}(t)\gg \,=\int_{-\infty}^{t} dt_o T \exp\left\{-i\int_{t_o}^{t}(\tilde{L}_{int}(t')+i\tilde{R})dt'\right\}|\,\tilde{\Lambda}(t_o)\gg \qquad (16)$$

Apart from some special cases where the T operator may be ignored, this solution is difficult to use and the only possibility is to write an expansion in successive powers of the interaction Liouvillian :

$$|\,\tilde{\rho}(t)\gg \,=\int_{-\infty}^{t} dt_o\, e^{\tilde{R}(t-t_o)}|\,\tilde{\Lambda}(t_o)\gg$$

$$+(-i)\int_{-\infty}^{t} dt_o\int_{t_o}^{t} dt'\, e^{\tilde{R}(t-t')}\tilde{L}_{int}(t')e^{\tilde{R}(t'-t_o)}|\,\tilde{\Lambda}(t_o)\gg$$

$$+(-i)^2\int_{-\infty}^{t} dt_o\int_{t_o}^{t} dt'\int_{t_o}^{t'} dt''\, e^{\tilde{R}(t-t')}\tilde{L}_{int}(t')e^{\tilde{R}(t'-t'')}$$

$$\tilde{L}_{int}(t'')e^{\tilde{R}(t''-t_o)}|\,\tilde{\Lambda}(t_o)\gg +\ldots \qquad (17)$$

We can introduce the solution in the absence of interaction :

$$|\,\tilde{\rho}^{(o)}(t)\gg \,=\int_{-\infty}^{t} dt_o\, e^{\tilde{R}(t-t_o)}|\,\tilde{\Lambda}(t_o)\gg \,=\int_{o}^{+\infty} d\tau\, e^{\tilde{R}\tau}|\,\tilde{\Lambda}(t-\tau)\gg \qquad (18)$$

By rearrangement of (17) we obtain :

$$|\,\tilde{\rho}(t)\gg \,=\,|\,\tilde{\rho}^{(o)}(t)\gg +\,(-i)\int_{-\infty}^{t} dt'\, e^{\tilde{R}(t-t')}\tilde{L}_{int}(t')|\,\tilde{\rho}^{(o)}(t')\gg$$

$$+(-i)^2\int_{-\infty}^{t} dt'\int_{-\infty}^{t'} dt''\, e^{\tilde{R}(t-t')}\tilde{L}_{int}(t')e^{\tilde{R}(t'-t'')}\tilde{L}_{int}(t'')|\,\tilde{\rho}^{(o)}(t'')\gg +\ldots \qquad (19)$$

(*) $$T \exp\left\{-i\int_{t_o}^{t} A(t')dt'\right\} = \lim_{\delta t'\to 0}\prod_{t_o}^{t}\exp\left\{-i\delta t' A(t')\right\}$$

which is equivalently obtained by iteration of the integral equation :

$$|\tilde{\rho}(t)\rangle\rangle = |\tilde{\rho}^{(o)}(t)\rangle\rangle - i\int_{-\infty}^{t} dt'\, e^{\tilde{R}(t-t')}\tilde{L}_{int}(t')|\tilde{\rho}(t')\rangle\rangle$$

$$= |\tilde{\rho}^{(o)}(t)\rangle\rangle - i\int_{o}^{+\infty} d\tau\, e^{\tilde{R}\tau}\tilde{L}_{int}(t-\tau)|\tilde{\rho}(t-\tau)\rangle\rangle \qquad (20)$$

which is an obvious solution of (15) since from (18) :

$$\frac{\partial|\tilde{\rho}^{(o)}(t)\rangle\rangle}{\partial t} = |\tilde{\Lambda}(t)\rangle\rangle + \tilde{R}|\tilde{\rho}^{(o)}(t)\rangle\rangle \qquad (21)$$

The formal solution (16) and the integral equation (20) would have much more complicated forms with the usual Hilbert space notation as we shall see later.

5 - Thermal equilibrium density matrix

If $|\tilde{\rho}^{(o)}\rangle\rangle$ is time independent, the source term can be rewritten as :

$$\tilde{\Lambda} = -\tilde{R}\tilde{\rho}^{(o)} \quad \text{or} \quad |\tilde{\Lambda}\rangle\rangle = -\tilde{R}|\tilde{\rho}^{(o)}\rangle\rangle \qquad (22)$$

As an example the equilibrium density operator $\tilde{\rho}^{(o)}$ for a canonical ensemble at temperature T is :

$$\tilde{\rho}^{(o)}(\alpha_o,\vec{p}_o) \equiv \rho^{(o)}(\alpha_o,\vec{p}_o) = |\alpha_o,\vec{p}_o\rangle\langle\alpha_o,\vec{p}_o|\,(N/Z)\, \exp\left[-(H_o/k_BT) - p_{op}^2/2Mk_BT\right]$$

where H_o is the Hamiltonian corresponding to this equilibrium, $\vec{p}_{op}$ is the momentum operator, k_B is the Boltzmann constant and Z is the partition function :

$$Z = \mathrm{Tr}\,\exp\left[-(H_o/k_BT) - (p_{op}^2/2Mk_BT)\right] \qquad (23)$$

which may be divided into its internal part $Z_{int} = \mathrm{Tr}\,\exp(-H_o/k_BT)$ and the translation partition function $Z_{tr} = V/\lambda_T^3$.
λ_T is the de Broglie wavelength $\lambda_T = h/Mu\sqrt{\pi}$ and u is the most probable velocity $u = (2k_BT/M)^{1/2}$
$\rho^{(o)}(\alpha_o,\vec{p}_o)$ may therefore be written :

$$\rho^{(o)}(\alpha_o,\vec{p}_o) = |\alpha_o,\vec{p}_o\rangle\langle\alpha_o,\vec{p}_o|\,(N/V)(h^3/Z_{int})F(\vec{p}_o)\exp(-E_{\alpha_o}/k_BT) \quad (24)$$

where $F(\vec{p}_o)$ is the normalized Maxwell-Boltzmann distribution :

$$F(\vec{p}_o) = (2\pi M k_B T)^{-3/2} \exp(-p_o^2/2Mk_BT) \tag{25}$$

If Tr_{int} is the trace operation over the internal degrees of freedom :

$$Tr_{int} < \vec{r}| \sum_{\alpha_o, \vec{p}_o} \rho^{(o)}(\alpha_o, \vec{p}_o) |\vec{r}> = N/V$$

is the total population per unit volume which remains finite as $N \rightarrow +\infty$ and $V \rightarrow +\infty$. In this limit the sums over discrete momentum states $\sum_{\vec{p}_o}$ can be replaced by integrations $\int d^3 p_o$ with the usual change in the normalization constant :

$$|<\vec{r}|\vec{p}_o>|^2 = V^{-1} \longrightarrow |<\vec{r}|\vec{p}_o>|^2 = h^{-3}$$

(It would be equivalent to keep the old normalization condition and to replace $\sum_{\vec{p}_o}$ by $Vh^{-3}\int d^3 p_o$).

6 - Interaction Hamiltonian with the electromagnetic field

We shall consider an interaction localized at the center of mass of the molecules that is such that(*) :

$$V_{int}(t) = \int d^3 r |\vec{r}> V(\vec{r},t) <\vec{r}| \tag{26}$$

In the $|\vec{r}>$ representation the matrix elements are :

$$<\vec{r}| V_{int}(t) | \vec{r}'> = V(\vec{r},t)\delta(\vec{r}-\vec{r}') \tag{27}$$

and in the momentum representation(*) :

$$<\vec{p}| V_{int}(t) | \vec{p}'> = \frac{1}{(2\pi)^{3/2}} \int d^3 k \; V(\vec{k},t)\delta(\vec{p}-\vec{p}'-\hbar\vec{k}) \tag{28}$$

From now on we shall assume more specifically an electric dipole interaction :

$$V(\vec{r},t) = -\vec{\mu}.\vec{\mathcal{E}}(\vec{r},t) \tag{29}$$

where $\vec{\mu}$ is the electric dipole moment operator.

(*) One may also introduce the position operator $\vec{R}_{op}$ and write :

$$V_{int}(\vec{R}_{op},t) = \frac{1}{(2\pi)^{3/2}} \int d^3 k \; V(\vec{k},t) \exp(i\vec{k}.\vec{R}_{op})$$

In a semiclassical theory $\vec{\mathcal{E}}(\vec{r},t)$ is the classical electric field that we shall write as the sum of two monochromatic counter-propagating fields (the theory can be generalized to a larger number of laser fields in a straightforward manner) :

$$\vec{\mathcal{E}}(\vec{r},t) = (\vec{E}_o^{+}/2)U_o^{+}(\vec{r})\exp\left[i(\omega t-\vec{k}.\vec{r}+\varphi^{+})\right]+ \\ +(\vec{E}_o^{-}/2)U_o^{-}(\vec{r})\exp\left[i(\omega t+\vec{k}.\vec{r}+\varphi^{-})\right] + \text{c.c.} \qquad (30)$$

where the functions $U_o^{\pm}(r)$ describe the transverse geometry of the light field. For Gaussian laser beams these functions and their Fourier transforms are given in reference (1).

A general method to obtain the matrix elements of the Liouvillian which takes into account the level degeneracy (Zeeman sublevels) is given in Appendix B.

7 - Quantities of interest for laser spectroscopy : macroscopic polarization, average absorbed power and transition rates

For most laser spectroscopy experiments the interesting quantity is the macroscopic polarization created per unit volume in the medium :

$$\vec{\mathscr{P}}(\vec{r},t) = \mathrm{Tr}_{int} <\vec{r}\,|\,\vec{\mu}\rho(t)|\,\vec{r}> =$$

$$= \sum_{\alpha_o}\int d^3p_o \sum_{\alpha,\beta} <\beta|\vec{\mu}|\alpha><\alpha|<\vec{r}|\rho(\alpha_o,\vec{p}_o,t)\,|\vec{r}>|\beta> \qquad (31)$$

The complex representation of $\vec{\mathscr{P}}$ may be written by introducing the positive frequency part $\rho^{(+)}_{\alpha\beta}$ of the off-diagonal density matrix elements :

$$\vec{P}(\vec{r},t) = 2\sum_{\alpha_o}\int d^3p_o \sum_{\alpha,\beta} \vec{\mu}_{\beta\alpha} <\vec{r}|\,\rho^{(+)}_{\alpha\beta}(\alpha_o,\vec{p}_o,t)|\,\vec{r}> \qquad (32)$$

where
$$\vec{\mu}_{\beta\alpha} \equiv <\beta|\,\vec{\mu}\,|\alpha>$$

If $\vec{E}$ is the complex representation of the electric field $\vec{\mathcal{E}}$, the average power absorbed per unit volume by the medium at the space-time point $(\vec{r},t)$ is :

$$\frac{d\overline{W}_{abs}}{dV} = (1/2)\mathrm{Re}\,\vec{E}^{*}\,.\,\frac{\partial\vec{P}}{\partial t} \simeq (\omega/2)\mathrm{Re}(i\vec{E}^{*}\,.\,\vec{P}) =$$

$$= 2\,\hbar\omega\;\mathrm{Re}\sum_{\alpha_o}\int d^3p_o \sum_{\alpha,\beta}(i\vec{\mu}_{\beta\alpha}\,.\,\vec{E}^{*}/2\hbar)<\vec{r}\,|\rho^{(+)}_{\alpha\beta}(\alpha_o,\vec{p}_o,t)|\,\vec{r}> \qquad (33)$$

Another way to derive this result from the transition rate to a given level is given in Appendix A. With the rotating-wave approximation, $\rho_{\alpha\beta}^{(+)}$ may be replaced by $\rho_{\alpha\beta}$ in (33) if the sum is restricted to states such that $E_\alpha < E_\beta$.

8 - Density matrix elements between energy-momentum states and corresponding equations

From the expressions derived in the previous section, it is clear that the matrix element that we need to calculate is $\langle \vec{r} |\, \rho_{\alpha\beta}(\alpha_0, \vec{p}_0, t)\, | \vec{r} \rangle$. Let us point out that the operator $\langle \vec{r} |\, \rho(\alpha_0, \vec{p}_0, t)\, | \vec{r} \rangle$ is the analog of the operator $\rho(\vec{r}, \vec{v}, t, \alpha_0)$ for molecules created with the velocity $\vec{v}$ in the state α_0 when the motion is treated classically.

In fact sub-Doppler techniques involve quasi-plane waves with almost discrete momentum exchanges :

when the molecules initially created in state α_0 with momentum $\vec{p}_0$ interact with electromagnetic waves having the wavevectors $\pm\vec{k}$ they are scattered into a superposition of narrow wave-packets centered around $\vec{p}_0 + m\hbar\vec{k}$ where m is an integer number (see figure 1).

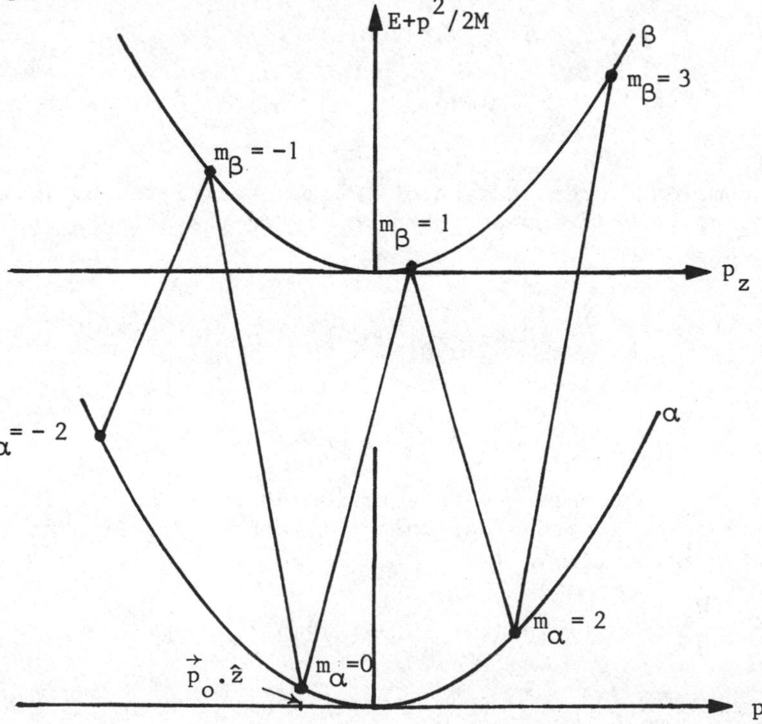

Figure 1

These wave-packets are well-resolved in the momentum domain since, if z is the coordinate along the common optical axis, the corresponding widths of wave-vector distributions are :

$$\Delta k_x \, , \Delta k_y \sim 1/w_o \quad \text{and} \quad \Delta k_z \sim 1/b \ll k$$

where w_o and b are respectively the beam radius and the confocal parameter for a Gaussian beam ($b = kw_o^2$).

As a result, a greater physical insight is obtained by working in the momentum representation rather than in the position representation. But this is at the expense of complicated integro-differential equations in the case of Gaussian laser beams. We can combine the advantages of both representations if we introduce the wave-packets centered at $\vec{p}_o + m\hbar\vec{k}$ and write :

$$< \vec{r} \mid \rho_{\alpha\beta}(\alpha_o,\vec{p}_o,t) \mid \vec{r} > = \int d^3p d^3p' < \vec{r} \mid \vec{p} > < \vec{p} \mid \rho_{\alpha\beta} \mid \vec{p}' > < \vec{p}' \mid \vec{r} >$$

$$= \sum_{m_\alpha m_\beta} < \vec{p}_o + m_\alpha \hbar\vec{k} \mid \rho_{\alpha\beta}(\alpha \, ,\vec{p}_o,\vec{r},t) \mid \vec{p}_o + m_\beta \hbar\vec{k} > / h^3 \qquad (34)$$

superposition of matrix elements between "energy-momentum states" which are functions of $\vec{r}$ given by :

$$< m_\alpha \mid \rho_{\alpha\beta} \mid m_\beta > = \int d^3p d^3p' < \vec{p} \mid e^{i\vec{p}_{op} \cdot \vec{r}/\hbar} \rho_{\alpha\beta}(\alpha_o,\vec{p}_o,t) e^{-i\vec{p}_{op} \cdot \vec{r}/\hbar} \mid \vec{p}' > \qquad (35)$$

where the integrations over $\vec{p}$ and $\vec{p}'$ are limited to intervals around $\vec{p}_o + m_\alpha \hbar\vec{k}$ and $\vec{p}_o + m_\beta \hbar\vec{k}$ small compared to $\hbar|\vec{k}|$ and large compared to $\hbar/b$ or $\hbar/w_o$.

This defines a new operator ρ which depends on space and time as parameters and has matrix elements between discrete energy-momentum states $E_\alpha, \vec{p}_o + m_\alpha \hbar\vec{k}$. These matrix elements are proportional to slow functions of $\vec{r}$, times $\exp i(m_\alpha - m_\beta)\vec{k}.\vec{r}$, and the expansion :

$$< \vec{r} \mid \rho_{\alpha\beta} \mid \vec{r} > = \sum_{m_\alpha, m_\beta} < m_\alpha \mid \rho_{\alpha\beta} \mid m_\beta > / h^3 \qquad (36)$$

generalizes the usual Fourier expansion of $\rho(\vec{r},\vec{v},t)$.

From the equations for the matrix elements $< \vec{p} \mid \rho_{\alpha\beta} \mid \vec{p}' >$, one can derive the following equations for $< m_\alpha \mid \rho_{\alpha\beta} \mid m_\beta >$ (see Appendix C).

$$\left(\frac{\partial}{\partial t}+(\vec{p}_o/M).\vec{\nabla}\right)<m_\alpha|\rho_{\alpha\beta}|m_\beta> = \lambda(\alpha_o,\vec{p}_o,t)\delta_{\alpha\alpha_o}\delta_{\beta\alpha_o}\delta_{m_\alpha 0}\delta_{m_\beta 0}$$

$$-(i\omega_{\alpha\beta}+i(m_\alpha^2-m_\beta^2)\delta+\gamma_{\alpha\beta})<m_\alpha|\rho_{\alpha\beta}|m_\beta>$$

$$+\frac{1}{i\hbar}\sum_{\beta'} V^+_{\alpha\beta'}(\vec{r},t)<m_\alpha-1|\rho_{\beta'\beta}|m_\beta>+\ldots \qquad (37)$$

where : $\delta = \hbar k^2/2M$ and

$$<\beta|<\vec{p}_o+(m_\alpha+1)\hbar\vec{k}|V|\vec{p}_o+m_\alpha\hbar\vec{k}>|\alpha> = V^+_{\beta\alpha}(\vec{r},t) =$$

$$= -(\vec{\mu}_{\beta\alpha}.\vec{E}_o^{+*}/2)U_o^{+*}(\vec{r})\exp\left[-i(\omega t-\vec{k}.\vec{r}+\varphi^+)\right]$$

$$-(\vec{\mu}_{\beta\alpha}.\vec{E}_o^{-}/2)U_o^{-}(\vec{r})\exp\left[i(\omega t+\vec{k}.\vec{r}+\varphi^-)\right] \qquad (38)$$

and :

$$<\beta|<\vec{p}_o+(m_\alpha-1)\hbar\vec{k}|V|\vec{p}_o+m_\alpha\hbar\vec{k}>|\alpha> = V^-_{\beta\alpha}(\vec{r},t) =$$

$$= -(\vec{\mu}_{\beta\alpha}.\vec{E}_o^{-*}/2)U_o^{-*}(\vec{r})\exp\left[-i(\omega t+\vec{k}.\vec{r}+\varphi^-)\right]$$

$$-(\vec{\mu}_{\beta\alpha}.\vec{E}_o^{+}/2)U_o^{+}(\vec{r})\exp\left[i(\omega t-\vec{k}.\vec{r}+\varphi^+)\right]$$

In operator form, one can write these equations[4]:

$$i\hbar\left\{\frac{\partial\rho}{\partial t}+\frac{1}{2M}\left[\vec{p}_{op}.\vec{\nabla}\rho+\vec{\nabla}\rho.\vec{p}_{op}\right]\right\} = i\hbar\Lambda+i\hbar R\rho+\left[H_o+V,\rho\right] \qquad (39)$$

where $\rho,\vec{p}_{op}$ and V are now operators in the sub-space of the discrete vectors $|\vec{p}_o+m\hbar\vec{k}>$

These equations generalize the usual density matrix equations for $\rho(\vec{r},\vec{v},t)$ and include the effects of the atomic recoil besides the laser beam geometry. If δ is neglected we recover the equations for $\rho(\vec{r},\vec{v},t)$ by summation of both sides over m_α and m_β (and multiplication by h^3).

9 - Reference frame problems and integral form of the density matrix equations

The previous formalism is invariant in a Galilean transformation of the coordinates. We have written the equations in the laboratory frame where the molecules have the velocity $\vec{v} = \vec{p}_o/M$ (if velocity-changes due to the recoil are neglected). These equations become simple differential equations in the molecular frame and can be formally integrated.

Let us consider the general transformation laws between the laboratory frame coordinates $(\vec{r},t)$ and the molecular frame coordinates $(\vec{r}\,',t')$:

$$\begin{aligned} \vec{r} &= \vec{R}(\vec{r}\,',t') \\ t &= T(\vec{r}\,',t') \end{aligned} \tag{40}$$

and :

$$\begin{aligned} \vec{r}\,' &= \vec{R}^{-1}(\vec{r},t) \\ t' &= T^{-1}(\vec{r},t) \end{aligned} \tag{41}$$

We shall assume that the interaction Hamiltonian is transformed as:

$$V'(\vec{r}\,',t') = V(\vec{r},t) = V\left[\vec{R}(\vec{r}\,',t')\ ,\ T(\vec{r}\,',t')\right] \tag{42}$$

The resulting equation for a molecular frame coherence between energy-momentum states is :

$$\begin{aligned} \frac{\partial}{\partial t'} &< m_\alpha | \rho'_{\alpha\beta}(\vec{r}\,',t',\vec{p}_o) | m_\beta > = \lambda(\alpha_o,\vec{p}_o,t')\delta_{\alpha\alpha_o}\delta_{\beta\alpha_o}\delta_{m_\alpha 0}\delta_{m_\beta 0} \\ &- \left[i\omega_{\alpha\beta} + i(m_\alpha^2 - m_\beta^2)\delta + \gamma_{\alpha\beta}\right] < m_\alpha | \rho'_{\alpha\beta}(\vec{r}\,',t',\vec{p}_o) | m_\beta > \\ &+ \sum_{\beta'}\Big[V^+_{\alpha\beta'}\left[\vec{R}(\vec{r}\,',t')\ ,\ T(\vec{r}\,',t')\right] < m_\alpha - 1 | \rho'_{\beta'\beta}(\vec{r}\,',t',\vec{p}_o) | m_\beta > \\ &+ V^-_{\alpha\beta'}\left[\vec{R}(\vec{r}\,',t')\ ,\ T(\vec{r}\,',t')\right] < m_\alpha + 1 | \rho'_{\beta'\beta}(\vec{r}\,',t',\vec{p}_o) | m_\beta > \Big] / i\hbar \\ &- \sum_{\alpha'}\Big[< m_\alpha | \rho'_{\alpha\alpha'}(\vec{r}\,',t',\vec{p}_o) | m_\beta + 1 > V^+_{\alpha'\beta}\left[\vec{R}(\vec{r}\,',t')\ ,\ T(\vec{r}\,',t')\right] \\ &+ < m_\alpha | \rho'_{\alpha\alpha'}(\vec{r}\,',t',\vec{p}_o) | m_\beta - 1 > V^-_{\alpha'\beta}\left[\vec{R}(\vec{r}\,',t')\ ,\ T(\vec{r}\,',t')\right]\Big] / i\hbar \end{aligned} \tag{43}$$

The integral solution of this equation is :

$$\begin{aligned} < m_\alpha | \rho'_{\alpha\beta}(\vec{r}\,',t',\vec{p}_o) | m_\beta > &= < m_\alpha | \rho'^{(o)}_{\alpha\beta}(\vec{p}_o) | m_\beta > \\ + \left(\frac{i}{\hbar}\right)\sum_{\alpha'}\int_{-\infty}^{+\infty} d\tau\ G_{\alpha\beta}(\tau) &< m_\alpha | \rho'_{\alpha\alpha'}(\vec{r}\,',t'-\tau,\vec{p}_o) | m_\beta + 1 > V^+_{\alpha'\beta}\Big[\vec{R}(\vec{r}\,',t'-\tau), \\ T(\vec{r}\,',t'-\tau)\Big] &+ \text{similar terms for } V^-_{\alpha'\beta'}\ ,\ V^+_{\alpha\beta'}\ ,\ V^-_{\alpha\beta'} \end{aligned} \tag{44}$$

with a propagator $G_{\alpha\beta}(\tau) = Y(\tau)\ \exp\left\{-\left[i\omega_{\alpha\beta}+i(m_\alpha^2-m_\beta^2)\delta+\gamma_{\alpha\beta}\right]\tau\right\}$ (45)

where $Y(\tau)$ is the Heaviside step function. We may now transform this solution to get the laboratory frame solution; a first step is :

$$<m_\alpha|\rho'_{\alpha\beta}(\vec{r}',t',\vec{p}_o)|m_\beta> = <m_\alpha|\rho'^{(o)}_{\alpha\beta}(\vec{p}_o)|m_\beta>$$

$$+(\frac{i}{\hbar})\sum_{\alpha'}\int_{-\infty}^{+\infty} d\tau\ G_{\alpha\beta}(\tau)<m_\alpha|\rho_{\alpha\alpha'}\left[\vec{R}(\vec{r}',t'-\tau),T(\vec{r}',t'-\tau),\vec{p}_o\right]|m_\beta+1>$$

$$V^+_{\alpha'\beta}\left[\vec{R}(\vec{r}',t'-\tau)\ ,\ T(\vec{r}',t'-\tau)\right]$$

$$+\text{ similar terms for } V^-_{\alpha'\beta}\ ,\ V^+_{\alpha\beta'}\ ,\ V^-_{\alpha\beta'}$$

finally :

$$<m_\alpha|\rho_{\alpha\beta}(\vec{r},t,\vec{p}_o)|m_\beta>= <m_\alpha|\rho^{(o)}_{\alpha\beta}(\vec{p}_o)|m_\beta>$$

$$+\frac{i}{\hbar}\sum_{\alpha'}\int_{-\infty}^{+\infty} d\tau\ G_{\alpha\beta}(\tau)<m_\alpha|\rho_{\alpha\alpha'}\left[\vec{R}(\overrightarrow{R^{-1}}(\vec{r},t),T^{-1}(\vec{r},t)-\tau),T(\overrightarrow{R^{-1}}(\vec{r},t),\right.$$

$$\left.T^{-1}(\vec{r},t)-\tau),\vec{p}_0\ |m_\beta+1> V^+_{\alpha'\beta}\left[\vec{R}(\overrightarrow{R^{-1}}(\vec{r},t),T^{-1}(\vec{r},t)-\tau),T(\overrightarrow{R^{-1}}(\vec{r},t),\ T^{-1}(\vec{r},t)-\tau)\right]\right.$$

$$+\text{ similar terms for } V^-_{\alpha'\beta}\ ,\ V^+_{\alpha\beta'}\ ,\ V^-_{\alpha\beta'} \qquad (46)$$

For the Galilean transformation :

$$\begin{cases} \vec{r} = \vec{r}' + \vec{v}t' & \vec{r}' = \vec{r} - \vec{v}t \\ t = t' & t' = t \end{cases} \qquad (47)$$

$$<m_\alpha|\rho_{\alpha\beta}(\vec{r},t,\vec{p}_o)|m_\beta>= <m_\alpha|\rho^{(o)}_{\alpha\beta}(\vec{p}_o)|m_\beta>$$

$$+\frac{i}{\hbar}\sum_{\alpha'}\int_{-\infty}^{+\infty} d\tau\ G_{\alpha\beta}(\tau)<m_\alpha|\rho_{\alpha\alpha'}(\vec{r}-\vec{v}\tau,t-\tau,\vec{p}_o)|m_\beta+1>$$

$$.\ V^+_{\alpha'\beta}(\vec{r}-\vec{v}\tau,t-\tau)$$

$$+\text{ similar terms for } V^-_{\alpha'\beta}\ ,\ V^+_{\alpha\beta'} \text{ and } V^-_{\alpha\beta'} \qquad (48)$$

A simpler way to derive this result is to use directly the integral equation (20) written in the Schroedinger representation:

$$|\rho(t)\gg = |\rho^{(o)}(t)\gg - i\int_0^{+\infty} d\tau \exp\left[-i(L_1+iR)\tau\right] L_{int}(t-\tau)|\rho(t-\tau)\gg \tag{49}$$

with
$$L_1 = L_o + L_{kinetic}$$

in the $|\vec{p}>$ representation in Hilbert space notations :

$$<\vec{p}|\rho_{\alpha\beta}(\vec{p}_o,t)|\vec{p}\,'> = <\vec{p}|\rho^{(o)}_{\alpha\beta}(\vec{p}_o)|\vec{p}\,'>$$

$$-\frac{i}{\hbar}\int_0^{+\infty} d\tau\, e^{-i(E_\alpha-E_\beta)\tau/\hbar-\gamma_{\alpha\beta}\tau-i[(p^2-p'^2)/2M]\tau/\hbar}$$

$$\sum_{\beta'}\frac{1}{(2\pi)^{3/2}}\int d^3k\, V^+_{\alpha\beta'}(\vec{k},t-\tau)<\vec{p}-\hbar\vec{k}|\rho_{\beta'\beta}(t-\tau)|\vec{p}\,'>$$

$$+\ 3 \text{ similar terms} \tag{50}$$

If we multiply both sides by $\exp i(\vec{p}-\vec{p}\,').\vec{r}/\hbar$ and integrate over $\vec{p}$ and $\vec{p}\,'$ over the same small intervals as before around $\vec{p}_o + m_\alpha\hbar\vec{k}$ and $\vec{p}_o + m_\beta\hbar\vec{k}$:

$$<m_\alpha|\rho_{\alpha\beta}(\vec{r},t,\vec{p}_o)|m_\beta> = <m_\alpha|\rho^{(o)}_{\alpha\beta}(\vec{p}_o)|m_\beta>$$

$$-\frac{i}{\hbar}\int_0^{+\infty} d\tau\, e^{-[i\omega_{\alpha\beta}+\gamma_{\alpha\beta}]\tau}\frac{1}{(2\pi)^{3/2}}\int d^3p\,d^3p'\,d^3k\; e^{-i[(p^2-p'^2)/2M]\tau/\hbar}$$

$$\sum_{\beta'} V^+_{\alpha\beta'}(\vec{k},t-\tau)\; e^{i\vec{k}.\vec{r}}\; e^{i(\vec{p}-\hbar\vec{k}-\vec{p}\,').\vec{r}/\hbar}<\vec{p}-\hbar\vec{k}|\rho_{\beta'\beta}(t-\tau)|\vec{p}\,'>$$

$$+\ 3 \text{ similar terms}$$

$$= <m_\alpha|\rho^{(o)}_{\alpha\beta}(\vec{p}_o)|m_\beta>$$

$$-\frac{i}{\hbar}\sum_{\beta'}\int_0^{+\infty} d\tau\; e^{-[i\omega_{\alpha\beta}+i(m_\alpha^2-m_\beta^2)\delta+\gamma_{\alpha\beta}]\tau}\; V^+_{\alpha\beta'}(\vec{r}-\vec{v}\tau,t-\tau)$$

$$<m_\alpha-1|\rho_{\beta'\beta}(\vec{r}-\vec{v}\tau,t-\tau,\vec{p}_o)|m_\beta>$$

$$+\ 3 \text{ similar terms} \tag{51}$$

If the recoil is neglected ($\delta = o$) we may use a summation of both sides over m_α and m_β and multiply by h^3 to recover the usual integral equation for $\rho_{\alpha\beta}(\vec{r},t,\vec{v})$:

$$\rho_{\alpha\beta}(\vec{r},t,\vec{v}) = \rho^{(o)}_{\alpha\beta}(\vec{r},t,\vec{v}) + (\frac{1}{i\hbar})\sum_{\beta'}\int_0^{+\infty} d\tau\, e^{-(i\omega_{\alpha\beta}+\gamma_{\alpha\beta})\tau}\, V_{\alpha\beta'}(\vec{r}-\vec{v}\tau,t-\tau)$$

$$\rho_{\beta'\beta}(\vec{r}-\vec{v}\tau,t-\tau,\vec{v})$$

$$+ \text{ similar terms} \tag{52}$$

This equation can be obtained directly from (49) written in the $|\vec{r}>$ representation. For classical motion $L_{kinetic} = -i\,\vec{v}.\vec{\nabla}$ and the operator $\exp(-i\,L_{kinetic}\,\tau)$ reduces to the translation operator $\exp(-\tau\,\vec{v}.\vec{\nabla})$ which has the following property :

$$\exp(-\tau\,\vec{v}.\vec{\nabla})g(\vec{r}) = g(\vec{r}-\vec{v}\tau)$$

easily established with the Fourier transform of g. This operator can also be used to solve directly an equation like (37) involving a hydrodynamic derivative, by treating it as an ordinary differential equation with respect to time. This is another way to recover (51). The quantities $(m_\alpha^2-m_\beta^2)\delta$ appear as recoil corrections to the eigenvalues $\omega_{\alpha\beta}$ of the proper Liouvillian L_o in a theory where the motion of each wave-packet is treated classically.

10 - Introduction of the second-order Doppler shift

The second-order Doppler shift is a relativistic effect which is naturally introduced when the equations are covariant in Lorentz transformations. In fact only the motion of the center of mass of the molecules requires a relativistic treatment and we are thus led to a hybrid formalism where the internal degrees of freedom may receive their usual treatment. A fully covariant formalism is outlined in the next section and we shall first limit ourselves to a few recipes towards a relativistic invariance of the equations in a change of reference frame. The first recipe which is applicable in the molecular frame is the following blue shift of the frequency seen by the molecules :

$$\omega' = \omega/\sqrt{1-v^2/c^2} \tag{53}$$

this law comes from the Lorentz transformation of the four-vector $(\omega,\vec{k})$. Owing to the relativistic invariance of the phase $\vec{k}.\vec{r}-\omega t$, this is equivalently obtained by a Lorentz transformation of the coordinates in the electric field expression :

$$\begin{cases} \vec{r} = \vec{r}'+(\gamma-1)(\vec{r}'.\vec{v})\vec{v}/v^2+\gamma\vec{v}t' = \vec{R}(\vec{r}',t') \\ t = \gamma(t'+\vec{r}'.\vec{v}/c^2) = T(\vec{r}',t') \end{cases} \quad (54)$$

and :

$$\begin{cases} \vec{r}' = \vec{r}+(\gamma-1)(\vec{r}.\vec{v})\vec{v}/v^2-\gamma\vec{v}t = \vec{R}^{-1}(\vec{r},t) \\ t' = \gamma(t-\vec{r}.\vec{v}/c^2) = T^{-1}(\vec{r},t) \end{cases} \quad (55)$$

where $\gamma = (1-v^2/c^2)^{-1/2}$ and $\vec{v} = \vec{p}_o/M\gamma$

The second recipe which is applicable in the laboratory frame is the following red shift of the Bohr frequencies :

$$\omega_{ba} \rightarrow \omega_{ba}\left(1-\frac{v^2}{2c^2}\right) \quad (56)$$

This second law comes from the relativistic expression of the energy in a frame where the molecule is moving :

$$E_a = [(Mc^2+E_a^o)^2 + p_a^2c^2]^{1/2} = \frac{Mc^2+E_a^o}{\sqrt{1-\beta_a^2}} \quad (57)$$

where the internal energy E_a^o is added to the rest mass energy Mc^2, where $\beta_a = \frac{|\vec{v}_a|}{c}$ and where the momentum $\vec{p}_a$ is given by :

$$\vec{p}_a = \frac{(M+E_a^o/c^2)\vec{v}_a}{\sqrt{1-\beta_a^2}} \quad (58)$$

Equation (57) is the equation of the hyperboloid which corresponds to the mass shell of a particle of mass $M+E_a^o/c^2$. The transition between two hyperboloids corresponding to states a and b is illustrated on fig. 2a and 2b in planes respectively parallel and perpendicular to the optical axis.

The absorbed or emitted frequencies can be obtained from the energy and momentum balances :

$$\hbar\omega = E_b - E_a \quad , \quad \hbar\vec{k} = \vec{p}_b - \vec{p}_a$$

If the rest mass M corresponds to halfway between the two

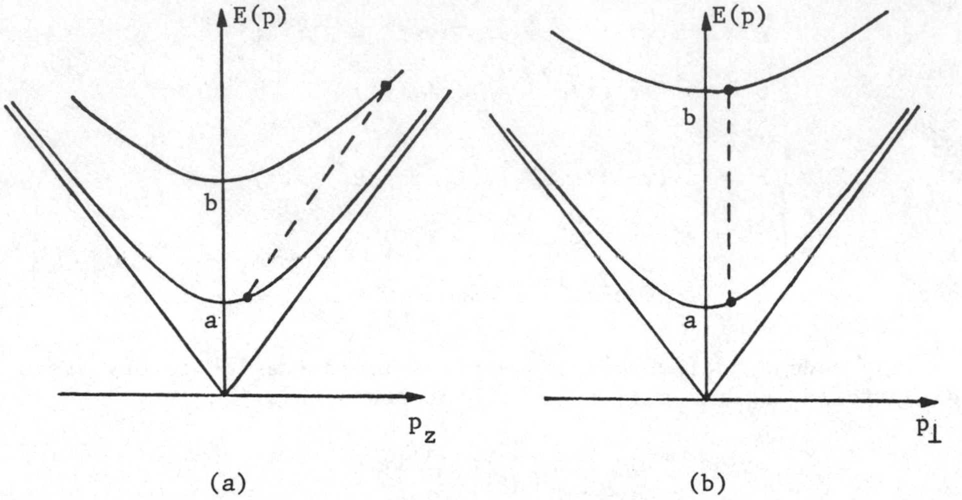

Figure 2

levels one finds :

$$\frac{\omega}{\omega_{ba}} = \frac{\sqrt{1-\beta_a^2}}{1-\hat{k}.\vec{v}_a/c} \cdot \frac{1}{1-\varepsilon} \quad \text{for the absorption frequency} \tag{59}$$

and :

$$\frac{\omega}{\omega_{ba}} = \frac{\sqrt{1-\beta_b^2}}{1-\hat{k}.\vec{v}_b/c} \cdot \frac{1}{1+\varepsilon} \quad \text{for the emission frequency} \tag{60}$$

with : $\hat{k} = \vec{k}c/\omega$, $\varepsilon = \frac{\hbar\omega}{2Mc^2}$ and $\omega_{ba} = \frac{E_b^o - E_a^o}{\hbar}$

These formulas give without any expansion the right combination of first order and transverse Doppler shifts with the recoil shift. They can easily be generalized to the case of the off-diagonal element $\langle m_b|\rho_{ba}|m_a\rangle$ with $\vec{p}_a \simeq \vec{p}_o + m_a\hbar\vec{k}$ and $\vec{p}_b \simeq \vec{p}_o + m_b\hbar\vec{k}$

$$[E_b - E_a - (\vec{p}_b - \vec{p}_a).\vec{v}_a]/\hbar = \omega_{ba}\sqrt{1-\beta_a^2}[1+(m_b-m_a)^2\varepsilon/(1-\varepsilon)] \tag{61}$$

and :

$$[E_b - E_a - (\vec{p}_b - \vec{p}_a).\vec{v}_b]/\hbar = \omega_{ba}\sqrt{1-\beta_b^2}[1-(m_b-m_a)^2\varepsilon/(1+\varepsilon)] \tag{62}$$

We recover the two previous formulas with respectively $m_a = 0$, $m_b = 1$ in the first case and $m_a = -1$, $m_b = 0$ in the second case.

With the approximations:

$$\frac{\vec{p}_o c^2 \sqrt{1-\beta_a^2}}{E_a^o + Mc^2} \simeq \frac{\vec{p}_o c^2 \sqrt{1-\beta_b^2}}{E_b^o + Mc^2} \simeq \frac{2\vec{p}_o c^2 \sqrt{1-\beta^2}}{E_a^o + E_b^o + 2Mc^2} = \vec{v} \tag{63}$$

$$\frac{\varepsilon}{1+\varepsilon} \simeq \frac{\varepsilon}{1-\varepsilon} \simeq \varepsilon$$

we can write the single formula :

$$[E_b - E_a - (\vec{p}_b - \vec{p}_a).\vec{v}]/\hbar = \omega_{ba} \sqrt{1-\beta^2}[1 + (m_b^2 - m_a^2)\,\varepsilon] \tag{64}$$

In the case of an arbitrary reference frame, both recipes have to be applied i.e. the transformation of the coordinates in the field expression and the relativistic expression of the eigenvalues of the Hamiltonian.

Furthermore it is possible to write a relativistically invariant Hamiltonian :

$$V = -\vec{\mu}.\vec{\mathcal{E}} + \gamma(\vec{v}\times\vec{\mu}^{(o)}).\vec{\mathcal{B}} \tag{65}$$

where :

$$\vec{\mu} = \gamma[\vec{\mu}^{(o)} - \vec{v}(\vec{v}.\vec{\mu}^{(o)})/v^2] + \vec{v}(\vec{v}.\vec{\mu}^{(o)})/v^2$$

and where $\vec{\mu}^{(o)}$ is the electric dipole operator in the rest frame of the molecule.

Let us now examine how the density matrix equations are modified by these rules. In the molecular frame, the equations (43) and their integral form remain obviously valid if the interaction Hamiltonian given in the laboratory frame is transformed according to (42) with the Lorentz transformation (54) of the coordinates.

The integral solution can be brought back into the laboratory frame provided that :

- first the density matrix elements transform as :

$$\gamma \langle m_\alpha | \rho'_{\alpha\beta}(\vec{r}\,',t',\vec{p}_o)| m_\beta \rangle = \langle m_\alpha | \rho_{\alpha\beta}(\vec{r},t,\vec{p}_o)| m_\beta \rangle \tag{66}$$

- second, the propagator $G_{\alpha\beta}(\tau)$ is invariant.

From the relations (64) we check that $[\omega_{ba} + (m_b^2 - m_a^2)\delta]\tau$ is invariant(*) since $(\vec{p},E)$ and $(\gamma\vec{v}\tau,\gamma\tau)$ are four-vectors and thus we may keep the same propagator in any reference frame and, in this propagator, τ is the proper time between two events which is defined in the molecular frame and becomes $\gamma\tau$ in any other frame.

We may thus use the solution (46) with R and T given by (54) and we find :

$$<m_\alpha|\rho_{\alpha\beta}(\vec{r},t,\vec{p}_o)|m_\beta> = <m_\alpha|\rho^{(o)}_{\alpha\beta}(\vec{p}_o)|m_\beta> +$$

$$+ \frac{i}{\hbar}\sum_{\alpha'} \gamma \int_{-\infty}^{+\infty} d\tau\, G_{\alpha\beta}(\tau) <m_\alpha|\rho_{\alpha\alpha'}(\vec{r}-\gamma\vec{v}\tau, t-\gamma\tau, \vec{p}_o)|m_\beta+1>$$

$$. V^+_{\alpha'\beta}\left[\vec{r}-\gamma\vec{v}\tau, t-\gamma\tau\right] \quad (66)$$

$$+ \text{ similar terms for } V^-_{\alpha'\beta}\ ,\ V^+_{\alpha\beta'} \text{ and } V^-_{\alpha\beta'} \quad (67)$$

In the laboratory frame the approximate equation :

$$(E_b - E_a)/\hbar = (\vec{p}_b - \vec{p}_a).\vec{v}/\hbar - \omega_{ba}\sqrt{1-\frac{v^2}{c^2}}\,[1+(m_b^2-m_a^2)\delta]$$

can be used to replace the non-relativistic expression $\omega_{ba} + (p_b^2 - p_a^2)/2\hbar M$ in C-3 to derive a "relativistic" equation equivalent to (C-6) and (43) where $\omega_{ba} + (m_b^2 - m_a^2)\delta - i\gamma_{ba}$ is simply multiplied by $\sqrt{1-\frac{v^2}{c^2}}$.

The same trick can be used in equation (50) :

$$<m_\alpha|\rho_{\alpha\beta}(\vec{r},t,\vec{p}_o)|m_\beta> = <m_\alpha|\rho^{(o)}_{\alpha\beta}(\vec{p}_o)|m_\beta> - \frac{i}{\hbar}\sum_{\beta'}\int_0^{+\infty} d\tau'$$

$$e^{-i\omega_{\alpha\beta}\sqrt{1-\frac{v^2}{c^2}}\,[1+(m_\alpha^2-m_\beta^2)\delta]\tau'} \quad \times$$

(*) $m_b^2\hbar\delta$ and $m_a^2\hbar\delta$ appear as corrections to the rest energy equal to the recoil kinetic energies.

$$\times\ e^{-\sqrt{1-\frac{v^2}{c^2}}\gamma_{\alpha\beta}\tau'} \frac{1}{(2\pi)^{3/2}}\int d^3p\,d^3p'\,d^3k\ V^+_{\alpha\beta'}(\vec{k},t-\tau')$$

$$e^{i\vec{k}(\vec{r}-\vec{v}\tau')}\ e^{i(\frac{\vec{p}-\hbar\vec{k}-\vec{p}'}{\hbar})(\vec{r}-\vec{v}\tau')} < \vec{p}-\hbar\vec{k}|\,\rho_{\beta'\beta}(t-\tau')\,|\,\vec{p}'> + \text{other terms} =$$

$$<m_\alpha|\,\rho^{(o)}_{\alpha\beta}(\vec{p}_o)\,|\,m_\beta> - \frac{i}{\hbar}\gamma\int_0^{+\infty} d\tau\ e^{-[\,i\omega_{\alpha\beta}+i(m_\alpha^2-m_\beta^2)\delta+\gamma_{\alpha\beta}]\tau}$$

$$\sum_{\beta'} V^+_{\alpha\beta'}(\vec{r}-\gamma\vec{v}\tau,t-\gamma\tau) <m_\alpha-1|\,\rho_{\beta'\beta}(\vec{r}-\gamma\vec{v}\tau,t-\gamma\tau,\vec{p}_o)\,|\,m_\beta> + \text{other terms} \tag{68}$$

11 - Outline of a fully covariant formalism

The previous treatment is not fully satisfactory neither with respect to relativity nor with respect to quantum mechanics.

1. Space and time do not play a symmetric role. They should always appear together as a single four-vector whereas in the previous theory, time is the only evolution parameter.

2. Relativistic expressions of the eigenvalues of the Hamiltonian have been artificially introduced whereas they should result from a relativistic expression of the Hamiltonian that is from a relativistic equation for the density operator itself.

One could try to introduce such an equation (e.g. the Klein-Gordon equation) but it is well known that there is no satisfactory relativistic wave equation in the presence of external fields and that the canonical answer to these problems in quantum electrodynamics is quantum field theory.[16,17]

We have therefore to introduce field operators for the molecules in the various states α as well as for the light field. Molecules in a given internal state α will be considered as elementary systems[43] with a mass corresponding to their total internal energy E^o_α. They will be described by creation and annihilation operators written in the configuration space as :

$$\psi_\alpha(\vec{r},t) = \frac{1}{(2\pi\hbar)^{3/2}}\int d^3p\sqrt{\frac{E^o_\alpha}{E_\alpha(p)}}\ c_\alpha(\vec{p})u^\alpha(p)e^{i(\vec{p}.\vec{r}-E_\alpha(p)t)/\hbar} \tag{69}$$

$$\psi^{\dagger}_{\alpha}(\vec{r},t) = \frac{1}{(2\pi\hbar)^{3/2}} \int d^3p \sqrt{\frac{E^o_\alpha}{E_\alpha(p)}}\; c^{\dagger}_{\alpha}(\vec{p})\, \bar{u}^{\alpha}(p)\gamma_o e^{-i(\vec{p}.\vec{r}-E(p)t)/\hbar} \quad (70)$$

with $E_\alpha(p) = (E^o_\alpha{}^2 + p^2c^2)^{1/2}$ and where $u^\alpha(p)$ are spinors (multi-spinors satisfying the Bargmann-Wigner equations in a more general theory) normalized such that :

$$\bar{u}^\alpha(p)\; u^\beta(p) = \delta_{\alpha\beta}$$

$$u^{\alpha\dagger}(p)\; u^\beta(p) = \frac{E_\alpha}{E^o_\alpha}\,\delta_{\alpha\beta} \quad (71)$$

These operators act on the occupation number in configuration space (Fock space) and for obvious reasons we restrict possible states to one particle states with positive energy. If $|\tilde{\psi}>$ is the state vector describing a one particle state, $<\tilde{\psi}|\psi^{\dagger}_{\alpha}(x)|o>$ is the amplitude for finding a molecule in state α at the point $x = (\vec{r},t)$.

The electromagnetic field is described by the tensor operators:

$$F_{\mu\nu}(x) = \partial_\mu A_\nu(x) - \partial_\nu A_\mu(x) \quad (72)$$

with :

$$A^\mu(x) = \sum_{\lambda=1}^{4} \frac{\sqrt{\hbar}}{(2\pi)^{3/2}} \int \frac{d^3k}{\sqrt{2\omega\varepsilon_o}}\, e^\mu_\lambda(\vec{k})\, a_\lambda(k)\, e^{i(\vec{k}.\vec{r}-\omega t)} + h.c. \quad (73)$$

The vector $|\tilde{\psi}>$, which describes the whole system (molecules + field) in the interaction representation, satisfies the Tomonaga-Schwinger equation :

$$i\hbar \frac{\delta|\tilde{\psi}(\sigma)>}{\delta\sigma(x)} = \tilde{\mathscr{H}}(x)|\tilde{\psi}(\sigma)> \quad (74)$$

where the functional derivative with respect to the space-like hypersurface σ at the space time point x plays the role of the time derivative in the Schroedinger equation and where $\tilde{\mathscr{H}}(x)$ is an invariant interaction Hamiltonian density.

We can define a density operator in the interaction representation :

$$\tilde{\rho} = \sum_{\psi} p_\psi |\tilde{\psi}><\tilde{\psi}| \quad (75)$$

which satisfies :

$$i\hbar \frac{\delta\tilde{\rho}(\sigma)}{\delta\sigma(x)} = [\tilde{\mathscr{H}}(x), \tilde{\rho}(\sigma)] \quad (76)$$

or in Liouville space notation :

$$i \frac{\delta |\tilde{\rho}(\sigma)\gg}{\delta\sigma(x)} = \tilde{\mathscr{L}}(x)|\tilde{\rho}(\sigma)\gg \tag{77}$$

where $\tilde{\mathscr{L}}(x)$ is a Liouvillian interaction density in interaction representation (which may be non-hermitian to represent relaxation).

Introducing the pumping operator $|\tilde{\Lambda}(x_o)\gg$ we shall write the general solution :

$$|\tilde{\rho}(\sigma)\gg = \int_{-\infty}^{\sigma} d^4x_o \, T \, e^{-i\int_{\sigma_o}^{\sigma} d^4x' \tilde{\mathscr{L}}(x')} |\tilde{\Lambda}(x_o)\gg \tag{78}$$

which is also solution of the integral equation :

$$|\tilde{\rho}(\sigma)\gg = |\tilde{\rho}^{(o)}(\sigma)\gg - i\int_{-\infty}^{\sigma} d^4x' \, \tilde{\mathscr{L}}(x')|\tilde{\rho}(\sigma')\gg \tag{79}$$

or in Hilbert space notations :

$$\tilde{\rho}(\sigma) = \tilde{\rho}^{(o)}(\sigma) - \frac{i}{\hbar}\int_{-\infty}^{\sigma} d^4x' [\tilde{\mathscr{H}}(x'),\tilde{\rho}(\sigma')] \tag{80}$$

The electric dipole interaction is written as :

$$\sum_{\alpha,\beta} g_{\alpha\beta}\bar{\psi}_\alpha(x)\gamma^5\sigma^{\mu\nu}\psi_\beta(x)F_{\mu\nu}(x) + \text{h.c.} \tag{81}$$

with[17] :

$$\sigma^{\mu\nu} = \frac{1}{2}(\gamma^\mu\gamma^\nu - \gamma^\nu\gamma^\mu) = (\vec{\alpha}, i\vec{\Sigma})$$

and :

$$\gamma^5\sigma^{\mu\nu} = (-\vec{\Sigma}, -i\vec{\alpha})$$

The density matrix elements that we need can be written(*) :

$$Tr_R<0|\psi_\alpha(x_1)\tilde{\rho}(\sigma)\bar{\psi}_\beta(x_2)|0> \text{ or } Tr_R<0|\psi_\alpha(x_1)\tilde{\rho}(\sigma)\psi^\dagger_\beta(x_2)|0>\gamma^o \tag{82}$$

where Tr_R is the trace operation with respect to the radiation field and where x_1 and x_2 belong to σ.

(*) We use a simplified notation where all bispinor indices are omitted.

From the integral equation for $\tilde{\rho}$ we may write :

$$\mathrm{Tr}_R \langle 0|\psi_\alpha(x_1)\tilde{\rho}(\sigma)\bar{\psi}_\beta(x_2)|0\rangle = \mathrm{Tr}_R \langle 0|\psi_\alpha(x_1)\tilde{\rho}^{(0)}(\sigma)\bar{\psi}_\beta(x_2)|0\rangle$$

$$+ \mathrm{Tr}_R \sum_{\alpha'} \frac{g_{\alpha\alpha'}}{i\hbar} \int_{-\infty}^{\sigma} d^4x_1' \int d\sigma'_{\mu'} \langle 0|\psi_\alpha(x_1)\bar{\psi}_\alpha(x_1')|0\rangle \gamma^5 \sigma^{\mu\nu} F_{\mu\nu}(x_1')$$

$$\langle 0|\psi_{\alpha'}(x_1')\tilde{\rho}(\sigma')\bar{\psi}_\beta(x_2')|0\rangle \gamma^{\mu'} \langle 0|\psi_\beta(x_2')\bar{\psi}_\beta(x_2)|0\rangle \tag{83}$$

+ 3 similar terms.

This equation is obviously covariant and displays the usual Feynman propagators.

We have used the closure relation for one particle states :

$$\sum_\beta \int d\sigma'_{\mu'}\, \bar{\psi}_\beta(x_2')|0\rangle \gamma^{\mu'} \langle 0|\psi_\beta(x_2')$$

$$= \sum_\beta \int d^3p\, c^\dagger_\beta(p)|0\rangle\langle 0|c_\beta(p) \tag{84}$$

To recover the equations of section 10 we write $\langle 0|\psi_\alpha(x_1)\tilde{\rho}(\sigma)\bar{\psi}_\beta(x_2)|0\rangle$ as a sum of density matrix elements between energy-momentum states :

$$\langle 0|\psi_\alpha(x_1)\tilde{\rho}(\sigma)\bar{\psi}_\beta(x_2)|0\rangle = \sum_{m_\alpha, m_\beta} \langle 0|\psi_{\alpha,m_\alpha}(x_1)\tilde{\rho}(\sigma,\vec{p}_o)\bar{\psi}_{\beta,m_\beta}(x_2)|0\rangle / h^3 \tag{85}$$

with :

$$\langle 0|\psi_{\alpha,m_\alpha}(x_1)\tilde{\rho}(\sigma,\vec{p}_o)\bar{\psi}_{\beta,m_\beta}(x_2)|0\rangle = \int d^3p_1 d^3p_2 \left(\frac{E^o_\alpha}{E_\alpha(p_1)} \frac{E^o_\beta}{E_\beta(p_2)} \right)^{1/2}$$

$$\langle 0|c_\alpha(p_1)u^\alpha(p_1)\tilde{\rho}(\sigma,\vec{p}_o)\bar{u}^\beta(p_2)c^\dagger_\beta(p_2)|0\rangle \exp\Big[i(\vec{p}_1.\vec{r}_1 - E_\alpha(p_1)t_1)/\hbar$$

$$- i(\vec{p}_2.\vec{r}_2 - E_\beta(p_2)t_2)/\hbar \Big] \tag{86}$$

where the integrations are performed over small intervals around $\vec{p}_o + m_\alpha \hbar\vec{k}$ and $\vec{p}_o + m_\beta \hbar\vec{k}$.

In a given reference frame we select the hyperplanes $t =$ constant for the hypersurfaces σ and we find :

$$<0|\psi_{\alpha,m_\alpha}(x_1)\tilde{\rho}(t,\vec{p}_0)\bar{\psi}_{\beta,m_\beta}(x_2)|0> = <0|\psi_{\alpha,m_\alpha}(x_1)\tilde{\rho}^{(0)}(\vec{p}_0)\bar{\psi}_{\beta,m_\beta}(x_2)|0>$$

$$+\sum_{\alpha'}\frac{g_{\alpha\alpha'}}{i\hbar}\int_{-\infty}^{t}dt'\,\frac{E^o_\alpha}{E_\alpha(m_\alpha)}\,u^\alpha(m_\alpha)\,\bar{u}^\alpha(m_\alpha)$$

$$\exp\left\{\frac{i}{\hbar}\left[(\vec{p}_0+m_\alpha\hbar\vec{k})(\vec{r}_1-\vec{r}'_1)-(\vec{p}_0+m_\beta\hbar\vec{k})(\vec{r}_2-\vec{r}'_2)-(E_\alpha(m_\alpha)-E_\beta(m_\beta))(t-t')\right]\right\}$$

$$\gamma^5\sigma^{\mu\nu}F^{(+)}_{\mu\nu}(x'_1)$$

$$<0|\psi_{\alpha',m_\alpha-1}(x'_1)\tilde{\rho}(t',\vec{p}_0)\bar{\psi}_{\beta,m_\beta}(x'_2)|0>\gamma^o\,\frac{E^o_\beta}{E_\beta(m_\beta)}\,u^\beta(m_\beta)\,\bar{u}^\beta(m_\beta) \quad (87)$$

+ 3 similar terms.

If we write (*)

$$\vec{r}_1-\vec{r}'_1 = \vec{r}_2-\vec{r}'_2 = \gamma\vec{v}\tau \quad , \quad t-t' = \gamma\tau \quad (88)$$

and use (64),we recover the same invariant propagator as before, apart from the rank-two bispinors :

$$\frac{E^o_\alpha}{E_\alpha(m_\alpha)}\,u^\alpha(m_\alpha)\,\bar{u}^\alpha(m_\alpha) \quad \text{and} \quad \frac{E^o_\beta}{E_\beta(m_\beta)}\,u^\beta(m_\beta)\,\bar{u}^\beta(m_\beta) \quad (89)$$

which reduce to unity only in the rest frame when basis spinors are eigenvectors of the z-component of the spin in that frame.

In any other frame the spin component along an arbitrary z-axis is not any more a good quantum number. This means that in a relativistic theory we cannot any more write a scalar propagator for Zeeman sub-levels of molecular levels. This might lead to interesting (however quite small) new effects.

Another point is that if we take the trace over the radiation field of both sides of the previous equation,we find the semi-classical theory with the classical electromagnetic field at point x'_1. But we could also introduce the density matrix elements between radiation states and a photon propagator just like we did for the matter field. The previous formalism can thus be easily extended to be fully quantized.

(*) For the relation between momentum and velocity see reference 42.

II - DENSITY MATRIX DIAGRAMS AND DIAGRAMMATIC RULES

The density matrix equations can be solved analytically only in the case of a single travelling monochromatic plane wave within the rotating-wave approximation, i.e. the case of a single Fourier component.(*) As soon as a second wave is added, solutions can be written only in terms of continued fractions[18] and practical expressions require either the rate equation approximation or perturbation expansions with respect to the second wave[19,4]. If the applied fields are either not monochromatic or not plane waves, the only choice left is between purely numerical solutions or perturbation expansions with respect to all fields. Numerical methods have been used with success in the case of strong fields with arbitrary spatial or temporal dependence. However, perturbative calculations which provide closed-form expressions bring a lot of insight and are preferred in many cases for a detailed physical discussion. It is well-known that diagrams are a useful tool to keep track of the various terms in perturbative expansions. This is illustrated by the wide use of Feynman diagrams. Furthermore besides the interest of diagrams in book-keeping terms, diagrammatic rules can be of great help to write down the solutions automatically without having to constantly refer to the basic equations. These reasons explain why density matrix diagrams have already appeared in a number of studies concerning laser spectroscopy[1,2,5,20-26]

1 - Semiclassical space-time diagrams

A perturbation approach and even the idea of a diagram are strongly suggested by the integral forms (48,52,66) of the density matrix equations. We can use them to relate the perturbation order ξ to the perturbation order $(\xi - 1)$:

$$\langle m_\alpha | \rho^{(\xi)}_{\alpha\beta}(\vec{r},t,\vec{p}_o) | m_\beta \rangle$$

$$= \sum_{\beta'} (i\Omega^+_{\alpha\beta'}) \int_0^{+\infty} d\tau G_{\alpha\beta}(\tau) U_o^{+*}(\vec{r} - \gamma\vec{v}\tau) \exp\left[-i[\omega(t-\gamma\tau) - \vec{k}.(\vec{r}-\gamma\vec{v}\tau) + \varphi^+]\right]$$

$$\times \langle m_\alpha - 1 | \rho^{(\xi-1)}_{\beta'\beta}(\vec{r}-\gamma\vec{v}\tau, t-\gamma\tau, \vec{p}_o) | m_\beta \rangle \qquad (90)$$

plus seven other terms with the same structure, where the Rabi

---

(*) At resonance, analytic solutions are known in the case of standing waves with arbitrary spatial dependence, or with travelling waves having very special time dependences.[27,28]

pulsations $\Omega^{\pm}_{\beta\alpha} = \langle \beta | \vec{\mu}.\hat{e}^{*} | \alpha \rangle E_o / 2\hbar$ can be found in reference [21] for degenerate levels and arbitrary laser polarization vectors $\hat{e}$.

$\langle m_\alpha | \rho_{\alpha\beta}(\vec{r},t,p_o) | m_\beta \rangle$ appears as the sum of contributions consisting of a source term, expressed at some previous epoch, times a propagator over the elapsed time τ. This structure suggests a diagram such as that of fig. 3a for each of these contributions. The vertical axis corresponds to the time flow

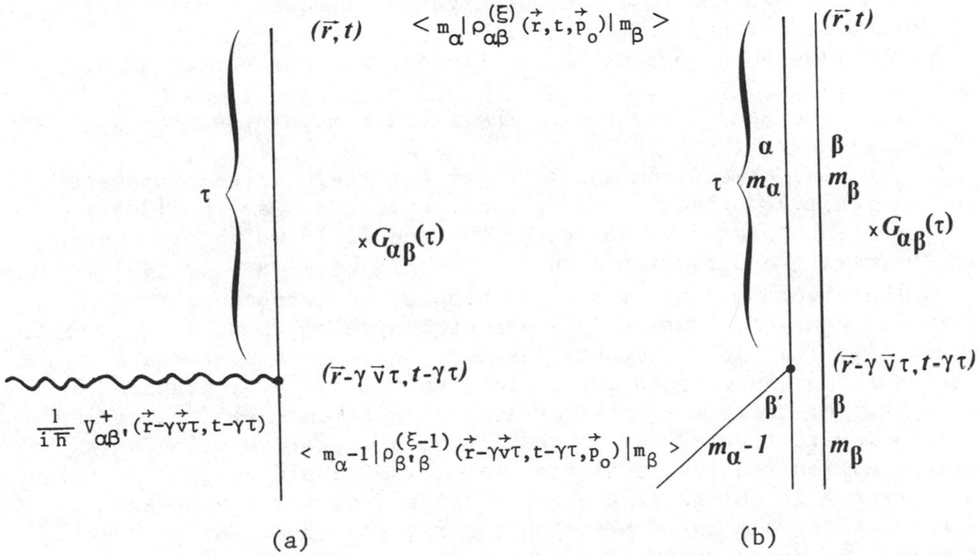

Figure 3

and the free propagator $G_{\alpha\beta}$ (retarded Green function) of the calculated matrix element is associated with the vertical bar. An interaction takes place at the vertex, and the electromagnetic field is represented by the wavy line coming from the side. Because of time dilation it is not surprising that, if τ is the proper time elapsed in the molecular frame, the two corresponding space-time points are separated by the 4-vector $(\gamma\vec{v}\tau, \gamma\tau)$ in the laboratory frame. The contribution to a density matrix element at the space-time point $(\vec{r},t)$ in the laboratory frame is therefore obtained from the corresponding diagram by the following rule : one first multiplies the previous density matrix element by the interaction matrix element, both taken at the laboratory space-time point $(\vec{r}-\gamma\vec{v}\tau, t-\gamma\tau)$; this quantity is then multiplied by the propagator of the calculated density matrix element during the time τ and integrated over all times τ. Such diagrams were first used used with a Galilean transformation of coordinates in reference [1]. It quickly appeared that more information could be carried by the diagrams and new rules were given in reference [2]. First, the vertical bar was replaced by a double bar to distinguish between both indices of a density matrix element. Vertices denoting the interactions with the electromagnetic field are of course on the vertical column associated with the subscript of ρ that changes in the interaction. Each segment is labeled by a letter α and an integer number m_α which indicate the energy level E_α and the linear momentum state $\vec{p}_\rho + m_\alpha \hbar \vec{k}$ on which the density operator is projected. Second, at each vertex the interaction with the field is represented by lines with four different orientations to distinguish between the positive and negative frequency parts and two different directions of propagation :

- the line goes upwards if the field component varies as $\exp(-i\omega t)$;
- the line goes downwards if the field component varies as $\exp(+i\omega t)$;
- the line comes from the left for the field which propagates in the positive direction along the $\hat{z}$-axis ((+) wave in (30));
- the line comes from the right for the field which propagates in the negative direction along the $\hat{z}$-axis ((-) wave in (30)).

The eight possible vertex topologies corresponding to the various terms of equation (90) are displayed on Table I. This table gives also the field component expression to be used at each vertex and an associated multiplicative factor (Rabi frequency).[21] Only half of these vertices correspond to a resonant interaction. For an interaction vertex on the left column, the rotating-wave approximation requires that the field line should go upwards for an increase in molecular energy and downwards for a decrease in energy. This rule is reversed on the right column. For each vertex topology we get a corresponding one-vertex diagram, of the type represented on fig.3b, which connects a density matrix element at a given perturbation order with a coupled density matrix element at the previous perturbation order. These elementary diagrams are then stacked on top of each other to represent a full perturbation chain, starting with equilibrium populations (zeroth order)[(*)] and ending with the desired perturbation order. Each of these diagrams represents a different non-linear process.

Different time intervals τ, τ', τ'' in the molecular frame separate adjacent vertices on the full diagram. This determines the laboratory space-time coordinates of the successive interaction vertices for a molecule with velocity $\vec{v}$:

$$(\vec{r},t) \quad ; \quad (\vec{r}-\gamma\vec{v}\tau, t-\gamma\tau) \quad ; \quad (\vec{r}-\gamma\vec{v}\tau-\gamma\vec{v}\tau' \, , \, t-\gamma\tau-\gamma\tau')\ldots$$

(*) Various forms of the zeroth order populations will be given in the applications (see also section I).

Table 1

Graphical representation	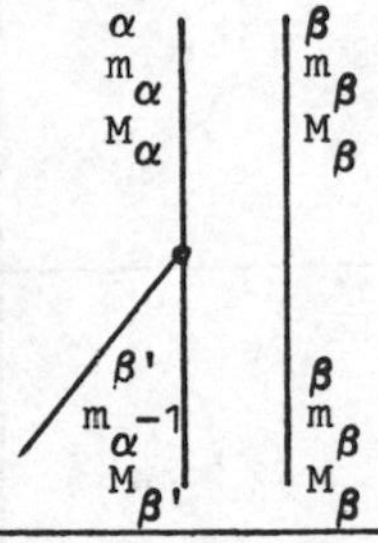	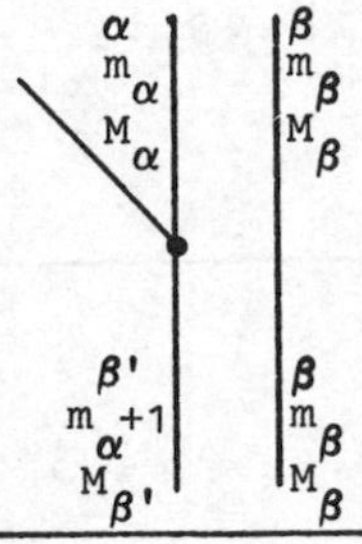	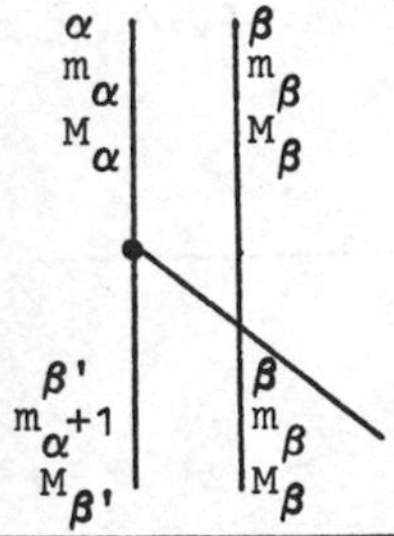	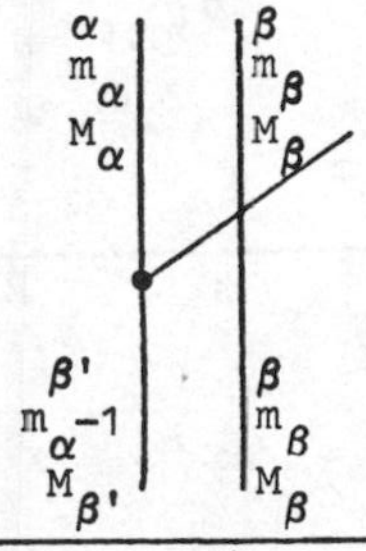
Consequence of the rotating-wave approximation (r.w.a.)	$E_\alpha > E_{\beta'}$	$E_\alpha < E_{\beta'}$	$E_\alpha > E_{\beta'}$	$E_\alpha < E_{\beta'}$
Field component $U^{\epsilon'}_{q\epsilon''}(\vec{r})\exp\left[\epsilon' i(\omega t - \epsilon''\vec{k}.\vec{r} + \varphi_{q\epsilon''}\right]$ $U^{+}_{q\epsilon''} \equiv U_{q\epsilon''}$; $U^{-}_{q\epsilon''} \equiv U^{*}_{q\epsilon''}$	$\epsilon' \equiv -$, $\epsilon'' \equiv +$	$\epsilon' \equiv +$, $\epsilon'' \equiv +$	$\epsilon' \equiv -$, $\epsilon'' \equiv -$	$\epsilon' \equiv +$, $\epsilon'' \equiv -$
Multiplicative factor at the vertex $\Omega_{\alpha\beta q\epsilon''} = <\alpha \| \mu_{q\epsilon''} \| \beta> \dfrac{E^{o}_{q\epsilon''}}{2\hbar}$	$i\,\Omega_{\alpha\beta' q^+}$ $(M_\alpha = M_{\beta'} + q^+)$	$i\,\Omega_{\beta'\alpha q^+}$ $(M_\alpha = M_{\beta'} - q^+)$	$i\,\Omega_{\alpha\beta' q^-}$ $(M_\alpha = M_{\beta'} + q^-)$	$i\,\Omega_{\beta'\alpha q^-}$ $(M_\alpha = M_{\beta'} - q^-)$

Table 1 (continued)

Graphical representation	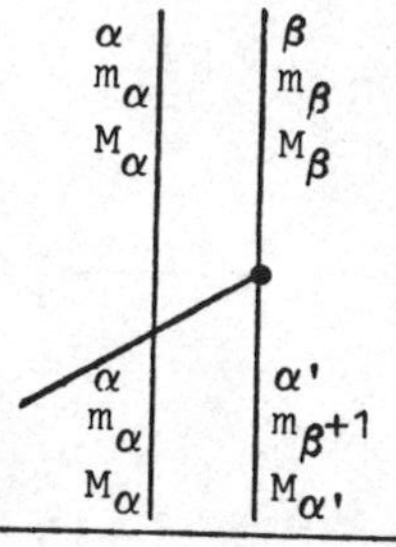	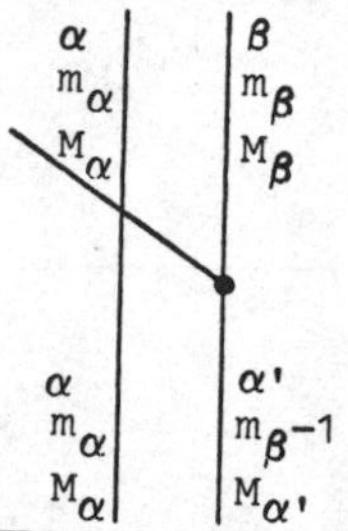	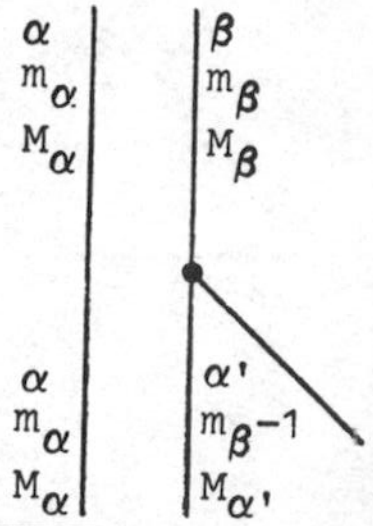	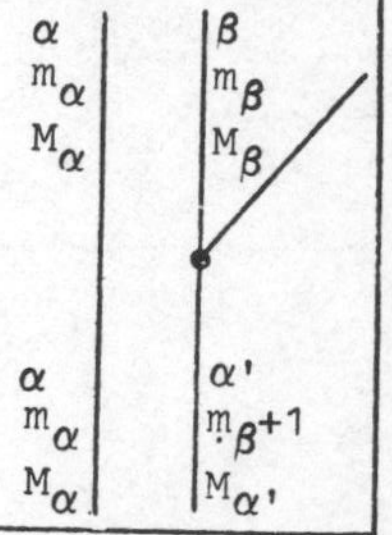
Consequence of the rotating-wave approximation (r.w.a.)	$E_{\alpha'} > E_\beta$	$E_{\alpha'} < E_\beta$	$E_{\alpha'} > E_\beta$	$E_{\alpha'} < E_\beta$
Field component $U^{\epsilon'}_{q\epsilon''}(\vec{r})\exp\left[\epsilon' i(\omega t-\epsilon''\vec{k}\cdot\vec{r}+\varphi_{q\epsilon''}\right]$ $U^{+}_{q\epsilon''} \equiv U_{q\epsilon''}$; $U^{-}_{q\epsilon''} \equiv U^{*}_{q\epsilon''}$	$\epsilon' \equiv -$, $\epsilon'' \equiv +$	$\epsilon' \equiv +$, $\epsilon'' \equiv +$	$\epsilon' \equiv -$, $\epsilon'' \equiv -$	$\epsilon' \equiv +$, $\epsilon'' \equiv -$
Multiplicative factor at the vertex $\Omega_{\alpha\beta q\epsilon''} = \langle\alpha\|\mu_{q\epsilon''}\|\beta\rangle \dfrac{E^{o}_{q\epsilon''}}{2\hbar}$	$-i\Omega_{\alpha'\beta q+}$ $(M_\beta = M_{\alpha'} - q^{+})$	$-i\Omega_{\beta\alpha' q+}$ $(M_\beta = M_{\alpha'} + q^{+})$	$-i\Omega_{\alpha'\beta q-}$ $(M_\beta = M_{\alpha'} - q^{-})$	$-i\Omega_{\beta\alpha' q-}$ $(M_\beta = M_{\alpha'} + q^{-})$

To each segment is associated the propagator of the corresponding density matrix element, and to each vertex is associated a field component and a multiplicative factor according to table I. The product of all these elements is then integrated over all times τ,τ',τ'' ... to give the desired density matrix element.

The contribution of the diagram to the macroscopic polarization (or equivalently the non-linear susceptibility) is obtained from this density matrix element and equations (32) and (34).

The contribution to the absorbed power is given by equations (33) and (34). Within the rotating-wave approximation, it is equivalent to calculate the energy absorbed per unit volume and per unit time in units of $\hbar\omega$, or the rate of change of the upper level population, as discussed in Appendix A.

$$\overline{W}_{abs} = \hbar\omega \sum_{\alpha_o} \int d^3p_o\, d^3r \left[\frac{\partial}{\partial t} < \vec{r}|\, \rho_{\beta\beta}(\alpha_o,\vec{p}_o,t)\,|\,\vec{r} > \right]_{rad}$$

$$= \hbar\omega \sum_{\alpha_o} \int \frac{d^3p_o\, d^3r}{h^3} \left[\frac{\partial}{\partial t} \sum_{m_\beta} < m_\beta|\, \rho_{\beta\beta}(\alpha_o,\vec{p}_o,t)\,|\,m_\beta > \right]_{rad} \tag{91}$$

The absorbed power is for each field :

$$\overline{W}^{\pm}_{abs} = 2\,\hbar\omega\mathrm{Re} \sum_{\alpha,\beta} (i\vec{\mu}_{\beta\alpha}.\vec{E}^{\pm *}/2\hbar)$$

$$\int d^3r \sum_{\alpha_o} \int d^3p_o \sum_{m_\alpha} <\vec{p}_o + m_\alpha \hbar\vec{k}|\, \rho_{\alpha\beta}(\alpha_o,\vec{p}_o,t)\,|\,\vec{p}_o + (m_\alpha \pm 1)\hbar\vec{k}>/h^3 \tag{92}$$

So that the absorbed power may be obtained by adding a last vertex to any diagram in order to terminate the diagram by an upper level population change (see fig. 4). The absorbed power is then

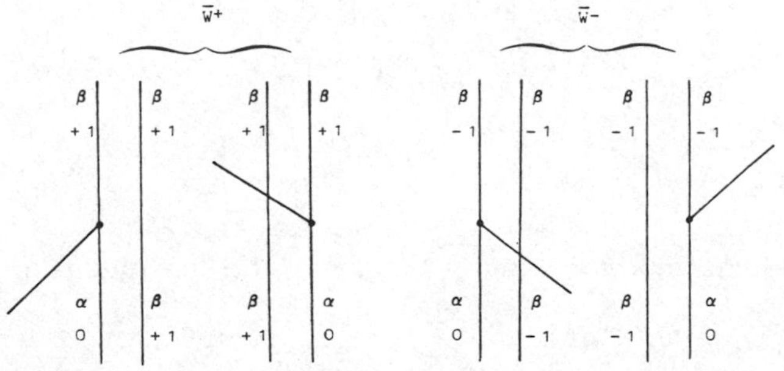

Fig. 4. Possible structures for the last vertex which corresponds to the absorbed power for a (+) wave coming from the left or a (-) wave coming from the right.

obtained :

1) by applying the usual diagrammatic rules to this last vertex,

2) by adding complex conjugate diagrams or by taking twice the real part (2Re) of one of them,

3) by multiplication by $\hbar\omega$ (if $E_\beta > E_\alpha$ and by $-\hbar\omega$ if $E_\beta < E_\alpha$),

4) by a final integration over the equilibrium momentum distribution and over space.

Here, two points of view may be adopted,

- first, all diagrams may start with a single momentum class $\vec{p}_o$:

$$<\vec{p}_o|\rho^{(o)}_{\alpha\alpha}|\vec{p}_o> \equiv <0|\rho^{(o)}_{\alpha\alpha}|0>$$

the distribution $F(\vec{p}_o)$ is then common to all diagrams;

- second, one can decide to have a last coherence of the form:

$$<\vec{p}'_o|\rho_{\alpha\beta}|\vec{p}'_o \pm \hbar\vec{k}> \equiv <0|\rho_{\alpha\beta}| \pm 1>$$

and all diagrams start with a different momentum class.

This means that, instead of calculating all the contributions between $\vec{p}_o + m_\alpha\hbar\vec{k}$ and $\vec{p}_o + (m_\alpha \pm 1)\hbar\vec{k}$ originating from a single $(\alpha_o,\vec{p}_o)$, we focus on all possible contributions to a single coherence $<\vec{p}'_o|\rho_{\alpha\beta}|\vec{p}'_o \pm \hbar\vec{k}>$ resulting from the injection of molecules in all the energy-momentum states coupled with $|E_\alpha,\vec{p}'_o>$ or $|E_\beta,\vec{p}'_o \pm \hbar\vec{k}>$.

This new point of view is obviously obtained by shifting to the new variable $\vec{p}'_o = \vec{p}_o + m_\alpha\hbar\vec{k}$:

$$\overline{W}^{\pm} = 2\hbar\omega\ \mathrm{Re}\sum_{\alpha,\beta} (i\vec{\mu}_{\beta\alpha}.\vec{E}^{\pm *}/2\hbar)$$

$$\int d^3r\int d^3p'_o <\vec{p}'_o|\sum_{\alpha_o,m_\alpha}\rho_{\alpha\beta}(\alpha_o,\vec{p}'_o - m_\alpha\hbar\vec{k},\vec{r},t)|\vec{p}'_o \pm \hbar\vec{k}>/h^3 =$$

$$2\hbar\omega\,\mathrm{Re}\sum_{\alpha,\beta}(i\vec{\mu}_{\beta\alpha}.\vec{E}^{\pm *}/2\hbar).\int d^3r\int d^3p'_o < \vec{p}'_o|\rho_{\alpha\beta}(\vec{p}'_o,\vec{r},t)|\vec{p}'_o \pm \hbar\vec{k}>/h^3 \tag{93}$$

The new density matrix $$\rho(\vec{p}'_o,\vec{r},t) = \sum_{\alpha_o,m_\alpha}\rho(\alpha_o,\vec{p}'_o - m_\alpha\hbar\vec{k},\vec{r},t) \tag{94}$$

satisfies equations (37) and (39) with the new pumping operator :

$$\Lambda(\vec{p}'_o,t) = \sum_{\alpha_o,m_\alpha}\Lambda(\alpha_o,\vec{p}'_o - m_\alpha\hbar\vec{k},t) \tag{95}$$

One can shift from one point of view to the other by adding a small integer to the m_α values on both sides of the diagrams.

2 - Second-quantized density matrix diagrams

The previous diagrammatic rules have the advantage of simplicity in their application to the computation of the various terms. They have a disadvantage which is that they do not display at first sight a simple interpretation in terms of emitted or absorbed photons. Within the second quantization formalism of section I-11, it is clear that density matrix diagrams can be interpreted as double Feynman diagrams. Each of these Feynman diagrams represents a probability amplitude for a given process in which particles are created and others are destroyed. The double diagram exhibits the correlations between processes and its contribution corresponds to the interference term between the probability amplitudes of the two processes (which may be either identical or different). It is therefore attractive to have identical rules for both sides of the diagrams, such that the emission or absorption of photons have an intuitive representation. This requirement leads us to the correspondence rules between semiclassical density matrix diagrams and double Feynman diagrams given in fig. 5.

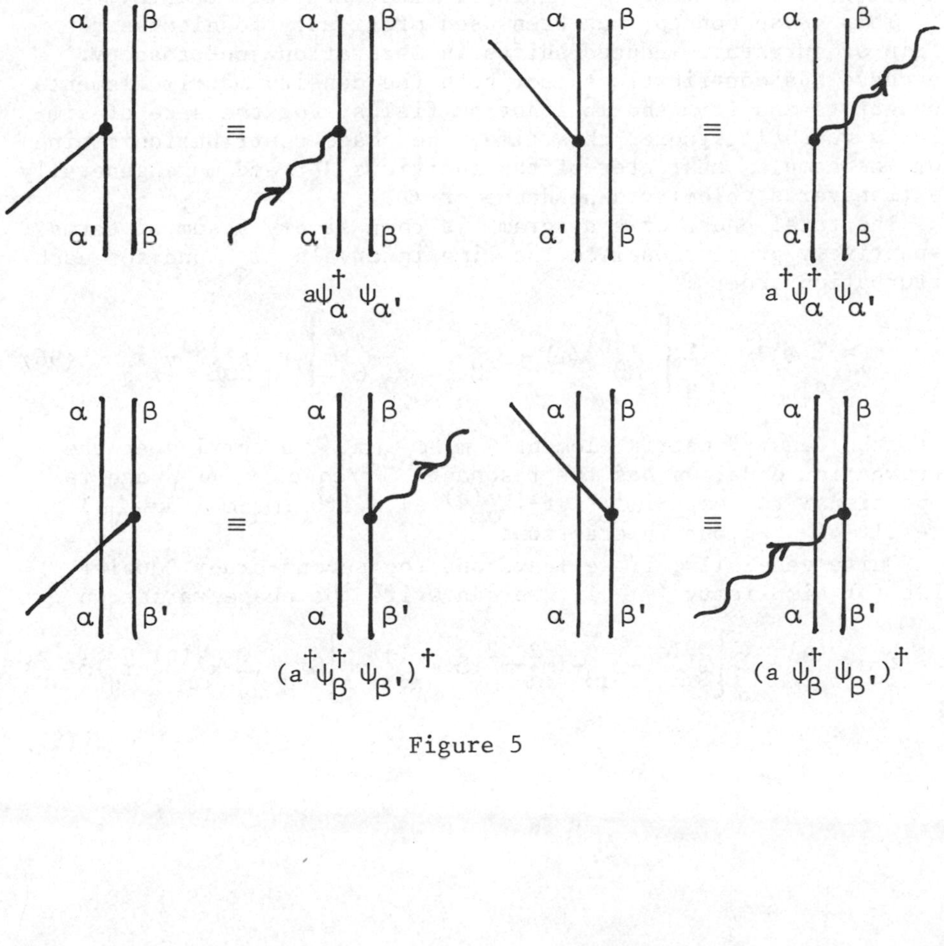

Figure 5

With these new diagrams we have the choice between two possibilities :

1) Use second-quantization for the electromagnetic field as well as for molecules. It is then necessary to introduce a propagator for the radiation density matrix elements.

2) Use the semiclassical approximation in which the classical electromagnetic field enters at each vertex with the corresponding space-time argument.

3 - Diagram phase

The contribution of a given diagram is a complex number with an amplitude A and a phase Φ which is then integrated over velocity and over the various time intervals τ_q :

$$\int d^3v d^n\tau_q \; A(\vec{v},\tau_q)\; e^{i\Phi(\vec{v},\tau_q)} \qquad (q=1,n)$$

The amplitude A is usually a slow function of velocity and of the τ'_q s , whereas the exponential factor exp $i\Phi$ oscillates rapidly except when the phase is stationary, and the greatest contribution to the velocity and τ_q integrals will come from the vicinity of stationary points, if there are such points.

This phase concept has been used previously to discuss the origin of curvature-induced shifts in saturation spectroscopy.[1] The phase has contributions from both the density matrix elements propagators and from the interaction fields. For the sake of simplicity we shall ignore, this time, the phase contribution coming from the complex character of the functions U_o and more generally the transverse velocity dependence of Φ.

The total phase of a diagram is then simply a sum of terms respectively proportional to the time intervals τ_q , one for each perturbation order :

$$\Phi = \sum_q \Phi^{(q)} = \sum_q \left[\omega'^{(q)}_{\alpha\beta}(\omega) - \omega_{\alpha\beta} - (m^2_\alpha - m^2_\beta)\delta \right] \tau_q + k'^{(q)}_{\alpha\beta} v_z \tau_q \qquad (96)$$

where the density matrix element $\langle m_\alpha | \rho_{\alpha\beta} | m_\beta \rangle$ occurring at the perturbation order q has the resonant frequency $\omega_{\alpha\beta}$ and is proportional to $\exp(-i\omega'^{(q)}_{\alpha\beta} t + ik'^{(q)}_{\alpha\beta} z)$ (with $(m_\alpha - m_\beta)k = k'_{\alpha\beta}$) as a result of previous interactions.

Quite generally, if we leave out the second-order Doppler shift for simplicity ($\gamma \equiv 1$), we can write the phase variation :

$$\delta\Phi = \sum_q \delta\Phi^{(q)} = \sum_q \left[\omega'^{(q)}_{\alpha\beta} - \omega_{\alpha\beta} - (m^2_\alpha - m^2_\beta)\delta + k'^{(q)}_{\alpha\beta} v_z \right] \delta\tau_q + \left(\sum_q k'^{(q)}_{\alpha\beta} u\tau_q\right)\delta\left(\frac{v_z}{u}\right)$$

(97)

from which we infer the stationarity conditions :

$$\begin{cases} \omega'^{(q)}_{\alpha\beta} - \omega_{\alpha\beta} - (m_\alpha^2 - m_\beta^2)\delta + k'^{(q)}_{\alpha\beta} v_z = 0 & \text{(A)} \\ \sum_q k'^{(q)}_{\alpha\beta} \tau_q = 0 & \text{(B)} \end{cases} \qquad (98)$$

In this simple case, if these conditions are fulfilled, the phase will be identically equal to zero.

If some of these conditions are not fulfilled, the corresponding ranges $\delta\tau_q$ or $\delta(v_z/u)$ over which there will be a significant contribution to the integrals, are such that $\delta\Phi \leqslant 1$ and are greatly reduced. Therefore, according to the number of conditions simultaneously satisfied, the process considered will have more or less importance.

First, let us consider the case where only one of the resonance conditions (A) is satisfied(*) (with $\omega > 0$ and values of v_z within the range of thermal velocities), for example that for $q = 2$. For each v_z in the range of values defined by the Maxwell-Boltzmann distribution, there will be a value of ω such that :

$$\omega'^{(2)}_{\alpha\beta} - \omega_{\alpha\beta} - (m_\alpha^2 - m_\beta^2)\delta + k'^{(2)}_{\alpha\beta} v_z = 0$$

is satisfied. The process is therefore Doppler-broadened with a maximum for $\omega'^{(2)}_{\alpha\beta} = \omega_{\alpha\beta} + (m_\alpha^2 - m_\beta^2)\delta$ <u>except</u> if $k^{(2)}_{\alpha\beta} = 0$ (case of Doppler-free two-photon spectroscopy). The range of values for τ_2 is limited by $(k^{(2)}_{\alpha\beta} u)^{-1}$, whereas the range of values for $\tau_{q \neq 2}$ is of the order of $1/\Delta\omega_q$, where $\Delta\omega_q$ is the detuning in the $\Phi_{q \neq 2}$.

The integral $\int d(v_z/u)\, d^n\tau_q \; A\, e^{i\Phi}$ is therefore of the order of :

$$\frac{1}{\Delta\omega_1} \, \frac{1}{k^{(2)}_{\alpha\beta} u} \, \frac{1}{\Delta\omega_3} \, \cdots\cdots \, \frac{1}{\Delta\omega_n}$$

If $k'^{(o)}_{\alpha\beta} = 0$ (e.g. DFTPS) the contribution domain for τ_2 is of the order of $1/\gamma^{(2)}_{\alpha\beta}$ and the process is of the order of :

$$\frac{1}{\Delta\omega_1} \, \frac{1}{\gamma^{(2)}_{\alpha\beta}} \, \frac{1}{\Delta\omega_3} \, \cdots\cdots \, \frac{1}{\Delta\omega_n}$$

(*) This might simply mean that the other conditions require $\omega < 0$ or values of $v_z \gg u$ for the tuning range of ω in which this first condition is satisfied for $v_z = 0$.

Second, let us consider the case where two of the resonance conditions (A) can be satisfied for a value v_o of v_z within the range of thermal velocities :

$$\omega'^{(1)}_{\alpha\beta} - \omega_{\alpha\beta} - (m_\alpha^2 - m_\beta^2)\delta + k'^{(1)}_{\alpha\beta}\, v_z = 0$$

$$\omega'^{(2)}_{\alpha'\beta'} - \omega_{\alpha'\beta'} - (m_{\alpha'}^2 - m_{\beta'}^2)\delta + k'^{(2)}_{\alpha'\beta'}\, v_z = 0$$

For the corresponding value of ω, we expect an enhanced contribution. Since this enhanced contribution would correspond to a well-defined velocity class,we expect the signal to be free of Doppler-broadening. Since A is a slow function of v_z, we may remove $A(v_o)$ from the v_z integral which reduces to :

$$\int dv_z \; \exp\left[i \sum_q k'^{(q)}_{\alpha\beta}\, v_z \tau_q \right] \tag{99}$$

This differs from zero only if condition (B) is satisfied. In the previous example this requires opposite signs for $k'^{(1)}_{\alpha\beta}$ and $k'^{(2)}_{\alpha'\beta'}$ otherwise all the τ's would have to be zero.

A complete calculation of the various possible diagrams in the case of plane waves (with multiple resonances) can be found in the Appendix of reference 22.

4 - Energy-momentum diagrams

To represent a given process,we may use a diagram in the energy-momentum domain instead of the space-time domain. (See fig. 6). The time axis is replaced by the energy axis and the optical $\hat{z}$-axis becomes the axis for the momentum component $\vec{p}.\hat{z}$. Energy-momentum states lie on a series of hyperbolas (parabolas in the non-relativistic limit), one for each discrete rest energy level. The matrix element of the interaction Hamiltonian connecting two such states is represented by an arrow :

- continuous arrow from (β, m_β) to (α, m_α) for $V_{\alpha\leftarrow\beta}$ acting before $\langle m_{\beta} | \rho_{\beta\beta'} | m_{\beta'} \rangle$ (left column of the space-time diagrams).

- dotted arrow from (α, m_α) to (β, m_β) for $V_{\alpha\rightarrow\beta}$ acting after $\langle m_{\alpha'} | \rho_{\alpha'\alpha} | m_{\alpha} \rangle$ (right column of the space-time diagrams).

(these rules differ from those used in references 22-24). Off-diagonal matrix elements (coherences) are represented by wavy arrows going from (β, m_β) to (α, m_α) for $\langle m_\beta | \rho_{\beta\alpha} | m_\alpha \rangle$

Unmodulated populations are represented by stars. Finally a perturbation order labels each element and indicates the time sequence.

Since the energy and momentum exchanges corresponding to each interaction can be read respectively on the vertical and on the

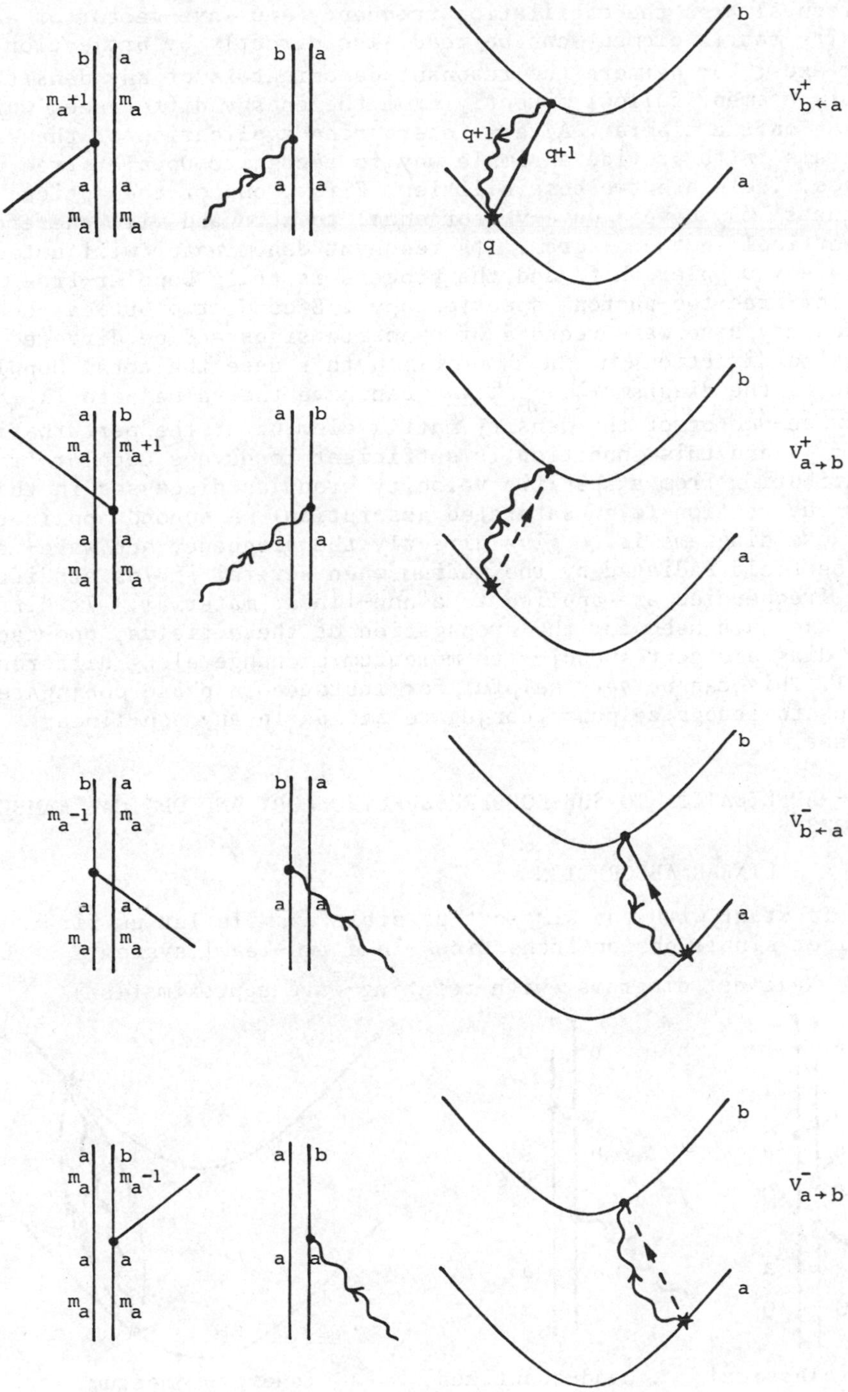
b
a
m_a+1
m_a
a
a
m_a
m_a
b
a
a
a
q+1
q+1
q
b
a
$V^+_{b\leftarrow a}$
a
b
m_a
m_a+1
a
a
m_a
m_a
a
b
a
a
b
a
$V^+_{a\rightarrow b}$
b
a
m_a-1
m_a
a
a
m_a
m_a
b
a
a
a
b
a
$V^-_{b\leftarrow a}$
a
b
m_a
m_a-1
a
a
m_a
m_a
a
b
a
a
b
a
$V^-_{a\rightarrow b}$

Figure 6

horizontal axes the oscillation frequency and wave-vector of each density matrix element can be read also directly by projection on both axes. Furthermore the resonant denominators of any density matrix element follows directly from the energy differences which appear on the diagram. A very interesting application of these diagrams is to provide a simple way to recognize Doppler-free processes. There are two possibilities. First, one of the optical coherences may have a wave-vector equal to zero and will therefore be vertical in the diagram. Its resonant denominator will not involve any Doppler shift and the process is truly Doppler-free (e.g. Doppler-free two-photon spectroscopy). Second, two optical coherences may have wave-vectors of opposite signs and be directed in opposite directions in the diagram. In this case the total Doppler phase of the diagram $\Sigma\, k_{\alpha\beta}^{(j)}\tau_j v_z$ can have the value zero ($k_{\alpha\beta}^{(j)}$ is the wave-vector of the density matrix element at the perturbation order j) and this condition is sufficient to have a Doppler-free contribution from a specific velocity group as discussed in the previous section (e.g. saturated absorption). A second application of these diagrams is to give directly the frequency and wave-vector of the field radiated by the medium when several fields at different frequencies are applied to a non-linear material. (If different axes are used for the propagation of these fields, one can draw diagrams corresponding to momentum exchange along different axes). This can be very helpful, for instance in phase conjugate optics, to recognize phase conjugate fields in any non-linear process.

III - APPLICATION TO SUB-DOPPLER SPECTROSCOPY AND OPTICAL RAMSEY FRINGES

A - LINEAR ABSORPTION

To start with the simplest possible example let us first consider single photon transitions in a two-level system $E_b > E_a$

Relevant diagrams (with rotating-wave approximation)

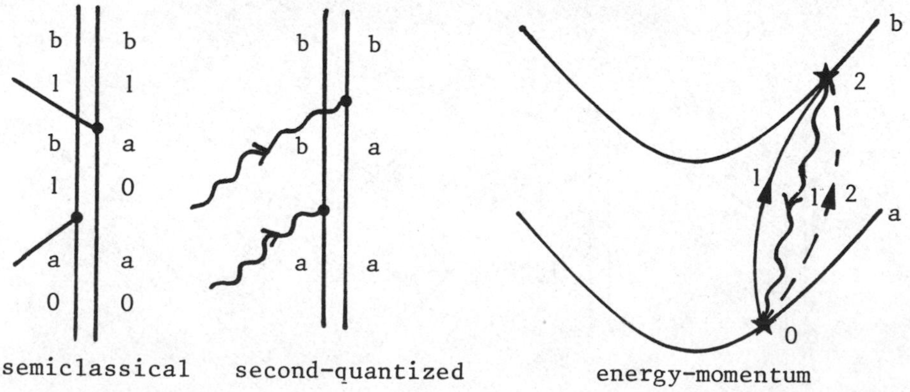

semiclassical second-quantized energy-momentum

+ similar diagrams starting with ρ_{bb} , + complex conjugate diagrams

Application of the diagrammatic rules to the 1st diagram (+c.c.)

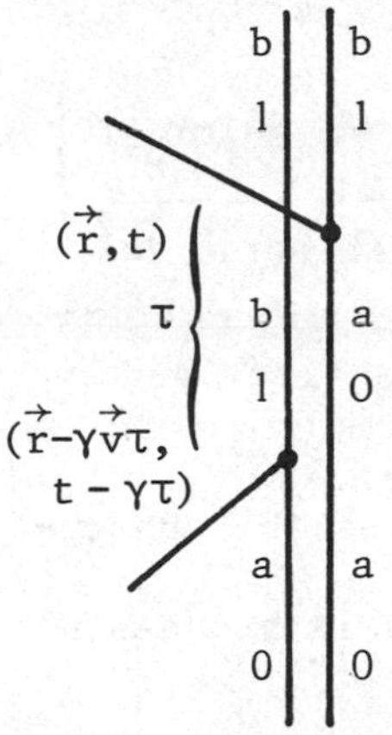

$$\left[\frac{\partial}{\partial t}<m_b=1|\,\rho_{bb}(\vec{r},t,\vec{p}_o)|\,m_b=1>\right]_{rad} = 2\,\mathrm{Re}\int_o^{+\infty} d\tau$$

$$-i\,[<a|\,\vec{\mu}.\hat{e}|\,b>E_o/2\hbar]\;U_o(\vec{r})\exp[i(\omega t-kz+\varphi)]$$

$$\exp[-(i\omega_{ba}+i\delta+\gamma_{ba})\tau]$$

$$i\,[<b\,|\vec{\mu}.\hat{e}^*|\,a>E_o/2\hbar]\;U_o^*(\vec{r}-\gamma\vec{v}\tau)\exp[-i(\omega(t-\gamma\tau)-k(z-\gamma v_z\tau)+\varphi)]$$

$$<m_a=0\,|\rho_{aa}^{(o)}(\vec{r}-\gamma\vec{v}\tau,\vec{p}_o)|\,m_a=0>$$

with
$$<o|\rho_{aa}^{(o)}(\vec{r}-\gamma\vec{v}\tau,\vec{p}_o)|\,o> \simeq n_a^{(o)}F(\vec{r},\vec{p}_o)h^3 \qquad (100)$$

The average absorbed power is :

$$\overline{W} = \hbar\omega\int\frac{d^3p_o\,d^3r}{h^3}\left[\frac{\partial}{\partial t}<1|\,\rho_{bb}(\vec{r},t,\vec{p}_o)|\,1>\right]_{rad} =$$

$$= \hbar\omega\; n_a^{(o)}\;\Omega_{ba}^2\int d^3p_o\,d^3r\;F(\vec{r},\vec{p}_o)$$

$$2\mathrm{Re}\int_o^{+\infty} d\tau\;\exp\left\{\left[i(\gamma\omega-\omega_{ba}-\delta-kv_z)-\gamma_{ba}\right]\tau\right\}U_o(\vec{r})U_o^*(\vec{r}-\vec{v}\tau) \qquad (101)$$

where :

$$\Omega_{ba}^2 = <b|\,\vec{\mu}.\hat{e}^*\,|\,a><a|\,\vec{\mu}.\hat{e}|\,b>E_o^2/4\hbar^2 \qquad (102)$$

Just for the illustration, let us consider first the cell case with a uniform Maxwell-Boltzmann velocity distribution :

$$F(\vec{r},\vec{p}_o)\longrightarrow F(\vec{v}) = \frac{1}{(\sqrt{\pi}\,u)^3}\;\exp(-\,v^2/u^2) \qquad (103)$$

and a TEM_{oo} laser mode :

$$U(\vec{r}) = L(z)\;\exp\left[-\,L(z)(x^2+y^2)/w_o^2\right] \qquad (104)$$

After x,y and velocity integrations, one finds :

$$\frac{\overline{dW}}{dz} = \hbar\omega\, n_a^{(o)} \frac{\pi w_o^2}{2} \Omega_{ba}^2 \left\{ 2\mathrm{Re} \int_o^{+\infty} d\tau \; \frac{\exp\left[\frac{-k^2u^2\tau^2/4}{(1-i\omega\tau u^2/2c^2)} + [i(\omega-\omega_{ba}-\delta)-\gamma_{ba}]\tau \right]}{(1+\tau^2u^2/2w_o^2 - i\omega\tau u^2/2c^2)(1-i\omega\tau u^2/2c^2)^{1/2}} \right\} \tag{105}$$

which includes in a single formula the effects of the first and 2nd order Doppler shifts, of transit broadening, of relaxation and of the recoil. (Multiplication by $1/(c.\frac{\varepsilon_o E_o^2}{2} . \frac{\pi w_o^2}{2})$ gives the absorption coefficient).

If the second order Doppler effect and the transit broadening are neglected, we recover the familiar Voigt profile(*) :

$$\left\{ \quad \right\} \longrightarrow \frac{2\sqrt{\pi}}{ku} \mathrm{Re}\, W\left(\frac{\omega-\omega_{ba}-\delta+i\gamma_{ba}}{ku} \right) \tag{106}$$

which reduces to normalized Doppler or Lorentz line shapes when γ_{ba}/ku is respectively very small or very large.

In the beam case (illustrated by fig. 9) $F(\vec{r},p_o)d^3\vec{p}_o$ is

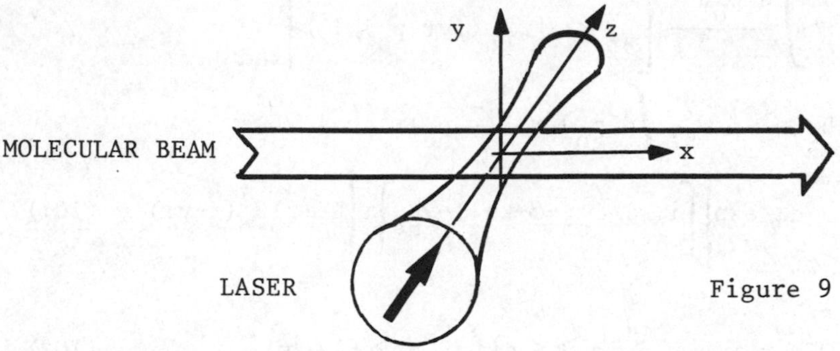

Figure 9

replaced by :

$$F(v)A(n_y, n_z, x, y, z)dv\, dn_y\, dn_z \tag{107}$$

where $F(v)$ is the distribution of velocity moduli and where $A(n_y, n_z, \vec{r})$ is the angular distribution of velocity directions $(\hat{n} = \vec{v}/v)$ at each point of the beam.

(*) $W(z) = \exp(-z^2)\,\mathrm{erfc}(-iz)$ is the probability function of complex argument.[29]

In the present problem, the x dependence of A may be ignored and, since the y direction is not critical for Doppler shifts, we may extract a $\delta(n_y)$ from A.

The laser field distribution $U_o(\vec{r})$ will be written as :

$$U_o(\vec{r}) = G(x)\, G(y) \qquad (108)$$

$$\overline{W} = \hbar\omega\, n_a^{(o)}\, \Omega_{ba}^2 \int dv\, dn_z\, dy\, dz\, |G(y)|^2 A(n_z, y, z) F(v)$$

$$2\mathrm{Re}\int_0^{+\infty} d\tau\, \exp\left\{\left[i(\gamma\omega - \omega_{ba} - \delta - kv_z) - \gamma_{ba}\right]\tau\right\}\int_{-\infty}^{+\infty} dx\, G(x) G^*(x - v\tau) \qquad (109)$$

For a TEM_{oo} mode :

$$G(x) = L^{1/2} \exp\left(-L\,\frac{x^2}{w_o^2}\right) \qquad (110)$$

$$\int_{-\infty}^{+\infty} dx\, G(x) G^*(x - v\tau) = \sqrt{\frac{\pi}{2}}\, w_o\, e^{-\frac{1}{2}\frac{v^2\tau^2}{w_o^2}} \qquad (111)$$

and, for each velocity class, the line shape is a Voigt profile :

$$2\mathrm{Re}\int_0^{+\infty} d\tau\, \exp\left\{\left[i(\gamma\omega - \omega_{ba} - \delta - kv_z) - \gamma_{ba}\right]\tau\right\}\int_{-\infty}^{+\infty} dx\, G(x) G^*(x - v\tau) =$$

$$\frac{\pi w_o^2}{v}\, \mathrm{Re}\, W\left[\frac{\omega_{ba} + \delta - \gamma\omega - kv_z + i\gamma_{ba}}{\underbrace{\sqrt{2}\, v/w_o}_{\text{transit width}}}\right] \qquad (112)$$

To go further, we shall assume a Gaussian dependence of the effective n_z distribution :

$$\int dy\, dz\, |G(y)|^2 A(n_z, y, z) = \sqrt{LL^*}\, S\, \frac{1}{\sqrt{\pi}\,\varepsilon} \exp\left[-\left(\frac{n_z}{\varepsilon}\right)^2\right] \qquad (113)$$

where S and ε are calculable or adjustable parameters (independent of the laser beam geometry when the laser beam diameter is large enough in comparison with the molecular beam transverse dimension).

One finds :

$$\overline{W} = n_a^{(o)} \hbar\omega\, \Omega_{ba}^2\, \pi w_o^2 \left(1 + \frac{k^2 w_o^2 \varepsilon^2}{2}\right)^{-1/2} \int_0^{+\infty} \frac{dv\, F(v)}{v} \sqrt{LL^*}\, S$$

$$\mathrm{Re}\, W \left[\frac{\omega_{ba} + \delta - \gamma\omega + i\gamma_{ba}}{\sqrt{2}(v/w_o)\left(1 + \frac{k^2 w_o^2 \varepsilon^2}{2}\right)^{1/2}} \right] \tag{114}$$

If γ_{ba} is negligible, we obtain a Gaussian line shape for each transverse velocity, with a width where the usual Doppler broadening and the transit broadening are combined quadratically :

$$\Delta\nu = \frac{\varepsilon v}{\lambda}\, 2\sqrt{\ln 2} \left[1 + \frac{2}{k^2 w_o^2 \varepsilon^2}\right]^{1/2} \tag{115}$$

In this limit of negligible relaxation, a very general expression for the line shape is obtained with the convolution theorem :

$$2\mathrm{Re} \int_0^{+\infty} d\tau\, \exp\left[\, i(\gamma\omega - \omega_{ba} - \delta - kv_z)\tau\right] \int_{-\infty}^{+\infty} dx\, G(x)\, G^*(x - v_x\tau)$$
$$= \frac{1}{v} \left| \int_{-\infty}^{+\infty} dx\; G(x)\; e^{i\frac{\Delta x}{v}} \right|^2 \tag{116}$$

where :

$$\Delta = \gamma\omega - \omega_{ba} - \delta - kv_z$$

That is, the line shape is the square of the modulus of the Fourier transform of the transverse field distribution. For a Gaussian G, we obtained a Gaussian line shape; for a square field distribution we would obtain a $(\mathrm{sinc})^2$ function of Δ.

Ramsey fringes

To reduce the transit broadening, it is necessary to increase the width of $G(x)$ as much as possible. To overcome this problem, N.F. Ramsey suggested in 1950 to split $G(x)$ into two parts[30,31] :

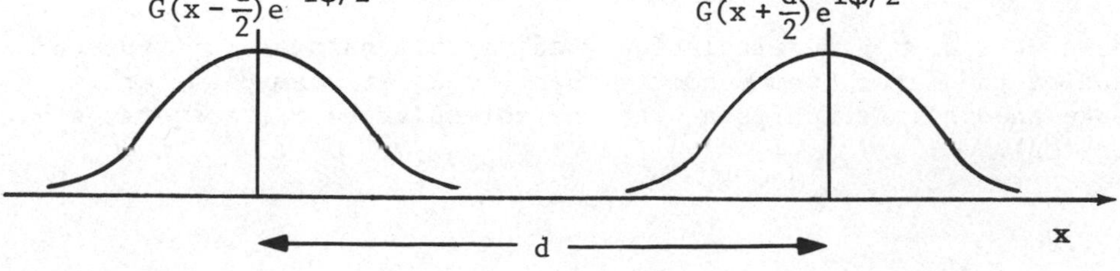

From the previous theorem (116), it is straightforward to get the line shape :

$$\left| \int_{-\infty}^{+\infty} dx\, e^{i\frac{\Delta x}{v}} \left[G(x - \frac{d}{2}) e^{-i\varphi/2} + G(x + \frac{d}{2}) e^{i\varphi/2} \right] \right|^2 =$$

$$= 2\,[1 + \cos(\Delta d/v - \varphi)\,] \,.\, \left| \int_{-\infty}^{+\infty} dx\, e^{i\frac{\Delta x}{v}} G(x) \right|^2 \qquad (117)$$

The line shape (corresponding to the squared modulus of the Fourier transform of each field zone) is now modulated by interference fringes and the fringe spacing is determined by d/v. The physics can be understood by inspection of the diagrams. The constant term corresponds to both one-photon processes occurring in the same field zone, whereas the oscillating term comes from an interference of two one-photon processes occurring in two different zones. On the diagram, we can also see that, at the first vertex in the first field zone, an optical dipole (off-diagonal matrix element or optical coherence) is created and precesses freely in the dark zone for the time duration d/v. When this dipole is projected on the field at the second interaction vertex, in the second zone, the phase difference, between the internal atomic clock and the laser field oscillation during the time d/v, gives rise to the interference term in the upper state population or in the absorbed power. Usually the integration over the $F(v)$ distribution will tend to average out the side fringes. It should also be noted that any uncontrolled phase difference φ will shift the observed line center. Finally in the optical domain, the first-order Doppler shift kv_z contained in Δ will destroy the fringes. Generally one considers standing waves in both field zones, in which case one should add the two corresponding diagrams, and $\cos(\Delta d/v - \varphi)$ in (117) is replaced by :

$$2 \cos [\,(\gamma\omega - \omega_{ba} - \delta)\, d/v - (\varphi^+ + \varphi^-)/2\,] \cos[\, kv_z d/v + (\varphi^- - \varphi^+)/2\,]$$

which averages to zero when the velocity distribution is large compared to v/kd. One way out of this problem is to introduce a modulation of the velocity distribution, preventing the previous cosine to give a zero average. This idea was proposed and demonstrated by G. Kramer using a grating to filter molecular velocities.[32,33] The other possibilities result from a combination of Ramsey's idea with Doppler-free spectroscopic methods that we shall examine next.

B - DOPPLER-FREE TWO-PHOTON SPECTROSCOPY

Our next illustration of the density matrix diagrams is one of the methods of non-linear Doppler-free spectroscopy. In this method, atoms undergo a two-photon transition from level a to level c under the influence of two counter-propagating waves.[41]

Relevant diagrams

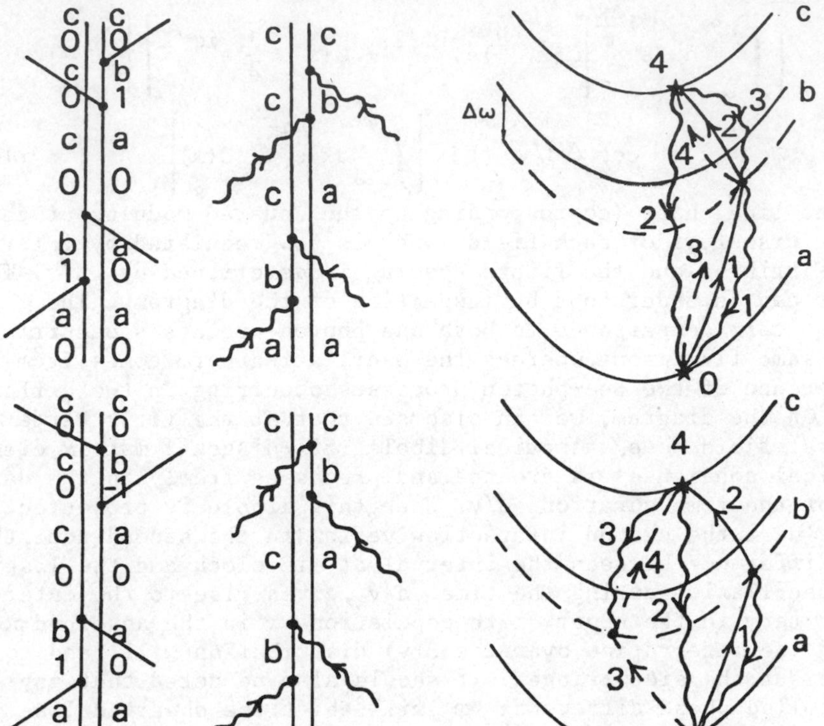

+ 2 diagrams obtained by exchange of the two first vertices.

To compute the power absorbed or emitted (e.g. in Doppler-free two-photon phase conjugation) one needs to consider also diagrams such as [2] :

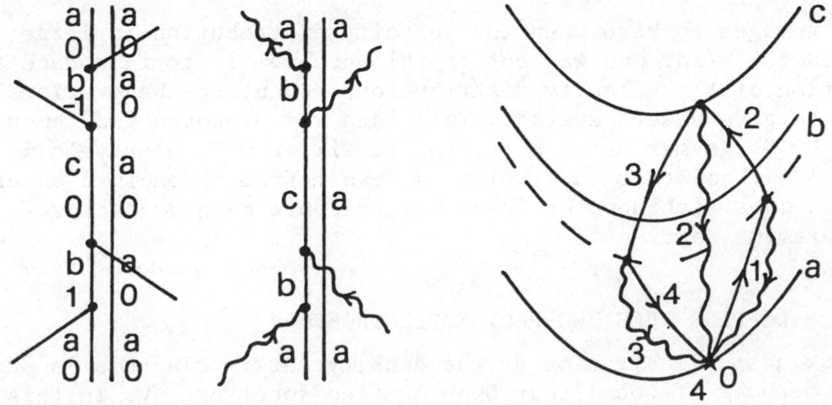

Figure 10

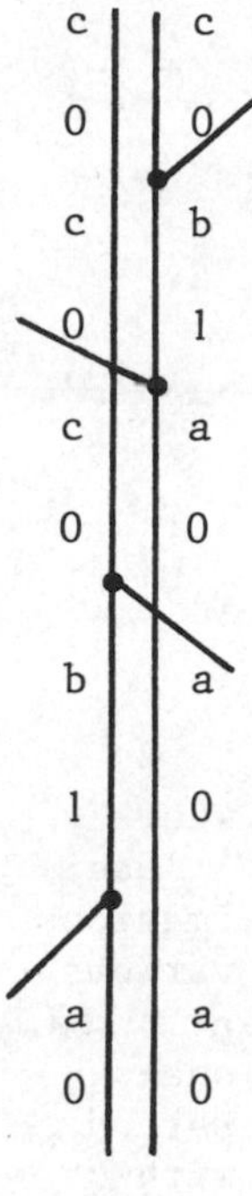

$$\left[\frac{\partial}{\partial t}\langle m_c=0|\,\rho_{cc}(\alpha_o=a,\vec{p}_o,\vec{r},t)\,|m_c=0\rangle\right]^{(4)}_{rad} = 2\,\mathrm{Re}\int_o^{+\infty} d\tau\, d\tau'\, d\tau''$$

$$(-i\,\vec{\mu}_{bc}\cdot\vec{E}_o^{-}/2\hbar)\,U_o^{-}(\vec{r},t)\exp[i(\omega t+kz+\varphi^{-})]$$

$$\exp[-(i\omega_{cb}-i\delta+\gamma_{cb})\tau]$$

$$(-i\,\vec{\mu}_{ab}\cdot\vec{E}_o^{+}/2\hbar)\,U_o^{+}(\vec{r}-\gamma\vec{v}\tau)\exp[i(\omega(t-\gamma\tau)-k(z-\gamma v_z\tau)+\varphi^{+})]$$

$$\exp[-(i\omega_{ca}+\gamma_{ca})\tau']$$

$$(+i\,\vec{\mu}_{cb}\cdot\vec{E}_o^{-*}/2\hbar)\,U_o^{-*}(\vec{r}-\gamma\vec{v}(\tau+\tau'))$$

$$\exp\{-i[\omega(t-\gamma(\tau+\tau'))+k(z-\gamma v_z(\tau+\tau'))+\varphi^{-}]\}$$

$$\exp[-(i\omega_{ba}+i\delta+\gamma_{ba})\tau'']$$

$$(+i\,\vec{\mu}_{ba}\cdot\vec{E}_o^{+*}/2\hbar)\,U_o^{+*}(\vec{r}-\gamma\vec{v}(\tau+\tau'+\tau''))$$

$$\exp\{-i[\omega(t-\gamma(\tau+\tau'+\tau''))-k(z-\gamma v_z(\tau+\tau'+\tau''))+\varphi^{+}]\}$$

$$h^3\, n_a^{(o)}\, F(\vec{p}_o,\vec{r}-\gamma\vec{v}(\tau+\tau'+\tau''))$$

Application of the diagrammatic rules to the first diagram of fig. 10

Figure 11

If the intermediate levels b are sufficiently far from resonance, the τ and τ'' integrations are straightforward and one finds :

$$\overline{W} \not\approx \hbar\omega\, n_a^{(o)} \sum_b \frac{(\Omega_{cb}^{-})^2(\Omega_{ba}^{+})^2}{(\Delta\omega)^2}\int d^3p_o$$

$$2\,\mathrm{Re}\int_o^{+\infty} d\tau\, \exp[(i\Delta-\gamma_{ca})\tau]$$

$$\int d^3r\, U_o^{-}(\vec{r})\,U_o^{+}(\vec{r})\,U_o^{-*}(\vec{r}-\vec{v}\tau)\,U_o^{+*}(\vec{r}-\vec{v}\tau)\,F(\vec{p}_o,\vec{r}-\vec{v}\tau)$$

$$+\ 7\ \text{similar expressions} \qquad (118)$$

with $\Delta\omega = \omega_{ba}-\omega \simeq \omega-\omega_{cb}$ and $\Delta = 2\gamma\omega-\omega_{ca}$ (*).

This formula is quite similar to the one-photon case. This is not too surprising since it is now well-known that the equations governing the atomic system in DFTPS can be reduced to those of an equivalent two-level system E_a, E_c with the following

---

(*) $\Delta\,\omega$ should not be confused with $\Delta\,.\,\omega$

correspondence rules :

- detuning $\gamma\omega - \omega_{ba} \longleftrightarrow \gamma\omega^+ + \gamma\omega^- - \omega_{ca}$ $(= 2\gamma\omega - \omega_{ca}$ for $\omega^+ = \omega^-)$
- Doppler shift $kv_z \longleftrightarrow (k^+ - k^-)v_z$ $(= 0$ if $k^+ = k^-)$
- Rabi pulsation $$\Omega_{ba} \longleftrightarrow \frac{\Omega^+_{cb}\,\Omega^-_{ba}}{\Delta\omega} + \frac{\Omega^-_{cb}\,\Omega^+_{ba}}{\Delta\omega} \qquad (119)$$
- Field geometry $U_o(\vec{r}) \longleftrightarrow U^-_o(\vec{r})\, U^+_o(\vec{r})$
- off-diagonal density matrix element $\rho_{ba} \longleftrightarrow \rho_{ca}$

In the cell case, $F(\vec{p}_o, \vec{r})d^3p_o$ is replaced by $F(\vec{v})d^3v$ where $F(\vec{v})$ is the Maxwell-Boltzmann distribution. For the TEM_{oo} laser mode, the integrals are easily calculated and various approximations for the line shape can be found in [2]. They correspond to various evaluations of the integral (105) with the previous replacement rules. Also, on the basis of the correspondence (119) between the one-photon and Doppler-free two-photon cases, it is easy to understand how Ramsey's idea can be transposed to the two-photon case with the major advantage of the absence of the Doppler shift kv_z.

<u>Beam case and Ramsey fringes</u> [34-37,3,5]

Beautiful experiments, where transit effects are studied in detail and where Ramsey fringes are demonstrated respectively with beams of Bi and Rb, have been reported by Lee, Helmcke and Hall in [36]. We shall have these experiments in mind for the calculations which follow.

We shall first repeat here the derivation of the two-photon Ramsey pattern, for the case of an ideal beam of atoms and the detection geometry described in [36].

In the absence of laser irradiation, the ground level population in the beam is written :(*)

$$\langle o|\rho^{(o)}_{aa}(\vec{r},\vec{v})|o\rangle/h^3 = n^{(o)}_a F(\vec{v})D(\vec{r})\exp[-\gamma_a(x+x_o)/v_x] \qquad (120)$$

where $F(\vec{v})$ is the velocity distribution at the oven exit slit $(x = -x_o)$:

$$F(\vec{v}) = (4v_x^2/\sqrt{\pi}u^3)\exp(-v_x^2/u^2)\delta(v_y)\delta(v_z) \qquad (121)$$

(*) In most cases the state a is the ground state and one has $\gamma_a = 0$

and where $D(\vec{r})$ is the unnormalized transverse spatial distribution of atoms in the beam, which we assume independent of x, $D(\vec{r}) = D(y,z)$ and such that $D(0,0) = 1$ ($n_a^{(0)}$ is the total population in the beam center).

$\langle 0|\rho_{aa}^{(0)}|0\rangle$ can be rewritten using the beam flux :

$$\Phi_a^{(0)} = (2n_a^{(0)} u/\sqrt{\pi}) \iint dy\, dz\, D(y,z)$$

as :

$$\langle 0|\rho_{aa}^{(0)}(\vec{r},\vec{v})|0\rangle = h^3 \Phi_a^{(0)} \frac{2v_x^2}{u^4} \frac{D(y,z)\delta(v_y)\delta(v_z)}{\iint dy\, dz\, D(y,z)}$$

$$\exp(-v_x^2/u^2 - \gamma_a (x+x_0)/v_x) \qquad (122)$$

In the two interaction regions, the laser field is written :

$$\vec{\mathcal{E}} = \text{Re} \sum_{j=1,2} \hat{e}_j^+ E_j^+ U_j^+(\vec{r}) \exp[i(\omega t - \vec{k}_j \cdot \vec{r} + \varphi_j^+)]$$

$$+ \hat{e}_j^- E_j^- U_j^-(\vec{r}) \exp[i(\omega t + \vec{k}_j \cdot \vec{r} + \varphi_j^-)] \qquad (123)$$

where we have made explicit the polarization vectors $\hat{e}_j^\pm$, the field amplitudes and phases and the transverse spatial distribution of the laser beams :

$$U_j^\pm(\vec{r}) = G_j^\pm(x - \varepsilon_j d/2, z) G_j^\pm(y,z) \qquad (124)$$

with : $\varepsilon_1 = -1$, $\varepsilon_2 = 1$

For the most general Gaussian TEM_{00} mode [1] :

$$G_j^\pm(x,z) = \left[L_j^\pm(z)\right]^{1/2} \exp\left[-L_j^\pm(z) x^2 / w_{j^\pm}^2\right]$$

but in this calculation we shall assume for simplicity that the interactions take place at the beam waists $L_j^\pm(z) = 1$ with a common beam radius $w_{j^\pm} = w_0$ as in [36] , that is :

$$G_j^\pm(x,z) = G(x) \quad \text{and} \quad G_j^\pm(y,z) = G(y)$$

with : $$G(x) = \exp(-x^2/w_0^2) \qquad (125)$$

We shall also assume the folded standing wave condition[2] :

$$\varphi_2^- + \varphi_2^+ - \varphi_1^- - \varphi_1^+ = 0 \qquad (126)$$

Finally we use the following notation for Rabi circular frequencies:

$$\Omega^{\pm}_{j\beta\alpha} = \frac{\langle\beta|\vec{\mu}\,.\,\hat{e}^{\pm *}_{j}|\alpha\rangle E^{\pm}_{j}}{2\hbar} \tag{127}$$

The diagrams of fig. 10-11 give the rate of formation of atoms in level c per unit-time under the influence of radiation (subscript rad) at fourth-order in the field :

$$\left[\frac{\partial}{\partial t}\langle 0|\rho_{cc}(\vec{r},\vec{v})|0\rangle\right]^{(4)}_{rad}$$

In the beam experiments described in [36], it is not exactly this rate which is actually measured, but rather a flux of excited atoms reaching a target detector at some distance ℓ :

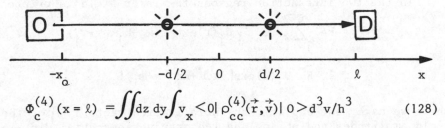

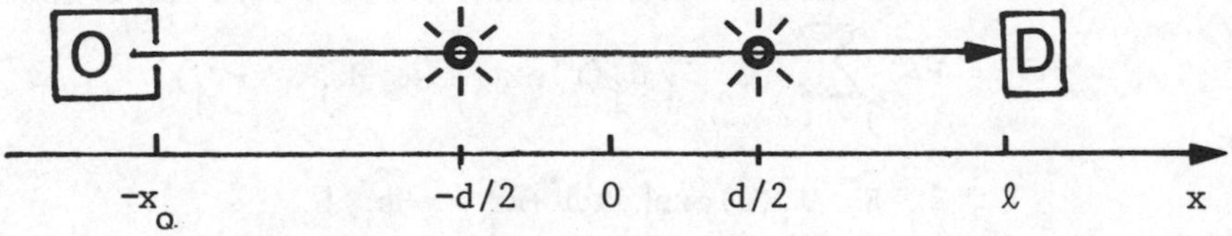

$$\Phi^{(4)}_{c}(x=\ell) = \iint dz\,dy \int v_x \langle 0|\rho^{(4)}_{cc}(\vec{r},\vec{v})|0\rangle d^3v/h^3 \tag{128}$$

The diagrams of fig. 10-11 should thus be extended to include the level c population propagator from the last field zone to the detector position. If τ is the proper time elapsed between the last interaction vertex and the detection point, the diagrammatic rules give :

$$\langle 0|\rho^{(4)}_{cc}(\vec{r},\vec{v})|0\rangle = \int_{0}^{+\infty} d\tau\, e^{-\gamma_c\tau}\left[\frac{\partial}{\partial t}\langle 0|\rho_{cc}(\vec{r}-\vec{v}\tau,\vec{v})|0\rangle\right]^{(4)}_{rad} \tag{129}$$

With the new variable $x = \ell - v_x\tau$

$$\Phi^{(4)}_{c} = \int d^3v \exp\left(-\frac{\gamma_c\ell}{v_x}\right)\iint_{-\infty}^{+\infty} dz\,dy \int_{-\infty}^{\ell} dx \exp\left(\frac{\gamma_c x}{v_x}\right)\left[\frac{\partial}{\partial t}\langle 0|\rho_{cc}(\vec{r},\vec{v})|0\rangle\right]^{(4)}_{rad}/h^3 \tag{130}$$

With a good approximation the upper bound of the x integral may be replaced by $+\infty$. If the relaxation of level c can be neglected during the transit time across the single zone radius w_o, the factor $\exp(\gamma_c x/v_x)$ may be replaced by $\exp(\gamma_c\,\varepsilon_j d/2v_x)$. With this approximation we recover a flux of excited atoms on the detector equal to their total rate of formation weighted, for each velocity class, by the relaxation factor $\exp[-\gamma_c(\ell-\varepsilon_j\,d/2)/v_x]$. The calculation is then, for each diagram, identical to what is presented in fig. 11, equation (118) and references [2,3] . In fact, as we shall see now, the factor $\exp(\gamma_c x/v_x)$ can be kept without any mathematical difficulty in (130) at the expense of slightly more cumbersome expressions.

The resulting expression for $\Phi_c^{(4)}$ is :

$$\Phi_c^{(4)} = n_a^{(o)} \sum_{\substack{\varepsilon,\varepsilon'=\pm \\ j,j'=1,2}} \frac{\Omega^{\varepsilon}_{jcb}\Omega^{-\varepsilon}_{jba}\Omega^{\varepsilon'}_{j'cb}\Omega^{-\varepsilon'}_{j'ba}}{(\Delta\omega)^2} \iint dy\, dz \exp\left(-\frac{4y^2}{w_o^2}\right) D(y,z)$$

$$\int_o^{+\infty} dv_x \frac{4v_x^2}{\sqrt{\pi}u^3} \exp\left(-\frac{v_x^2}{u^2} - \frac{\gamma_c \ell + \gamma_a x_o}{v_x}\right) 2\,\mathrm{Re} \int_o^{+\infty} d\tau \exp[(i\Delta - \gamma_{ca} + \gamma_a)\tau]$$

$$\times \int_{-\infty}^{+\infty} dx \exp[-(\gamma_a - \gamma_c)x/v_x]\, G^2(x-\varepsilon_{j'}d/2) G^2(x-\varepsilon_j d/2 - v_x\tau) \qquad (131)$$

The x and τ integrals are easily performed :

$$\int_o^{+\infty} d\tau \exp\Big[(i\Delta - \gamma_{ca} + \gamma_a)\tau\Big] \int_{-\infty}^{+\infty} dx \exp\Big[-(\gamma_a - \gamma_c)x/v_x\Big] G^2(x-\varepsilon_{j'}d/2)$$

$$G^2(x-\varepsilon_j d/2 - v_x\tau) =$$

$$= \frac{\pi}{4}\frac{w_o^2}{v_x} \exp\left\{-\left[1-\delta_{jj'}\right]\frac{d^2}{w_o^2} + \frac{(\gamma_c-\gamma_a)(\varepsilon_j+\varepsilon_{j'})d}{4v_x} + \frac{(\gamma_c-\gamma_a)^2 w_o^2}{16v_x^2}\right\}$$

$$W\left[\frac{\Delta + i\left(\gamma_{ca} - \frac{\gamma_c}{2} - \frac{\gamma_a}{2} + (\varepsilon_j - \varepsilon_{j'})\frac{dv_x}{w_o^2}\right)}{2v_x/w_o}\right] \qquad (132)$$

where W is the probability function of complex arguments (see page 40). As pointed out in [3] the fringe pattern can be extracted with the following relationship [29] :

$$W(x-iy) = 2\exp(y^2-x^2+2ixy) - W^*(x+iy) \qquad (133)$$

with :

$$x = \frac{\Delta}{2v_x/w_o} \quad \text{and} \quad y = \frac{(\gamma_a+\gamma_c)/2 - \gamma_{ca} + 2dv_x/w_o^2}{2v_x/w_o}$$

To a good approximation, $\mathrm{Re}\, W\left[\dfrac{\Delta + i(\gamma_{ca} - (\gamma_a+\gamma_c)/2 + 2\, dv_x/w_o^2)}{2v_x/w_o}\right]$ and $-\mathrm{Re}\, W^*\left[\dfrac{\Delta + i(2dv_x/w_o^2 - \gamma_{ca} + (\gamma_a+\gamma_c)/2}{2v_x/w_o}\right]$ cancel each other, and in any case disappear as d/w_o increases.

The signal is finally obtained as the sum of a background and of a fringe pattern :

$$\Phi_c^{(4)} = \Phi_c^{(4)'} + \Phi_c^{(4)''} \tag{134}$$

with :

$$\Phi_c^{(4)'} = \Phi_a^{(0)} \xi^{(4)} \sum_{\substack{\varepsilon,\varepsilon'=\pm \\ j=1,2}} \eta_{jj}^{\varepsilon\varepsilon'} f_{jj}(\omega) \tag{135}$$

and :

$$\Phi_c^{(4)''} = \Phi_a^{(0)} \xi^{(4)} \sum_{\varepsilon,\varepsilon'=\pm} \eta_{12}^{\varepsilon\varepsilon'} f_{12}(\omega) \tag{136}$$

where :

$$\eta_{jj'}^{\varepsilon\varepsilon'} = \frac{\Omega_{jcb}^{\varepsilon} \Omega_{jba}^{-\varepsilon} \Omega_{j'cb}^{\varepsilon'} \Omega_{j'ba}^{-\varepsilon'}}{(\omega_{cb} - \omega)(\omega - \omega_{ba})} \frac{\pi w_o^2}{2u^2} \tag{137}$$

is a two-photon pumping efficiency

$$\xi^{(4)} = \frac{\iint dy\, dz\, \exp(-4y^2/w_o^2)\, D(y,z)}{\iint dy\, dz\, D(y,z)} \tag{138}$$

is a filling factor(*), and where f_{jj} and f_{12} are the following line shapes:

$$f_{jj}(\omega) = 2\mathrm{Re}\int_0^{+\infty} \frac{v_x dv_x}{u^2} \exp\left[-\frac{v_x^2}{u^2} - \frac{\gamma_a x_o + \gamma_c \ell}{v_x} + \frac{\varepsilon_j(\gamma_c - \gamma_a)d}{2v_x} + \frac{(\gamma_c-\gamma_a)^2 w_o^2}{16 v_x^2}\right]$$

$$W\left[\frac{\Delta + i\left(\gamma_{ca} - \frac{\gamma_c+\gamma_a}{2}\right)}{2v_x/w_o}\right] \tag{140}$$

(*) In a more realistic theory one should take the molecular beam divergence into account and replace $F(\vec{v})D(\vec{r})d^3v$ as defined by (121) with $F(v)A(n_y,n_z,y,z)dv\, dn_y\, dn_z$ as defined in (107). For the fringes, the filling factor $\xi^{(4)}$ is then replaced by :

$$\xi_{12}^{(4)} = \frac{\int G^2(y) G^2(y-n_y d) A(n_y,n_z,y,z)\, dy\, dz\, dn_y\, dn_z}{\int A(n_y,n_z,y,z)\, dy\, dz\, dn_y\, dn_z} \tag{139}$$

for the background signals and

$$f_{12}(\omega) = 2\int_{0}^{+\infty} \frac{2v_x dv_x}{u^2} \exp\left[-\frac{v_x^2}{u^2} - \frac{\gamma_a x_o + \gamma_c \ell}{v_x} + \frac{(\gamma_c-\gamma_a)^2 w_o^2}{16 v_x^2}\right]$$

$$\times \exp\left\{\frac{\left(\gamma_{ca} - \frac{\gamma_c+\gamma_a}{2} - \frac{2dv_x}{w_o^2}\right)^2}{4v_x^2/w_o^2} - \frac{d^2}{w_o^2} - \frac{\Delta^2}{4v_x^2/w_o^2}\right\} \cos \frac{\left(\frac{2dv_x}{w_o^2} + \frac{\gamma_c+\gamma_a}{2} - \gamma_{ca}\right)\Delta}{2v_x^2/w_o^2} \tag{141}$$

for the fringes.

When the relaxation is negligible during the single zone transit they reduce to the approximate forms :

$$f_{jj}(\omega) = \int_{0}^{+\infty} \frac{2v_x dv_x}{u^2} \exp\left[-\frac{v_x^2}{u^2} - \frac{\gamma_c(\ell-\varepsilon_j d/2)+\gamma_a(x_o+\varepsilon_j d/2)}{v_x} - \frac{\Delta^2 w_o^2}{4v_x^2}\right] \tag{142}$$

and :

$$f_{12}(\omega) = 2\int_{0}^{+\infty} \frac{2v_x dv_x}{u^2} \exp\left[-\frac{v_x^2}{u^2} - \frac{\gamma_c(\ell-d/2)+\gamma_a(x_o-d/2)+\gamma_{ca} d}{v_x} - \frac{\Delta^2 w_o^2}{4v_x^2}\right] \cos(\Delta d/v_x) \tag{143}$$

When the relaxation is totally neglected, we can use the convolution theorem to evaluate the double integrals (132) as products of Fourier transform of the transverse field distribution :

$$2\mathrm{Re}\int_{0}^{+\infty} d\tau \exp(i\Delta\tau)\int_{-\infty}^{+\infty} dx\ G^2(x-\varepsilon_j d/2)G^2(x-\varepsilon_j d/2-v_x\tau)$$

$$= \left|\int_{-\infty}^{+\infty} dx\ G^2(x) \exp i(x\Delta/v_x)\right|^2 /v_x \tag{144-a}$$

$$\sum_{j\neq j'} 2\mathrm{Re}\int_{0}^{+\infty} d\tau \exp(i\Delta\tau)\int_{-\infty}^{+\infty} dx\ G^2(x-\varepsilon_{j'} d/2)G^2(x-\varepsilon_j d/2-v_x\tau)$$

$$= 2\cos(d\Delta/v_x)\left|\int_{-\infty}^{+\infty} dx\ G^2(x)\exp i(x\Delta/v_x)\right|^2 /v_x \tag{144-b}$$

with :

$$\left|\int_{-\infty}^{+\infty} dx\ G^2(x)\exp i(x\Delta/v_x)\right|^2 = \frac{\pi w_o^2}{2}\exp\left(-\frac{\Delta^2 w_o^2}{4v_x^2}\right) \tag{145}$$

for Gaussian beams,

or:
$$= a^2\left(\frac{\sin\frac{a\Delta}{2v_x}}{\frac{a\Delta}{2v_x}}\right)^2 \qquad (146)$$

for square pulses of width a and unit height.

We can then reintroduce average relaxation weight factors to recover (142) and (143) from (145) and (144).

An important question, connected with the use of Ramsey fringes in Doppler-free two-photon spectroscopy, is the calculation of the residual light shifts associated with this technique. It has been suggested that these shifts might be reduced by the ratio w_o/d of the laser beam waist radius to the zone separation[36]. We shall see now that this question can also be answered by using density matrix diagrams. Light shifts occur whenever two levels are coupled by a non-resonant field. This situation is intrinsic in two-photon spectroscopy where both the ground level a and the excited level c are shifted through their non-resonant coupling with the intermediate level b. These couplings give rise to higher order terms which are not symmetric with respect to the detuning and thus lead to a shift. These higher order terms are obtained by dressing either state a or c of the off-diagonal density matrix element ρ_{ca} with two or more interactions which are non-resonant at the coupled transitions frequencies ω_{ba} or ω_{cb}. To the lowest order, two additional vertices are necessary. These two interactions are quasi-simultaneous, owing to the non-resonant character of the interactions. The corresponding diagrams are displayed on fig. 12. For strong fields, it may be easily shown from the

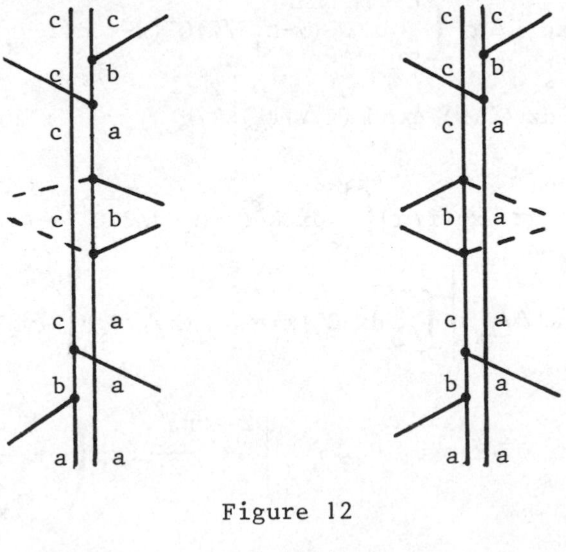

Figure 12

density matrix equations (or by resummation of diagrams) that the light shift can be taken into account, at any order, through a modified propagator for ρ_{ca} :

$$\exp\left\{-(i\omega_{ca}+\gamma_{ca})\tau - i\sum_{\pm,j}\left(\frac{(\Omega^{\pm}_{jba})^2}{\Delta\omega}-\frac{(\Omega^{\pm}_{jcb})^2}{\Delta\omega}\right)\int_0^{\tau} d\tau_1 \left|U^{\pm}_j(\vec{r}-\gamma\vec{v}\tau_1, t-\gamma\tau_1)\right|^2\right\} \tag{147}$$

When the field is independent of space and time : $U_j \equiv 1$, and we recover Cohen Tannoudji's familiar result [38]. In our case, this strong field propagator is too cumbersome and will be expanded to first order in Ω^2. This leads to results identical to those obtained from the diagrams of fig. 12. We shall therefore limit ourselves to the application of the usual diagrammatic rules to these diagrams in order to get the light shift contributions. This calculation will be presented here only for the fringes, and with the approximation that the relaxation of level c may be neglected during the transit time across the last zone and for $\gamma_a = o$:

$$\begin{aligned}
\Phi_c^{(6)''} = n_a^{(o)} 2\,\mathrm{Re} \sum_{\substack{\varepsilon_1,\varepsilon_2,\varepsilon=\pm \\ j=1,2}} \int d^3r d^3v \exp(-\gamma_c(\ell-d/2)/v_x) F(\vec{r},\vec{v}) \\
\left(-i\Omega_{2cb}^{-\varepsilon_2}\right) U_2^{-\varepsilon_2}(\vec{r}) \exp[i(\omega t+\varepsilon_2 \vec{k}_2.\vec{r})] \\
(-i\Omega_{2ba}^{\varepsilon_2}) \int_0^{+\infty} d\tau_5 \exp\left[-(i\omega_{cb}+\gamma_{cb})\tau_5\right] U_2^{\varepsilon_2}(\vec{r}-\vec{v}\tau_5) \\
\exp i\left[\omega(t-\gamma\tau_5)-\varepsilon_2\vec{k}_2.(\vec{r}-\vec{v}\tau_5)\right] \\
(-i\Omega_{jba}^{\varepsilon}) \int_0^{+\infty} d\tau_4 \exp\left[-(i\omega_{ca}+\gamma_{ca})\tau_4\right] U_j^{\varepsilon *}(\vec{r}-\vec{v}(\tau_4+\tau_5)) \\
\exp\left\{-i\left[\omega(t-\gamma\tau_4-\gamma\tau_5)-\varepsilon\vec{k}_j.\left(\vec{r}-\vec{v}(\tau_4+\tau_5)\right)\right]\right\} \\
(-i\Omega_{jba}^{\varepsilon}) \int_0^{+\infty} d\tau_3 \exp\left[-(i\omega_{cb}+\gamma_{cb})\tau_3\right] U_j^{\varepsilon}\left(\vec{r}-\vec{v}(\tau_3+\tau_4+\tau_5)\right) \\
\exp\left\{i\left[\omega\left(t-\gamma(\tau_3+\tau_4+\tau_5)\right)-\varepsilon\vec{k}_j.\left(\vec{r}-\vec{v}(\tau_3+\tau_4+\tau_5)\right)\right]\right\} \\
(i\Omega_{1cb}^{-\varepsilon_1}) \int_0^{+\infty} d\tau_2 \exp\left[-(i\omega_{ca}+\gamma_{ca})\tau_2\right] U_1^{-\varepsilon_1^*}\left(\vec{r}-\vec{v}(\tau_2+\tau_3+\tau_4+\tau_5)\right) \\
\exp\left\{-i\left[\omega\left(t-\gamma(\tau_2+\tau_3+\tau_4+\tau_5)\right)+\varepsilon_1\vec{k}_1.\left(\vec{r}-\vec{v}(\tau_2+\tau_3+\tau_4+\tau_5)\right)\right]\right\}
\end{aligned}$$

$$(i\,\Omega_{1ba}^{\varepsilon_1}) \int_0^{+\infty} d\tau_1 \,\exp\left[-(i\omega_{ba}+\gamma_{ba})\tau_1\right] U_1^{\varepsilon_1^*}\left(\vec{r}-\vec{v}\,(\tau_1+\tau_2+\tau_3+\tau_4+\tau_5)\right)$$

$$\exp\left\{-i\left[\omega\left(t-\gamma(\tau_1+\tau_2+\tau_3+\tau_4+\tau_5)\right) - \varepsilon_1\vec{k}_1.\left(\vec{r}-\vec{v}\,(\tau_1+\tau_2+\tau_3+\tau_4+\tau_5)\right)\right]\right\}$$

+ a similar term where ρ_{cb} is replaced by ρ_{ba} during τ_3 (148)

If the levels b are sufficiently off-resonant, the time intervals τ_1 , τ_3 and τ_5 collapse and the corresponding integrals are trivial :

$$\Phi_c^{(6)''} = \Phi_a^{(o)} \sum_{\varepsilon_1,\varepsilon_2} \eta_{12}^{\varepsilon_1\varepsilon_2} \sum_{\varepsilon,j} \xi_j^{(6)} \delta_j^{\varepsilon}\, 2\,\mathrm{Im} \int_0^{+\infty} \frac{v^2 dv}{u^2}\, e^{-\gamma_c(\ell-d/2)/v}\, e^{-v^2/u^2}$$

$$\frac{4}{\pi w_0^2} \int_0^{+\infty} d\tau_2 \,\exp\left[(i\Delta-\gamma_{ca})\tau_2\right] \int_0^{+\infty} d\tau_4 \,\exp\left[(i\Delta-\gamma_{ca})\tau_4\right]$$

$$\int_{-\infty}^{+\infty} dx\, G^2(x-\tfrac{d}{2})G^2\,(x-v\tau_4-\varepsilon_j d/2)G^2(x-v(\tau_2+\tau_4)+\tfrac{d}{2}) \qquad (149)$$

where $\Phi_a^{(o)}$ and $\eta_{12}^{\varepsilon_1\varepsilon_2}$ have been defined before ;

$$\text{where} \quad \xi_j^{(6)} = \frac{\int dn_y dn_z dydz\, G^{2j}(y)\, G^{4/j}(y-n_y d) A(n_y,n_z,y,z)}{\int dn_y dn_z dydz\, A(n_y,n_z,y,z)} \qquad (150)$$

are new filling factors, and where :

$$\delta_j^{\varepsilon} = \frac{(\Omega_{jba}^{\varepsilon})^2}{\Delta\omega} - \frac{(\Omega_{jcb}^{\varepsilon})^2}{\Delta\omega} \qquad (151)$$

are plane wave light shifts.

The integrations are then performed on x and then first on τ_2 or τ_4 depending whether j=1 or 2. The same decomposition of the W function is performed as before to extract the oscillating part of the signal and the final τ_4 or τ_2 integration is possible. In our limited space, we shall give the final result only in the approximation where the relaxation over one field zone is neglected :

$$\Phi_c^{(6)''} = \Phi_a^{(o)} \sum_{\varepsilon_1\varepsilon_2} \eta_{12}^{\varepsilon_1\varepsilon_2} \sum_{\varepsilon,j} \xi_j^{(6)} \delta_j^{\varepsilon} \sqrt{\frac{\pi}{2}} \int_0^{+\infty} \frac{2v dv}{u^2}$$

$$\exp\left[-\frac{\gamma_c(\ell-d/2)}{v} - \frac{\gamma_{ca} d}{v} - \frac{\Delta^2 w_0^2}{4 v_x^2}\right] \left(\frac{w_0}{v}\right) \sin\frac{\Delta d}{v} \qquad (152)$$

The fringe pattern including the light shift contribution is the sum of a cosine and of a sine, which can be rewritten as a single cosine with a frequency shift δ equal to :

$$\delta = \frac{1}{2}\sqrt{\frac{\pi}{2}}\,\frac{w_o}{d}\sum_{\varepsilon j}\frac{\xi_j^{(6)}}{\xi^{(4)}}\,\delta_j^{\varepsilon} \tag{153}$$

We end up with the very important conclusion that the light shift for two-photon Ramsey fringes is reduced by the factor d/w_o, which is consistent with the experimental observations in [36].

The same result can be obtained, if the Gaussian dependence of the field is approximated by a rectangle function, by using Ramsey's formula for the equivalent two-level system with a shifted atomic frequency in each field zone.

C - RAMSEY FRINGES IN SATURATION SPECTROSCOPY

For the general case of closely spaced sublevels (e.g. hyperfine sublevels) noted a,a' for the lower level, b,b' for the upper level, saturation spectroscopy diagrams are displayed in fig.13.[21,24]

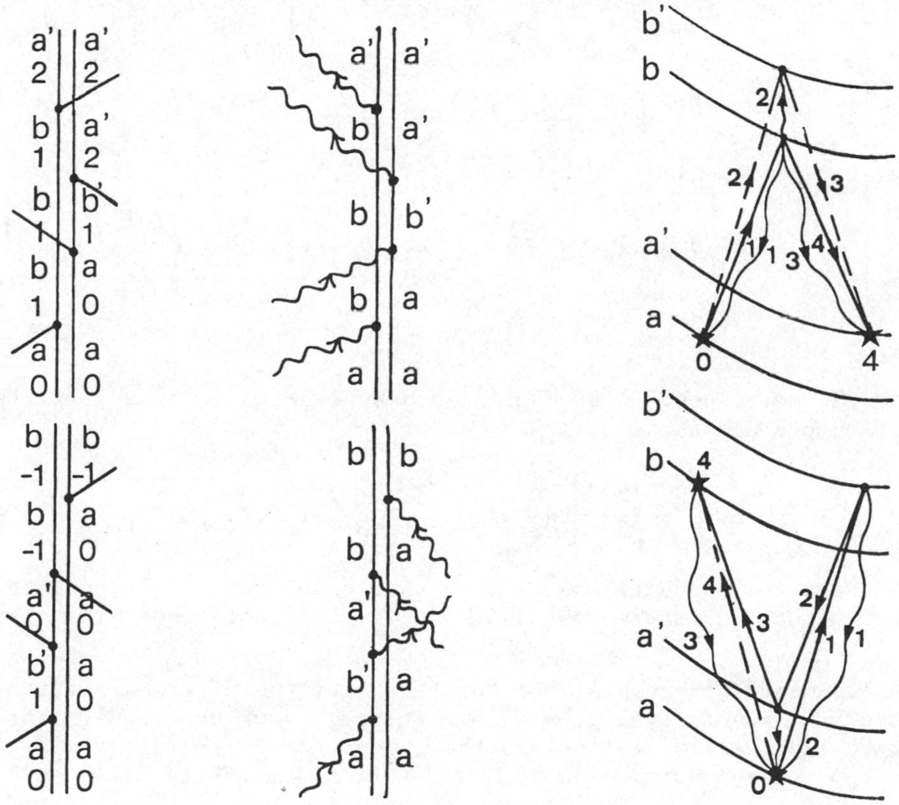

Figure 13

Only the processes which do not average out after velocity integration have been represented. We check that they all involve two optical coherences with arrows pointing in opposite directions.[24]

We illustrate the application of the diagrammatic rules in the case of the first diagram on fig. 14.

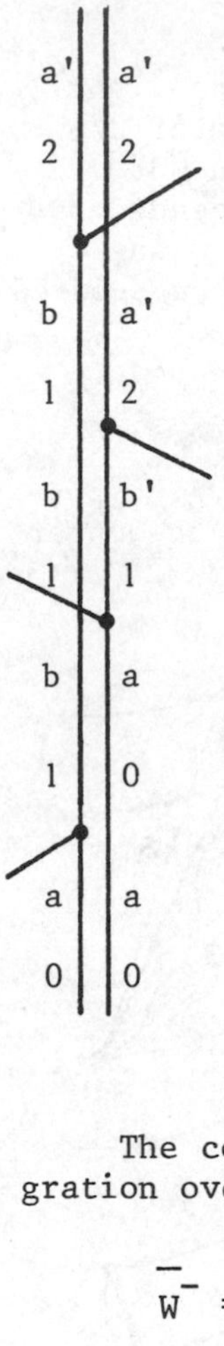

$$\left[\frac{\partial}{\partial t}\langle m_{a'}=2|\rho_{a'a'}(\alpha_o=a,\vec{p}_o,\vec{r},t)|m_{a'}=2\rangle\right]^{(4)}_{rad} = 2\,\mathrm{Re}\iiint d\tau d\tau' d\tau''$$

$$(+i\vec{\mu}_{a'b}.\vec{E}_o^{-}/2\hbar)U_o^{-}(\vec{r},t)\exp\left[i(\omega t+kz+\varphi^{-})\right]$$

$$\exp\left[-(i\omega_{ba'}-3i\delta+\gamma_{ba'})\tau\right]$$

$$(-i\vec{\mu}_{b'a'}.\vec{E}_o^{-*}/2\hbar)U_o^{-*}(\vec{r}-\gamma\vec{v}\tau)\exp\left[-i(\omega(t-\gamma\tau)+k(z-\gamma v_z\tau)+\varphi^{-})\right]$$

$$\exp\left[-(i\omega_{bb'}+\gamma_{bb'})\tau'\right]$$

$$(-i\vec{\mu}_{ab'}.\vec{E}_o^{+}/2\hbar)\,U_o^{+}(\vec{r}-\gamma\vec{v}(\tau+\tau'))\exp\left[i(\omega(t-\gamma(\tau+\tau'))-k(z-\gamma v_z(\tau+\tau'))+\varphi^{+})\right]$$

$$\exp\left[-(i\omega_{ba}+i\delta+\gamma_{ba})\tau''\right]$$

$$(i\vec{\mu}_{ba}.\vec{E}_o^{+*}/2\hbar)U_o^{+*}(\vec{r}-\gamma\vec{v}(\tau+\tau'+\tau''))\exp\left\{-i\left[\omega(t-\gamma(\tau+\tau'+\tau''))-k(z-\gamma v_z(\tau+\tau'+\tau''))+\varphi^{+}\right]\right\}$$

$$\times\; h^3\, n_a^{(o)} F(\vec{r}-\gamma\vec{v}(\tau+\tau'+\tau''),\vec{p}_o)$$

Figure 14

The corresponding average absorbed power is obtained by integration over phase space(*):

$$\overline{W^{-}} = -\hbar\omega\int\frac{d^3p_o\,d^3r}{h^3}\left[\frac{\partial}{\partial t}\langle 2|\rho_{a'a'}(\alpha_o=a,\vec{p}_o,\vec{r},t)|2\rangle\right]_{rad}$$

Detailed calculations of line shapes for the single zone case can be found in references[1,21] and we shall not repeat them here.

(*) The minus sign comes from the fact that the diagram is terminated by a lower level population. It would have been equivalent for the computation of the absorbed power to end the diagram with the upper level population $\langle 1|\rho_{bb}|1\rangle$ and multiply its contribution by $\hbar\omega$.[21]

We shall rather focus on the computation of the fringes in the case of three field zones and a simple two-level system ($a \equiv a'$, $b \equiv b'$):

$\hat{e}\, E_o^{\pm}\, U_o^{\pm}\, e^{i\varphi^{\pm}}$ is replaced by :

$$\begin{aligned}&\hat{e}_1^{\pm}\, E_1^{\pm}\, G_1^{\pm}(y)\, G_1^{\pm}(x+d)\, e^{i\varphi_1^{\pm}}\\ +\,&\hat{e}_2^{\pm}\, E_2^{\pm}\, G_2^{\pm}(y)\, G_2^{\pm}(x)\, e^{i\varphi_2^{\pm}}\\ +\,&\hat{e}_3^{\pm}\, E_3^{\pm}\, G_3^{\pm}(y)\, G_3^{\pm}(x-d)\, e^{i\varphi_3^{\pm}}\end{aligned} \tag{154}$$

and :

$$F(\vec{r},\vec{p}_o)\,d^3p_o \quad \text{by} \quad \frac{4v^2}{\sqrt{\pi}\,u^3}\exp\left(-\frac{v^2}{u^2}\right) A(n_y,n_z,y,z)\,dv\,dn_y\,dn_z \tag{155}$$

$$\text{with} \quad \hat{n} = \vec{v}/v$$

Neglecting transit effects and the 2nd order Doppler shift in the z direction, the v_z (or n_z) integration can be performed first in the limit of infinite Doppler width :

$$\int_{-1}^{+1} dn_z\, A(n_y,n_z,y,z)\,e^{ikvn_z(\tau-\tau'')} \Longrightarrow \frac{2\pi}{kv}\,\delta(\tau-\tau'')\,A(n_y,o,y,z) \tag{156}$$

Then, the calculation is simple only if either one of the two following assumptions is made :

1) A independent of y : $A(n_y,n_z,z)$. This applies to the cell case or when the laser beam transverse extension is small compared to the molecular beam.

2) The n_y distribution can be approximated by a Dirac distribution $\delta(n_y)$. This applies when the laser beam diameter is large compared to the molecular beam width.

In the first case the x and y integrations can both be performed. In the absence of wave-front curvature and for equal beam waists w_o :

$$G_1^{\pm} \equiv G_2^{\pm} \equiv G_3^{\pm} = \exp\left(-\frac{x^2}{w_o^2}\right)$$

$$\int_{-\infty}^{+\infty} dx\; G_3^{-}(x)\,G_2^{-*}(x-v_x\tau)\,G_2^{+}\big(x-v_x(\tau+\tau')\big)\,G_1^{+*}\big(x-v_x(\tau'+2\tau)\big)$$

$$= \frac{\sqrt{\pi}\,w_o}{2}\exp\left\{-2\frac{v_x^2\tau^2}{w_o^2} - 2\frac{v_x^2\tau'\tau}{w_o^2} + 4\frac{v_x d}{w_o^2}\tau - \frac{v_x^2\tau'^2}{w_o^2} + 2\frac{v_x d\tau'}{w_o^2} - \frac{2d^2}{w_o^2}\right\}$$

and the same result for $\int dy$ with $v_x \rightarrow v_y$ and $d = o$.

Finally the τ and τ' integrations can be performed as in reference [3] by using the property (133) to extract the oscillating part after the τ integration :

$$\overline{W}^{(4)-} = -2\pi^{(5/2)}\hbar\omega\left[\Omega^-_{3ba}\Omega^-_{2ba}\Omega^+_{2ba}\Omega^+_{1ba}\left(\frac{w_o}{u}\right)^4\right]\frac{n_a^{(o)}u}{kw_o}\int dz\, dn_y A(n_y, n_z = o, z)$$

$$\exp\left(-\frac{2d^2}{w_o^2}\right)\int_0^{+\infty}\frac{dv}{v}\exp\left(-\frac{v^2}{u^2}\right)\exp\left[\frac{w_o^2}{2v^2}\left(\frac{2v_x d}{w_o^2}-\gamma_{ba}\right)^2-\frac{(\gamma\omega-\omega_{ba}+\delta)^2}{2v^2/w_o^2}\right]$$

$$\mathrm{Re}\left\{\exp\left[i(\varphi_3^- - \varphi_2^- + \varphi_2^+ - \varphi_1^+)\right]\exp\left[\frac{i(\gamma\omega-\omega_{ba}+\delta)}{2v^2/w_o^2}\left(\frac{2v_x d}{w_o^2}-\gamma_{ba}\right)\right]\right.$$

$$\left. W\left[\left(\omega_{ba}-\delta-\gamma\omega+i(\gamma_b-\gamma_{ba})\right)w_o/\sqrt{2}\,v\right]\right\} \qquad (157)$$

We can make the approximation $v_x \simeq v$ except for a factor $\exp\left[-n_y^2\,\frac{2d^2}{w_o^2}\right]$. The n_y and z integrations give simply the multiplicative factor :

$$\int dz\, dn_y \exp\left(-n_y^2\,\frac{2d^2}{w_o^2}\right)A(n_y, n_z = o, z) \qquad (158\text{-a})$$

In the second case this factor is replaced by :

$$\frac{2}{\sqrt{\pi}\,w_o}\int dy\; dz\; G^4(y)A(n_z = o, y, z)\left(\int dn_y\;\delta(n_y)\right) \qquad (158\text{-b})$$

A good approximation for the same factor in the general case, which includes the two previous limits, is obtained by setting $v_y\tau \simeq n_y d$, $v_y\tau' \simeq o$ and $v_x \simeq v$:

$$\frac{2}{\sqrt{\pi}\,w_o}\int dy\, dz\, dn_y\; G^4(y-n_y d)\exp\left(-2\,\frac{n_y^2 d}{w_o^2}\right)A(n_y, n_z = o, y, z) \qquad (159)$$

The final result is :

$$\frac{\overline{W}^{(4)-}}{\hbar\omega} = -4\pi^2\,\eta\,\Phi_{eff}\,.\,\mathcal{F}(\omega) \qquad (160)$$

where :

$$\eta = (\langle a|\vec{\mu}.\hat{e}_3^-|b\rangle\langle b|\vec{\mu}.\hat{e}_2^{-*}|a\rangle\langle a|\vec{\mu}.\hat{e}_2^+|b\rangle\langle b|\vec{\mu}.\hat{e}_1^{+*}|a\rangle \; E_3^- E_2^- E_2^+ E_1^+/16\hbar^4)(w_o/u)^4 \quad (161)$$

is an excitation efficiency,

$$\Phi_{eff} = \left[n_a^{(o)} \frac{u}{kw_o}\int dy\, dz\, dn_y\, G^4(y-n_y d)\exp(-2n_y^2\frac{d^2}{w_o^2})A(n_y, n_z = o, y, z)\right]^{(*)} \quad (162)$$

is an effective flux of atoms ($1/kw_o$ is the fraction of atoms used in the n_z space) and :

$$\mathcal{F}(\omega) = \exp\left(-\frac{2d^2}{w_o^2}\right)\int_o^{+\infty}\frac{dv}{v}\exp\left[-\frac{v^2}{u^2}+\frac{(\frac{2vd}{w_o^2}-\gamma_{ba})^2}{2v^2/w_o^2}-\frac{(\gamma\omega-\omega_{ba}+\delta)^2}{2v^2/w_o^2}\right]$$

$$\mathrm{Re}\left\{\exp\left[i(\varphi_3^- - \varphi_2^- + \varphi_2^+ - \varphi_1^+)\right]\exp\left[i\frac{(\gamma\omega-\omega_{ba}+\delta)}{v^2/w_o^2}(\frac{2vd}{w_o^2}-\gamma_{ba})\right]\right.$$

$$\left. W\left(i\frac{w_o}{\sqrt{2}\,v}\left[i(\gamma\omega-\omega_{ba}+\delta)+\gamma_b-\gamma_{ba}\right]\right)\right\} \quad (163)$$

is a line shape averaged over the effective velocity distribution and exhibiting the Ramsey fringes.

If relaxation may be neglected during the transit time across each zone, this line shape simplifies into :

$$\mathcal{F}(\omega) = \mathrm{Re}\int_o^{+\infty}\frac{dv}{v}\exp(-\frac{v^2}{u^2})\exp(-2\gamma_{ba}\frac{d}{v})$$

(effective velocity distribution)(relaxation)

$$\exp\left[-\frac{(\omega-\omega_{ba})^2}{2v^2}w_o^2\right]W\left(\frac{w_o}{\sqrt{2}v}(\omega_{ba}-\omega)\right)$$

(envelope)

$$\exp\left[i(\varphi_3^- - \varphi_2^- + \varphi_2^+ - \varphi_1^+)\right]\exp\left[i\frac{(\gamma\omega-\omega_{ba}+\delta)}{v}2d\right] \quad (164)$$

(phases) (oscillating term)

(*) A couple of typographical errors have appeared in expressions (3) and (5) of reference [8] and should be corrected according to (162) and (164)

with[29] :

$$W\left(\frac{w_o}{\sqrt{2}v}(\omega_{ba}-\omega)\right) = \exp\left[-\frac{(\omega_{ba}-\omega)^2}{2v^2}w_o^2\right] + \frac{2i}{\sqrt{\pi}}D\left(\frac{w_o}{\sqrt{2}v}(\omega_{ba}-\omega)\right)$$

where D is Dawson's integral.

These expressions are valid both for beam and cell experiments and should allow a direct comparison between these two cases. (For a cell $A(n_y, n_z, y, z) \equiv 1/4$).

For a beam experiment, the flux of excited atoms reaching a given point at a distance ℓ from the last zone is also given by $\overline{W}/\hbar\omega$ in which the relaxation factor $\exp(-\gamma_b \ell/v)$ is introduced as in the two-photon case. Finally expressions (160-164) correspond to a single diagram and one should add three similar terms corresponding to $\overline{W}^{(4)+}$ ($+ \longleftrightarrow -$ in η) and to the other recoil peak ($\delta \longrightarrow -\delta$ and $\gamma_b \longrightarrow \gamma_a$).

Power contraction of the recoil splitting[4,5]

One of the difficulties associated with the recoil splitting in saturation spectroscopy is the light shift of the recoil peaks towards each other. The origin of this light shift is again the existence of non-resonant channels coupled with the energy-momentum states involved in either resonance and is illustrated by the energy-momentum diagrams of fig. 15 a . The detuning, which can be

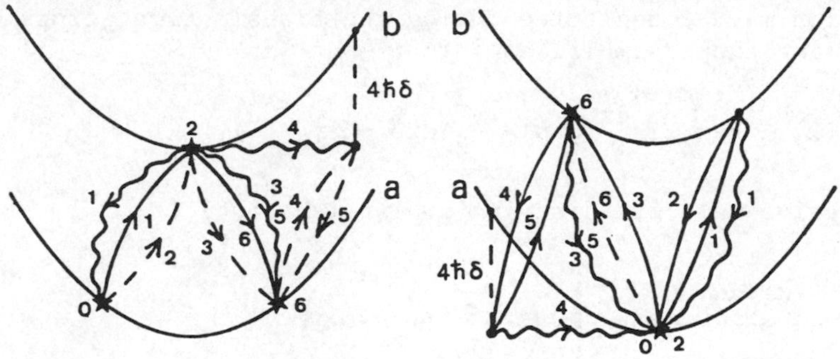

Figure 15-a

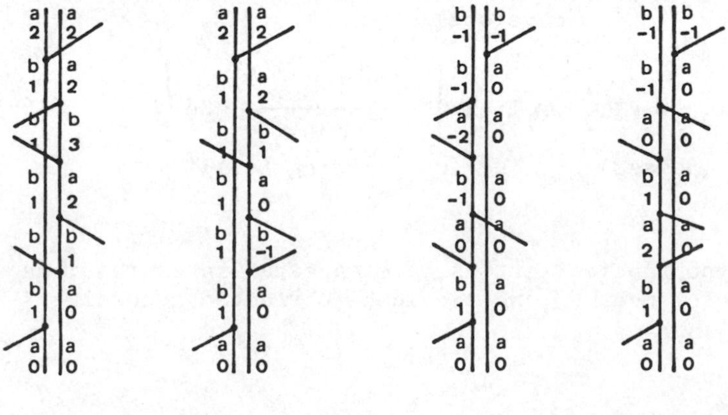

Figure 15-b

read directly on the figure, amounts to 4δ and is therefore responsible for a light shift equal to $[(\Omega^+_{ba})^2 + (\Omega^-_{ba})^2]/8\delta$. (Here 2δ is the recoil splitting.) This is precisely the low-field limit found with the plane wave theory of reference [4] (for high fields this light shift levels off at a constant value). One would like to know what this shift becomes for the Ramsey fringes. The answer to this question can be found with the sixth-order diagrams of fig.15b. The corresponding calculations are too cumbersome to be given here but they lead to the same simple conclusion as for the two-photon case : the above shift is reduced by the ratio d/w_o and this result has been confirmed by the calcium experiments[7,8]. Also, it is clear from the above diagrams that this light shift is suppressed when the three standing waves are replaced by four travelling waves as in [27]. This is an illustration of the physical insight provided by the density matrix diagrams and of their predictive value.

APPENDIX A

In this appendix we wish to discuss the relationship between the absorbed laser power and the transition rate from one level to another. The number R_β of molecules transferred per unit time to the level β, under the influence of the interaction Hamiltonian V, can be written :

$$R_\beta = \left[\frac{\partial}{\partial t} \mathrm{Tr}_{ext}\, \rho_{\beta\beta}(t)\right]_{rad} = \sum_{\alpha_o}\int d^3 p_o \left[\frac{\partial}{\partial t} \mathrm{Tr}_{ext}\, \rho_{\beta\beta}(\alpha_o,\vec{p}_o,t)\right]_{rad} \tag{A-1}$$

where Tr_{ext} is the trace operation over translation degrees of freedom. From the density matrix equations we find :

$$R_\beta = \sum_{\alpha_o}\int d^3 p_o\, \mathrm{Tr}_{ext}\left[V,\rho(\alpha_o,\vec{p}_o,t)\right]_{\beta\beta} /i\hbar \tag{A-2}$$

where :

$$\mathrm{Tr}_{ext}\left[V,\rho\right]_{\beta\beta} /i\hbar = \int d^3r \sum_\alpha \Big[V_{\beta\alpha}(\vec{r},t) <\vec{r}|\, \rho_{\alpha\beta}(\alpha_o,\vec{p}_o,t)|\,\vec{r}>$$

$$- <\vec{r}|\,\rho_{\beta\alpha}(\alpha_o,\vec{p}_o,t)|\,\vec{r}> V_{\alpha\beta}(\vec{r},t)\Big] /i\hbar =$$

$$= 2\mathrm{Re}\int d^3 r \sum_\alpha V_{\beta\alpha}(\vec{r},t) <\vec{r}\,|\rho_{\alpha\beta}(\alpha_o,\vec{p}_o,t)|\,\vec{r}>/i\hbar \tag{A-3}$$

with $V_{\beta\alpha}(\vec{r},t) = <\beta|<\vec{r}|\,V(t)|\vec{r}>|\,\alpha> = -<\beta|\vec{\mu}|\,\alpha>.\,\vec{\mathcal{E}}(\vec{r},t)$

If we keep only terms varying slowly with time :

$$\overline{R_\beta} = \overline{R^+_\beta} + \overline{R^-_\beta} \tag{A-4}$$

with :

$$\bar{R}_\beta^+ = 2\mathrm{Re}\int d^3r \sum_{\alpha_o}\int d^3p_o \sum_\alpha (i\vec{\mu}_{\beta\alpha}.\vec{E}^*/2\hbar) <\vec{r}|\rho_{\alpha\beta}^{(+)}(\alpha_o,\vec{p}_o,t)|\vec{r}> \quad \text{(A-5)}$$

and :

$$\bar{R}_\beta^- = 2\mathrm{Re}\int d^3r \sum_{\alpha_o}\int d^3p_o \sum_\alpha (i\vec{\mu}_{\beta\alpha}.\vec{E}/2\hbar) <\vec{r}|\rho_{\alpha\beta}^{(-)}(\alpha_o,\vec{p}_o,t)|\vec{r}> \quad \text{(A-6)}$$

where $\rho_{\alpha\beta}^{(+)}$ and $\rho_{\alpha\beta}^{(-)}$ are the $\exp(i\omega t)$ and $\exp(-i\omega t)$ parts of $\rho_{\alpha\beta}$.

For $E_\beta > E_\alpha$ the first term ($\bar{R}_\beta^+$) corresponds to resonant processes and the second ($\bar{R}_\beta^-$) to antiresonant ones, and vice-versa if $E_\beta < E_\alpha$. We shall therefore split these two terms into a part for which $E_\alpha < E_\beta$ and one for which $E_\alpha > E_\beta$

$$\bar{R}_\beta^+ = \bar{R}_{\beta>}^+ + \bar{R}_{\beta<}^+$$
$$\bar{R}_\beta^- = \bar{R}_{\beta>}^- + \bar{R}_{\beta<}^- \quad \text{(A-7)}$$

with the following properties :

$$\sum_\beta \bar{R}_{\beta<}^- = -\sum_\beta \bar{R}_{\beta>}^+ \quad \text{for resonant processes}$$

and :

$$\sum_\beta \bar{R}_{\beta>}^- = -\sum_\beta \bar{R}_{\beta<}^+ \quad \text{for antiresonant processes} \quad \text{(A-8)}$$

The power absorbed at the circular frequency ω is given by :

$$\bar{W}_{abs} = \frac{\hbar\omega}{2}\sum_\beta (\bar{R}_{\beta>}^+ - \bar{R}_{\beta<}^- - \bar{R}_{\beta>}^- + \bar{R}_{\beta<}^+) \quad \text{(A-9)}$$

where the factor 1/2 comes from the fact that each level is counted twice (once as an arrival level and once as a departure level). If we use (A-8) we find :

$$\bar{W}_{abs} = \hbar\omega\sum_\beta \bar{R}_\beta^+ =$$

$$2\hbar\omega\mathrm{Re}\int d^3r \sum_{\alpha_o}\int d^3p_o \sum_{\alpha,\beta} (i\vec{\mu}_{\alpha\beta}.\vec{E}^*/2\hbar)<\vec{r}|\rho_{\alpha\beta}^{(+)}(\alpha_o,\vec{p}_o,t)|\vec{r}> \quad \text{(A-10)}$$

which is identical to the integral of $d\bar{W}_{abs}/dV$ given by (33). To obtain this result we have had to assume opposite signs for the energy quanta exchanged in resonant and antiresonant processes. If we consider a two-level system a,b such that $E_b > E_a$, antiresonant processes correspond to the emission of one quantum from the lower level a and to the absorption of one quantum from the upper level b.

APPENDIX B : LIOUVILLE SPACE DERIVATION OF INTENSITIES OF NON-LINEAR PROCESSES

The atomic level degeneracies and the polarization characteristics of the optical fields can be handled using standard techniques of summation over magnetic quantum numbers such as those used in [21] for saturation spectroscopy. However in the absence of any external field lifting this degeneracy, a much more powerful tool is the Racah algebra in Liouville space. This approach implies that the density operator $|\rho\gg$, considered as a vector of the Liouville space,is expanded on a basis of irreducible tensors [39] :

$$|\,\alpha F_\alpha (\beta F_\beta)^\dagger \;;\; KQ \gg \qquad \text{(B-1)}$$

From the physicist's point of view,Liouville space may be considered as the tensor product of the usual Hilbert space by its dual space, and operators in Liouville space (such as the the interaction Liouvillian) can be written as sums of tensor products of Hilbert space operators, acting in the Hilbert space part of $|\rho\gg$,by hermitian conjugate operators of Hilbert space operators,acting on the dual space part of $|\rho\gg$. These superoperators are written as irreducible tensorial sets,and we use the notations and conventions of Fano and Racah [40] to deal with the corresponding coupling algebra :

$$\mathscr{M}^{[k]} = \left[\mathscr{S}^{[k_1]} \otimes \mathscr{U}^{(k_2)} \mathscr{T}^{[k_2]\dagger}\right]^{[k]} \qquad \text{(B-2)}$$

It can be shown through a recoupling analog to equ.(15.1) and (15.2) of [40] that :

$$\mathscr{M}_q^{[k]} = \sum_{\substack{\alpha,\beta,K_1,K_2\\ Q_1,Q_2}} <\alpha_2 F_{\alpha_2} \| S^{[k_1]} \| \alpha_1 F_{\alpha_1}><\beta_2 F_{\beta_2} \| T^{[k_2]} \| \beta_1 F_{\beta_1}>^*$$

$$\sqrt{(2K_2+1)(2K_1+1)(2k+1)} \begin{Bmatrix} F_{\alpha_2} & F_{\alpha_1} & k_1 \\ F_{\beta_2} & F_{\beta_1} & k_2 \\ K_2 & K_1 & k \end{Bmatrix} (-1)^{K_2-Q_2} \begin{pmatrix} K_2 & k & K_1 \\ -Q_2 & q & Q_1 \end{pmatrix}$$

$$|\,\alpha_2 F_{\alpha_2} (\beta_2 F_{\beta_2})^\dagger \;;\; K_2 Q_2 \gg \ll \alpha_1 F_{\alpha_1} (\beta_1 F_{\beta_1})^\dagger \;;\; K_1 Q_1 \,| \qquad \text{(B-3)}$$

which can be considered as the expression of Wigner-Eckart theorem in Liouville space with the following reduced matrix element in the coupled basis :

$$\ll \alpha_2 F_{\alpha_2} (\beta_2 F_{\beta_2})^{\dagger} \;;\; K_2 \| M^{[k]} \| \alpha_1 F_{\alpha_1} (\beta_1 F_{\beta_1})^{\dagger} \;;\; K_1 \gg$$

$$= \sqrt{(2k+1)(2K_2+1)(2K_1+1)} \begin{Bmatrix} F_{\alpha_2} & F_{\alpha_1} & k_1 \\ F_{\beta_2} & F_{\beta_1} & k_2 \\ K_2 & K_1 & k \end{Bmatrix} \langle \alpha_2 F_{\alpha_2} \| S^{[k_1]} \| \alpha_1 F_{\alpha_1} \rangle$$

$$\langle \beta_2 F_{\beta_2} \| T^{[k_2]} \| \beta_1 F_{\beta_1} \rangle^{*} \qquad \text{(B-4)}$$

For electric dipole interactions we take :

$$\mathscr{M}^{[1]} = \left[(i\mu)^{[1]} \otimes 1^{[0]} \right]^{[1]} - \left[1^{[0]} \otimes (i\mu)^{[1]} \right]^{[1]} \qquad \text{(B-5)}$$

where $(i\mu)^{[1]}$ is the contrastandard set defined from the dipole moment operator with the standardization convention of [40] as :

$$(i\mu)_1^{[1]} = \frac{-i\mu_x + \mu_y}{\sqrt{2}}, \quad (i\mu)_o^{[1]} = i\mu_z, \quad (i\mu)_{-1}^{[1]} = \frac{i\mu_x + \mu_y}{\sqrt{2}} \qquad \text{(B-6)}$$

and which is self-conjugate. With a similar definition for the sets associated with the electric field $\vec{\mathscr{E}}$ and with the light polarization vector $\hat{e} = \Sigma\, (-)^q \hat{u}_{-q}\, e_q$, the interaction Liouvillian is :

$$\mathscr{L}_{int} = \mathscr{M}^{[1]} \cdot (i\,\mathscr{E})^{[1]} / \hbar \qquad \text{(B-7)}$$

the required matrix elements are those of :

$\mathscr{M}^{[1]} \cdot (ie)^{[1]}$ for the interactions involving the $\exp(i\omega t)$ part of the field ($\varepsilon' = +1$),

and $\mathscr{M}^{[1]} \cdot (ie)^{(1)*}$ for those involving $\exp(-i\omega t)$. ($\varepsilon' = -1$)

From the Wigner-Eckart theorem in Liouville space we get :

$$\ll \alpha_2 F_{\alpha_2} (\beta_2 F_{\beta_2})^{\dagger} \;;\; K_2 Q_2 | \mathscr{M}_q^{[1]} | \alpha_1 F_{\alpha_1} (\beta_1 F_{\beta_1})^{\dagger} \;;\; K_1 Q_1 \gg =$$

$$i\delta_{\beta_2\beta_1} \delta_{F_{\beta_2} F_{\beta_1}} \langle \alpha_2 F_{\alpha_2} \| \mu \| \alpha_1 F_{\alpha_1} \rangle (-1)^{F_{\alpha_2} + F_{\beta_2} - Q_2} \begin{pmatrix} K_2 & K_1 & 1 \\ -Q_2 & Q_1 & q \end{pmatrix} \times$$

$$\times\sqrt{(2K_2+1)(2K_1+1)}\begin{Bmatrix} K_2 & K_1 & 1 \\ F_{\alpha_1} & F_{\alpha_2} & F_{\beta_2}\end{Bmatrix} - i\,\delta_{\alpha_2\alpha_1}\,\delta_{F_{\alpha_2}F_{\alpha_1}} <\beta_1 F_{\beta_1}\|\mu\|\beta_2 F_{\beta_2}>$$

$$(-1)^{F_{\beta_2}+F_{\alpha_2}-Q_2}\begin{pmatrix} K_2 & 1 & K_1 \\ -Q_2 & q & Q_1\end{pmatrix}\sqrt{(2K_2+1)(2K_1+1)}\begin{Bmatrix} K_1 & K_2 & 1 \\ F_{\beta 2} & F_{\beta_1} & F_{\alpha_2}\end{Bmatrix}$$

$$= i(-1)^{F_{\alpha_2}+F_{\beta_2}-Q_2}\sqrt{(2K_2+1)(2K_1+1)}\begin{pmatrix} K_2 & K_1 & 1 \\ -Q_2 & Q_1 & q\end{pmatrix}$$

$$\Bigg[\delta_{\beta_2\beta_1}\,\delta_{F_{\beta_2}F_{\beta_1}} <\alpha_2 F_{\alpha_2}\|\mu\|\alpha_1 F_{\alpha_1}>$$

$$\begin{Bmatrix} K_2 & K_1 & 1 \\ F_{\alpha_1} & F_{\alpha_2} & F_{\beta_2}\end{Bmatrix} - \delta_{\alpha_2\alpha_1}\,\delta_{F_{\alpha_2}F_{\alpha_1}} <\beta_1 F_{\beta_1}\|\mu\|\beta_2 F_{\beta_2}>$$

$$(-1)^{K_1+K_2+1}\begin{Bmatrix} K_1 & K_2 & 1 \\ F_{\beta_2} & F_{\beta_1} & F_{\alpha_2}\end{Bmatrix}\Bigg] \qquad \text{(B-8)}$$

The intensity factor for any n^{th} order non-linear process will be: (we assume a single q at each j otherwise there is a corresponding summation)

$$\sum_{\substack{K_jQ_j\\ q_j}} \prod_{j=1}^{j=n} (-\varepsilon_j')^{q_j} \ll \alpha_j F_{\alpha_j}(\beta_j F_{\beta_j})^{\dagger}; K_jQ_j|\mathscr{M}^{[1]}_{-\varepsilon_j' q_j}|\alpha_{j-1}F_{\alpha_{j-1}}(\beta_{j-1}F_{\beta_{j-1}})^{\dagger};$$

$$K_{j-1}Q_{j-1}\gg\sqrt{(2F_{\alpha_n}+1)(2F_{\alpha_o}+1)}\,(n_o)\,i^n\left(\frac{1}{2i\hbar}\right)^n E_j^o\, e_{q_j} \qquad \text{(B-9)}$$

where the factors $\sqrt{2F_{\alpha_o}+1}$ and $\sqrt{2F_{\alpha_n}+1}$ come from the conversion of populations from the spherical tensor basis and where n_o is the population of each Zeeman sublevel.

For example, the general formula for the intensity of 4th order processes is :

$$(-\varepsilon_4')^{q_4}(-\varepsilon_3')^{q_3}(-\varepsilon_2')^{q_2}(-\varepsilon_1')^{q_1}(-1)^{F_{\alpha_2}+F_{\beta_2}} <\beta_3 F_{\beta_3}\|\mu\|\alpha_3 F_{\alpha_3}> \quad \times$$

$$\times \sum_{\substack{K_2,Q_2\\ q_1q_2q_3q_4}} (2K_2+1)(-1)^{-Q_2} \begin{pmatrix} K_2 & 1 & 1 \\ -Q_2 & -\varepsilon_1' q_1 & -\varepsilon_2' q_2 \end{pmatrix} \begin{pmatrix} K_2 & 1 & 1 \\ Q_2 & -\varepsilon_3' q_3 & -\varepsilon_4' q_4 \end{pmatrix}$$

$$\left[<\alpha_3 F_{\alpha_3} \| \mu \| \alpha_2 F_{\alpha_2}> \delta_{\beta_3\beta_2} \begin{Bmatrix} 1 & K_2 & 1 \\ F_{\alpha_2} & F_{\alpha_3} & F_{\beta_3} \end{Bmatrix} - <\beta_2 F_{\beta_2} \| \mu \| \beta_3 F_{\beta_3}> \right.$$

$$\left. (-1)^{K_2} \delta_{\alpha_3\alpha_2} \begin{Bmatrix} K_2 & 1 & 1 \\ F_{\beta_3} & F_{\beta_2} & F_{\alpha_3} \end{Bmatrix} \right] \left[<\alpha_2 F_{\alpha_2} \| \mu \| \alpha_1 F_{\alpha_1}> \delta_{\beta_2\beta_1} \begin{Bmatrix} K_2 & 1 & 1 \\ F_{\alpha_1} & F_{\alpha_2} & F_{\beta_2} \end{Bmatrix} \right.$$

$$\left. -(-1)^{K_2} <\beta_1 F_{\beta_1} \| \mu \| \beta_2 F_{\beta_2}> \delta_{\alpha_2\alpha_1} \begin{Bmatrix} K_2 & 1 & 1 \\ F_{\beta_1} & F_{\alpha_2} & F_{\beta_2} \end{Bmatrix} \right]$$

$$(<\alpha_1 F_{\alpha_1} \| \mu \| \alpha_0 F_{\alpha_0}> \delta_{\beta_1\beta_0} - <\beta_0 F_{\beta_0} \| \mu \| \beta_1 F_{\beta_1}> \delta_{\alpha_1\alpha_0}) \delta_{\alpha_0\beta_0}$$

$$(E_1^o e_{q_1} \; E_2^o e_{q_2} \; E_3^o e_{q_3} \; E_4^o e_{q_4} \,/\, 16\, \hbar^4) n_o \qquad \text{(B-10)}$$

This general result applies to saturation spectroscopy, two-photon spectroscopy, CARS, stimulated Raman or any four-wave mixing process. For example the formulas of [21] for saturation spectroscopy are recovered with :

$$\varepsilon_1' = -1 \;,\; \varepsilon_2' = +1 \;,\; \varepsilon_3' = -1 \;,\; \varepsilon_4' = +1$$

and either :

$$\alpha_o = \beta_o = \beta_1 \equiv a \;,\; \alpha_1 = \alpha_2 = \alpha_3 \equiv b \;,\; \beta_2 \equiv b' \;,\; \beta_3 \equiv a'$$

or :

$$\alpha_o = \beta_o = \beta_1 = \beta_2 = \beta_3 \equiv a \;,\; \alpha_1 \equiv b \;,\; \alpha_2 \equiv a' \;,\; \alpha_3 \equiv b'$$

In the two-photon case,we read on the diagrams the correspondence :

$$\alpha_o = \beta_o = \beta_1 = \beta_2 \equiv a \;,\; \alpha_1 = \beta_3 \equiv b \;,\; \alpha_2 = \alpha_3 \equiv c$$

$$\varepsilon_1' = \varepsilon_2' = -1 \qquad \varepsilon_3' = \varepsilon_4' = +1$$

which gives for the intensity of each diagram :

$$\sum_{K,Q} (2K+1) \langle cF_c \| \mu \| bF_b \rangle^2 \langle bF_b \| \mu \| aF_a \rangle^2$$

$$\begin{pmatrix} K & 1 & 1 \\ Q & q_1 & q_2 \end{pmatrix} \begin{pmatrix} K & 1 & 1 \\ Q & q_3 & q_4 \end{pmatrix} \begin{Bmatrix} K & 1 & 1 \\ F_b & F_a & F_c \end{Bmatrix}^2 E_4^o e_{q_4} \; E_3^o e_{q_3} \; E_2^o e_{q_2} \; E_1^o e_{q_1} \quad \text{(B-11)}$$

which is consistent with the formulas given in [41] but allows for 4 different polarizations as long as $q_1 + q_2 = q_3 + q_4$.

APPENDIX C

The density matrix elements in representation $|\vec{r}\rangle$ and $|\vec{p}\rangle$ are related by :

$$\langle \vec{r} | \rho_{\alpha\beta}(\alpha_o, \vec{p}_o, t) | \vec{r}\,' \rangle =$$

$$\int d^3p\, d^3p' \langle \vec{p} | \rho_{\alpha\beta}(\alpha_o, \vec{p}_o, t) | \vec{p}\,' \rangle \exp\left[i\,(\vec{p}.\vec{r} - \vec{p}\,'.\vec{r}\,')/\hbar \right] / h^3 \qquad \text{(C-1)}$$

If the integrations are limited to intervals small compared to $\hbar k$ but large compared to $\hbar/w_o$ (C-1) can be written as a discrete sum:

$$\langle \vec{r} | \rho_{\alpha\beta}(\alpha_o, \vec{p}_o, t) | \vec{r}\,' \rangle = \sum_{m_\alpha, m_\beta} \langle \vec{p}_o + m_\alpha \hbar\vec{k} | \rho_{\alpha\beta}(\alpha_o, \vec{p}_o, \vec{r}, \vec{r}\,', t) | \vec{p}_o + m_\beta \hbar\vec{k} \rangle / h^3 \qquad \text{(C-2)}$$

$\langle \vec{p}_o + m_\alpha \hbar\vec{k} | \rho_{\alpha\beta}(\alpha_o, \vec{p}_o, \vec{r}, \vec{r}\,', t) | \vec{p}_o + m_\beta \hbar\vec{k} \rangle$ will be written in short:

$$\langle m_\alpha | \rho_{\alpha\beta}(\vec{r}, \vec{r}\,') | m_\beta \rangle$$

In the momentum representation the density matrix equations can be written :

$$i\hbar \frac{\partial}{\partial t} \langle \vec{p} | \rho_{\alpha\beta}(\alpha_o, \vec{p}_o, t) | \vec{p}\,' \rangle = i\hbar\lambda(\alpha_o, \vec{p}_o, t)\,\delta(\vec{p} - \vec{p}_o)\,\delta(\vec{p}\,' - \vec{p}_o)\,\delta_{\alpha\alpha_o}\,\delta_{\beta\alpha_o}$$

$$- i\hbar\left[(i\omega_{\alpha\beta} + \gamma_{\alpha\beta}) \langle \vec{p} | \rho_{\alpha\beta}(\alpha_o, \vec{p}_o, t) | \vec{p}\,' \rangle\right] + \left[(\vec{p}^{\,2} - \vec{p}\,'^2)/2M\right] \langle \vec{p} | \rho_{\alpha\beta}(\alpha_o, \vec{p}_o, t) | \vec{p}\,' \rangle$$

$$+ \sum_{\beta'} \int d^3k\, V_{\alpha\beta'}(\vec{k}, t) \langle \vec{p} - \hbar\vec{k} | \rho_{\beta'\beta}(\alpha_o, \vec{p}_o, t) | \vec{p}\,' \rangle$$

$$- \sum_{\alpha'} \int d^3k \langle \vec{p} | \rho_{\alpha\alpha'}(\alpha_o, \vec{p}_o, t) | \vec{p}\,' + \hbar\vec{k} \rangle V_{\alpha'\beta}(\vec{k}, t) \qquad \text{(C-3)}$$

To get the corresponding equations for $<m_\alpha|\rho_{\alpha\beta}(\vec{r},\vec{r}\,')|m_\beta>$, both sides of equation (C-3) are multiplied by $\exp[i(\vec{p}.\vec{r}-\vec{p}\,'.\vec{r}\,')/\hbar]$ and integrated over the intervals defined previously :

$$i\hbar\frac{\partial}{\partial t}<m_\alpha|\rho_{\alpha\beta}(\vec{r},\vec{r}\,')|m_\beta>=i\hbar\lambda(\alpha_o,\vec{p}_o,t)\delta_{\alpha\alpha_o}\delta_{\beta\alpha_o}\delta_{m_\alpha o}\delta_{m_\beta o}e^{i\vec{p}_o.(\vec{r}-\vec{r}\,')}$$

$$-\,i\hbar(i\omega_{\alpha\beta}+\gamma_{\alpha\beta})<m_\alpha|\rho_{\alpha\beta}(\vec{r},\vec{r}\,')|m_\beta>+(1/2M)\int d^3p\,d^3p'(p^2-p'^2)$$

$$\exp\left[i(\vec{p}.\vec{r}-\vec{p}\,'.\vec{r}\,')/\hbar\right]<\vec{p}|\rho_{\alpha\beta}(\alpha_o,\vec{p}_o,t)|\vec{p}\,'>$$

$$+\sum_{\beta'}\int d^3k\,V_{\alpha\beta'}(\vec{k},t)\exp i\vec{k}.\vec{r}\int d^3p\,d^3p'<\vec{p}-\hbar\vec{k}|\rho_{\beta'\beta}(\alpha_o,\vec{p}_o,t)|\vec{p}\,'>$$

$$\exp\left\{i\left[(\vec{p}-\hbar\vec{k}).\vec{r}-\vec{p}\,'.\vec{r}\,'\right]/\hbar\right\}$$

$$-\sum_{\alpha'}\int d^3k\,V_{\alpha'\beta}(\vec{k},t)\exp i\,\vec{k}.\vec{r}\,'\int d^3p\,d^3p'<\vec{p}|\rho_{\alpha\alpha'}(\alpha_o,\vec{p}_o,t)|\vec{p}\,'+\hbar\vec{k}>$$

$$\exp\left\{i\left[\vec{p}.\vec{r}-(\vec{p}\,'+\hbar\vec{k}).\vec{r}\,'\right]/\hbar\right\} \qquad \text{(C-4)}$$

We shall use the following approximation :

$$(1/2M)\int d^3p\,d^3p'(p^2-p'^2)<\vec{p}|\rho_{\alpha\beta}(\alpha_o,\vec{p}_o,t)|\vec{p}\,'>\exp\left[i(\vec{p}.\vec{r}-\vec{p}\,'.\vec{r}\,')/\hbar\right]$$

$$\simeq-\frac{i\hbar}{2M}\left[2\vec{p}_o+(m_\alpha+m_\beta)\hbar k\right].(\vec{\nabla}_{\vec{r}}+\vec{\nabla}_{\vec{r}\,'})<m_\alpha|\rho_{\alpha\beta}(\vec{r},\vec{r}\,')|m_\beta>$$

$$\simeq-\,i\hbar(\vec{p}_o/M).(\vec{\nabla}_{\vec{r}}+\vec{\nabla}_{\vec{r}\,'})<m_\alpha|\rho_{\alpha\beta}(\vec{r},\vec{r}\,')|m_\beta>$$

$$+\,\hbar(m_\alpha^2-m_\beta^2)\delta<m_\alpha|\rho_{\alpha\beta}(\vec{r},\vec{r}\,')|m_\beta> \qquad \text{(C-5)}$$

with : $\delta=\hbar k^2/2M$

We finally get the equations :

$$i\hbar\left[\partial/\partial t+(\vec{p}_o/M).(\vec{\nabla}_{\vec{r}}+\vec{\nabla}_{\vec{r}\,'})\right]<m_\alpha|\rho_{\alpha\beta}(\vec{r},\vec{r}\,')|m_\beta>=$$

$$i\hbar\lambda(\alpha_o,\vec{p}_o,t)\delta_{\alpha\alpha_o}\delta_{\beta\alpha_o}\delta_{m_\alpha o}\delta_{m_\beta o}e^{i\vec{p}_o.(\vec{r}-\vec{r}\,')}-i\hbar\left[i\omega_{\alpha\beta}+i(m_\alpha^2-m_\beta^2)\delta+\gamma_{\alpha\beta}\right]\times$$

$$\times \langle m_\alpha | \rho_{\alpha\beta}(\vec{r},\vec{r}\,')| m_\beta \rangle + \sum_{\beta'} \Big[V^+_{\alpha\beta'}(\vec{r},t) \langle m_\alpha -1 | \rho_{\beta'\beta}(\vec{r},\vec{r}\,')\, m_\beta \rangle$$

$$+ V^-_{\alpha\beta'}(\vec{r},t)\langle m_\alpha+1|\rho_{\beta'\beta}(\vec{r},\vec{r}\,')|m_\beta\rangle\Big]$$

$$-\sum_{\alpha'}\Big[\langle m_\alpha|\rho_{\alpha\alpha'}(\vec{r},\vec{r}\,')|m_\beta+1\rangle V^+_{\alpha'\beta}(\vec{r}\,',t)$$

$$+\langle m_\alpha|\rho_{\alpha\alpha'}(\vec{r},\vec{r}\,')|m_\beta-1\rangle V^-_{\alpha'\beta}(\vec{r}\,',t)\Big] \qquad \text{(C-6)}$$

For $\vec{r} \equiv \vec{r}\,'$ the operator $\vec{\nabla}_{\vec{r}\,'}$ simply disappears and these equations can be put in the operator form given in the text.

REFERENCES

1. Ch.J. Bordé, J.L. Hall, C.V. Kunasz and D.G. Hummer, Phys. Rev. A 14 , 236 (1976).
2. Ch.J. Bordé, C.R. Acad. Sc. Paris 282 B, 341 (1976).
3. Ch.J. Bordé, C.R. Acad. Sc. Paris 284 B, 101 (1977).
4. Ch.J. Bordé, C.R. Acad. Sc. Paris 283 B, 181 (1976).
5. Ch.J. Bordé, in Laser Spectroscopy III, p. 121, Eds. J.L. Hall and J.L. Carlsten, Springer Verlag 1977.
6. J.L. Hall and Ch.J. Bordé, Bull. Am. Phys. Soc. 19, 1196 (1974). J.L. Hall, Ch.J. Bordé and K. Uehara, Phys. Rev. Lett. 37, 1339 (1976).
7. R.L. Barger, Opt. Lett. 6, 145 (1981) and references therein.
8. Ch. Salomon, Ch. Bréant, Ch.J. Bordé and R.L. Barger, Journal de Physique Colloques 42, C8 (1981) and references therein.
9. U. Fano, Rev. of Mod. Phys. 29, 74 (1957). E. Wigner, Phys. Rev. 40, 749 (1932).
10. A.P. Kol'chenko, S.G. Rautian and R.I. Sokolovskii, Sov. Phys. JETP 28, 986 (1969).
11. J.H. Shirley, J. Phys. B 13, 1537 (1980). J.H. Shirley and S. Stenholm, J. Phys. A 10, 613 (1977).
12. S. Stenholm, J. Phys. B. 7, 1235 (1974).
13. V.A. Alekseev, T.L. Andreeva and I.I. Sobel'man, Sov. Phys. JETP 37, 413 (1973).
14. Ch.J. Bordé, S. Avrillier and M. Gorlicki, J. de Physique Lettres 38, L-249 (1977); 40, L-35 (1979).
15. A. Ben-Reuven, Phys. Rev. 141, 34 (1966); 145, 7 (1966). U. Fano, Phys. Rev. 131, 259 (1963). M. Baranger, Phys. Rev. 111, 481 (1958); 111, 494 (1958); 112, 855 (1958). R. Zwanzig, J. Chem. Phys. 33, 1338 (1960).

16. J.D. Bjorken and S.D. Drell, Relativistic Quantum fields, Mc Graw-Hill (1965).
P.Roman, Introduction to quantum Field theory, J. Wiley and Sons (1969).
O. Costa de Beauregard, Précis de mécanique quantique relativiste, Dunod, Paris (1967).
17. V.B. Berestetski, E.M. Lifshitz and L.P. Pitayevski, Relativistic Quantum theory, Pergamon (1971).
18. B.J. Feldman and M.S. Feld, Phys. Rev. A 1, 1375 (1970).
19. S. Haroche and F. Hartmann, Phys. Rev. A 6, 1280 (1970).
20. J. Bordé and Ch.J. Bordé, C.R. Acad. Sc. Paris 285 B, 287(1977).
21. J. Bordé and Ch.J. Bordé, J. of Mol. Spectrosc. 78, 353 (1979).
22. S.A.J. Druet, J.-P. E. Taran and Ch.J. Bordé, J. de Physique 40, 819 (1979).
23. S.A.J. Druet, J.-P. E. Taran and Ch.J. Bordé, J. de Physique 41, 183 (1980).
24. Ch.J. Bordé, G. Camy and B. Decomps, Phys. Rev. A 20, 254(1979).
25. S.A.J. Druet, B. Attal, T.K. Gustafson and J.-P.E. Taran, Phys. Rev. A 18, 1529 (1978).
26. S.Y. Yee, T.K. Gustafson, S.A.J. Druet and J.-P.E. Taran, Opt. Comm. 23, 1 (1977).
27. Ch.J. Bordé, S. Avrillier, A. Van Lerberghe, Ch. Salomon, D. Bassi and G. Scoles, J. de Phys. Colloques 42, C8 (1981).
28. L. Allen and J.H. Eberly, Optical resonance and two-level atoms, J. Wiley and Sons (1975).
29. M. Abramowitz and I.A. Stegun, Handbook of mathematical functions, Dover, New York 1965.
30. N.F. Ramsey, Phys. Rev. 78, 695 (1950).
31. N.F. Ramsey, Physics Today (July 1980) p. 25.
32. G. Kramer, J. Opt. Soc. Am. 68, 1634 (1978).
33. G. Kramer and D.N. Ghosh Roy, CPEM Digest (1980).
34. V. Baklanov, V.P. Chebotayev and B.Ya. Dubetsky, Appl. Phys. 11, 201 (1976).
35. V.P. Chebotayev, A.V. Shishayev, B.Ya. Yurshin and L.S. Vasilenko, Appl. Phys. 15, 43 (1978).
36. S.A. Lee, J. Helmcke and J.L. Hall in Laser Spectroscopy IV, Edited by H. Walther and K.W. Rothe, Springer-Verlag 1979, p. 130.
37. E.V. Baklanov and B.Ya. Dubetsky, Sov. J. Qu. Electr. 8, 51 (1978).
38. C. Cohen-Tannoudji, Metrologia 13, 161 (1977) and references therein.
39. A. Omont, Prog. Quantum Electronics 5, 69 (1977).
40. U. Fano and G. Racah, Irreducible Tensorial Sets, Academic Press, New York (1959).
41. G. Grynberg, F. Biraben, E. Giacobino and B. Cagnac, J. de Physique 38, 629 (1977) and references therein.
42. R.F. O'Connell and E.P. Wigner, Phys. Lett. 61 A, 353 (1977).
43. T.D. Newton and E.P. Wigner, Rev. Mod. Phys. 21, 400 (1949).

RYDBERG ATOMS IN MAGNETIC AND ELECTRIC FIELDS

Serge Feneuille

Laboratoire Aimé Cotton
C.N.R.S.II, bat 505, Université de Paris-Sud
91405 Orsay Cedex, France

During the last five years, numerous experimental and theoretical studies have been carried out on highly excited atoms in the presence of an uniform field, magnetic or electric. The aim of these four lectures was to describe our present knowledge of this subject which is of a great theoretical interest and which recently reaped a large profit from laser spectroscopy.

RYDBERG ATOMS IN MAGNETIC FIELDS

Very well known qualitative results were first recalled[1,2]: Hamiltonian of an electron interacting simultaneously with a Coulomb field and an uniform magnetic field; evolution of the various energy terms with increasing the principal quantum number n ; constants of motion (parity, magnetic quantum numbers); introduction of a critical magnetic field $B_0 = 2.35\ 10^5$T ; non separable character of the problem.

Then, the various following approaches were considered : perturbation theory to first order in the very low field limit ($B/B_0 \ll n^{-3}$) ; higher orders of perturbation theory ; Bohr models and WKB approximation in the strong field limit ($B/B_0 \gtrsim n^{-3}$).

Concerning perturbation theory to first order, very well known results were recalled[2] : 1-mixing induced by the diamagnetic contribution, necessity of introducing an additional label κ to distinguish the various levels belonging to the same hydrogenic

---

New address : Lafarge Coppée, 28, ruc Emile Menier
75782 PARIS CEDEX 16, France

manifold (by convention, κ being minimum for the level which, in the low-magnetic-field limit, rises fastest in energy with increasing the magnetic field). A most recent contribution allowing an interpretation of κ in terms of SO(4) symmetry were also briefly discussed[3].

Of course, higher orders of perturbation theory induce n-mixing. However, in despite of this inter-n mixing, it was recently shown that anticrossings between levels with same parity and same magnetic quantum numbers become very weak for highly excited hydrogenic states[4,5]. In the corresponding lecture, this result was strongly emphasized for it is very important : on one hand, it could suggest the existence of an approximate dynamical symmetry and, on the other hand, it shows that one can easily follow "diabatic" levels labelled by the same quantum numbers as in the low magnetic field limit.

The interest of more or less simplified Bohr models[6] and, in particular, of the O'Connell[7] approximation is to easily explain the equally spaced structures (so called quasi Landau resonances) which extend across the zero field series limit into the continuum with a spacing of about 1.5 $\hbar\omega_c$, ω_c being the cyclotron frequency[8]. The corresponding results were recalled and it was explained why Bohr models are valid for states corresponding the minimum value of κ, $\kappa_<$, only[9]. It was also shown that, for this particular value of κ, WKB approximation provides good predictions[7,10-12] and that the corresponding "diabatic" levels obey some scaling laws, valid for the whole range of the magnetic field[11,13].

Finally, it was explained why the minimum value of κ plays a major role in the interpretation of the quasi-Landau resonances[12,14,15] such resonances emerge by a concentration of oscillator strength into the diabatic level which, in the low-magnetic-field-limit, rises fastest in energy with increasing the magnetic field. So, the appearance of quasi-Landau resonances can be seen as a direct consequence of optical excitation . Remaining open problems, such as ionization properties of Rydberg atoms in magnetic fields (autoionization, observation of Fano profiles), three-dimensional solutions, influence of the atomic core, were also briefly discussed.

RYDBERG ATOMS IN ELECTRIC FIELDS

Hydrogenic atoms in an uniform electric field were first considered, and the various consequences of separability, in parabolic coordinates, of the corresponding problem were described in detail[1]. Then, the results of perturbation theory to first order, which is valid in the low-electric-field-limit, were briefly recalled before underlining the asymptotic character of perturbation expansion when high orders are taken into account[16,17].

Still in the hydrogenic approximation, ionization properties were qualitatively investigated by using parabolic coordinates. This approach allows one to predict that, according to the excitation energy, E, and the strength of the electric field, F, three situations can be distinguished : for negative energies, there is a low-electric-field-limit in which only stable Stark levels exist (region I); in an intermediate region (II), quasi-stable Stark levels are superimposed to ionization continua while in the third region (III) only broad ionization continua can be found (high electric field limit)[18] . Then, the various approaches, which have been used to calculate ionization probabilities by tunnelling, were described. Special attention was paid to WKB approximation[19,20]. Predictions of one dimensional models,[21-24] which are valid in the strong field limit only, were also given before describing, in greater detail, recent "exact" calculations[25-27]. These exact calculations concern not only state density but also photoionization cross sections from the ground state[25,26] or from weakly excited states[27]. They provide the first complete explanation of the resonances observed above the zero field ionization limit in atomic photoionization spectra in the presence of an uniform electrid field.

Secondly, alkali atoms were considered . The additional phenomena resulting from the influence of the atomic core and from the lack of separability were successively discussed in a qualitative way : anticrossing of Stark levels[28], autoionization of quasi stable Stark states in the intermediate region[29-31] defined previously, strong perturbation of oscillator strengths from the ground state[32]. Various possibilities for describing quantitatively these non hydrogenic phenomena were also briefly discussed (diagonalization of the Hamiltonian within a truncated spherical basis[28], configuration interaction[33] and quantum defect theory[34] in parabolic coordinates).

RYDBERG ATOMS IN CROSSED FIELDS

Finally, the properties of Rydbergatoms in crossed fiels were briefly analyzed. When the magnetic field is much more intense than the electric one ($Bn^3 \gg Fn^4$ in a.u.), new resonances appear because of the mixing between different magnetic states induced by the electric field. This mixing leads, in the vicinity of the field free ionization limit, to the observation of quasi-Landau resonances with a spacing now equal to $0.5\ \hbar\omega_c$.[35] In the opposite case ($Fn^4 \gg Bn^3$ in a.u.), the observed phenomena are basically the same ones but the spacings are now essentially governed by the electric field[36].

In the intermediate case, very amazing predictions can be qualitatively derived from the potential energy surface which exhibits, in addition to the inner Coulomb valley, an outer well whose position, depth and width depend essentially of the respective strengths of the two fields. Therefore, a possibility appears of creating atoms with an electron temporarily trapped in an eccentric orbit far from the nucleus[37,38]. Observation of such asymmetry atoms has not yet been achieved however.

REFERENCES

This list of references is not intended to be exhaustive but only representative.

1. See for example : Bethe H.A., and Salpter E.E. (1957) Quantum Mechanics of One- and Two-Electron Atoms, Springer Verlag, Berlin.
2. Garstang R.H. (1977) Rep. Prog. Phys. 40, 105.
3. Labarthe J-J.(1981) J. Phys. B14, L467.
4. Zimmerman M.L., Kash M.M. and Kleppner D. (1980) Phys. Rev. Lett. 45, 1092.
5. Delande D. and Gay J-C. (1981) Phys. Lett. 82A, 399.
6. Canuto V. and Kelly D.C. (1972) Astrophys and Space Sci 17, 277.
7. O'Connell R.F. (1974) Astrophys. J. 187, 275.
8. Garton W.R.S. and Tomkins F.S. (1969) Astrophys. J. 158, 839.
9. Delande D. and Gay J.C. (1981) Phys. Lett. 82A, 393.
10. Rau, A.R.P. (1977) Phys. Rev. A16, 613.
11. Fonck R.J., Roesler F.L., Tracy D.H. and Tomkins F.S. (1980) Phys. Rev. A21, 861.
12. Gay J-C. Delande D. and Biraben F. (1980) J. Phys. B13, L729.
13. Feneuille S. (1982) Phys. Rev. A. (to appear).
14. Castro J.B., Zimmerman M.L., Hulet, R.G., Kleppner D. and Freeman R.R. (1980) Phys. Rev. Lett. 45, 1780.
15. Clark C.W. and Taylor K.T. (1980) J. Phys. B13, L737.
16. Koch P.M. (1978) Phys. Rev. Lett. 41, 99.
17. Silverstone H.J. (1978) Phys. Rev. A18, 1853.
18. Feneuille S. and Jacquinot P. (1982) Adv. in Atom. and Mol. Phys. (to appear).
19. Lanczos C. (1930) Z. Phys. 62, 518 ; 65 (431) ; (1931) Z. Phys. 68, 204.
20. Many references can be found in : Yamale T., Tachibana A. and Silverstone H.J. (1977) Phys. Rev. A16, 877. See also : Harmin D.A. (1981) Phys. Rev. A24, 2491.
21. O'Connell R.F. (1977) Phys. Lett. 60A, 481.
22. Freeman R.R. and Bjorklund C.C. (1978) Phys. Rev. Lett. 40,118.
23. Rau A.R.P. (1979) J. Phys. B12, L193.
24. Freeman R.R. and Economu N.P. (1979) Phys.Rev. A.20, 2356.
25. Luc-Koenig E. and Bachelier A. (1979) Phys.Rev. Lett. 43, 921.
26. Luc-Koenig E. and Bachelier A. (1980) J. Phys. B13, 1743, 1769.
27. Harmin D.A. (1981) private communication.
28. Zimmerman M.L., Littman M.G., Kash M.M. and Kleppner D. (1979) Phys. Rev. A20, 2251.
29. Feneuille S., Liberman S., Pinard J. and Jacquinot P. (1977) C.R. Acad. Sci. B284, 291.

30. Littman M.G., Kash M.M. and Kleppner D. (1978) Phys. Rev. Lett. 41, 103.
31. Feneuille S. Liberman S., Pinard J. and Taleb A. (1979) Phys. Rev. Lett. 42, 1404.
32. Luc-Koenig E., Liberman S. and Pinard J. (1979) Phys. Rev. A20, 519.
33. Luc-Koenig E. and Bachelier A., private communication.
34. Fano U. (1981) Phys. Rev. A24, 619.
35. Crosswhite H., Fano U., Lu K.T. and Rau A.R.P. (1979) Phys. Rev. Lett. 42, 963.
36. Feneuille S., Luc-Koenig E., S. Liberman, Pinard J. and Taleb A. (1982) Phys. Rev. A. (to be published).
37. Rau A.R.P. (1979) J. Phys. B12, L193.
38. Gay J.C., Pendrill L.R. and Cagnac B. (1979) Phys. Lett. 72A 315.

DOPPLER FREE SPECTROSCOPY

A. Javan

Massachusetts Institute of Technology
Cambridge, Mass.

These notes will give an elementary formalism to acquaint the students with the simplest aspects of the theory. It will provide background theoretical material for the applications to high resolution spectroscopy. The presentation will be tutorial. The lectures will then be followed by a review of the recent work of the author's laboratory in the area of Doppler free spectroscopy.

BACKGROUND FORMULATION

Consider a two level system, a and b, with n_a and n_b as the number densities of the molecules in a and b respectively. Assume a and b to be connected by a dipole matrix element, μ. Assume an applied field at a frequency ω close to the resonance frequency ω_o, given by $E = E_o e^{i(\omega t-\phi)}$.

Ignore for now the Doppler effect. The rate of transition (per molecule) induced by the applied field is a Lorentzian function of $(\omega-\omega_o)$ and is given by

$$R = \frac{\frac{1}{2}\left(\frac{\mu E_o}{\hbar}\right)^2 T}{(\omega-\omega_o)^2 T^2 + 1}$$

where T is a lifetime parameter and determines the half width of the Lorentzian response:

$$\Delta\omega = \frac{1}{T}$$

For collision dominated lifetime, $\frac{1}{T} = \frac{1}{2}\left(\frac{1}{T_a} + \frac{1}{T_b}\right)$, where T_a is lifetime of the upper level and T_b is lifetime of the lower level. Assume for simplicity $T_a = T_b = T$.

For $I = \frac{c}{8\pi} E_o^2$, the incident intensity, the change ΔI by molecules per unit volume is given by

$$\Delta I = \left(n_a - n_b\right)\hbar\omega R$$

Note $\Delta I > 0$, for $n_a > n_b$ and < 0 for $n_a < n_b$.

SATURATION

In the high field limit, n_a and n_b differ from their equilibrium values in the absence of the field. Designate the latter by n_a^o and n_b^o.

The rate equations determining n_a and n_b are:

$$\frac{dn_a}{dt} = -\frac{n_a - n_a^o}{T_a} - n_a R + n_b R$$

$$\frac{dn_b}{dt} = -\frac{n_b - n_b^o}{T_b} + n_a R - n_b R$$

In the steady state, $\frac{dn_a}{dt} = \frac{dn_b}{dt} = 0$. The solution for $n_a - n_b$ will be:

$$n_a - n_b = \frac{n_a^o - n_b^o}{1 + 2RT} .$$

We have assumed $T_a = T_b = T$.

For $\omega = \omega_o$, $R = \frac{1}{2}\left(\frac{\mu E_o}{\hbar}\right)^2 T$. In the weak field limit, $\frac{\mu E_o}{\hbar} T << 1$, i.e., $R << \frac{1}{T}$, have $n_a - n_b = n_a^o - n_b^o$. For high field, $R >> \frac{1}{T}$ and $(n_a - n_b) \longrightarrow 0$.

Consider now ΔI:

$$\Delta I = \left(n_a - n_b\right)\hbar\omega R$$

$$= \frac{\left(n_a^o - n_b^o\right)\hbar\omega R}{1 + 2RT}$$

Substitute the expression for R. We obtain

$$\Delta I = \frac{\frac{1}{2}\left(n_a^o - n_b^o\right)\hbar\omega\left(\frac{\mu E_o}{\hbar}\right)^2 T}{\left(\omega-\omega_o\right)^2 T^2 + 1 + \left(\frac{\mu E_o}{\hbar}\right)^2 T^2}$$

The absorption coefficient (or gain factor) is defined by:

$$\alpha = \frac{\Delta I}{I} \; ; \qquad I = \frac{cE_o^2}{8\pi}$$

We have $\left(\text{recognizing that } \omega = \frac{2\pi c}{\lambda}\right)$:

$$\alpha = \frac{\alpha_o}{\left(\omega-\omega_o\right)^2 T^2 + 1 + \left(\frac{\mu E_o}{\hbar}\right)^2 T^2}$$

where

$$\alpha_o = \left(n_a^o - n_b^o\right) \frac{8\pi^2}{\lambda} \frac{\mu^2}{\hbar} T$$

α_o is the peak absorption coefficient in the low field limit

$$\left(\text{low field: } \frac{\mu E_o}{\hbar} T << 1\right)$$

The saturated absorption coefficient is sometimes written as:

$$\alpha = \frac{\alpha_o}{\left(\omega-\omega_o\right)^2 T^2 + 1 + \frac{I}{I_s}}$$

where $I_s = \frac{c\hbar^2}{8\pi\mu^2 T^2}$. It depends on μ and T. Saturation occurs when I approaches I_s. The shorter is the T and the smaller is μ, the harder it is to saturate.

Considering that

$$\frac{dI}{dz} = \alpha I$$

in the low field limit ($I << I_s$), we have

$$I = I_o e^{\alpha z} \left(\alpha < 0 \quad \text{for} \quad n_a^o < n_b^o\right)$$

In the high field limit ($I >> I_s$), we have

$$\frac{dI}{dz} = \alpha_o I_s \quad \int_i^o \frac{dI}{dz}\, dz = \int_i^o \alpha_o I_s\, dz$$

$$I_o - I_i = \alpha_o I_s (z_o - z_i)$$

output, o

input, i

In the fully saturated limit, no longer have we an exponential in z; in fact, we have a linear function of z.

Substitute the expressions for α_o and I_s, and obtain

$$\frac{dI}{dz} = \frac{1}{2}\left(n_a^o - n_b^o\right)\frac{\hbar\omega}{T} .$$

<u>Homework</u>:

Solve $\frac{dI}{dz} = \alpha I$ for $\omega = \omega_o$, keeping the saturation terms. Assume $\alpha > 0$ (i.e., $n_a > n_b$), showing that in the asymptotic limit your expression agrees with mine. Interpret your result in that limit, in terms of the act of emission or absorption of photons (of energy $\hbar\omega$) by individual molecules in the states a or b. What does the $\frac{1}{2}$ factor mean?

DOPPLER EFFECT

Assume the applied field propagating in the +z-direction:

$$E = E_o e^{i(\omega t - kz)} ; \qquad k = \frac{\omega}{c}$$

Consider a molecule with velocity component along the z-direction given by $v_z = v$; (v can be negative or positive). For that molecule, the applied field will be Doppler shifted to a frequency $\omega - kv$.

We can write

$$\omega - kv = \omega - \omega\frac{v}{c}$$

$$= \omega\left(1 - \frac{v}{c}\right)$$

For $v > 0$, i.e., the molecule moving along the direction propagation, ω will be down-shifted. For $v < 0$, i.e., the molecule moving opposite to the propagating field, ω will be up-shifted.

The transition rate for molecules with $v_z = v$ will be given by the expression for R on page 1 with the substitution $\omega \to \omega\left(1 - \frac{v}{c}\right)$, or

$$R(v) = \frac{\frac{1}{2}\left(\frac{\mu E_o}{\hbar}\right)^2 T}{\left[\omega\left(1 - \frac{v}{c}\right) - \omega_o\right]^2 T^2 + 1}$$

Assume the molecular velocity distribution to be thermal at a temperature T. The distribution will be proportional to $e^{-\frac{1}{2}mv^2/kT}$ or e^{-v^2/u^2}, where $u = \sqrt{\frac{2kT}{m}}$ is the mean velocity.

Consider molecules with v lying in the range v to v + dv:

$$dn_a = \frac{n_a}{u\sqrt{\pi}} e^{-v^2/u^2} dv$$

$$dn_b = \frac{n_b}{u\sqrt{\pi}} e^{-v^2/u^2} dv$$

Note $\int_{-\infty}^{+\infty} e^{-v^2/u^2} dv = \sqrt{\pi}\, u$; hence $\int dn_a = n_a$, and $\int dn_b = n_b$

For dn_a and dn_b molecules (with velocity component $v_z = v$), the absorbed (or emitted) radiation, designated by δI, will be;

$$\delta I = \frac{1}{\sqrt{\pi}\,u}(n_a - n_b)e^{-v^2/u^2}\ R(v)\ \hbar\omega\ dv$$

Integrating over v will give ΔI as follows: $\left(\int_{v=-\infty}^{v=+\infty}\delta I = \Delta I\right)$

$$\Delta I = \frac{1}{\sqrt{\pi}\,u}(n_a - n_b)\hbar\omega\int_{-\infty}^{+\infty} e^{-v^2/u^2}\ R(v)\ dv.$$

Let us write R(v) as follows:

$$R(v) = \frac{\frac{1}{2}\left(\frac{\mu E_o}{\hbar}\right)^2 T}{(v_o - v)^2\ q^2 + 1}$$

where

$$v_o = \frac{\omega - \omega_o}{\omega}\ c$$

$$q = \frac{\omega T}{c}$$

R(v) is a Lorentzian function of v centered at $v = v_o$, with half velocity width $\delta v = \frac{1}{q} = \frac{c}{\omega T}$.

Write $v_o = \frac{\omega - \omega_o}{\omega}\ c$ as:

$$\omega_o = \omega - \omega\,\frac{v_o}{c}$$

Accordingly, v_o is the velocity of the molecules which Doppler shift ω to ω_o.

Similar manipulation shows that $\delta v = \frac{c}{\omega T}$, written as $\omega \frac{\delta v}{c} = \frac{1}{T}$, will be a velocity change corresponding to a Doppler shift equal to the half width $\frac{1}{T}$.

$$\Delta I = \frac{1}{2\sqrt{\pi}\, u} \left(n_a - n_b\right) \hbar\omega \left(\frac{\mu E_o}{\hbar}\right)^2 T$$

$$\times \int_{-\infty}^{+\infty} \frac{e^{-v^2/u^2}\, dv}{(v-v_o)^2 q^2 + 1}$$

For $\delta v = \frac{1}{q} << u$, have:

$$\int_{-\infty}^{+\infty} \frac{e^{-v^2/u^2}\, dv}{(v-v_o)^2 q^2 + 1} \simeq e^{-v_0^2/u^2} \int_{-\infty}^{+\infty} \frac{dv}{(v-v_o)^2 q^2 + 1}$$

$$= \frac{\pi}{q}\, e^{-v_0^2/u^2}$$

Substituting for $v_o = \frac{\omega - \omega_o}{\omega} c$, and define, $\Delta\omega_D = \omega \frac{u}{c}$, we have finally:

$$\Delta I = \frac{\sqrt{\pi}}{2\Delta\omega_D} (n_a - n_b)\, \hbar\omega \left(\frac{\mu E_o}{\hbar}\right)^2 e^{-(\omega - \omega_o)^2/\Delta\omega_D^2}$$

or

$$\alpha = \frac{\Delta I}{I}$$

$$= \alpha_o\, e^{-(\omega - \omega_o)^2/\Delta\omega_D^2}$$

$$\alpha_o = \frac{8\pi^2}{\lambda} (n_a - n_b) \frac{\mu^2 \sqrt{\pi}}{\hbar\, \Delta\omega_D}$$

SATURATION EFFECT

Consider molecules with velocity in the range v, v+dv. Proceed as before:

$$\frac{d}{dt}(dn_a) = -\frac{dn_a - dn_a^o}{T_a} - (dn_a - dn_b)R(v)$$

$$\frac{d}{dt}(dn_b) = -\frac{dn_b - dn_b^o}{T_b} + (dn_a - dn_b)R(v)$$

Therefore, in the steady state we have

$$dn_a - dn_b = \frac{dn_a^o - dn_b^o}{1+2RT}$$

assuming $T_a = T_b = T$.

In the absence of the field, we have

$$dn_a^o = \frac{n_a^o}{u\sqrt{\pi}} e^{-v^2/u^2} dv \; ; \; dn_b^o = \frac{n_b^o}{u\sqrt{\pi}} e^{-v^2/u^2} dv$$

Hence

$$d(n_a - n_b) = \frac{1}{u\sqrt{\pi}} \frac{(n_a^o - n_b^o)\, e^{-v^2/u^2}\, dv}{1 + 2RT}$$

For $E_0^2 = 0$, i.e., in the absence of the applied field, when $R(v) = 0$, the population difference is a Boltzmann distribution. In the presence of the field, it will differ by the factor $\frac{1}{1+2RT}$.

Let us inspect this for the simple case where: $(\frac{\mu E_0}{\hbar})^2 T^2 << 1$, for which we have

$$\frac{1}{1+2RT} \approx 1-2RT$$

We need to inspect the product $(1-2RT)e^{-v^2/u^2}$:

$$1-2RT = 1 - \frac{\left(\frac{\mu E_o}{\hbar}\right)^2 T^2}{(v-v_o)^2 q^2 + 1}$$

Plot e^{-v^2/u^2} :

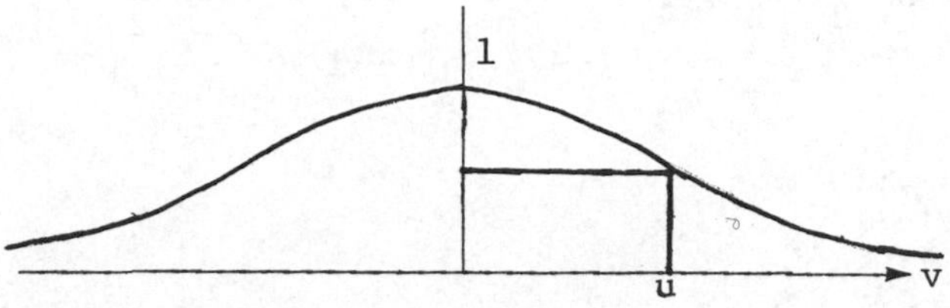

Plot 1-2RT, for $v_o > 0$ (i.e., $\omega > \omega_o$):

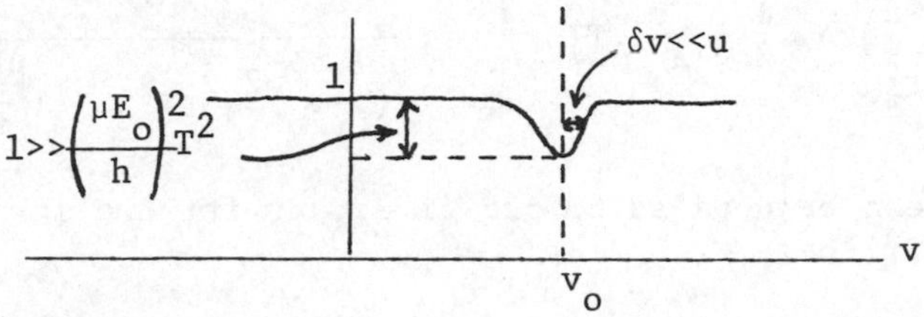

Plot the product

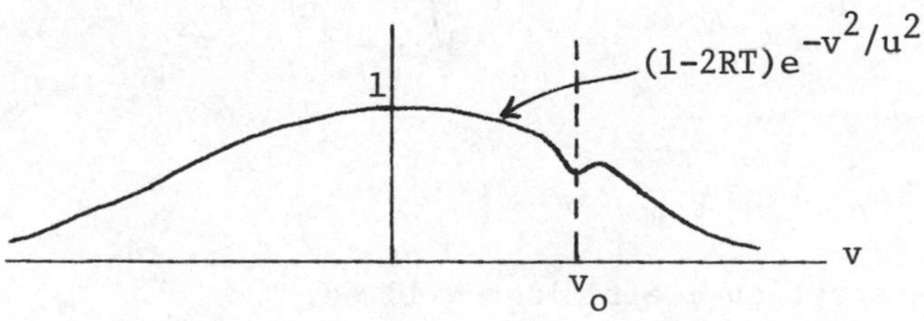

Note that as ω is tuned, the dip will be tuned. For $\omega = \omega_o$, it will be centered on $v = 0$.

Saturation effect will appear via a change in population difference:

Designate as before the intensity change due to molecules with $v_z = v$ in a range dv by δI:

$$\delta I = (dn_a - dn_b)\, \hbar\omega\, R(v)$$

$$= \frac{n_a^o - n_b^o}{u\sqrt{\pi}} \; \frac{e^{-v^2/u^2} dv}{1+2RT} R\hbar\omega$$

Substitute R:

$$\delta I = \frac{n_a^o - n_b^o}{u\sqrt{\pi}} \hbar\omega \frac{1}{2} \left(\frac{\mu E_o}{\hbar}\right)^2 T^2 \frac{e^{-v^2/u^2} dv}{(v-v_o)^2 q^2 + 1 + \left(\frac{\mu E_o}{\hbar}\right)^2 T^2}$$

$$\Delta I = \int_{v=-\infty}^{+\infty} \delta I$$

or

$$\Delta I = \frac{n_a^o - n_b^o}{u\sqrt{\pi}} \hbar\omega \frac{1}{2} \left(\frac{\mu E_o}{\hbar}\right)^2 T^2 \int_{-\infty}^{+\infty} \frac{e^{-v^2/u^2} dv}{(v-v_o)^2 q^2 + 1 + \left(\frac{\mu E_o}{\hbar}\right)^2 T^2}$$

Integration can be done as before (i.e., as for the low field case). The result will be:

$$\Delta I = \frac{\sqrt{\pi}}{2\Delta\omega_D} (n_a^o - n_b^o)\, \hbar\omega \left(\frac{\mu E_o}{\hbar}\right)^2 \frac{e^{-(\omega-\omega_o)^2/\Delta\omega_D^2}}{\sqrt{1 + \left(\frac{\mu E_o}{\hbar}\right)^2 T^2}}$$

where $\Delta\omega_D = \omega\frac{u}{c}$; $(u = \sqrt{\frac{2kT}{m}})$.

The absorption coefficient will be:

$$\alpha = \frac{\Delta I}{I} = \frac{\alpha_o\, e^{-(\omega-\omega_o)^2/\Delta\omega_D^2}}{\sqrt{1 + \left(\frac{\mu E_o}{\hbar}\right)^2 T^2}}$$

with

$$\alpha_o = \frac{8\Pi^2}{\lambda}(n_a^o - n_b^o)\frac{\mu^2\sqrt{\pi}}{\hbar\,\Delta\omega_D}$$

For $(\frac{\mu E_o}{\hbar})^2 T^2 << 1$, we will have the same results as before.

As $(\frac{\mu E_o}{\hbar})^2 T^2$ approaches unity, α decreases as $\frac{1}{\sqrt{1+\left(\frac{\mu E_o}{\hbar}\right)^2 T^2}}$

DOPPLER FREE RESONANCES

With a single propagating field, there will be no Doppler free resonance. The nonlinear dependence on the applied field would appear as a decrease in absorption coefficient, with the line shape remaining as a Gaussian Doppler profile, $e^{-(\omega-\omega_o)^2/\Delta\omega_D^2}$.

Assume now an <u>intense</u> applied field at a fixed frequency ω, saturating the Doppler broadened resonance. The direction of propagation of this field will define the position z-direction.

$\rightarrow + z$

$$E = E_o\, e^{-i(\omega t - kz)}$$

Let us at the same time apply a <u>weak</u> probe field at the frequency ω_p, propagating <u>opposite</u> to the intense field.

+z

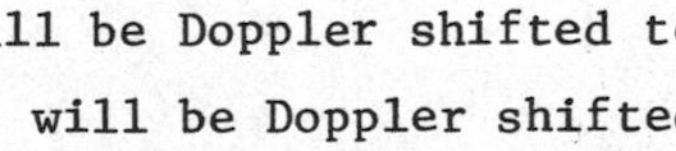

$$E = E_o\, e^{i(\omega t - kz)}$$

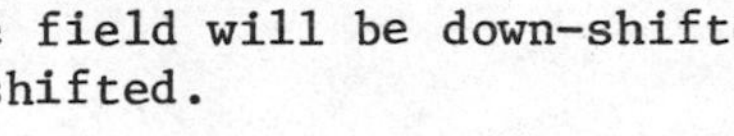

$$E_p = E_p^o\, e^{i(\omega_p t + k_p z)}$$

For molecules with a velocity component $v_z = v$, the intense field will be Doppler shifted to $\omega \rightarrow \omega(1-\frac{v}{c})$. The weak probe field, however, will be Doppler shifted to $\omega_p \rightarrow \omega_p(1+\frac{v}{c})$; e.g., if $v > 0$, the intense field will be down-shifted, but the weak probe field will be up-shifted.

Consider now the absorption of the probe field in the presence of an intense field at a fixed frequency ω.

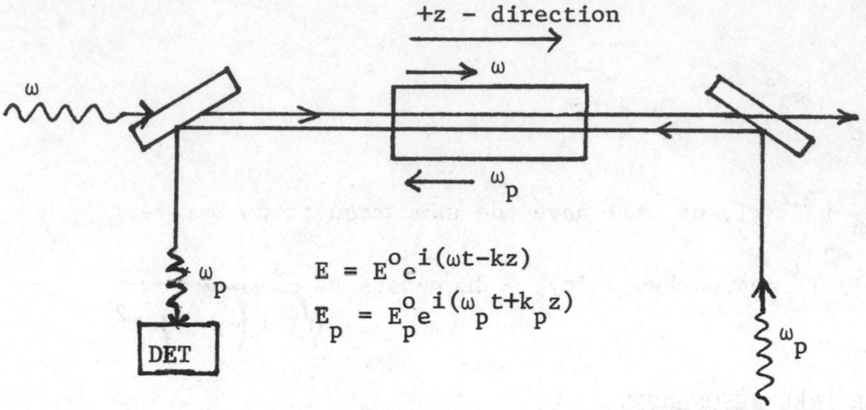

As before, the expression for δI_p due to molecules with a velocity v in a narrow range dv will be

$$\delta I_p = (dn_a - dn_b)\,\hbar\omega_p\, R(\omega_p)$$

The expression for $R(\omega_p)$ will differ from $R(\omega)$, because of the sign of the Doppler shift:

$$R(\omega) = \frac{\frac{1}{2}\left(\frac{\mu E^o}{\hbar}\right)^2 T}{\left[\omega\left(1 - \frac{v}{c}\right) - \omega_o\right]^2 T^2 + 1}$$

$$R(\omega_p) = \frac{\frac{1}{2}\left(\frac{\mu E_p^o}{\hbar}\right)^2 T}{\left[\omega_p\left(1 + \frac{v}{c}\right) - \omega_o\right]^2 T^2 + 1}$$

Let us write $R(\omega_p)$ as follows:

Define, as before, $v_p^o = \frac{\omega_p - \omega_o}{\omega} c$, and $q_p = \omega_p T/c$. We recognize that

$$q_p \simeq q = \omega T/C$$

$$\therefore \quad R_p(v) = \frac{\frac{1}{2}\left(\frac{\mu E_p^o}{\hbar}\right)^2 T}{(v + v_p^o)^2 q^2 + 1}$$

This is the expression for the transition rate induced by the probe field on the molecules with velocity v.

Compare this with the previous expression for the transition rate induced on the same molecules (with velocity v) by the strong field:

$$R(v) = \frac{\frac{1}{2}\left(\frac{\mu E^o}{\hbar}\right)^2 T}{(v-v^o)^2 q^2 + 1}$$

where $v^o = \frac{\omega - \omega_o}{\omega} c.$

Let us inspect δI_p:

Since the probe field is weak, it will not affect $(dn_a - dn_b)$. The population difference, however, will be saturated by the strong field. From page 8, have

$$(dn_a - dn_b) = \frac{1}{u\sqrt{\pi}} (n_a^o - n_b^o)\, e^{-v^2/u^2} dv \, \frac{1}{1+2RT}$$

It follows:

$$\delta I_p = \frac{(n_a^o - n_b^o)}{u\sqrt{\pi}} e^{-v^2/u^2} dv \, \frac{R_p(v)}{1+2R(v)T} \hbar\omega_p$$

Let us inspect this to the lowest order in $\left(\frac{\mu E_o}{\hbar}\right)^2 T^2$, where:

$$\frac{1}{1+2R(v)T} \simeq 1-2R(v)T \; \left[\text{for } \left(\frac{\mu E_o}{\hbar}\right)^2 T^2 \ll 1,\; 2R(v)T \ll 1\right]$$

$$\delta I_p = \frac{\hbar\omega_p (n_a^o - n_b^o)\, e^{-v^2/u^2} dv}{u\sqrt{\pi}} (1-2R(v)T) R_p(v)$$

Substituting for R(v) and $R_p(v)$, have:

$$\delta I_p = \frac{(n_a^o - n_b^o)}{u\sqrt{\pi}} \frac{1}{2} \hbar\omega \left(\frac{\mu E_p^o}{\hbar}\right)^2 T \left[1 - \frac{\left(\frac{\mu E_o}{\hbar}\right)^2 T^2}{(v - v_o)^2 q^2 + 1}\right] \frac{e^{-v^2/u^2} dv}{(v + v_p^o)^2 q^2 + 1}$$

Write the above as:

$$\delta I_p = A\left[1 - \frac{\beta^2 E_o^2}{(v - v_o)^2 q^2 + 1}\right] \frac{e^{-v^2/u^2}\, dv}{(v + v_p^o)^2 q^2 + 1}$$

where

$$A = \frac{n_a^o - n_b^o}{u\sqrt{\pi}}\, \hbar\omega\, \frac{1}{2} \left(\frac{\mu E_p^o}{\hbar}\right)^2 T$$

$$\beta^2 = \frac{\mu^2}{\hbar^2} T^2$$

$$\Delta I_p = \int_{v=-\infty}^{v=+\infty} \delta I_p = A\left[\int_{-\infty}^{+\infty} \frac{e^{-v^2/u^2}\, dv}{(v + v_p^o)^2 q^2 + 1} - \beta^2 E_o^2 \int_{-\infty}^{+\infty} \frac{e^{-v^2/u^2}\, dv}{[(v + v_p^o)^2 q^2 + 1][(v - v_o)^2 q^2 + 1]}\right]$$

In the Doppler limit, have

$$\delta v = \frac{1}{q}$$

$$= \frac{c}{\omega T} << u$$

In the first integral, $[(v + v_p^o)^2 q^2 + 1]^{-1}$ is a Lorentzian centered at $v = -v_p^o$, with a half width $\delta v << u$; hence:

$$\int_{-\infty}^{+\infty} \frac{1}{(v + v_p^o)^2 q^2 + 1}\, e^{-v^2/u^2} dv \simeq e^{-{v_p^o}^2/u^2} \quad dv \frac{1}{(v + v_p^o)^2 q^2 + 1}$$

$$\int_{-\infty}^{+\infty} \frac{dv}{(v + v_p^o)^2 q^2 + 1} = \frac{\pi}{q}$$

(integrate this by contour)

Hence the first integral is $\frac{\Pi}{q}\, e^{-v_p^2/u^2}$, or

$$\Delta I_p = A\left[\frac{\pi}{q}\, e^{-v_p^2/u^2} - \beta E_o^2\, Q\right]$$

where

$$Q = \int_{-\infty}^{+\infty} \frac{e^{-v^2/u^2}\,dv}{[(v+v_p)^2q^2+1]\;[(v-v_o)^2q^2+1]}$$

$$\begin{cases} v_p = \dfrac{\omega_p - \omega_o}{\omega_p} \\[2ex] v_o = \dfrac{\omega - \omega_o}{\omega} \end{cases}$$

Before integrating, inspect Q. Q is a function of v_o and v_p:

$Q(v_o, v_p)$.

If $v_p = -v_o$, then the product of the two Lorentzians will peak when $v = v_p (= -v_o)$.

If v_p is appreciably different from v_o, however, the product will be nearly zero for all v's.

HOMEWORK:

Prove the above, by inspecting the product of the two Lorentzians.

From above, it follows:

$$Q \simeq e^{-v_p^2/u^2} \int_{-\infty}^{+\infty} \frac{dv}{[(v+v_p)^2q^2+1]\;[(v-v_o)^2q^2+1]}$$

By contour integration, we obtain:

$$Q = \frac{\pi}{2q} \frac{1}{\left(\frac{v_o + v_p}{2}\right)^2 q^2 + 1} e^{-v_p^2/u^2}$$

Note that for $v_p = -v_o$, there will be a maximum, for v_p appreciably different from $-v_o$, the Lorentzian will fall to zero, as per our inspection.

Finally:

$$\Delta I_p = A \frac{\pi}{q} e^{-v_p^2/u^2} \left[1 - \frac{1}{2} \frac{\beta^2 E_o^2}{\left(\frac{v_o + v_p}{2}\right)^2 q^2 + 1} \right]$$

Substituting v_p, v_o and q, have

$$\Delta I_p = B\, e^{-(\omega_p - \omega_o)^2/\Delta\omega_D^2} \left[1 - \frac{1}{2} \beta^2 E_o^2 \frac{1}{\left(\frac{\omega + \omega_p}{2} - \omega_o\right)^2 T^2 + 1} \right]$$

This is our first expression for Doppler free resonance. I define this as the Displaced Doppler Free Resonance, as follows:

The frequency response of ΔI_p will be:

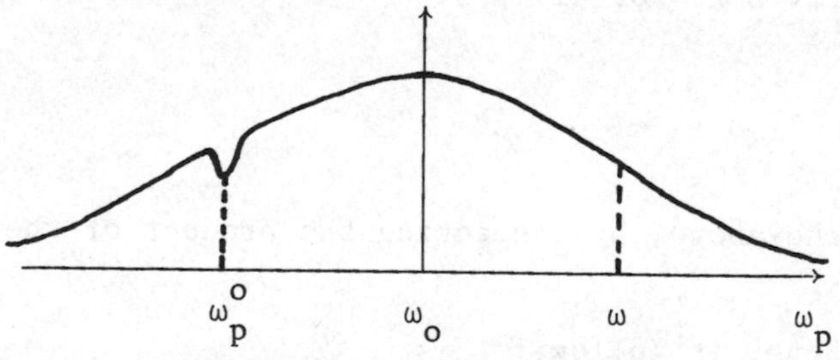

The displaced dip is centered at $\omega = \omega_p^o$.

CENTRAL DIP

Assume now the probe field to be at the same frequency as the saturating field:

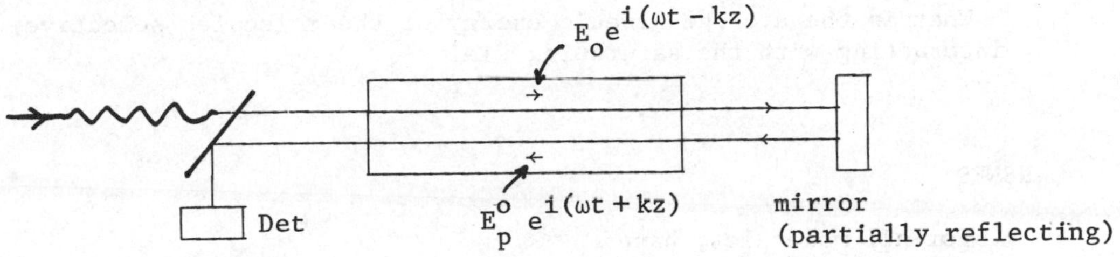

In this case we set $\omega = \omega_p$ in the expression for ΔI_p. We will have:

$$\Delta I_p = B\, e^{-(\omega-\omega_o)^2/\Delta\omega_D^2}\left[1 - \frac{1}{2}\beta^2 E_o^2 \frac{1}{(\omega-\omega_o)^2 T^2 + 1}\right]$$

In this case the dip will appear centered at ω_o, with a half width $\frac{1}{T}$.

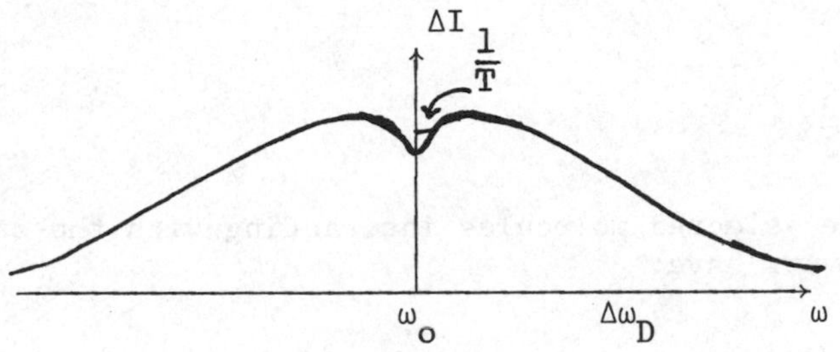

$$\frac{1}{T} << \Delta\omega_D$$

$1/T$ can be less than $\Delta\omega_D$ by several orders of magnitude.

Applications to be discussed later relate to orders of magnitude and improved resolution in spectroscopic observations.

REMARKS ABOUT DISPLACED DOPPLER FREE RESONANCE

The saturating field at the frequency ω selectively interacts with molecules having $v_z = v = \frac{\omega - \omega_o}{\omega} c$. The probe field counter-propagating) will pick up the resonance due to these molecules at a frequency $\omega_p = \omega_o - (\omega - \omega_o)$.

QUESTION:

What is the average kinetic energy of the molecules selectively interacting with the saturating field?

ANSWER:

For all molecules, have:

$$\overline{K.E.} = \overline{\frac{1}{2}mv_x^2} + \overline{\frac{1}{2}mv_y^2} + \overline{\frac{1}{2}mv_z^2}$$

with:

$$\overline{\frac{1}{2}mv_x^2} = \overline{\frac{1}{2}mv_y^2} = \overline{\frac{1}{2}mv_z^2} = \frac{1}{2}kT$$

Hence

$$\overline{K.E.} = \frac{3}{2}kT$$

For the selected molecules interacting with the saturating field, however, have:

$$\overline{\frac{1}{2}mv_x^2} = \overline{\frac{1}{2}mv_y^2} = \frac{1}{2}kT$$

$$\overline{\frac{1}{2}mv_z^2} = \frac{1}{2}mv^2; \quad v = \frac{\omega - \omega_o}{\omega} c$$

$\therefore$

$\overline{K.E}$. for the selected molecules will be:

$$(\overline{K.E.})_s = \frac{1}{2}kT + \frac{1}{2}kT + \frac{1}{2}m\left(\frac{\omega - \omega_o}{\omega}\right)^2 c^2$$

Consider the last term:

$\frac{1}{2}m\left(\frac{\omega - \omega_o}{\omega^2}\right)^2 c^2$, multiply the numerator and denominator by u^2:

$\frac{1}{2}mu^2\,(\omega - \omega_o)^2 c^2/\omega^2 u^2$

Since $u = \sqrt{\frac{2kT}{m}}$, and $\Delta\omega_D = \omega\frac{u}{c}$, have

$$\frac{1}{2}mu^2\,\frac{(\omega - \omega_o)^2}{\omega^2 u^2}\,c^2 = kT\,\frac{(\omega - \omega_o)^2}{\Delta\omega_D^2}$$

Hence:

$$\overline{(K.E.)_s} = kT\left(1 + \frac{(\omega - \omega_o)^2}{\Delta\omega_D^2}\right).$$

For central dip $\omega = \omega_o$, the average $\overline{K.E.}$ is kT, (instead of $\frac{3}{2}kT$ for all molecules).

For $\omega - \omega_o = \Delta\omega_D$, have

$$\overline{(K.E.)_s} = 3kT$$

For $\omega - \omega_o = 2\Delta\omega_D$

$$\overline{(K.E.)_s} = 5kT$$

By tuning the saturating field away from ω_o, we select molecules from the room temperature gas.

Define an effective temperature for the selected molecules:

$$\overline{(K.E.)_s} = \frac{3}{2}k\,T_e$$

or

$$T_e = \frac{2}{3}T\left(1 + \frac{(\omega - \omega_o)^2}{\Delta\omega_D^2}\right)$$

For $\omega = \omega_o$, have $\frac{2}{3}$ of room temperature molecules, i.e., for $t = 300$ K , have $T_e = 200$ K .

For $\omega - \omega_o = 2\Delta\omega_D$, have $T_e = \frac{10}{3}T$, or for $T = 300$ K (room temperature), $T_e = 1000$ K .

This is manifested in the collision line width of the displaced resonances. Study of it gives detailed information on the dynamics of intermolecular forces resulting in collision broadening, to be discussed in the lecture.

* * * * * * * *

The last lecture was devoted to a review of the work of the author's laboratory in the area of Doppler Free Spectroscopy. Historically, the first Doppler-free spectroscopy was performed by the author[1] and his students and MIT staff in 1963. In that experiment, the precise isotope shift of an optical transition of Ne^{20} and Ne^{22} was measured, using the Lamb-dip at the outputs of two He-Ne lasers, one of which had pure Ne^{20} and the other pure Ne^{22} isotope. The lasers were each oscillating at a single frequency and reset at their respective Lamb-dip centers. The measured difference frequency (in the 400 MHz range) gave the precise isotope shift. In a number of other[2-13] early MIT experiments extended over several years, several other approaches were devised and applied to obtain optical Doppler-free spectra of atoms. The methods devised were used to determine precise sub-Doppler splittings caused by hyperfine interaction, both in the zero-field and in the presence of an external magnetic field. These MIT experiments constituted the early demonstrations and utilizations of the different approaches to Doppler-free spectroscopy.

In the late sixties and the seventies, the attention in the author's laboratory was shifted to the observation of the effect in molecules and their high resolution studies[14-24].

In an independent series of experiments, Jan Hall at Boulder, see his lecture in this issue, succeeded in obtaining the first molecular Doppler-free resonance in the 3.39 μm line of methane, using a 3.39 μm He-Ne laser. This resonance, in the expert hands of Jan Hall, has offered the highest resolution and accuracy offered by Doppler free spectroscopy.

The MIT efforts in the nineteen-seventies have pursued a search to uncover the various ways in which Doppler-free spectroscopy can be used to obtain high resolution spectroscopic information unobtainable by other means. For this, instead of a transmission-type experiment, as described in the above account of the theory, a different method was applied[15] to observe weakly absorbing transitions. The method is uniquely suitable for hot-band molecular spectroscopy, or the transitions due to a pair of high-lying atomic levels. In this case, the spontaneous emission from an upper (or a lower) level of an optical transition is observed in the presence of a laser radiation in the form of two oppositely propagating running waves, interacting with the optical transition. The change in the intensity of the emitted spontaneous radiation versus the laser frequency, will show the Doppler-free resonance as the laser frequency is tuned across the center of the optical transition. The first demonstration of this effect was done by Freed and Javan, to observe Doppler-free resonances in a low-pressure CO_2 absorber gas. In the experiment, a line tunable CO_2 laser was used to observe the effect in the various ro-vibrational lines of the CO_2 10.6 μm and 9.4 μm bands (both the P and R transitions).

Observation of Doppler-free resonances over an entire molecular rotational-vibrational band offers a great flexibility to study the nature and the details of a spectroscopic effect (or a perturbation) causing weak sub-Doppler shifts or splittings of the various ro-vibrational lines. Such effects are usually J-dependent. Comparison of the splittings (or the shift) of a low-J with a high J-line, for instance, can reveal details not obtainable from the study of a single line. The first observation of this type has been done[18,20] at the author's Lab in the hot bands of CO_2 and N_2O molecules. In the experiment, the weak rotational magnetic moment of these molecules (g_J) and their dependence on vibrational quantum numbers were observed and precisely measured. The latter effect gives rise to an anomalous Zeeman effect, evident only when Doppler-free resonances are studied over an entire band. Along with this, the Doppler-free resonances are used to determine accurate g_J values for the individual ro-vibration lines. In a recent experiment at the author's laboratory, the effect of diamagnetism, giving rise to H^2 dependence in the weak Zeeman splitting is observed[24] in a low-J line of CO_2. From the diamagnetic susceptivity, it has been possible to determine molecular electrical quadrupole moment of the molecules.

Doppler-free spectroscopy has also offered accurate determination of the center of a ro-vibrational band, hyperfine structure of molecular ro-vibrational lines, and others.

In a recently reported[25] precision MIT experiment, a Doppler-free resonance of a selected CO_2 line is used to obtain an accurate measure of the speed of light[25,26]; this is done via simultaneous frequency and wavelength determination of the center of a narrow CO_2 resonance. For a review of the early work of the author in extending frequency measuring technology into the optical region, see references[27-30].

For the work of the author's laboratory in Doppler-free spectroscopy of the selected kinetically hot molecules (the displaced Doppler-free resonances presented above), the reader is referred to references 22 and 23. These references give the first observation of the effect and its utilization to obtain detailed information on intermolecular forces. The experiment was done in NH_3 molecules. The work is currently in progress to extend the method to the other molecules.

REFERENCES

Early MIT Publications; Sub-Doppler Spectroscopy -- Mainly Atoms

1 - Phys. Rev. Letters. 10, 521 (1963). 2 - Appl. Phys. Lett. 7, 322 (1965). 3 - Physics of Quantum Electronics, McGraw-Hill Book Co., p. 567 (1965). 4 - Phys. Rev. Lett. 17, 1242 (1966). 5 - Phys. Rev. 150, 267 (1966). 6 - Phys. Rev. Lett. 18, 730 (1967). 7 - Phys. Rev. Lett. 20, 578 (1968). 8 - Appl. Phys. Lett. 13, 142 (1968). 9 - Phys. Rev. Lett. 22, 267 (1969). 10 - Phys. Rev. 177, 540 (1969). 11 - J. of Mol. Spec. 29, 502 (1969). 12 - Phys. Rev. A 5 (1972). 13 - Fund. and Appl. Laser Physics, Proc. of Esfahan Symp., Wiley-Interscience, page 295, 1973.

Later MIT Publications; Molecular Sub-Doppler Spectroscopy

14 - Phys. Rev. Lett. 23, 559 (1969). 15 - App. Phys. Lett. 17, 53 (1970). 16 - J. Chem. Phys. 56, 4028 (1972). 17 - App. Phys. Lett. 21, 303 (1972). 18 - Laser Spectroscopy, Proc. of Megeve Conf., Springer-Verlag, Page 439, 1975. 19 - J. Mol. Spec. 64, 491 (1977). 20 - Phys. Rev. Lett. 37, 686 (1976), 21 - Phys. Rev. A, 15, 2356 (1977). 22 - Chem. Phys. Lett. 38, 176 (1976). 23 - App. Phys. Lett. 23, 675 (1973). 24 - Optics Lett. 5, 123 (1980).

Absolute Frequency and Wavelength Measurements

For a review of early MIT work, see: Fund. and App. Laser Phys., Proc. of Esfahan Symp. Wiley-Interscience, p. 295, 1973. 25 - Appl. Optics 20, 736 (1981). 26 - Optics Lett. 1, 5 (1977). 27 - App. Phys. Lett. 15, 398 (1969). 28 - Appl. Phys. Lett. 17, 53 (1970). 29 - Appl. Phys. Lett. 10, 147 (1967). 30 - Laser Spectroscopy. Plenum Press, Page, 11, 1973.

THE LINESHAPES OF SUBDOPPLER RESONANCES OBSERVABLE WITH FM SIDE-BAND (OPTICAL HETERODYNE) LASER TECHNIQUES

J. L. Hall,* H. G. Robinson, T. Baer, and L. Hollberg

Joint Institute for Laboratory Astrophysics, National Bureau of Standards and University of Colorado
Boulder, CO 80309

I. BACKGROUND AND OVERVIEW

One of the most powerful spectroscopic techniques available for the measurement of weak absorption is the use of frequency modulation of the source complemented by phase sensitive detection. This approach is highly effective in separating the interesting narrow resonance features from the broad background profile. While optimum amplitude of the frequency modulation leads to recovery of nearly the full signal component, it also leads to broadened resonance profiles which are not immediately related to the physical resonances of interest. Furthermore, it is attractive to employ high modulation frequencies to take advantage of the nearly universal situation that one finds experimentally, namely the concentration of excess noise toward low frequencies. Ultimately then, the resonance profiles of interest are further "distorted" by the modulation process when the modulation frequency is comparable to the resonance width. This constraint is particularly painful in contemporary ultrahigh resolution optical experiments when the measured lines may only be of ~ kiloHertz width.

In this lecture we describe the "new" technique of FM spectroscopy in which the modulation frequency is greater than the linewidths of interest. With this approach we may almost completely avoid the low frequency noise of a technical nature while recovering undistorted resonance profiles of a simple and characteristic shape with excellent signal/noise ratio. It is the theory of these profiles which forms the main subject for this lecture/paper.

---

*Staff Member, Quantum Physics Division, NBS.

II. REVIEW OF MODULATION SPECTROSCOPY AND INTRODUCTION TO RF SIDEBAND OPTICAL HETERODYNE SPECTROSCOPY

The fundamental goal of modulation techniques in spectroscopy is to encode the resonance information of interest into a frequency or time pattern optimally chosen to avoid the source and/or detector noise and drift. Before discussing the "new" rf sideband techniques, it is appropriate to review conventional modulation methods. For example, the weak absorption of an intracavity atomic beam might be made more conspicuous by chopping the atomic beam and looking for a modulated component of the laser power level synchronous with the chopping rate. In gas cell experiments, a small frequency modulation superimposed on a slow sweep may be used to preferentially detect narrow (weak) resonances by emphasizing their higher slope relative to a broader background resonance. This technique works because a purely frequency-modulated signal does not have any intrinsic amplitude change or modulation. Ideally, only the spectrally-sharp resonances of interest can perform the FM to AM conversion.

For a Lorentzian resonance of unit height and unit fullwidth at half maximum, a peak-peak modulation excursion of 0.707 full linewidths is attractive as this modulation value leads to a maximal slope (1.21) near the center of the derivative-like response of the lockin detector. The antisymmetric peak signals in this case are ±0.57, separated by 0.78 fullwidths rather than the $1/\sqrt{3} = 0.577$ separation of the Lorentz derivative peaks. Thus we have a 1.35-fold broadening produced by the chosen value of frequency modulated amplitude. A larger modulation (2.0 fullwidths peak-peak) leads to a maximal recovery of the signal peaks, ±0.78, but their separation is then further broadened to 1.74 full linewidths. This "modulation broadening" problem has been studied analytically and graphically.[1,2]

Although this first derivative signal recovery technique provides excellent isolation from baseline dc drifts in the original absorption signal without serious loss of signal to noise performance, it still will pick up baseline slope directly. Thus one is led to consider frequency modulation of the source as before, but with signal recovery at the second harmonic of the modulation frequency to form an approximation to second derivative signal detection. This signal has a pronounced peak at the resonance center, and two smaller peaks of the other sign located symmetrically around the center. The central peak amplitude has a soft maximum of 0.54 for a modulation excursion of 2.3 fullwidths peak-peak. The second-harmonic-detected signal is not influenced by the resonance's baseline tilt but is contaminated still by curvature of the background resonance. Further the signal has an even symmetry around the central resonance frequency, making it unsuitable for servo control applications with usual techniques.

Coherent detection at the third harmonic frequency is a widely employed scheme to further enhance selectivity against the background.[3,4] Roughly speaking, one recovers the third derivative of the resonance profile in this manner. Further, this signal now possesses the odd-symmetry around line center that is needed for servo applications. Thus problems of baseline dc stability, tilt, and curvature are essentially eliminated by third harmonic detection.

However, the signal-to-noise ratio loss does begin to be noticeable in this case. For example, a peak-peak modulation excursion of 1.64 fullwidths gives an optimum slope at the resonance center of 0.530 using third harmonic detection. Under these conditions the peak signals are ±0.26 and are separated by 0.77 fullwidths, in contrast to the 0.44 fullwidth separation of the corresponding true third-derivative peaks which would be obtained with small modulation.

In practice one needs to be careful that pure frequency modulation is achieved without any spurious amplitude modulation. For example, spurious AM would feed a portion of the background dc level directly into a first harmonic detection system. Distortion of the FM waveform will also lead to spurious frequency offsets.

The general pattern of this type of modulation spectroscopy is the trade-off of a little signal-to-random-noise ratio to eliminate the often enormously-larger problems arising from dc instability, baseline tilt, and curvature. However, from the examples given, we can see that another important "cost" of this type of modulation spectroscopy is broadening and distortion of the resonance profile. Larger modulation amplitude leads to a stronger resonance signal from the lockin detector, but the shape and width are less transparently related to the intrinsic resonance. Another problem arises in trying to choose the modulation frequency. As noted before, the source and detector noise are typically smaller at higher frequencies, and so there is a signal/noise advantage in using high modulation frequencies. However for very narrow lines — say of width comparable to or less than the modulation frequency — there is severe distortion of the spectral profile. Ultimately the narrow resonance serves to analyze the frequency modulation wave form into its spectral components ("sidebands") and the resonance profile becomes rather complex.

In NMR experiments, Smaller[5] and later Acrivos[6] were led by these considerations to the use of a modulation frequency which was much greater than the line width of the resonant transition of interest. In this way they separated from the main resonance the auxiliary resonances associated with modulation-induced spectral sidebands. The excitation of the resonant system becomes effectively single frequency and transient excitation phenomena are avoided. At the same time, the effective frequency modulation excursion could be large enough to provide approximately optimum signal recovery. Instructive lineshape results have been derived for the homogeneously-broadened lines typically encountered in NMR.[7] Closely related and basically equivalent results and techniques have been recently and independently developed for optical spectroscopy by Bjorklund[8] and colleagues, Drever and Hough,[9] and the present authors.[10] We have shown that with these FM techniques, very simple and precise resonance profiles can be obtained with a signal/noise level within 3.4 dB of the optimum value limited by shot noise.[11] These low noise results are available if the laser source has some (high frequency) spectral region in which its amplitude noise level approaches the quantum (shot noise) limit — even though at typical modulation frequencies (~kHz) the noise density may be 4 or more orders of magnitude above the shot noise level.

It is highly instructive to represent the pure, phase (frequency) modulated wave by a Fourier spectral series. We have the mathematical relation

$$E = E_0 \sin(\Omega t + \beta \sin \omega t) \tag{1}$$

$$= E_0\left\{ \sum_{n=0}^{\infty} J_n(\beta)\sin(\Omega+n\omega)t + \sum_{n=1}^{\infty} (-1)^n J_n(\beta)\sin(\Omega-n\omega)t \right\} ,$$

where $J_n(\beta)$ is the Bessel function of order n and the peak phase excursion, β, is called the modulation index. We will often speak of Eq. (1) as representing a frequency-modulated wave. The basic equivalence can be easily shown by identifying the time derivative of the phase as a (generalized) frequency. From Eq. (1) we have $\phi = \Omega t + \beta \sin \omega t$, and $f \equiv (1/2\pi)(d\phi/dt) = (\Omega + \beta\omega \cos \omega t)/2\pi$. Thus the peak frequency excursion is $\beta\omega/2\pi$. It is important to note the sign alternation of the lower frequency sidebands in Eq. (1). For the cases of most interest to us, the modulation index β is $\lesssim 1$, for which the infinite sideband spectrum collapses mainly into the carrier at angular frequency Ω and two first-order sidebands at $\Omega \pm \omega$. A small percentage of the power is in the second-order sidebands at $\Omega \pm 2\omega$.

This particular sum of optical frequencies has the remarkable property that an ideal amplitude photodetector will generate no time-dependent photocurrent. Physically, the photodiode will have two main sources of photocurrent at frequency ω, namely cross terms of the optical carrier with the two first-order sidebands. However, as may be determined by inspection of Eq. (1), the two currents are of equal magnitude and opposite sign. Thus they will (ideally) exactly cancel, leaving no net current variation at frequency ω. However, rather powerful beat signals at frequency ω would arise if the cancellation were no longer perfect. For example, if we pass the beam represented by Eq. (1) through a medium possessing a sharp resonant absorption/dispersion feature, the sideband amplitudes may become unequal if there are different absorption coefficients at the two sideband frequencies. Alternatively, the phase of any of the three principal spectral components may be shifted by near-resonant interaction. Either process will thus lead to a net photocurrent at the modulation frequency, ω. This is the basic idea behind "rf-sideband optical heterodyne spectroscopy." In the following we first develop general expressions for the observable signals at ω and 2ω in terms of the phase shifts and amplitude absorptions produced by the medium. We then model the medium in microscopic terms and obtain detailed line profiles for the cases of interest. A final section discusses transient effects.

III. DEVELOPMENT OF GENERAL LINE SHAPE RESULTS

Suppose we have incident on a resonant medium the phase-modulated signal as described by Eq. (1). The wave emerging from the medium will in general have a different absorption, $\alpha(\Omega')$, and phase shift, $\phi(\Omega')$, for each spectral component. If ℓ is the absorption cell length, we may write the output field

$$E_{out}(\Omega') = E_{in}(\Omega')\, e^{-\alpha(\Omega')\ell/2}\, e^{-i\phi(\Omega')} \tag{2}$$

where Ω' can be Ω, $\Omega + \omega$, and $\Omega - \omega$. A convenient notation is o, +, - for these frequency indices. In this way the attenuation (Beers' Law) factors may be written $T^o = e^{-\alpha(\Omega)\ell/2}$, $T^+ = e^{-\alpha(\Omega+\omega)\ell/2}$ and $T^- = e^{-\alpha(\Omega-\omega)\ell/2}$. Similarly the phase shift for the three main spectral components may be denoted by ϕ^o, ϕ^+, ϕ^-. In Eq. (2) we have suppressed an overall phase shift due to propagation through the cell.

Amplitude and phase factors for the second order sidebands may be denoted A^{++}, ϕ^{++}, A^{--}, ϕ^{--}. In these terms the output of the cell is of the form

$$\begin{aligned} E_{out} \simeq\ & T^o J_0(\beta)\sin(\Omega t-\phi^o) + T^+ J_1(\beta)\sin\big((\Omega+\omega)t-\phi^+\big) \\ & - T^- J_1(\beta)\sin\big((\Omega-\omega)t-\phi^-\big) + T^{++} J_2(\beta)\sin\big((\Omega+2\omega)t-\phi^{++}\big) \\ & + T^{--} J_2(\beta)\sin\big((\Omega-2\omega)t - \phi^{--}\big) \quad . \end{aligned} \tag{3}$$

The photodiode responds to the incident light intensity, that is, to the square of the field represented by Eq. (3). For signal detection of the photodiode current at the modulation frequency ω, we are interested in cross-terms of the first sideband frequencies with the neighboring second sidebands and with the carrier. It is convenient to introduce the current responsivity of the diode, η, defined so that $i \equiv \eta E^2$, where i is the photodiode current resulting from placing it in an optical electric field E. In terms of the intrinsic quantum efficiency η_0, $\eta = e\eta_0(c\varepsilon/2)A/\hbar\Omega$ where A represents an effective area of the diode, and ε is the permitivity of free space. The photodiode currents at rf frequency ω can now be derived from the following cross terms in $(E_{out})^2$:

$$\frac{i(\omega)}{\eta E_0^2} = T^+ J_1(\beta)\sin(\Omega^+ t-\phi^+) \times$$
$$[T^o J_0(\beta)\sin(\Omega t-\phi^o) + T^{++} J_2(\beta)\sin(\Omega^{++} t-\phi^{++})]$$
$$- T^- J_1(\beta)\sin(\Omega^- t-\phi^-) \times$$
$$[T^o J_0(\beta)\sin(\Omega t-\phi^o) + T^{--} J_2(\beta)\sin(\Omega^{--} t-\phi^{--})] \quad . \qquad (4)$$

Simplification is possible using the identity sinAsinB = 1/2[cos(A-B)-cos(A+B)] and by averaging over the optical sum frequency. One finds

$$\frac{i(\omega)}{\eta E_0^2} = \frac{T^+}{2} J_1(\beta)\left[T^o J_0(\beta)\cos\left(\omega t-(\phi^+-\phi^o)\right) + T^{++} J_2(\beta)\cos\left(\omega t-(\phi^{++}-\phi^+)\right)\right]$$
$$- \frac{T^-}{2} J_1(\beta)\left[T^o J_0(\beta)\cos\left(\omega t-(\phi^o-\phi^-)\right)\right.$$
$$\left. + T^{--} J_2(\beta)\cos\left(\omega t-(\phi^- -\phi^{--})\right)\right] \quad . \qquad (5)$$

Expanding again, and collecting related terms we obtain

$$\frac{i(\omega)}{\eta E_0^2 J_1(\beta)/2} = \cos\omega t\left\{J_0(\beta)T^o\left[T^+\cos(\phi^+-\phi^o) - T^-\cos(\phi^o-\phi^-)\right]\right.$$
$$+ J_2(\beta)T^+T^{++}\cos(\phi^{++}-\phi^+) - \left. J_2(\beta)T^-T^{--}\cos(\phi^- -\phi^{--})\right\}$$
$$+ \sin\omega t\left\{J_0(\beta)T^o\left[T^+\sin(\phi^+-\phi^o) - T^-\sin(\phi^o-\phi^-)\right]\right.$$
$$+ J_2(\beta)T^+T^{++}\sin(\phi^{++}-\phi^+) - \left. J_2(\beta)T^-T^{--}\sin(\phi^- -\phi^{--})\right\} \quad . \qquad (6)$$

Phase-sensitive detection at frequency ω will allow us to select either the rf sine or cosine-dependent term in Eq. (6). Even so, it can easily be seen that this function contains the possibility to produce a rather complicated spectral profile in the general case. In any case the resonances have complete odd symmetry around the spectral center.

Before making contact with a microscopic model of the resonances, it is interesting to derive the signal that can be observed by rf detection at twice the modulation frequency. Such terms will arise as beats between the two first-order sidebands and also from beats of the carrier with either second-order sideband. From Eq. (3) we have

$$\frac{i(2\omega)}{\eta E_0^2} = \{-T^+T^-J_1^2(\beta)\sin(\Omega^+t-\phi^+)\sin(\Omega^-t-\phi^-)$$
$$+ T_0J_0(\beta)J_2(\beta)\sin(\Omega t-\phi^o)[T^{++}\sin(\Omega^{++}t-\phi^{++})$$
$$+ T^{--}\sin(\Omega^{--}t-\phi^{--})]\} \quad . \qquad (7)$$

Proceeding as before, we obtain for the signal at the rf second harmonic frequency,

$$\frac{i(2\omega)}{\eta E_0^2/2} = \sin 2\omega t[T^oJ_0(\beta)J_2(\beta)(T^{++}\sin(\phi^{++}-\phi^o)$$
$$+ T^{--}\sin(\phi^o-\phi^{--})) - T^+T^-J_1^2(\beta)\sin(\phi^+-\phi^-)]$$
$$+ \cos 2\omega t[(T^oJ_0(\beta)J_2(\beta))(T^{++}\cos(\phi^{++}-\phi^o)$$
$$+ T^{--}\cos(\phi^o-\phi^{--})) - T^+T^-J_1^2(\beta)\cos(\phi^+-\phi^-)] \quad . \qquad (8)$$

IV. LINE SHAPES IN THE LOW ABSORPTION LIMIT

It is useful and highly instructive to recover the first terms of an expansion of these line shape functions ($i(\omega)$ and $i(2\omega)$) under the assumption that the absorption and phase shifts are small. Using this approximation

$$T^j \equiv e^{-\alpha^j \ell/2} \simeq 1 - \alpha^j \frac{\ell}{2} + \ldots \rightarrow 1-A^j \quad , \qquad (9)$$

where $j = ++,+,0,-,--$ is the sideband frequency index. Similarly we will take $\cos(\phi j-\phi j') \rightarrow 1$ and $\sin(\phi j-\phi j') \rightarrow \phi j-\phi j'$. Higher-order expansion terms are relatively small in the case that is of most interest for spectroscopy, namely the detection of weak absorption and dispersion resonances. In the low absorption limit we find

$$\frac{i(\omega)}{\eta E_0^2 J_1(\beta)/2} = -\cos\omega t\left[\left(J_0(\beta)+J_2(\beta)\right)(A^+-A^-) + J_2(\beta)(A^{++}-A^{--})\right]$$

$$+ \sin\omega t\left[\left(J_0(\beta)-J_2(\beta)\right)(\phi^+-2\phi^0+\phi^-)\right.$$

$$\left.+ J_2(\beta)(\phi^{++}-2\phi^0+\phi^{--})\right] \quad . \qquad (10)$$

This form makes apparent, even more clearly than Eq. (6), the intrinsic antisymmetry about the carrier frequency $\Omega^0 = \Omega$ of the absorption signal (coefficient of $\cos\omega t$). Furthermore, remembering that the phase shift ϕ of a resonant medium has odd symmetry, we can see that the dispersion signal (coefficient of $\sin\omega t$) will also have pure odd symmetry. This property will be of fundamental importance in using these resonances for precision spectroscopy or for laser locking. Ultimately the antisymmetry of signals at frequency ω arises from the antisymmetry intrinsic to the FM probe signal, Eq. (1). [One interesting case where the absorption and dispersion are not very small arises in locking a laser to an optical resonance cavity. To investigate in detail the symmetry of the expected resonance profiles, we have in fact carried out the calculation completely to order of terms like A^2, ϕ^2 and $A\phi$ (22 for the $\cos\omega t$ term, 19 for the $\sin\omega t$ term). However, for brevity we do not display the results here; it is much more convenient to evaluate Eq. (6) by numerical methods.]

We now turn again to the signal observable at the second harmonic frequency. In the low absorption limit, this signal (Eq. (8)) becomes

$$\frac{i(2\omega)}{\eta E_0^2/2} = \sin 2\omega t\left[J_0(\beta)J_2(\beta)(\phi^{++}-\phi^{--}) - J_1^2(\beta)(\phi^+-\phi^-)\right]$$

$$- \cos 2\omega t\left[J_0(\beta)J_2(\beta)(A^{++}+2A^0+A^{--}) - J_1^2(\beta)(A^++A^-)\right.$$

$$\left.+ J_1^2(\beta) - 2J_0(\beta)J_2(\beta)\right] \quad . \qquad (11)$$

For usual values of the modulation index (say $\beta = 1$), the $J_1^2(\beta)$ coefficient (0.194) is about twice as large as $J_0(\beta)J_2(\beta)$ (0.089). In fact over the whole β range from 0 to 1.1, the ratio $J_1^2(\beta)/J_0(\beta)J_2(0)$ increases slowly from 2.00 to 2.26. Thus the nonresonant coefficient in the cosine term is rather small, and in fact only arises from our truncation of the series. As before, one can obtain two separate spectral lineshapes by choice of the rf reference phase. These resonances are (individually) even-symmetric about the central resonance frequency, consistent with our experimental observations. Calculated profiles will be presented after we obtain the characteristic resonance functions ϕ and A.

V. CONNECTING WITH MICROSCOPIC PHYSICS

In this section we suppose that a purely FM-modulated laser beam (of the form given by Eq. (1)) is incident on a cell containing an absorbing gas. To avoid unessential complications we assume the gas has only a single nondegenerate resonant transition within the frequency range of interest. Sargent and Toschek[12] have given some interesting results for the case of co-running beams where the laser fields are strong enough to produce saturation effects. Here, we are more interested in low opacity and the ultimate sensitivity limits of the sideband techniques. For example, a typical application might be the monitoring of the absorption/dispersion produced by weak transitions in a gas cell or by dilute radical ions in a flowing gas environment, or by trace constituents in the atmosphere. Our expectation is that the use of a high rf modulation frequency can encode the sample's resonance information to such a high frequency that laser noise of a technical nature and/or gas-cell turbulence etc. do not degrade the sensitivity by introducing noise in excess of the basic quantum-noise limit.

Our first task is to make the connection between the cell-integrated absorption and phase shift and the intrinsic absorption/dispersion response of each absorber. We can avoid many complications by restricting our attention to the thin absorbing cell limit. In this limit each molecule interacts with the same strength fields and makes its own independent contribution to the accumulated absorption and phase shift which are assumed to be small. We will take the incident field in the form

$$E = E_0\left((-i)\,\mathrm{Im}\; e^{i(\Omega t+\beta\sin\omega t-kz)}\right) \quad , \tag{12}$$

where our somewhat unconventional choice to use the sine (rather than cosine) part is dictated by the structure of the identity in Eq. (1). The wave enters the cell at $z = 0$, and exits at $z = \ell$. In the medium, the propagation constant k is shifted from its free space value $k = \Omega/c$ to

$$k' = k\left[1 + \frac{\chi'(\Omega)}{2n^2}\right] - i\,\frac{k\chi''(\Omega)}{2n^2} \quad . \tag{13}$$

Here n is the index of refraction. The complex susceptibility of the medium χ may be conveniently written explicitly in terms of the two real functions $\chi'(\Omega)$ and $\chi''(\Omega)$ which describe index-of-refraction effects and absorption effects respectively. Their specification provides a useful interface between our previous macroscopic discusion and the more detailed microscopic models which follow. The susceptibility language is also the natural one when thick cell (propagation) effects must be considered. For our case $n \simeq 1$, and the propagation constant becomes

$$k' \to k\left(1 + \frac{\chi'}{2}(\Omega)\right) - ik\frac{\chi''}{2}(\Omega) \quad . \tag{14}$$

Integrating through the absorption cell leads via Eqs. (12) and (13) to an output field of the form

$$E_{out} = E_{in} \times e^{-ik\ell}\, e^{-ik\chi'\ell/2}\, e^{-k\chi''\ell/2} \quad .$$

This equation is of the form postulated in Eq. (2), so we may make the identifications

$$A = \frac{\alpha\ell}{2} = \frac{k\chi''\ell}{2} \quad \text{and} \quad \phi = \frac{k\chi'\ell}{2} \quad . \tag{15}$$

For simplicity of notation in Eqs. (14) and (15), the frequency indices on the quantities A, α, ϕ, χ'', and χ' are suppressed. The quantity k in Eq. (14) is taken to be the (free space) wave vector of the carrier: the frequency splitting of the sidebands is very small and the medium-induced effects are small in any case. [However, in Eq. (12) it is important to keep the proper frequency-dependent k in the main exponential propagation term: otherwise a spurious and confusing FM-to-AM sideband conversion appears at some definite cell length related to the modulation frequency ω. Clearly, if the wave is launched as an FM wave it must remain so in the absence of appreciable dispersion.]

VI. THE LINESHAPE OF A SINGLE FM BEAM INTERACTING WITH AN ABSORBING MEDIUM

Our task now is to derive specific lineshapes in several cases of physical interest. For example, we know that to a good approximation the line shape resulting from pressure broadening is pure Lorentzian in form. Thus we may take as our first pedagogical example, and as a good model for atmospheric absorption produced by trace constituents — the case of strong collision damping and associated Lorentz line shapes. Several authors present useful density matrix discussions of the response of a two level system with (exponential) damping characterized by rates $(T_1)^{-1}$ for the population difference and $(T_2)^{-1}$ for the coherent dipole moment. For brevity we will only quote those results here.[13] Taking $E = E_0 \sin\Omega t$ as the driving field leads to the following macroscopic dipole moment P for the ensemble of unperturbed population difference ΔN_0:

$$P = \frac{\mu^2 T_2 \Delta N_0}{\hbar} E_0 \left(\frac{\cos\Omega t + (\Omega_0 - \Omega) T_2 \sin\Omega t}{1 + (\Omega - \Omega_0)^2 T_2^2 + 4\Omega_R^2 T_1 T_2}\right) \tag{16a}$$

and

$$\Delta N = \Delta N_0 \frac{1 + (\Omega-\Omega_0)^2 T_2^2}{1 + (\Omega-\Omega_0)^2\ T_2^2 + 4\Omega_R^2 T_1 T_2}\ , \tag{16b}$$

where μ is the transition moment, and $\Omega_R = \mu E_0/2\hbar$ is the Rabi transition frequency at resonance. Using the complex susceptibility $\chi = \chi' - i\chi''$ we can rewrite (16) in the form

$$\chi''(\Omega) = \frac{\mu^2 T_2 \Delta N_0}{\varepsilon_0 \hbar} \frac{1}{1 + (\Omega-\Omega_0)^2\ T_2^2 + 4\Omega_R^2 T_1 T_2}$$

$$\chi'(\Omega) = \frac{\mu^2 T_2 \Delta N_0}{\varepsilon_0 \hbar} \frac{(\Omega_0-\Omega) T_2}{1 + (\Omega-\Omega_0)^2 T_2^2 + 4\Omega_R^2 T_1 T_2}\ . \tag{17}$$

The actual driving field, Eqs. (1) and (12), contains a sum of Fourier frequencies and so we could have a rather complicated response. However if the fields are too weak to induce appreciable saturation — which will be the case for our prototypical atmospheric pressure absorption experiment — we may obtain their joint effect by superposition of the individual responses. The saturation parameter $4\Omega_R{}^2 T_1 T_2 \ll 1$ for each component.

To avoid having our equations become so clumsy that we cannot readily discern their meaning, it is attractive to introduce some simplifying notation. We have from Eqs. (15) and (17)

$$\chi''(\Omega{=}\Omega_0) = \frac{\alpha_0}{k} = \frac{\mu^2 T_2 \Delta N_0}{\varepsilon_0 \hbar}\ , \quad \text{in the limit } \Omega_R \to 0\ .$$

We introduce

$$L^j \equiv \frac{1}{1 + (\Omega^j-\Omega_0)^2 T_2^2 + S}$$

$$D^j \equiv \frac{(\Omega_0-\Omega^j) T_2}{1 + (\Omega^j-\Omega_0)^2 T_2^2 + S}\ , \tag{18}$$

$j = {+}{+}, +, 0, -, {-}{-}$ corresponding to $\Omega^j = \Omega+2\omega$, $\Omega+\omega$, Ω, $\Omega-\omega$, and $\Omega-2\omega$ respectively. The quantity $S \equiv 4\Omega_R^2 T_1 T_2 = (\mu E_0/\hbar)^2 T_1 T_2$ is the saturation parameter and is assumed to be $\ll 1$ in the present discussion. In later sections we will give S an index also. Collecting these results together we can rewrite Eq. (15) in the form

$$A^j = \frac{\alpha_0 \ell}{2} \cdot L^j$$

$$\phi^j = \frac{\alpha_0 \ell}{2} D^j \quad . \qquad (19)$$

Using this notation we can obtain a very compact expression by rewriting Eq. (10)

$$\frac{i(\omega)}{\eta E_0^2 J_1(\beta)\alpha_0 \ell/4} = -\cos\omega t[(J_0+J_2)(L^+-L^-) + J_2(L^{++}-L^{--})]$$

$$+ \sin\omega t[(J_0-J_2)(D^+-2D^o+D^-) + J_2(D^{++}-2D^o+D^{--})] \quad . \qquad (20)$$

To make these functions concrete, in Fig. 1a and b we have plotted the absorption ($\propto\cos\omega t$) and dispersion ($\propto\sin\omega t$) lineshapes. The figure contains four curves for each lineshape corresponding to different choices of the resolution, $R \equiv \omega T_2$. [This resolution parameter amounts to the ratio of the phase-modulation frequency to the resonance line width (HWHM).] Curves are given for R = 5, 2.5, 1.5, 0.5 and 0.375. Experimentally, the phase of the detection rf is not known a priori and so it is useful to see the evolution of the line shape for several phases around the pure absorption and pure dispersion cases. Such curves are shown in Fig. 2a and b, for phase steps of ±6° and ±12° for the case R = 7.5.

An important experimental consideration in our model case is that atmospheric pressure broadening can easily lead to line widths of several GHz. Effective phase modulation at such frequencies, however, represents a serious technical challenge. Thus it is important to know the loss of the signal if the modulation frequency is too low for the modulation sidebands to be well resolved by the resonance. Under these conditions it is easy to see from Eq. (20) that serious cancellation will occur and the peak-to-peak signal observable by tuning the laser over the resonance will diminish rapidly. Figure 3, calculated from Eq. (20), shows the scaled peak-peak signals plotted for a variable rf modulation frequency. The abscissa is labeled by the resolution parameter $R = \omega T_2$, i.e. by the modulation frequency measured in (HWHM) linewidths. Inspection of Fig. 3 shows that the sensitivity loss does not exceed 20 dB if the modulation frequency is above 0.1 linewidths HWHM. Development of efficient phase modulators suitable for multi-GHz operation is clearly in order if we are to have the ultimate high sensitivity detection of atmospheric absorbers.

Knowing that the resonances are Lorentzian in form allows us to give an immediate concrete form for the signals at 2ω as given by Eq. (11). We plot the two double-frequency signals separately

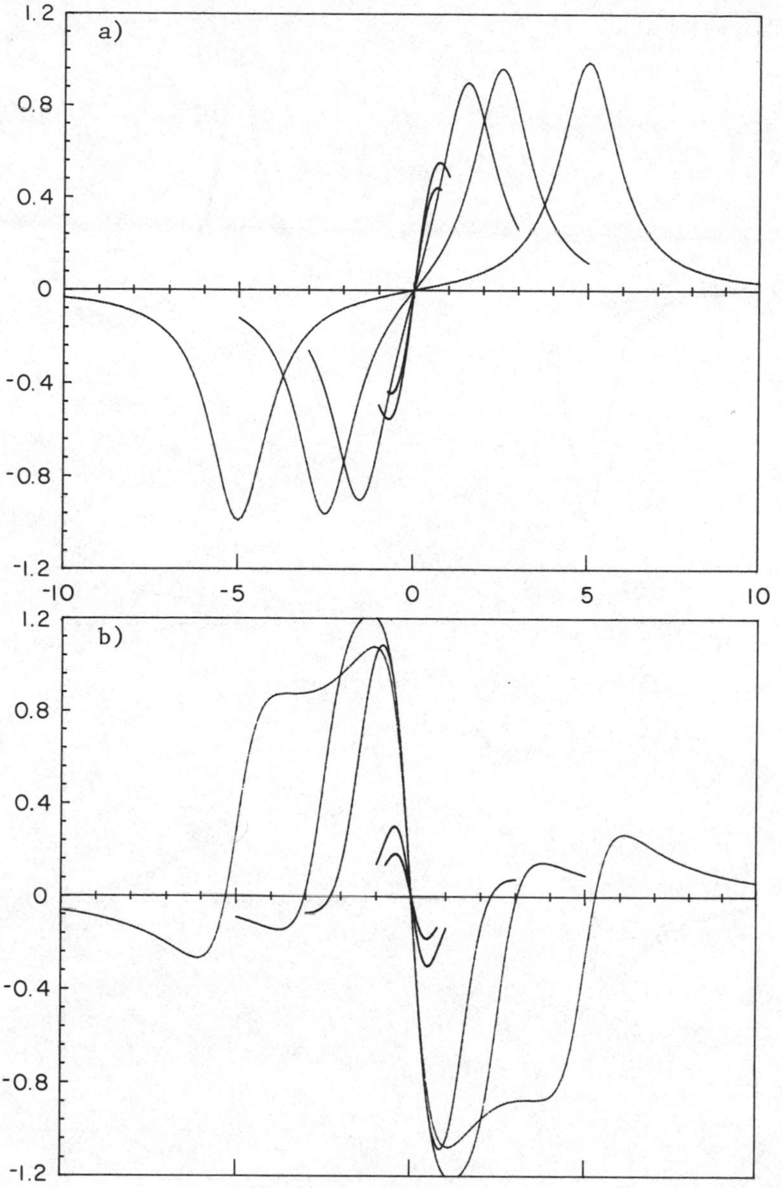

Fig. 1. Calculated resonance profiles versus laser detuning for several modulation frequencies. a) Absorption curves for $R_l = 5$, 2.5, 1.5, 0.5, and 0.375. See text for definition of R_l. b) Dispersion curves for same resolution values. Curves are calculated out to twice the modulation frequency. For nonlinear sub-Doppler resonances, the corresponding modulation frequencies are twice as large, but the response curves are the same. All frequencies are normalized by the linewidth (HWHM). The $J_2(\beta)$ contributions are not included here.

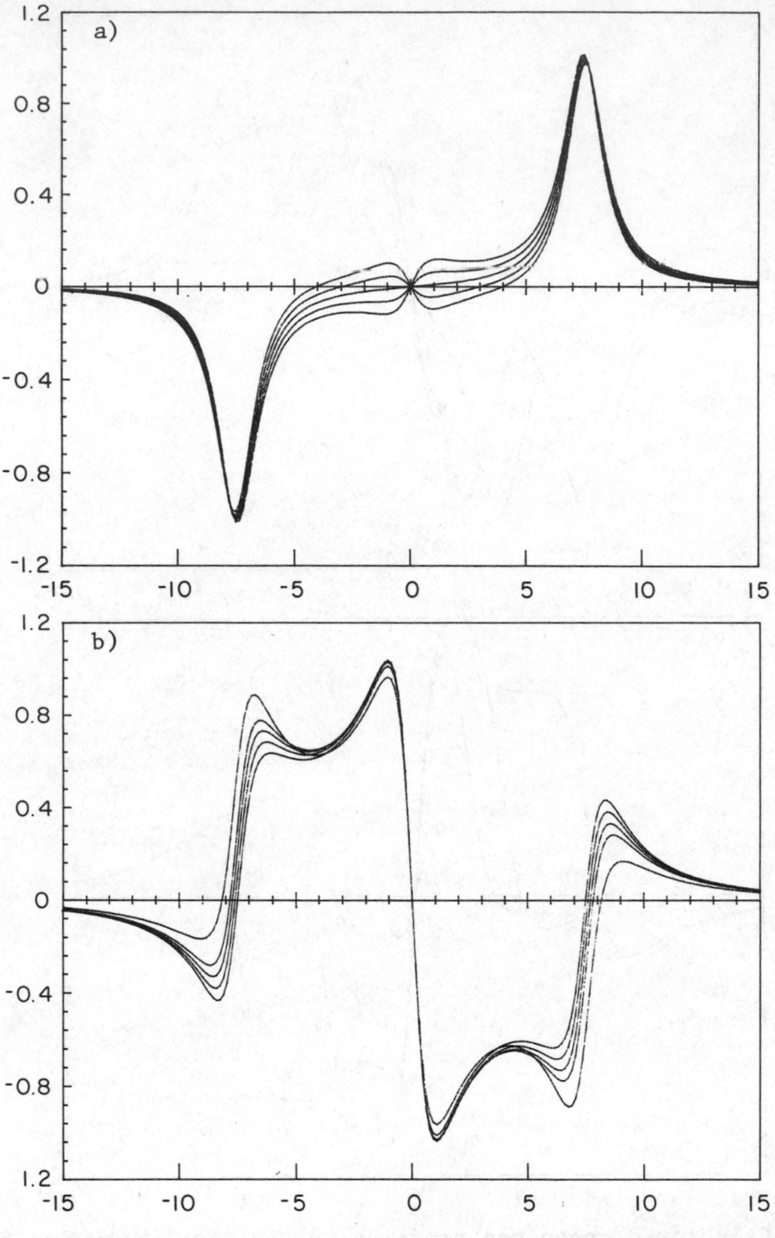

Fig. 2. Effect of phase variation around pure absorption (a) and pure dispersion (b). Phase increments are 0°, ±6°, ±12°. Positive phase steps give negative absorption signals for small positive detunings. Uppermost dispersion signal at positive detunings corresponds to 78° (-12° from pure dispersion). Modulation frequency is 7.5 regarding these curves as linear resonances; 15 if they are nonlinear sub-Doppler signals. The $J_2(\beta)$ contributions are omitted here.

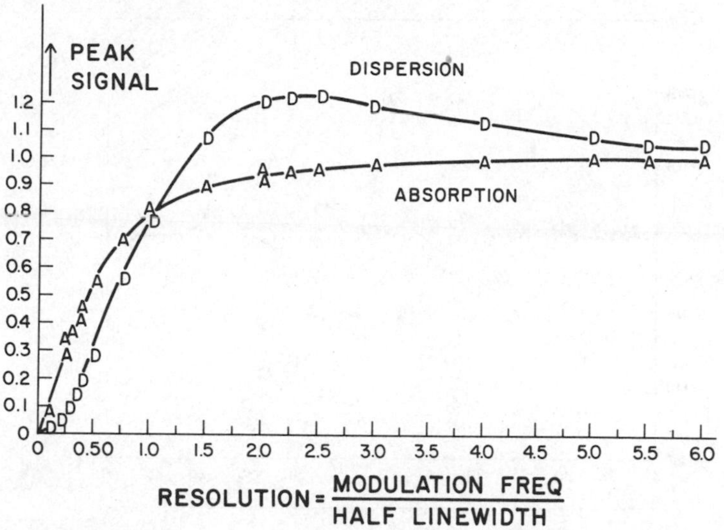

Fig. 3. Peak signal size versus modulation frequency. Abscissa scale assumes linear spectroscopy. Note that there is an optimum resolution value to maximize the dispersion slope near the origin, for servo-control purposes for example. At low resolution the absorption signal decreases less rapidly than the dispersion.

in Fig. 4, corresponding to absorption and dispersion phases. It may be seen that absorption resonances occur at $\Omega-\Omega_0 = 0, \pm\omega$, and $\pm 2\omega$, while dispersion resonances occur at $\Omega-\Omega_0 = \pm\omega$ and $\pm 2\omega$. Both curves have overall even symmetry.

VII. SUBDOPPLER LINE SHAPES VIA SATURATED ABSORPTION

We turn now to the low pressure limit in which pressure-broadening is much smaller than the Doppler broadening arising from the absorber's thermal motion. With only one beam probing the absorption, we will again have the same kind of sensitivity loss just calculated unless the modulation frequency substantially exceeds the Doppler broadened line width. The associated Gaussian profile will lead to a slightly different dependence on modulation frequency from that shown in Fig. 3, but the overall trend will be just the same. Instead of calculation of such results in detail, what we want to concentrate on here is a two beam experiment in which saturated absorption effects can provide resolution of spectral features much sharper than the Doppler-broadened line. The essential idea is that under suitable resolution conditions the second, counter-running beam can influence the absorption/dispersion for individual probe beam FM spectral components if the counter-running beam is strong enough to cause saturation.

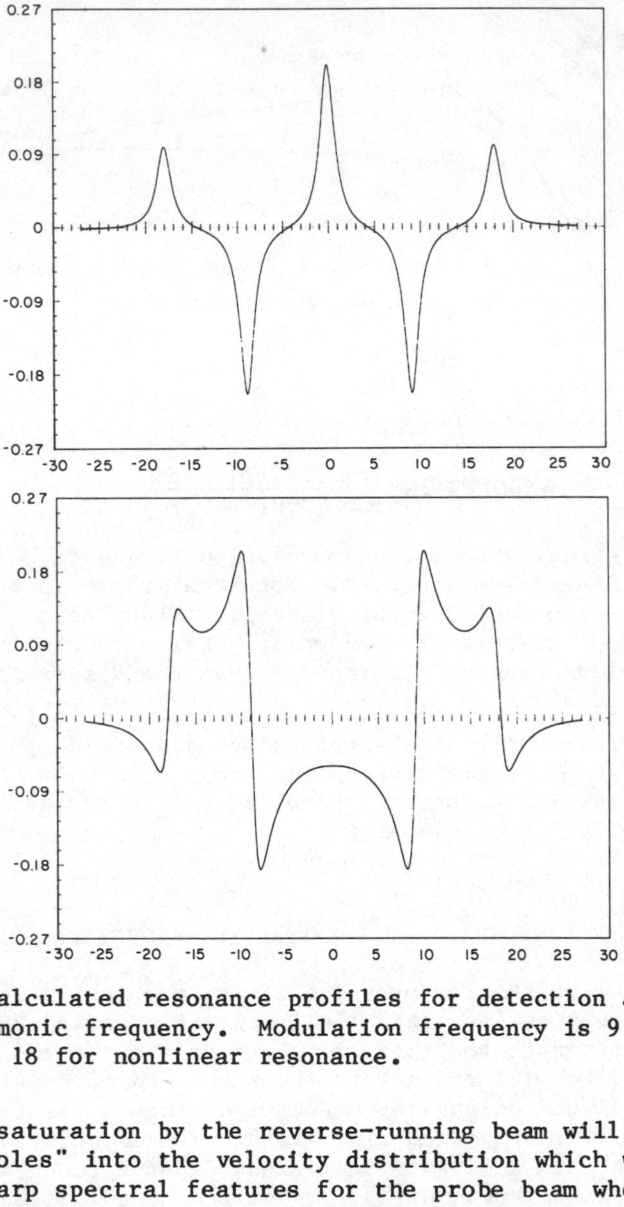

Fig. 4. Calculated resonance profiles for detection at the rf second harmonic frequency. Modulation frequency is 9 for a linear resonance, 18 for nonlinear resonance.

Such saturation by the reverse-running beam will burn "Bennett holes" into the velocity distribution which will lead in turn to sharp spectral features for the probe beam when one of its spectral components interacts with the affected part of the velocity distribution. The line width over which the mutual interaction is appreciable will be the very narrow spectral interval over which the interaction may be said to be "homogeneous," limited typically by residual pressure or lifetime broadening. Ultimately, for long-lived absorbers, the resolution limit will be set by the finite time of interaction as the absorber atoms move across the Gaussian laser mode. For simplicity of analysis we will take the homogeneous broadening to be that resulting from a relaxation process. This will lead to a Lorentzian response of linewidth (HWHM) $\Delta\omega = 1/T_2$, where — as before — T_2^{-1} is the decay rate of the coherent molecular dipole moment.

A complete physical description of the observable saturated absorption/dispersion resonances could be developed using the powerful density matrix diagram techniques discussed by Professor Bordé elsewhere in this volume. These methods make it rather simple to make a perturbation theory expansion of the complete density matrix treatment. We would calculate a polarization produced by a first interaction with the saturating beam. A second interaction would produce a velocity-resolved (nonlinear) peak in the excited state population. Interaction with this population by the three probe frequency components would lead to nonlinear polarizations at the probe frequencies. Finally, in a fourth interaction, the nonlinear polarizations would modify the propagation for the probe beam components to produce the sub-Doppler nonlinear attenuation and phase shift. However, for our discussion this approach has several disadvantages. For one, with four applied frequencies and four interactions there are a number of terms to be considered. Secondly, the usual plethora of mathematical complexity surrounding Lamb dip spectroscopy is fully reproduced here, including weak convergence of the higher-order terms associated with reflection of the saturating beam into the probe beam direction by the spatially-modulated density and phase gratings induced in the medium.[14,15] With three probe frequencies it does not seem reasonable at this time for us to make a theory of the optical-heterodyne-detected resonances along these lines.

Instead we will recall Shirley's result[16] for the single frequency Lamb dip case that the phenomena is surprisingly well-described by a population-type argument, the so-called "rate equation approximation" to the full density matrix treatment. This language gives us a simple and extremely powerful physical basis for understanding the resonance lineshapes which will be very useful when we have so many frequencies present. Also, this description is accurate at very low saturation and useful even up to $S \simeq 1$. A further useful approximation for us to make could be called the "high-resolution approximation" in which the modulation frequency is substantially larger than the homogeneous line width. In this limit the effects of the three probe frequencies will be essentially independent.

Our basic task is to take the Doppler distribution of velocities into account. In the lab frame, the absorber's natural frequency is shifted from Ω_0 to $\Omega' = \Omega_0 + \bar{k}\cdot\bar{v}$. The velocity distribution — in the absence of saturation — is given by the Doppler distribution

$$G_0(v) = \frac{1}{\sqrt{\pi}}\,\frac{1}{u}\,e^{-v^2/u^2} \tag{21}$$

where $u = (2kT/M)^{1/2}$ is the Maxwell velocity. When the field is strong enough to cause some saturation, the velocity distribution globally will still be like that given by Eq. (21). However, near the velocity which leads — via the Doppler shift — to resonance with the externally-applied field, we will have a local "Bennett hole" of the term given by Eq. (16b). The Doppler velocity distribution, including saturation will thus take the form

$$G(v) = \frac{1 + (\Omega-\Omega_0-\bar{k}\cdot\bar{v})^2 T_2^2}{1 + (\Omega-\Omega_0-\bar{k}\cdot\bar{v})^2 T_2^2 + S} \left(\frac{1}{\pi}\right)^{1/2} \frac{1}{u} e^{-v^2/u^2} \quad . \tag{22}$$

For the presence of a saturating beam to lead to sharp spectral profiles, we must have the saturating and probe beams counter-running. We choose the saturating beam, s, to be running in the (−)z direction. The frequency-modulated probe beam, p, is running in the (+)z direction. Writing $\Omega_0' = \Omega_0 + kv_z$ as the resonant frequency in the lab frame, we can rewrite Eq. (17) to give the susceptibility of a single velocity packet as a function of velocity and frequency:

$$\chi''(\Omega,v_z) = \frac{\alpha_0}{k} \frac{1}{1 + (\Omega-\Omega_0-kv_z)^2 T_2^2 + S_p}$$

$$\chi'(\Omega,v_z) = -\frac{\alpha_0}{k} \frac{(\Omega-\Omega_0-kv_z)T_2}{1 + (\Omega-\Omega_0-kv_z)^2 T_2^2 + S_p} \quad . \tag{23}$$

We can now perform the velocity integral to obtain the observable susceptibility of the gas sample. We have

$$\chi(\Omega) = \int_{-\infty}^{\infty} \chi(\Omega,v_z) G(v_z) dv_z \quad . \tag{24}$$

In more detail the susceptibility has the form

$$\chi''(\Omega) = \frac{\alpha_0}{k}\left(\frac{1}{u}\right)\left(\frac{1}{\pi}\right)^{1/2} \int_{-\infty}^{\infty} \frac{1}{1 + (\Omega-\Omega_0-kv_z)^2 T_2^2 + S_p}$$

$$\times \frac{1 + (\Omega-\Omega_0+kv_z)^2 T_2^2}{1 + (\Omega-\Omega_0+kv_z)^2 T_2^2 + S_s} e^{-v_z^2/u^2} dv_z$$

$$\chi'(\Omega) = -\frac{\alpha_0}{k}\left(\frac{1}{u}\right)\left(\frac{1}{\pi}\right)^{1/2} \int_{-\infty}^{\infty} \frac{(\Omega-\Omega_0-kv_z)T_2}{1 + (\Omega-\Omega_0-kv_z)^2 T_2^2 + S_p}$$

$$\times \frac{1 + (\Omega-\Omega_0+kv_z)^2 T_2^2}{1 + (\Omega-\Omega_0+kv_z)^2 T_2^2 + S_s} e^{-v_z^2/u^2} dv_z \quad . \tag{25}$$

These equations can be simplified in form by introducing $x = (\Omega-\Omega_0)T_2$ and $y = kv_zT_2$. Substituting into (25) we obtain

$$\chi''(\Omega) = \frac{\alpha_0}{k}\left(\frac{1}{\pi}\right)^{1/2}\frac{1}{kuT_2}\int_{-\infty}^{\infty}\frac{1}{1+(x-y)^2+S_p}$$

$$\times\frac{1+(x+y)^2}{1+(x+y)^2+S_s}e^{-y^2/k^2u^2T_2^2}dy$$

$$\chi'(\Omega) = -\frac{\alpha_0}{k}\left(\frac{1}{\pi}\right)^{1/2}\frac{1}{kuT_2}\int_{-\infty}^{\infty}\frac{(x-y)}{1+(x-y)^2+S_p}$$

$$\times\frac{1+(x+y)^2}{1+(x+y)^2+S_s}e^{-y^2/k^2u^2T_2^2}dy \quad . \qquad (26)$$

We can check the equation for χ'' for example by integrating Eq. (26) with the saturating beam intensity $S_s = 0$. Using the fact that the major contribution to the integral comes for a small range in $y \approx x$ and that $kuT_2 \gg 1$, we may reasonably estimate the integral by taking the more slowly-varying terms outside

$$\chi''(\Omega) = \frac{\alpha_0}{k}\left(\frac{1}{\pi}\right)^{1/2}\frac{1}{kuT_2}e^{-x^2/k^2u^2T_2^2}\int_{-\infty}^{\infty}\frac{1}{1+(x-y)^2+S_p}dy$$

$$= \frac{\alpha_0}{k}(\pi)^{1/2}\frac{1}{kuT_2}\left(\frac{1}{1+S_p}\right)^{1/2}e^{-(\Omega-\Omega_0)^2/(ku)^2}$$

$$\equiv \frac{(\alpha_{pk})}{k}\left(\frac{1}{1+S_p}\right)^{1/2}e^{-(\Omega-\Omega_0)^2/(ku)^2} \quad . \qquad (27)$$

Equation (27) shows four physically-interesting aspects. First we can see the Doppler profile term $\exp\{-[(\Omega-\Omega_0)/ku]^2\}$ which arises from the velocity distribution: large detuning from the rest frequency implies a large velocity for the resonantly interacting atoms. The factor $1/kuT_2$ amounts to a comparison of the homogeneous and Doppler widths. Only those particles whose velocity lies near the resonant velocity can contribute effectively. However the wings of a Lorentzian resonance are rather wide so that particles with a velocity near the resonance velocity can contribute a little: the total amounts to a $(\pi)^{1/2}$ increase over the contribution of a single velocity class. Thus $\alpha_{pk} = \alpha_0\sqrt{\pi}/(kuT_2)$, introduced in Eq. (27), is indeed the peak absorption coefficient of the Doppler-broadened but unsaturated absorption line. Finally the factor $1/(1+S_p)^{1/2}$ shows that if the probe beam is sufficiently intense, then probe-induced saturation can deplete the ground state absorbing population somewhat, thus leading to a reduction of the peak absorption coefficient, α_{pk}.

The really interesting case concerns the subDoppler resonances observable when the saturating beam is present, i.e. where $S_s > 0$ in Eq. (25). Then we will have a narrow saturation-induced hole in the population distribution which can lead to differential absorption and/or dispersion of the three frequency components making up the probe beam. The original organization of this beam as a pure frequency-modulated beam will then be upset as the narrow velocity hole is scanned sequentially over the probe beam components by tuning the laser. This physical picture is implicit in our Eq. (25), but it takes a little work with contour integration to make the results apparent. Also we should explicitly note that we intend to substitute for Ω the several frequencies present in the probe beam. Thus in Eq. (25) we really have an implicit sum over Ω^j, and a question could arise about the validity of superposition. As shown by J. Shirley,[16] our present approach is equivalent to the rate equation and so is reasonable for modest intensity of the probe beam. The higher terms recoverable with a full density matrix treatment turn out to be small corrections and correspond physically to reflection of the saturation beam into the probe direction by interaction with a standing-wave grating of modulated ground state population produced by a previous interaction of the probe and saturating beams with the medium. As noted above, the density matrix diagrams discussed in this volume by Bordé are ideal tools for investigating these higher order corrections. Some rather interesting effects occur in these higher orders due to the multiplicity of frequencies in the probe beam. Several time-dependent population grating terms have been identified and confirmed experimentally,[17] but we will not dwell on them here.

To integrate our rate-equation line shapes, Eq. (26), we will as before introduce the dimensionless variables $x = (\Omega-\Omega_0)T_2$, $y = kv_z\, T_2$. We have

$$\chi''(\Omega) = \frac{\alpha_{pk}}{k\pi} \int_{-\infty}^{\infty} \frac{1}{1 + (x-y)^2 + S_p} \times \frac{1 + (x+y)^2}{1 + (x+y)^2 + S_s}\, e^{-y^2/k^2u^2T_2^2}\, dy \tag{28}$$

and

$$\chi'(\Omega) = -\frac{\alpha_{pk}}{k\pi}\int_{-\infty}^{\infty}\frac{(x-y)}{1+(x-y)^2+S_p} \times \frac{1+(x+y)^2}{1+(x+y)^2+S_s}\, e^{-y^2/k^2u^2T_2^2}\, dy \quad .$$

It is useful to rewrite the second term in the integrand of Eq. (28). We have $[1+(x+y)^2]/(1+(x+y)^2+S_s) = 1 - S_s/(1+(x+y)^2+S_s)$. In this form we can easily identify the first term as leading to the linear susceptibility integral previously discussed. The second term clearly represents the depletion notch burned into the absorbers' velocity distribution by the saturating beam. For the present purpose then we will keep only the second term. Let $c = (kuT_2)^{-2}$

$$\chi''(\Omega) \to -\frac{\alpha_{pk}}{k\pi} S_s \int_{-\infty}^{\infty}\frac{1}{1+(x-y)^2+S_p}\cdot\frac{1}{1+(x+y)^2+S_s}\, e^{-cy^2}\, dy$$

$$\chi'(\Omega) \to \frac{\alpha_{pk}}{k\pi} S_s \int_{-\infty}^{\infty}\frac{(x-y)}{1+(x-y)^2+S_p}\cdot\frac{1}{1+(x+y)^2+S_s}\, e^{-cy^2}\, dy \quad . \qquad (29)$$

By noting that the resonance denominators are very sharp compared with the interval for appreciable change of the exponential, we can obtain simple results by contour integration. One finds

$$\chi''(\Omega^j) = -e^{-(\Omega_0-\Omega^j)^2/(ku)^2}\, \frac{\alpha_{pk}}{4k}\left(\frac{\sqrt{1+S_s}+\sqrt{1+S_p}}{\sqrt{1+S_s}\,\sqrt{1+S_p}}\right) \times S_s \times \left[\frac{1}{\left(\frac{\sqrt{1+S_s}+\sqrt{1+S_p}}{2}\right)^2 + (\Omega^j-\Omega)^2\, T_2^2)}\right]$$

$$\chi'(\Omega^j) = e^{-(\Omega_0-\Omega^j)^2/(ku)^2}\, \frac{\alpha_{pk}S_s}{2k\sqrt{1+S_s}}\, \frac{(\Omega^j-\Omega)T_2}{\left\{\frac{\sqrt{1+S_s}+\sqrt{1+S_p}}{2}\right\}^2 + (\Omega^j-\Omega)^2\, T_2^2} \qquad (30)$$

where we have re-introduced the frequency index j = ++, +, o, −, −− for the several applied frequencies. Thus we find Lorentzian dispersion and absorption resonances again from the influence of the saturating beam: it should be borne in mind that Eqs. (29) and (30) give only the nonlinear part of the susceptibility. It seems physically reasonable that the line widths of the resonances are power broadened equally by the probe and saturating beams. It is interesting that the strength of the dispersion χ' is not explicitly reduced by self-saturation. This is reasonable since the dispersion arises mainly from absorbers that are near — but not at — resonance. The absorption strength, on the other hand, does show the influence of self-saturation by the probe beam.

Another important point concerns the frequency axis. In the linear spectroscopy studied in part IV we have a simple direct relationship between sideband offset frequency and resolution of the resonance. However, in the sub-Doppler nonlinear resonances discussed here we need to remember that both the saturating and probe beam are assumed to be tuned, but that the sidebands are only present on the one beam. Thus in the convolution integral, Eq. (29), the width and detuning will each be increased two-fold while the sideband offset frequency will only be reproduced. The conclusion is that the apparent resonance width will be the "true" width, but that the sideband-related features will appear with only 1/2 the "expected" detuning. This minor difference may be taken into account in the definition of the resolution parameter R_i = modulation frequency/(linewidth) by using the halfwidth-at-half maximum linewidth for both linear, R_1, and nonlinear spectroscopy, R_2. However, for nonlinear spectroscopy one can use 1/2 of the modulation frequency in the definition of R_2. Thus if we now regard Figs. 1 and 2 as being nonlinear sub-Doppler resonances, the assumed resolution parameter, R_2, would be 15 (rather than 7.5).

One useful point of contact with experiment is the ratio, P, of the peak dispersion signal to the peak of the absorption signal. Some calculation based on Eq. (30) leads to the result

$$P = \frac{J_0(\beta) - J_2(\beta)}{J_0(\beta) + J_2(\beta)} \times \sqrt{1+S_{probe}} \quad .$$

This relation is found to be in agreement with our experiments. For low modulation depth and small saturation this reduces to the value unity. This is a factor of two increase of the peak dispersion/peak absorption ratio over the value familiar in Kramers-Kronig relations. It arises from a doubling of the central dispersion resonance due to contributions from the two first-order sidebands.

Finally, to close this section it may be of interest to present Fig. 5 which shows experimental profiles obtained with the rf sideband optical heterodyne techniques we have been discussing. The data are the points, while the solid line represents the theoretical functions Eq. (30) appropriately added to give profiles basically equivalent to Eq. (20). The agreement between experimental and theoretical line shapes is seen to be truly remarkable.

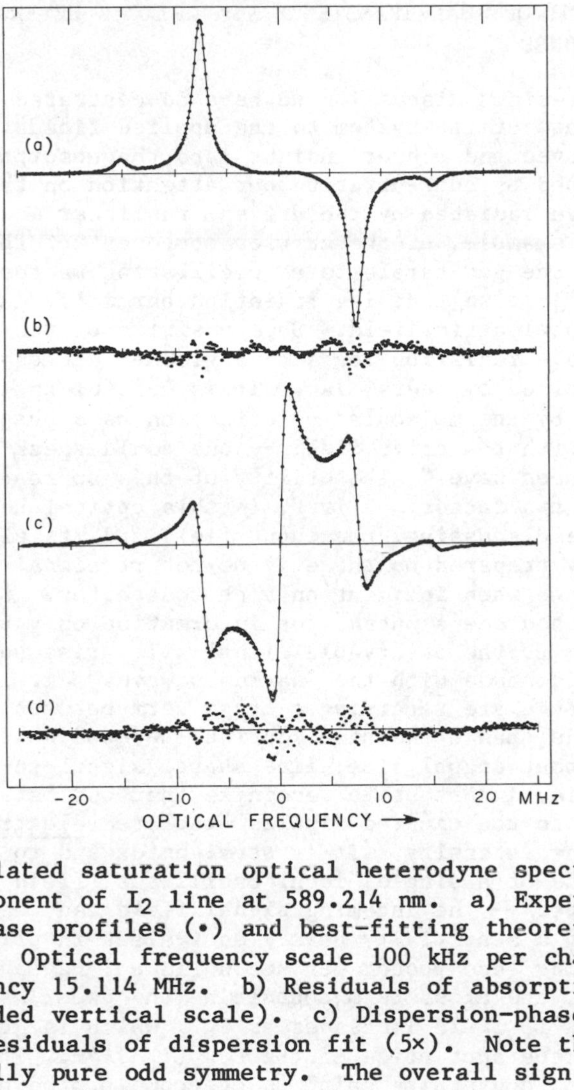

Fig. 5. Modulated saturation optical heterodyne spectrum of central hfs component of I_2 line at 589.214 nm. a) Experimental absorption-phase profiles (•) and best-fitting theoretical shape (solid line). Optical frequency scale 100 kHz per channel, modulation frequency 15.114 MHz. b) Residuals of absorption-phase fit (5× expanded vertical scale). c) Dispersion-phase data and theory. d) Residuals of dispersion fit (5×). Note that residuals have essentially pure odd symmetry. The overall sign of the resonances depends on rf phase-delays and so is arbitrary.

Because the data were taken with small probe intensity, we took $S_p = 0$ in the fitting functions. Differential power broadening associated with the several spectral components of the probe beam has not yet been included in our fitting program. However, we did show that spatial variation of the saturating beam saturation parameter can qualitatively account for the shape of the residuals in Fig. 5.

VIII. COHERENT OPTICAL TRANSIENTS VIA OPTICAL HETERODYNE RF SIDEBAND RESPONSE

In the previous discussion we have concentrated on the equilibrium response of the system to the applied fields. In some ways, an improved and deeper insight into the absorption process can be developed by concentrating our attention on the (possibly transient) wave radiated by the driven, nonlinear medium. The presence, for example, of the carrier component of the probe beam, gives rise in the gas sample to an oscillating macroscopic dipole moment phased just so that its radiation partially cancels the applied wave's electric field. Superposition of the input field with this dipole radiation leads to a gradual decrease in the total, as required by Beers' law. It is helpful to think of this wave radiated by the molecular polarization as a phase-locked wave in antiphase with the drive field — one could speak of it as being a "darkness wave." The utility of this approach stems from the following two factors. First, in this optical heterodyne spectroscopy we're discussing, the input field and its sidebands have been expressly prepared so there is no net rf signal term from the probe beam alone when incident on a photodetector. Since we will filter the photodiode's output for information only at the modulation frequency ω, the observable signal will arise purely from the beat of the sidebands with the "darkness wave." It is quickly appreciated that we are receiving a cross-term between two optical electric fields, hence the name "optical heterodyne spectroscopy." In thinking about signal size, line shape, signal-to-noise ratio, etc., it is also important to recognize that our heterodyne system is responding to the excited dipoles' radiated <u>electric field</u>, rather than the intensity. It is straightforward to show that — in the presence of a single "local oscillator" field for the heterodyne receiver — the incoming signal field can be optimally detected, with a sensitivity nearly equivalent in principle to quantum counting (one photon per second in a 1 Hz bandwidth). Fundamentally, the cross term containing the (weak) signal heterodyned with the LO field increases as E_{LO}, which is just the same dependence as the shot noise of the LO $\propto (i_{LO})^{1/2}$. The advantage of the implied "conversion gain" to overcome photodiode noise has been clearly demonstrated in the infrared heterodyne astronomy experiments of Betz, Townes <u>et al</u>.[18] For dispersion signals the presence of two sidebands rather than the one local oscillator field is of no consequence, as the two rf signals resulting from the optical heterodyne are mutually phase-coherent.

The second utility of the "darkness-wave-radiated-by-the-molecular-polarization" viewpoint becomes evident when we consider the nonequilibrium phenomena. The molecular polarization will not be able to follow very rapid phase or frequency changes of the laser. Thus for a time $\sim\gamma^{-1}$ we will have a "darkness wave" carrying the previous laser phase/frequency information against which we may compare the new laser sidebands' phases. This beat term will again lie at ω and can be synchronously detected as before.[19] We now will do just enough analysis to recover the interesting dependences of the transient signals.

It is convenient to work in a reference frame rotating with the original laser phase and frequency. In this frame we have $\bar{E}_{laser}$ and $\bar{E}_{rad}$, the molecular radiation field, in antiphase. The magnitude of E_{rad} may be obtained via Beers' Law. For the power P_{out} remaining of the original power P_{in}, we have

$$P_{out} = P^{o}_{in} e^{-\alpha L}$$

or

$$E_{out} = E_{in} e^{-\alpha L/2} \quad .$$

By definition of E_{rad}, $E_{out} = E_{in} + E_{rad}$, so that under equilibrium conditions we have

$$E_{rad} = -E_{in}(1-e^{-\alpha L/2}) \simeq -E_{in}\left(\frac{\alpha L}{2}\right) \quad .$$

Now suppose that at time t' the laser phase jumps by an angle θ, carrying with it the newly-produced sideband structure. After the phase jump we suppose the laser frequency returns to its original value. The dipole polarization cannot respond instantly, so for a brief moment we have a phase angle θ between the new laser field and the (decaying) dipolar radiation field. The simplest representation in the laboratory frame is

$$E_{out} = E_{in} \sin(\Omega t+\theta) + E_{rad} \sin\Omega t \quad .$$

Before the phase step, $\bar{E}_{rad}$ had contributions from two distinguishable sorts of molecules: those saturated molecules in resonance with the field contributed an "opacity" $\alpha_{res} = \alpha_0 \, S/(1+S)$ with the rest of the "opacity" being contributed by other parts of the Doppler velocity distribution. After the phase jump the polarization of the latter group will rapidly dephase, roughly within a time given by one inverse Doppler line width. The resonantly-saturated group however, will maintain their mutual coherence a much longer time, approximately the inverse of the homogeneous velocity packet width, $\gamma = 1/T_2$. Thus for a first appreciation of the behavior of the system after the laser phase jump at time t' we may take

$$E_{rad} \sin\Omega t \rightarrow E_{in}\left(\frac{\alpha_0 \ell}{2}\right)\cdot\left(\frac{S}{1+S}\right) \sin\Omega t \; e^{-2\gamma(t-t')} \quad \text{for } t > t' \quad .$$

In this discussion, S represents the saturation parameters produced by the probe frequency component of interest, taken here to be the laser carrier.

Recalling the sideband expansion from Eq. (1), we can now write the total field incident on the photodetector just after the laser phase jump

$$E = E_0\{J_0[\sin(\Omega t+\theta) - (\frac{\alpha L}{2})(\frac{S}{1+S})\sin\Omega t\ e^{-2\gamma(t-t')}]$$

$$+ J_1[\sin((\Omega+\omega)t+\theta) - \sin((\Omega-\omega)t+\theta)]\}$$

where the rapidly decaying linear polarization term has been neglected. In the photocurrent $i = \eta E^2$ we look for the rf term at the modulation frequency ω, obtaining

$$i(\omega) \propto E_0^2 \frac{J_0 J_1}{2} (\frac{\alpha L}{2})(\frac{S}{1+S})e^{-2\gamma(t-t')}[\cos(\omega t-\theta) - \cos(\omega t+\theta)]$$

$$= E_0^2 J_0 J_1 (\frac{\alpha L}{2})(\frac{S}{1+S})\ e^{-2\gamma(t-t')}\ \sin\omega t \sin\theta \quad .$$

Thus for the duration of the dipolar decay time, the molecular system provides an optical phase reference which enables us to obtain a signal directly proportional to the laser's change in optical phase. This optical phase detector phenomena offers some interesting and amusing possibilities. First we note that this signal proportional to optical phase change arises in the same rf phase ($\sin\omega t$) in which we would obtain the dispersion-shaped resonances in the presence of the reverse-running saturating beam. Thus when we are stabilizing the laser to the center of one of these dispersion resonances, the high-speed laser phase variations are directly and additionally read out by the one-beam saturation process just described. The availability of a signal proportional to <u>phase</u> error (rather than <u>frequency</u> error) is welcome since at short times the signal-to-noise of the saturated absorption signals may not be sufficient for high precision locking to the molecular/atomic resonance.[19] Second we can notice that the phase-change signal arises from the FM probe beam alone and so is available for laser phase servo purposes no matter where the laser is tuned within the Doppler profile. Preliminary experiments to investigate this point were made with an F-center laser and HF absorbers and with a dye-jet ring laser and I_2 absorbers. Rather profound decreases of the laser free-running phase noise were possible in both cases, by servo-controlling to this phase error signal.

Some interesting additional effects observable at longer times are presently under investigation. For example, if the laser is deliberately frequency swept about one linewidth in one lifetime, rather large transient signals are expected and observed. If instead a slow sinusoidal frequency modulation is imposed, one finds experimental transient signals which scale proportional to the (frequency of FM × T_2) × (peak-peak frequency excursion × T_2).

This result is readily obtained from a density matrix treatment.

As a third point, we observe that the laser phase sensitivity of the molecular polarization gives rise to a very large increase (>10x) in the noise in the dispersion-phase signal if the laser stabilization is inadequate or is deliberately deactivated. This latter mode is experimentally useful since there is a well-defined dip in this excess phase noise signal when the rf phase shift is set precisely to receive pure absorption phase signals ($\alpha\cos\omega t$). The availablity of an "absolute" phase setting reference is interesting for precise comparison of theoretical and experimental resonance profiles, for example, since a calibrated phase shifter can be used to step from this pure absorption phase to any other. However, we emphasize that the resonance profiles produced in optical heterodyne saturation spectroscopy are purely odd symmetric in frequency for all reference phases, so high accuracy stabilization to line center does not in fact require the corresponding high accuracy phase reference (even though it is available).

In summary we have discussed some of the interesting new possibilities for laser spectroscopy and stabilization using rf sideband techniques. These FM techniques when suitably applied to saturable systems can provide narrow optical resonance profiles of unprecedented symmetry and S/N. Furthermore by working at high modulation frequencies where there is little laser noise we may, in this optical heterodyne saturation spectroscopy method, closely approach the fundamental limit set by the quantum noise. We can foresee many useful applications of the method in precision spectroscopy (stabilized lasers, optical frequency standards, etc.), in trace constituent spectroscopy (such as molecular ions and free radicals), and perhaps in the overlap set (radiatively-cooled trapped ions, longitudinal interaction with supersonic molecular beams).

The authors are grateful to G. C. Bjorklund and to M. Levenson for preprints of their closely-related work. One of us (JLH) is happy to acknowledge and thank R. W. P. Drever for many stimulating conversations and for his collaboration in the laser stabilization experiments. We also thank J. H. Shirley for several useful discussions and for sharing his unpublished results with us. We thank K. Burnett for several interesting discussions about the transient response signals. H.G.R. has been a JILA Visiting Fellow, 1980-81, on leave from Duke University. The work has been supported in part by the National Bureau of Standards under its program of precision measurement research for possible application to basic standards, and in part by the Office of Naval Research and the National Science Foundation through grants to the University of Colorado.

References

1. M. Wahlquist, J. Chem. Phys. 35, 1708 (1961).
2. R. L. Smith, J. Opt. Soc. Am. 61, 1015 (1971).
3. A. J. Wallard, J. Physics E: Sci. Instrum. 5, 926 (1972).
4. J. L. Hall, in Physics of Quantum Electronics Lectures, 1970. Unpublished calculations performed with C. V. Kunasz give the broadening and sensitivity of signal recovery for arbitrary resonance lineshapes.
5. B. Smaller, Phys. Rev. 83, 812 (1951).
6. J. V. Acrivos, J. Chem. Phys. 36, 1097 (1962).
7. R. Karplus, Phys. Rev. 73, 1027 (1948); B. A. Jacobsohn and R. K. Wangness, Phys. Rev. 73, 942 (1948).
8. G. C. Bjorklund, Opt. Lett. 5, 15 (1980); G. C. Bjorklund and M. D. Levenson, Phys. Rev. A 24, 167 (1981); G. C. Bjorklund, K. Jain, J. D. Hope, Appl. Phys. Lett. 38, 747 (1981).
9. R. W. P. Drever and J. Hough, University of Glasgow, private communication, August 1979.
10. J. L. Hall, L. Hollberg, T. Baer and H. G. Robinson, Appl. Phys. Lett. 39, 680 (1981); and in Laser Spectroscopy 5, eds. A. R. W. McKellar, T. Oka, and B. P. Stoicheff (Springer-Heidelberg, 1981), p. 178.
11. J. L. Hall, L. Hollberg, Ma Long-Sheng, T. Baer, H. G. Robinson, J. Physique Special Publication, 3rd Symposium on Frequency Metrology and Standards, Aussois, October 1981.
12. M. Sargent and P. E. Toschek, Appl. Phys. 11, 107 (1976).
13. Useful discussions are given in e.g. Laser Physics by M. Sargent, M. O. Scully and W. E. Lamb, Jr. (Addison-Wesley, Reading, Mass, 1974); and in Quantum Electronics by A. Yariv (Wiley, New York, 1975) Section 8.
14. The physics of these processes is equivalent to four-wave mixing ("wavefront conjugation") but is somewhat complicated by the presence of three frequencies in the probe beam. See Ref. 12 and references therein. See also R. K. Raj, D. Bloch, J. J. Snyder, G. Camy, and M. Ducloy, Phys. Rev. Lett. 44, 1251 (1980).
15. J. H. Shirley has recently given the first few correction terms due to such coherence effects. Unpublished, 1981, private communication.
16. J. H. Shirley, J. Phys. B: Atom. Molec. Phys. 13, 1537 (1980).
17. L. Hollberg and J. H. Shirley, to be published.
18. See, e.g., A. L. Betz, in Laser Spectroscopy V, eds. A. R. W. McKellar, T. Oka and B. P. Stoicheff (Springer-Verlag, Heidelberg, 1981).
19. This idea is the molecular equivalent of the original phase-locking proposal made by Drever, Ref. 9.

HIGH RESOLUTION LASER SPECTROSCOPY*

T. W. Hänsch

Department of Physics
Stanford University
Stanford, California 94305, U.S.A.

ABSTRACT

Methods of Doppler-free laser spectroscopy are reviewed, including several recent advances such as FM sideband or polarization intermodulation techniques. New approaches to laser frequency stabilization are also discussed. Precision studies of the simple hydrogen atom serve to illustrate the power of high resolution laser spectroscopy, and to point out future challenges and opportunities.

1. INTRODUCTION: LASER SPECTROSCOPY WITHOUT DOPPLER BROADENING

Lasers have revolutionized the field of optical high resolution spectroscopy.[1,2] Several types of lasers, notably dye lasers[3] in the visible spectral region, can be tuned continuously over wide wavelength ranges while at the same time providing extreme spectral purity.

To take advantage of the very narrow instrumental linewidth of such laser sources, however, it is necessary to overcome the Doppler-broadening of spectral lines. Atoms or molecules which are relatively free and undisturbed are almost inevitably moving with high thermal velocity. Those atoms moving towards an observer appear to absorb or emit light at higher frequencies than atoms at rest, and atoms moving away appear to absorb or emit at lower frequencies. In a gas, with atoms moving at random in all directions, the lines appear blurred, and important details are often obscured.

In the oldest approach to eliminate Doppler-broadening, a well collimated atomic beam is used to select just a group of atoms that move nearly perpendicularly to the observer's line of sight. Rather spectacular results are now being achieved by laser-excitation of atomic or molecular beams. However, the method has its limitations. For instance, it is difficult to observe rare species or short living excited states in this way.

Doppler-broadening can also be reduced by cooling of a gas sample. Because the velocity spread diminishes only with the square root of the temperature, this approach tends to be much less effective, as long as conventional cooling schemes are employed. On the other hand, the light pressure of near-resonant laser radiation can cool atoms rapidly to very low temperatures,[4] and radiation cooling of trapped ions promises extremely narrow spectral lines.[5] However, while such methods are of interest for the development of optical frequency standards, the many special requirements tend to preclude their use for general spectroscopy.

Fortunately, the high intensity of laser light makes it possible to eliminate Doppler-broadening in gas samples without any need for cooling. Two conceptually different approaches have so far been used very successfully: 1) In saturation spectroscopy, a monochromatic laser beam "labels" a group of atoms with a narrow range of axial velocities through excitation or optical pumping, and a Doppler-free spectrum of these selected atoms is then observed. 2) In two-photon spectroscopy, spectra free of Doppler broadening can be recorded without any need for velocity selection by excitation with two counterpropagating laser beams whose first order Doppler shifts cancel.

Although excellent text books on laser spectroscopy have recently become available,[1,2] the field is still in rapid development, and new approaches and techniques continue to emerge. In the first lecture, we will try to give an overview of the different methods of Doppler-free saturation spectroscopy, including some recent advances. Polarization Intermodulated Excitation (POLINEX) will serve as an example for a particularly sensitive and versatile new method. In the second lecture, we will show how FM sideband and polarization techniques have led to effective new methods for laser frequency stabilization. In the third and final lecture we will review precision studies of the simple hydrogen atom in order to illustrate the power of high resolution laser spectroscopy and to point out future challenges and opportunities.

2. SATURATION SPECTROSCOPY

Saturation spectroscopy started when Macfarlane, et al.[6] and Szoke, et al.[7] demonstrated the Lamb dip caused by gain saturation at the center of the tuning curve of a single mode He-Ne laser. Lee and Skolnik[8] showed, that an absorbing gas inside the laser resonator can produce "inverse Lamb dips," which could be used for spectroscopy and for laser frequency stabilization.

2.1 Saturated Absorption Spectroscopy

Saturation spectroscopy became more widely useful when C. Bordé[9] and, independently, T. W. Hänsch, et al.[10] introduced a modulation method, which gives clean Doppler-free spectra of gas samples external to the laser. As shown in Fig. 1, the light from the laser is divided by a partial mirror into two beams which pass through the sample in nearly opposite directions. The stronger "pump" beam is chopped at an audio frequency. When it is on it party "saturates" the absorption of the gas by exciting the atoms or molecules and so removing them from the absorbing lower level. As a result, it bleaches a path for the probe beam. The probe intensity will hence be modulated at the chopper frequency. However, this modulation occurs only when the laser is tuned to interact with atoms at rest, or at least with zero velocity along the direction of the beams. Any atom moving along the beams sees one wave shifted up in frequency and the other wave shifted down, and so a moving atom cannot be simultaneously in resonance with both beams.

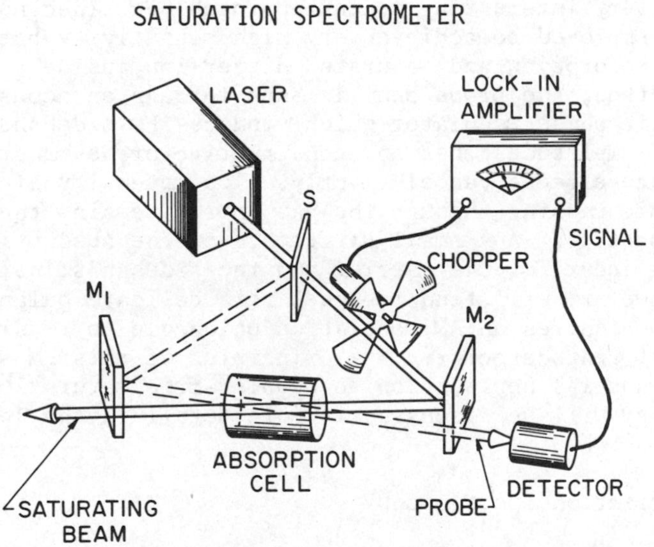

Fig. 1. Apparatus for Doppler-free saturated absorption spectroscopy of a gas sample.

It has been pointed out recently by C. Borde and others that a seemingly different interpretation can be given to this type of probe modulation: The saturating beam acquires closely spaced modulation sidebands, when passing through the chopper. Resonant four-wave mixing in the sample will then produce new frequency components travelling in the same direction as the probe, and by monitoring an intensity modulation of the probe, we are in effect observing these new frequency components in an optical heterodyne detection scheme. The influence of laser amplitude fluctuations can be reduced by proper choice of the heterodyne frequency, and M. Ducloy and collaborators[11,12] have achieved nearly shot noise limited detection sensitivity in this way.

2.2 Saturated Dispersion and FM Sideband Spectroscopy

Many variants of Doppler-free saturation spectroscopy have since been developed. The different techniques are distinguished mainly by the way the nonlinear coupling of the counterpropagating laser beams is detected.

Rather than monitoring saturation-induced changes in absorption, it is also possible to monitor changes in dispersion. Doppler-free saturated dispersion spectra have been recorded with the help of interferometers[13,14] or even via a deflection of the probe beam.[15]

In a very interesting recent approach, FM sideband techniques have been employed to achieve very high sensitivity both to saturated absorption and saturated dispersion inside a gas sample.[17] In this method, the probe beam is sent through an acoustooptic or electrooptic phase modulator which produces FM sidebands of such phases and amplitudes that any constructive or destructive interference effects cancel exactly. The intensity of the probe beam, before passing through the gas smple, remains therefore exactly constant. Any small difference in the absorption or refractive index for the carrier and the sidebands inside the sample, however, will tend to upset this delicate balance, and the probe beam acquires an AM modulation which can be readily detected with a fast photodetector. The principles of this FM sideband technique and its application to Doppler-free saturation spectroscopy will be discussed in more detail in the lectures of J. Hall.

2.3 Polarization Spectroscopy

High detection sensitivities are also achieved by monitoring the laser induced birefringence and dicroism in the absorbing gas. This method of polarization spectroscopy[18] is illustrated in Fig. 2. The probe beam "sees" the sample placed between nearly crossed linear polarizers so that only very little light arrives at the

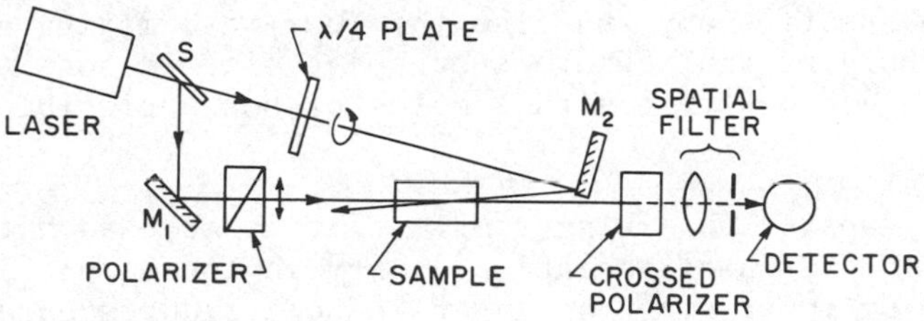

Fig. 2. Setup for Dopple-free polarization spectroscopy

photodetector. The saturating beam is made circularly polarized with a birefringent plate. Alternatively, a linearly polarized beam is used with its polarization axis rotated by 45°.

Normally, in a gas, the atoms or molecules have their rotation axes distributed at random in all directions. But the probability for absorbing polarized light depends on the molecular orientation. Thus, the saturating beam depletes preferentially molecules with a particular orientation, leaving the remaining ones polarized. As a result the sample becomes birefringent and dicroic and can change the polarization of the probe. The probe beam acquires thus a component that can pass through the crossed polarizer, but again this happens only at the line center, where both beams are interacting with the same molecules. The highest sensitivity is actually obtained with the analyzing polarizer slightly uncrossed, so that some probe light always passes into the detector. The new polarization components can then be observed through their interference with this "carrier," i.e. by degenerate optical heterodyne detection.

Aside from its higher sensitivity, polarization spectroscopy offers another interesting advantage. Large Doppler-broadened pedestals are often observed in the older probe modulation technique. These pedestals are caused by velocity changing elastic collisions, which redistribute the atoms over the Maxwellian velocity distribution, thus defeating the velocity selection by spectral hole burning.[19] In polarization spectroscopy, the signal is due to oriented atoms, and the pedestals are absent whenever there is a high probability for disorientation in an elastic collision.

2.4 Intermodulated Fluorescence Spectroscopy and Optogalvanic Spectroscopy

All techniques of saturation spectroscopy discussed so far monitor directly some change in the probe beam. Although polarization spectroscopy and also the four-wave mixing and FM sideband techniques can achieve shot noise limited sensitivity, all these techniques work best with samples of non-negligible absorption.

For very weakly absorbing samples, it is more advantageous to detect the deposition of radiation energy in the sample indirectly, for instance by monitoring the laser-induced fluorescence. Doppler-free spectra can be recorded with good selectivity by the method of intermodulated fluorescence spectroscopy first proposed by M. S. Sorem and A. L. Schawlow.[21] In this technique, illustrated in Fig. 3, the two counterpropagating beams are chopped at two different frequencies f_1 and f_2, and the signal is detected as a modulation of the total excitation rate at the sum- or difference frequency $f_1 \pm f_2$. Such an intermodulation occurs again only when the two beams are interacting with the same atoms so that they can saturate each other's absorption.

As the number of absorbing atoms becomes smaller, both signal and background are reduced proportionately, and a respectable signal-to-noise ratio can be maintained down to very low concentrations. It should even be feasible to apply the technique to a single ion in a trap.

The same intermodulation method can also be used with other indirect detection schemes, and both optogalvanic detection and optoacoustic detection have been demonstrated.[22-25]

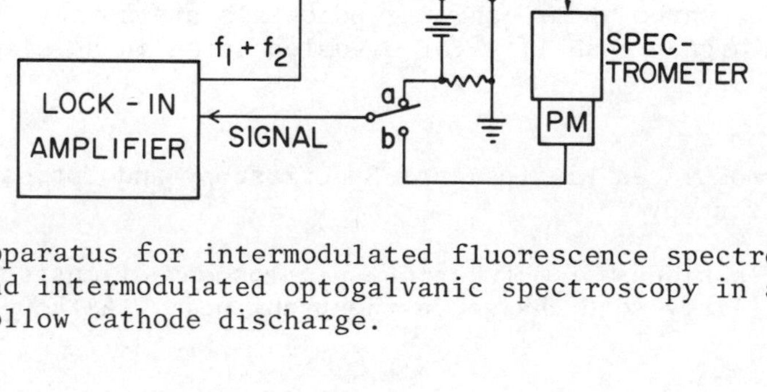

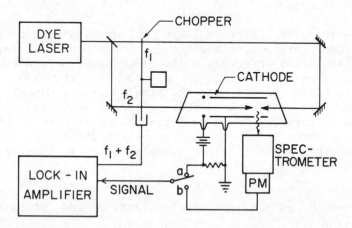

Fig. 3. Apparatus for intermodulated fluorescence spectroscopy and intermodulated optogalvanic spectroscopy in a hollow cathode discharge.

Optogalvanic spectroscopy is particularly convenient for studies of excited atoms inside gas discharges. The absorption of resonant laser light changes the populations of levels with different ionization probabilities, and the signal is observed simply as a resulting change in the discharge current, without any need for a photodetector. Although it is difficult to predict the signal magnitude quantitatively, very high sensitivities have been observed experimentally.[26] It is even possible to observe optogalvanic signals from a radiofrequency discharge, by monitoring changes in the impedance of the exciting coil.[27-28]

Intermodulated optogalvanic spectroscopy and intermodulated fluorescence spectroscopy have recently been used in our laboratory to record Doppler-free saturation spectra of metastable molybdenum and copper atoms sputtered inside a special hollow cathode lamp designed by J. E. Lawler.[29-30] The top part of Fig. 4 shows an

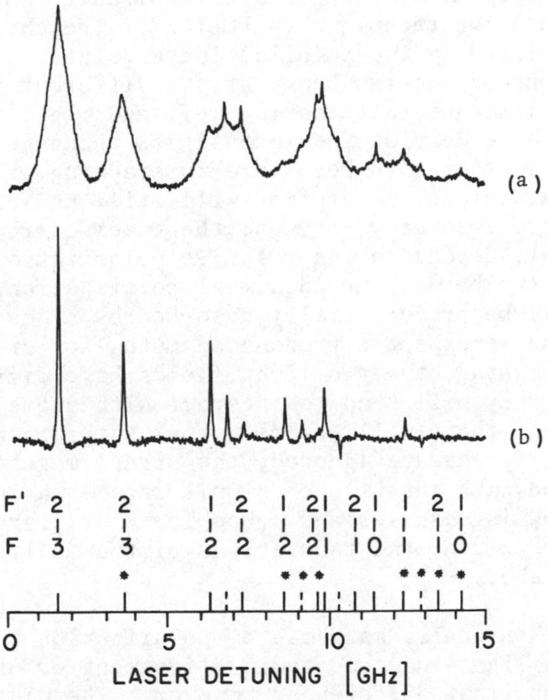

Fig. 4. Hyperfine spectrum of the 578.2 nm transition of Cu.
(a) Intermodulated fluorescence spectrum
(b) POLINEX spectrum.

intermodulated fluorescence spectrum of the Cu ($3d^{10}$ $4p$ $^2P_{1/2}$ - $3d^9$ $4s^2$ $^2D_{3/2}$) transition at 578.2 nm (the yellow Cu laser line), recorded in this way. Despite the relatively low pressure, the narrow Doppler-free resonances appear on top of very pronounced pedestals, which are ascribed to velocity changing elastic collisions.

Spurious signals similar to these collision induced pedestals can sometimes be produced by imperfections in the chopper or by nonlinear frequency mixing in the detector or amplifiers, and considerable experimental care is necessary to avoid such artifacts.

2.5 Polarization Intermodulated Excitation (POLINEX) Spectrocopy

In order to overcome some of these drawbacks and limitations of the older intermodulation methods, we have recently developed a sensitive and versatile new technique of saturation spectroscopy, which takes advantage of Polarization Intermodulated Excitation (POLINEX).[31]

Only seemingly minor changes are required to convert from the older intermodulation technique to POLINEX: The chopper is simply removed and replaced by two polarization modulators, which modulate the polarizations of the two beams at two different frequencies, producing, for instance, alternating left and right hand circular polarization, while leaving the intensities unchanged. When the laser is tuned so that both beams are interacting with the same atoms, the total rate of absorption will still be modulated at the sum or difference frequency, because the combined rate of excitation will, in general, depend on the relative polarization of the two beams. If the two beams have identical polarization, both lightfields will be preferentially absorbed by atoms of the same orientation, and we expect a pronounced mutual or cross saturation. If, on the other hand, the two light fields have different polarizations, they will tend to interact with atoms of different orientations, and there will be less cross saturation. If multiple optical pumping cycles are ignored, the signal magnitude can be easily predicted with the help of simple rate equations, valid in the limit of low intensities and a Doppler width large compared to the natural line width, and tabulations of the analytical results have been published.[31-32]

The POLINEX signals, as those of polarization spectroscopy, are entirely due to light-induced atomic alignement or orientation. There is one important difference, however. The older polarization technique is sensitive to both the dicroism and the birefringence of the sample, and the combination of these two effects generally produces asymmetric line shapes. The POLINEX lines, on the other hand, remain symmetric, since the signal is determined only by the total light absorption, i.e. by the imaginary part of the nonlinear susceptibility tensor.

At the same time, the POLINEX method maintains the advantages of indirect detection, offered by the older intermodulation techniques. But again, we find a rather important difference: In POLINEX spectroscopy, neither beam alone is capable of producing a modulated signal, because the total rate of (steady state) absorption in an isotropic medium does not depend on the sign or direction of the light polarization, even at high intensities. Consequently, any modulation of the excitation rate immediately gives the desired nonlinear signal, and nonlinear mixing in detector or amplifiers will no longer produce spurious signals. In fact, good selectivity for the Doppler-free signals can be maintained even if one of the polarization modulators is removed, so that $f_2 = 0$.

Such a simple setup, as illustrated in Fig. 5, was used for POLINEX spectroscopy of the 578 nm Cu line in a Lawler-type hollow cathode lamp.[32] An electrooptic modulator was inserted into one of the beams to produce alternating left and right hand circular polarization with a modulation frequency of about 800 Hz, while the circular polarization of the other beam remained fixed. The signal was observed as a modulation of the discharge current, or, alternatively, as a modulation of the fluorescence light emitted by the upper level. Figure 4b shows a POLINEX spectrum recorded via fluorescence detection. The Doppler broadened background is almost

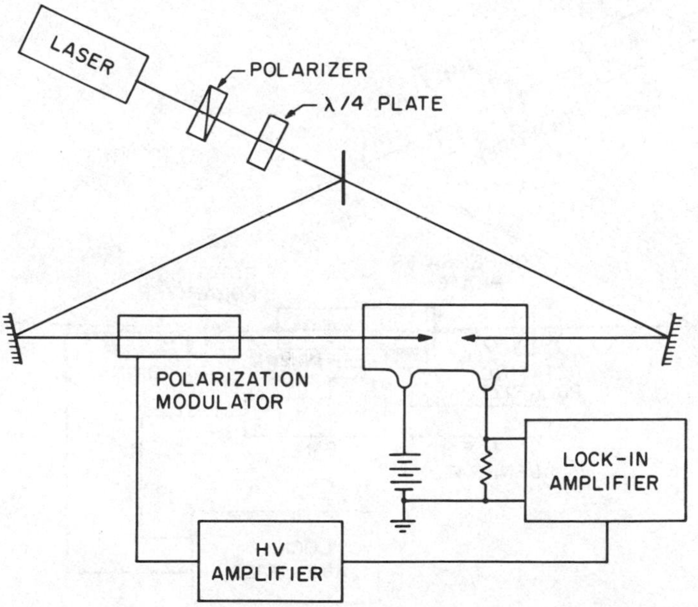

Fig. 5. Apparatus for Doppler-free POLINEX spectroscopy with a single electrooptic polarization modulator.

completely suppressed, indicating that the metastable Cu atoms are readily depolarized in elastic collisions.

A somewhat different setup was used for POLINEX studies of several of the yellow 1s - 2p transitions of neon.[30] As shown in Fig. 6, the two laser beams were first made circularly polarized and then sent through two mechanically rotating Polaroid filters, which produce linearly polarized light with the polarization axis rotating around the direction of light propagation. The signal was observed as a modulation in the fluorescent sidelight. If the two filters are rotating in the same sense, the Doppler-free signal is modulated at the difference of the two modulation frequencies, and if they are spinning in opposite directions, the signal appears at the sum frequency. The proper reference signal for phase sensitive detection can be obtained by simply monitoring the intensity of a laser beam that has passed through both polarizers.

Doppler-broadened pedestals were again clearly suppressed in the POLINEX spectra of neon. However, some residual Doppler-broadening due to velocity changing elastic collisions will always remain even in a POLINEX spectrum, because not all elastic collisions will destroy the atomic orientation. This broadening could be substantial for certain collision partners, such as molecules in high angular momentum states, which behave like rapidly spinning tops, whose angular momentum cannot be easily disoriented.

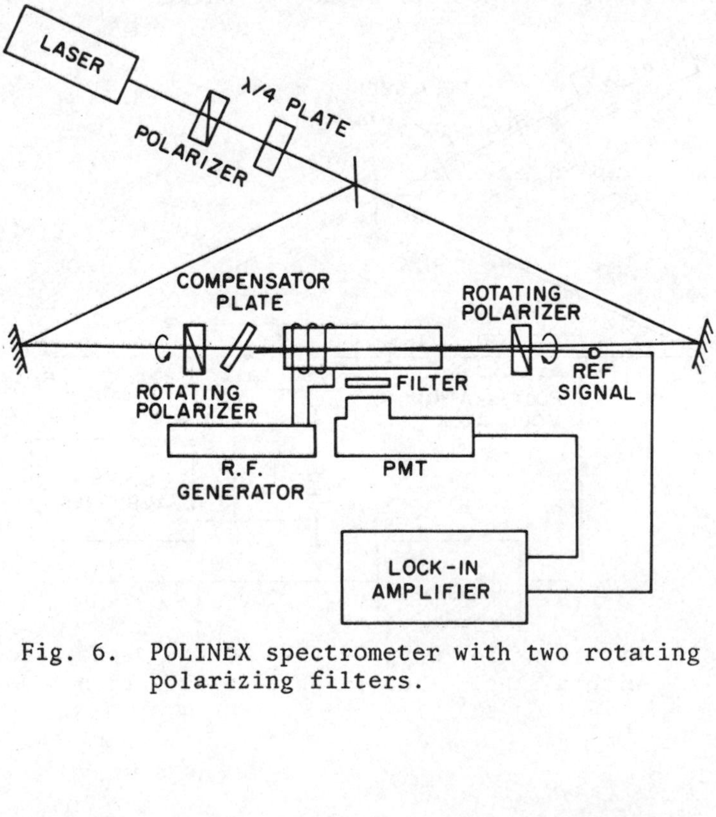

Fig. 6. POLINEX spectrometer with two rotating polarizing filters.

3. ADVANCES IN THE STABILIZATION OF LASER FREQUENCIES

Some of the same tricks which are permitting very sensitive Doppler-free saturation spectroscopy have recently been exploited for improved wavelength or frequency stabilization of tunable lasers.

The cavity of a laser is subject to various external and internal perturbations, and the frequency stability of a laser can generally be improved by actively locking its frequency to some passive reference cavity. The intensity transmitted by such a cavity as a function of the incoming laser frequency exhibits the familiar comb of equidistant maxima or fringes, as described by the Airy function. In the standard approach, employed in current commercial cw dye lasers, the transmitted intensity is monitored together with the incoming intensity, and a fast differencing method is used to lock the laser frequency to the side of one of the fringes.[33] Unfortunately, it is difficult to use this scheme with long reference cavities of high finesse, because any accidental frequency jump of as little as one fringe width can confuse the electronics and throw the system out of lock. Moreover, it is undesirable to operate at the side of a fringe, away from exact resonance, because any sudden change in laser frequency or phase produces a transient "ringing" response, which makes fast corrections difficult.[34]

In 1946 Pound has proposed two different methods for frequency stabilization of a microwave oscillator, which produce much more desirable almost truly dispersion shaped error signals by reflecting a wave off a reference cavity and monitoring a component with a phase shift of 90°.[35]

One of these methods relies on FM sidebands of the incoming wave. Away from exact resonance, the central "carrier" suffers a phase shift relative to the sidebands, and the FM modulation is translated into an AM modulation of the reflected wave. The sensitive techniques of FM spectroscopy mentioned earlier are, of course, closely related to this approach. An optical analog of this type of microwave stabilizer has recently been developed by J. L. Hall and R. W. P. Drever,[17] and details are given in the lectures by J. Hall.

The second method does not require any modulation, and an optical analog of such a DC Pound stabilizer has been demonstrated in our laboratory.[36] The setup is illustrated in Fig. 7. Linearly polarized laser light is reflected by a confocal reference cavity with an internal Brewster plate or linear polarizer. The incoming light can be decomposed into two orthogonal linearly polarized components, with the field vector parallel and perpendicular to the transmission axis of the intracavity polarizer. The parallel

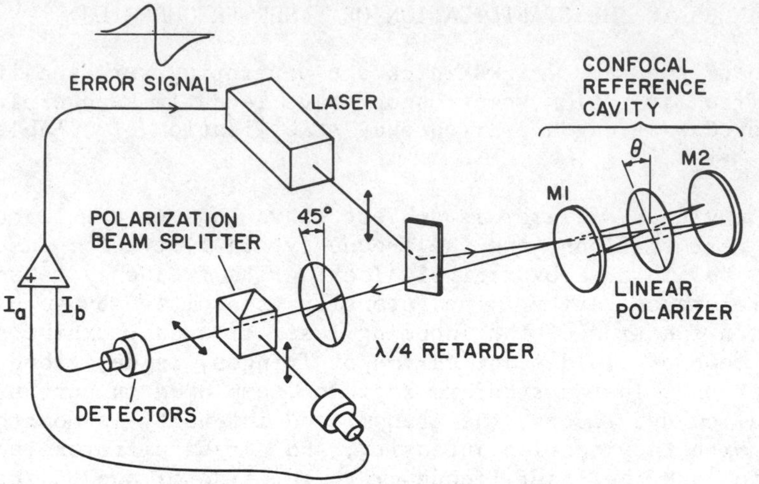

Fig. 7. Optical analog of the DC Pound frequency stabilizer.

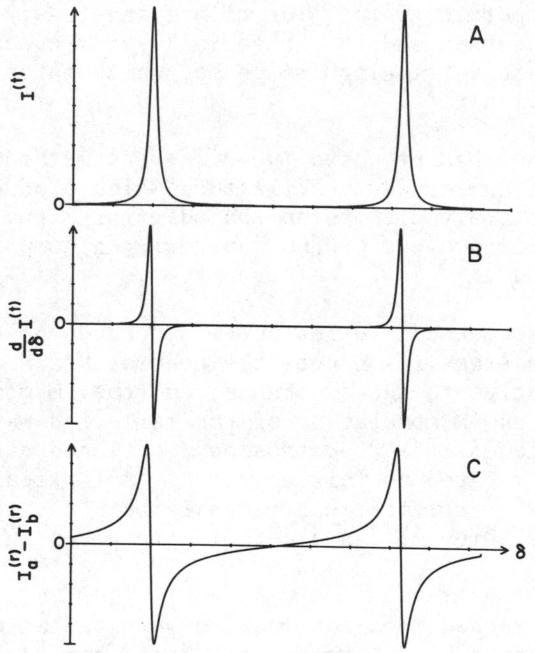

Fig. 8. A: Spectrum of the intensity transmitted by a passive cavity. B: First derivative of the transmission spectrum spectrum. C: Dispersive resonances obtained from the polarization of the reflected radiation.

component sees a cavity of low loss and experiences a frequency dependent phase shift in reflection. The perpendicular component is, to first approximation, simply reflected by the entrance mirror M_1 and serves as a reference. At exact resonance, both reflected components are in phase, and the reflected beam remains linearly polarized. Away from resonance, however, the parallel component acquires a phase shift relative to the perpendicular component, and the reflected beam becomes elliptically polarized. This ellipticity can be detected with a simple polarization analyzer, consisting of a $\lambda/4$ retarder and a polarization beamsplitter. The elliptically polarized light can be considered a superposition of left and right hand circularly polarized waves. These components are separated by the analyzer, so that their intensities can be measured by two separate photodetectors. A differential amplifier yields a dispersion shaped error signal with far reaching wings, as illustrated in Fig. 8. Such a signal is ideally suited for frequency locking to exact resonance, without the drawbacks of the conventional approach.

The same setup should, in fact, make it possible to phase lock the laser to the radiation field stored inside the passive cavity, because any sudden jump in the phase of the incident laser beam will immediately alter the polarization of the reflected light. A similar phase locking scheme, using the FM method, has been demonstrated by Hall and collaborators, and extremely narrow laser linewidths have indeed been achieved in this way.

Our polarization method has also proven very useful for the active frequency locking of a passive enhancement cavity to a tunable laser. Intensities many times larger than the laser output can be maintained inside such a cavity, permitting efficient second harmonic generation, nonlinear frequency mixing, or nonlinear spectroscopy.

4. PRECISION LASER SPECTROSCOPY OF HYDROGEN

This final lecture will illustrate the power of Doppler-free laser spectroscopy by reviewing a series precision studies of hydrogen, the simplest of the stable atoms. Spectroscopy of atomic hydrogen has played a crucial role in the development of atomic physics and quantum mechanics.[37] More than once, seemingly minute discrepancies between experiment and theory have led to major revolutions in our understanding of quantum physics, from Bohr and Sommerfeld to Schrödinger, Dirac, and modern QED. The advent of laser techniques has created exciting new opportunities for spectroscopic precision measurements, and future work may well lead once more to some surprising fundamental discoveries.

4.1 Laser Spectroscopy of Hydrogen Balmer Lines

In the laboratory, the visible Balmer lines of hydrogen can be observed most readily in the light emitted by an electric glow discharge. The regular Balmer spectrum has been studied intensively for more than 100 years. However, no classical spectroscopic observation has ever succeeded in fully resolving the intricate fine structure of these lines. The spectra remained always blurred by Doppler-broadening due to the rapid thermal motion of the light hydrogen atoms. Dramatic progress in spectral resolution became possible only with the advent of highly monochromatic tunable dye lasers together with the techniques of Doppler-free laser spectroscopy.

The prominent red Balmer-α line was the first hydrogen line to be studied by Doppler-free saturated absorption spectroscopy. Even though only a relatively simple pulsed dye laser was available in our initial experiments at Stanford, we were thrilled by the spectra (Fig. 9) that could be recorded in a simple Wood-type gas discharge. For the first time, we could resolve single fine structure components, and we could observe the n = 2 Lamb shift directly in the optical spectrum.

It was immediately clear that the new spectra would permit a more precise absolute wavelength measurement than any of the earlier experiments. And a comparison with theoretical computations could provide a more accurate value of the Rydberg constant, which describes the binding energy between electron and nucleus and provides one of the cornerstones in the determination of other fundamental constants. In 1974, M. H. Nayfeh, et al.[39] completed a first absolute wavelength measurement of the strong $P_{3/2}$ - $D_{5/2}$ component of hydrogen and deuterium which yielded an eightfold improved value of the Rydberg constant.

In 1978, J. E. M. Goldsmith, et al. at Stanford[41] undertook a new measurement of the Rydberg constant, observing the hydrogen Balmer-α line in a mild helium-hydrogen discharge by polarization spectroscopy with a cw dye laser. Because of its higher sensitivity, the method permitted measurements at lower atom densities and lower laser intensities, so that pressure broadening, power broadening, and related problems were much reduced. Examples of the new spectra are shown in Fig. 10. The resolution of the weak but narrow $S_{1/2}$ - $P_{1/2}$ fine structure component was at least 5 times better than in the earlier pulsed experiments. The absolute wavelength of this component was measured relative to the i-th hyperfine component of the $^{127}I_2$ B-X R(127) 11-5 transition of molecular iodine at 632.8 nm, using a near coincident I_2 line (the i-th hyperfine component of the $^{127}I_2$ B-X R(73) 5-5 transition) as an intermediate reference.[42]

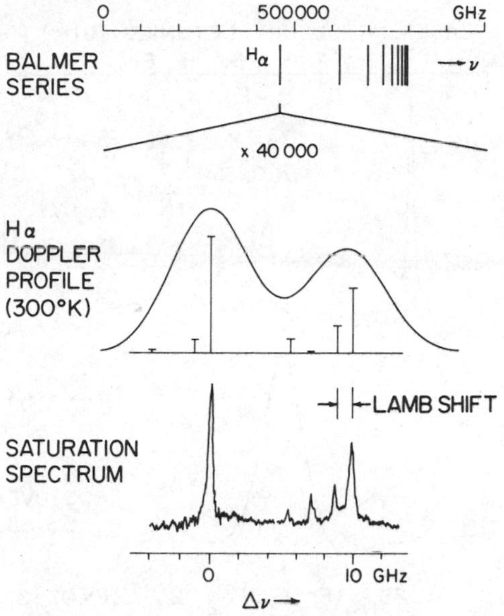

Fig. 9. Top: Balmer spectrum of atomic hydrogen. Center: Doppler-profile of the red Balmer-α line at room temperature and theoretical fine structure components. Bottom: Doppler-free spectrum of Balmer-α, recorded by saturated absorption spectroscopy with a pulsed dye laser.

Systematic line shifts due to the discharge plasma were studied very carefully in a series of measurements which yielded several results interesting in their own right.[43,44] For instance, anomaleous pressure shifts have been observed which could be explained in terms of collisional decoupling of the 3P hyperfine structure.[43] The final evaluation of all measurements gave a threefold improvement in the accuracy of the Rydberg value. The results, as shown in Fig. 11, are in good agreement with the 1974 measurement. The values given in Table 1 have been slightly adjusted to be consistent with the rounded value of the iodine reference wavelength, 632.991 339 nm, as recommended by the Committee for the Definition of the Meter.[45]

Table 1 and Fig. 11 also give the results of an independent measurement of the Rydberg constant, reported in 1980 by B. W. Petley et al.[46] The Balmer-α line of hydrogen was observed in a Wood type gas discharge by saturated absorption spectroscopy with a cw dye

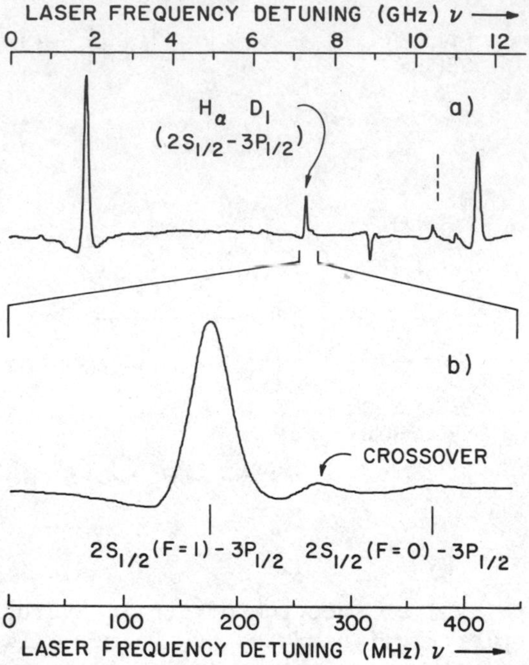

Fig. 10. Polarization spectrum of the Balmer-α line.

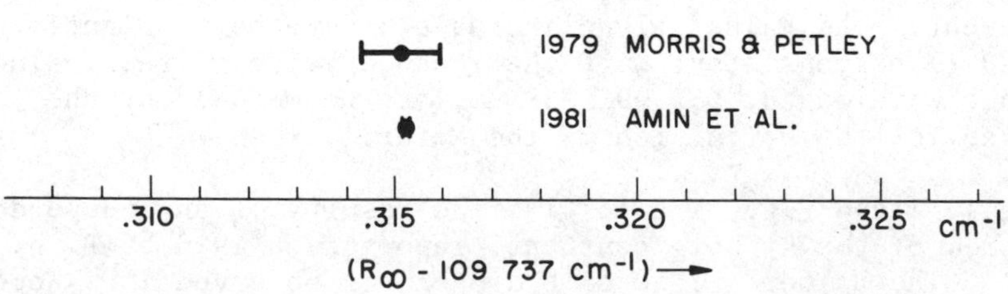

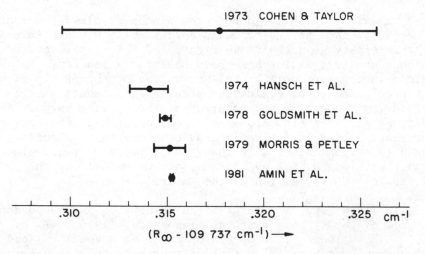

Fig. 11. Recent measurements of the Rydberg constant.

Table 1. Measurements of the Rydberg Constant

		R_∞ [cm^{-1}]
73	COHEN, TAYLOR (6)	109 737.317 70 ± 0.008 30
74	HANSCH NAYFEH, LEE, CURRY, SHAHIN (50	109 737.314 10 ± 0.001 00
78	GOLDSMITH, WEBER, HANSCH (10)	109 737.314 90 ± 0.000 32
79	PETLEY, MORRIS, SHAWYER (15)	109 737.315 13 ± 0.000 85
81	AMIN, CALDWELL, LICHTEN (16)	109 737.315 21 ± 0.000 11

laser. The result is in good agreement with the earlier Stanford values.

4.2 Laser Spectroscopy of a Metastable Hydrogen Beam

The accuracy of the best measurements in hydrogen discharges[41] does not appear to be limited by pressure shifts or Stark effect in the discharge plasma. Nonetheless, it has long been obvious that a collisionless beam of metastable hydrogen atoms would be a more ideal sample for precision laser spectroscopy.[39]

S. R. Amin et al. of Yale University are the first who have succeeded with such an atomic beam experiment, and they have reported very recently on a new Rydberg measurement.[47] The Balmer-α line of hydrogen and deuterium is observed by exciting the metastable 2S (F = 1) atoms with cw dye laser beams which cross the atomic beam at a right angle. Most of the excited 3P atoms quickly decay into the 1S ground state, and the resulting quenching of the 2S state can be observed with a detector for metastable atoms. Such linear atomic beam spectroscopy requires fewer systematic corrections than nonlinear saturation spectroscopy, and a wavelength measurement by direct comparison with an iodine stabilized He-Ne laser has yielded a new Rydberg value accurate to one part in 10^9, as shown in Table 1 and Fig. 11.

Although the accuracy of linear laser spectroscopy of a metastable hydrogen beam appears amenable to further improvements, the resolution of the narrowest Balmer-α components will always be limited by the short lifetime of the upper 3P state to no better than 29 MHz.

4.3 Radiofrequency-Optical Double-Quantum Spectroscopy

D. E. Roberts and E. N. Fortson[48] were the first to point out that narrower lines can be obtained if an additional radiofrequency field is applied so that radiofrequency optical double quantum transitions are induced from the $2S_{1/2}$ level to the longer living $3S_{1/2}$ and $3D_{1/2}$ level. E. W. Weber and J. E. M. Goldsmith[49] have observed lines as narrow as 20 MHz by applying this technique to hydrogen atoms in a gas discharge, and they have been able to measure the small $3P_{3/2}$ - $3D_{3/2}$ Lamb shift directly by comparing single- and double-quantum signals. C. E. Wieman and collaborators[50] have recently begun to apply the same technique to a beam of metastable hydrogen atoms, and they expect to reach a resolution better than 1 MHz, corresponding to the natural width of the 3S level.

4.4 Doppler-Free Two-Photon Spectroscopy

The same narrow lines could also be observed by excitation with two laser photons of equal frequency. If the two photons come from opposite directions (Fig. 12), their first order Doppler shifts cancel.[52] Although this method of Doppler-free two-photon excitation has become a widely used technique of high resolution laser spectroscopy, its application to the Balmer-α transition has so far been stifled by the lack of a suitable highly monochromatic tunable lasers in the near infrared. But visible dye lasers should make it possible to study transitions from 2S to high Rydberg levels by this technique.

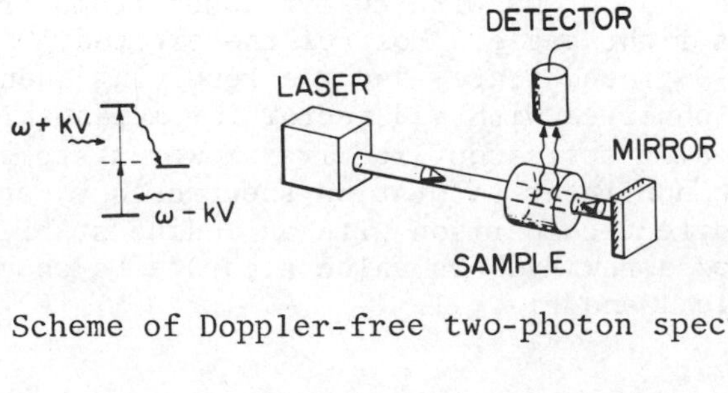

Fig. 12. Scheme of Doppler-free two-photon spectroscopy.

There is another, even more intriguing transition in hydrogen which can be studied by Doppler-free two-photon spectroscopy: The transition from the 1S ground state to the metastable 2S state. The 1/7 sec lifetime of the upper level implies an ultimate natural linewidth as narrow as 1Hz or a spectral resolution better than 1 part in 10^{15}.

There is no intermediate near resonant level which would enhance the two-photon transition rate. However, even small numbers of excited 2S hydrogen atoms can be detected with high sensitivity, by monitoring the vacuum ultraviolet Lyman-α radiation emitted after conversion to the 2P state by collisions or external fields, or by photoionizing the 2S atoms and observing charged particles.

Unfortunately, however, two-photon excitation of 1S-2S requires monochromatic ultraviolet radiation near 243 nm, where there are still no good tunable laser sources available. Intense coherent radiation at this wavelength can be generated by frequency doubling of a pulsed dye laser in a nonlinear optical crystal, and 1S-2S two-photon spectra have been observed at Stanford with such sources.[53-55] But the resolution remained limited by bandwidth of the pulsed lasers.

The best 1S-2S spectra so far have been recorded by C. E. Wieman,[55] who reached a resolution of 120 MHz (fwhm at 243 nm) with the help of a blue single mode cw dye laser oscillator with nitrogen-pumped pulsed dye laser amplifier chain and lithium niobate frequency doubler. The hydrogen atoms were generated in a Wood-type discharge tube and carried by gas flow and diffusion into the

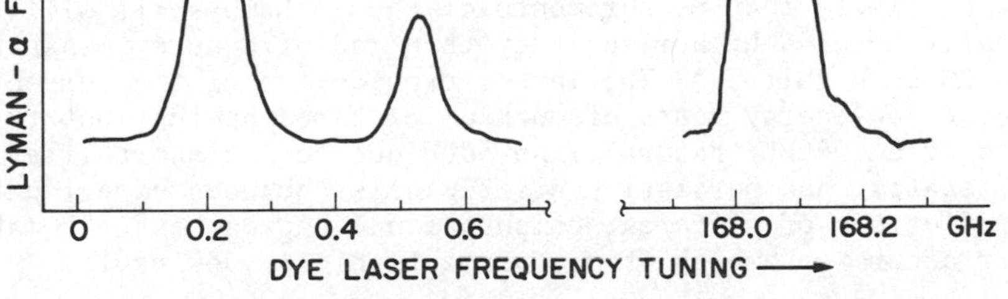

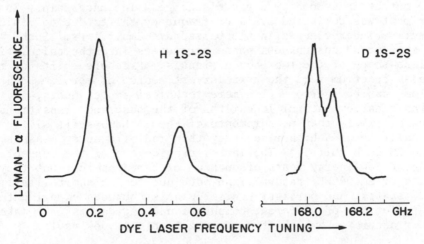

Fig. 13. Doppler-free two-photon spectrum of the 1S-2S transition in hydrogen and deuterium with resolved hyperfine splittings.

observation chamber, where they were excited by two counter-propagating beams from the laser system. The emitted vacuum ultraviolet Lyman-α photons were observed through a magnesium fluoride side window by a solar blind photomultiplier. Although the spectral resolution remained short of the envisioned ultimate limit by a factor of 100 million, it was sufficient to resolve the hyperfine doublets (Fig. 13). And even these crude spectra permitted a measurement of the 611 GHz H-D isotope shift to within 6.3 MHz and provided a first qualitative confirmation of the predicted small 11.9 MHz relativistic correction due to nuclear recoil.

In the same series of experiments.[53-55] the 1S-2S energy interval was compared with the n = 2-4 interval, by simultaneously observing the Balmer-β line with the visible dye laser output. If the simple Bohr theory were correct, the n = 1-2 interval would be exactly four times the n = 2-4 interval, and both transitions would be observed at exactly the same laser frequency. In reality, this degeneracy is lifted by relativistic and quantum electrodynamic corrections, and we expect line splittings and displacements as illustrated in Fig. 14. By measuring the separation of the 1S-2S resonance from one of the Balmer components, one can determine an experimental value of the Lamb shift of the 1S ground state, which cannot be measured by radiofrequency techniques, because there is no 1P state which could serve as a reference.

In the most accurate of the past Stanford experiment[55] the predicted 8149.43 + 0.08 MHz Lamb shift of the hydrogen 1S state has been confirmed within 0.4%, by comparing the 1S-2S spectrum with a polarization spectrum of the Balmer-β line, observed with the cw dye laser output in a Wood-type gas discharge. The uncertainty in this experiment was dominated by laser frequency shifts due to rapid refractive index changes in the pulsed dye amplifiers. Such chirping introduces unknown phase parameters into the calculation of the line-shape of the two-photon signals. Pressure shifts of the Balmer-β spectrum were the next largest source of error. Both problems can be overcome with mere technical improvements, and the intrinsic narrow natural linewidths of the observed transitions make it appear likely that measurements of the 1S Lamb shift will eventually reach a higher accuracy than radiofrequency measurements of the 2S Lamb shift.[56] The latter provide one of the current most stringent low-energy tests of quantum electrodynamics, but they are plagued by a 100 MHz natural linewidth due to the short lifetime of the 2P state. And persisting discrepancies between experiment and the predictions of different computational approaches[57,58] make further accurate Lamb shift measurements highly desirable.

In order to avoid the limitations of a pulsed laser source, A. I. Ferguson, J. E. M. Goldsmith, B. Couillaud, A. Siegel, J. E. Lawlwer, and other collaborators at Stanford have invested

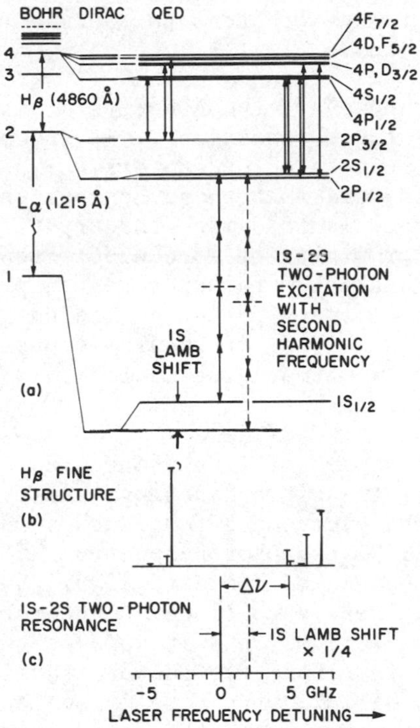

Fig. 14. Top: Simplified diagram of hydrogen energy levels and transitions. The Dirac fine structure and QED corrections are shown on an enlarged scale, hyperfine structure has been ignored. Bottom: Fine structure spectrum of the Balmer-β line and relative position of the 1S-2S two-photon resonance, as recorded with the second harmonic frequency. The dashed line gives the hypothetical position of the 1S-2S resonance if there were no 1S Lamb shift.

considerable efforts into an experiment designed to observe the hydrogen 1S-2S two-photon transition with low power continuous wave ultraviolet radiation. While known nonlinear optical crystals do not permit efficient 90° phase matched second harmonic generation down to 243 nm, cw ultraviolet radiation at this wavelength can be produced as the sum frequency of a blue krypton ion laser and a yellow rhodamine 6G dye laser in a crystal of ammonium dihydrogen phosphate (ADP), cooled close to liquid nitrogen temperature.[59] With frequency stabilized single frequency lasers of 0.6 and 2.5 watts power, respectively, focused to a waist diameter of 100 μm inside a 5 cm long ADP crystal, about 700 μW of tunable cw ultraviolet power have been produced. A frequency locked external passive enhancement cavity increases this power to several mW at the sample. Such a power should be sufficient for a resonant signal of of several hundred Lyman-α photons per second under the chosen experimental conditions.

Unfortunately, however, the ADP crystal is damaged within less than a minute under these conditions. We speculate that the ultraviolet light produces color centers, perhaps associated with heavy ion impurities, and that these centers then absorb enough of the intense primary beams to damage the crystal. We are presently investigating whether mixing of more widely separated primary wavelengths in ADP near room temperature can provide a cure for this problem, and we are actively pursuing alternative approaches. Once the damage problem is solved, it should be possible reduce the bandwidth of such a cw ultraviolet source to a few kHz with the help of fast servo controls.

However, other causes of line broadening, in particular pressure broadening, transit broadening, and transverse Doppler-broadening have to be overcome before such a resolution can be approached in the 1S-2S two-photon spectrum.[59] A beam of ground state hydrogen atoms, cooled close to liquid helium temperature, and interacting with nearly collinear counterpropagating laser beams could minimize such problems, and it appears technically quite feasible to observe the 1S-2S two-photon transition with a line width of a few tens of kHz, or a resolution approaching one part in 10^{11}. The line center could then be determined to within 1 part in 10^{13} or better, once accurate frequency standards become available in the visible and ultraviolet region.

Such a precise measurement of the 1S-2S frequency could, of course, be used to determine a still better value of the Rydberg constant. However, the current uncertainty of the electron/proton mass ratio[61] (about 0.14 ppm) limits the accuracy of such a Rydberg value to about 1 part in 10^{10}. Considerable improvements of direct measurements of the electron/proton mass ratio have been predicted.[61] Alternatively, a better mass ratio could be determined from a precision measurement of the 611 GHz H-D isotope shift of the 1S-2S frequency. However, uncertainties of the fine structure constant and of the mean square radii of the nuclear charge distributions would still impose error limits of about 4 parts in 10^{11} for the Rydberg.

If the electron/proton mass ratio and the fine structure constant can be measured independently with improved accuracy, then a precise measurement of the 1S-2S H-D isotope shift could provide an accurate probe for nuclear structure and recoil shifts.

To determine a precise Rydberg value that is not limited by nuclear structure corrections, one could combine the 1S-2S measurement with a precise measurement of a two photon transition from 2S to one of the higher nS levels. Taking advantage of the fact that the lowest order nuclear structure corrections scale with the inverse cube of the principal quantum number n, one can easily construct differences of transition frequencies which are no longer sensitive to the exact nuclear sizes.

Obviously, neither a measurement of the 1S-2S frequency nor of the isotope shift by itself can provide a very stringent test of quantum electrodynamics, because we are free to adjust the values of fundamental constants until the calculations agree with the observations. However, if we form the ratio of the 1S-2S frequency to the frequency of a different hydrogenic transition, such as a Balmer transition, or a transition to or between Rydberg states, we arrive at a dimensionless quantity, which, to lowest order, no longer depends on the Rydberg constant, and which can be calculated with very high precision. An accurate measurement of such a frequency ratio would permit a very interesting test of theory. In this way, one might detect, for instance, some small deviations from Coulomb's law, which may exist within atomic dimensions, but which may have escaped detection in the past.

---

* Work supported by the National Science Foundation under Grant PHY-80-10689, and by the U.S. Office of Naval Research under Contract ONR N00014-78-C-0403.

REFERENCES

1. Demtröder, "Laser Spectroscopy," Springer Series in *Chemical Physics, Vol. 5*, Springer-Verlag, Berlin, Heidelberg, New York 1981.
2. M. D. Levenson, *Introduction to Nonlinear Laser Spectroscopy*, Academic Press, New York, 1982.
3. "Dye Lasers," F.P.Schäfer, ed., in *Topics in Applied Physics Vol. 1*, Springer-Verlag, Berlin, Heidelberg, New York 1977.
4. T. W. Hänsch, A. L. Schawlow, Opt. Comm. 13, 68 (1975).
5. H. Dehmelt in *Laser Spectroscopy V*, A.R.W.McKellar, T.Oka, B.P.Stoicheff, eds., Springer Series in Optical Sciences, Vol. 30, Springer-Verlag, Berlin, Heidelberg, New York (1981), p. 353.
6. R. A. Macfarlane, W. R. Bennett, W. E. Lamb, Jr., Appl. Phys. Letter 2, 189 (1963).
7. A. Szöke, A. Javan, Phys. Rev. Letters 10, 521 (1963).
8. P. H. Lee and M. L. Skolnick, Appl. Phys. Letters, 10, 303 (1967).
9. C. Bordé, Compt. Rend. 271, 371 (1970).
10. T. W. Hänsch, M. D. Levenson, A. L. Schawlow, Phys. Rev. Letters 26, 946 (1971).

11. R. K. Raj, D. Bloch, J. J. Snyder, G. Carmy, M. Ducloy, Phys. Rev. Letters 19, 1251 (1980).
12. D. Bloch, R. K. Raj, M. Ducloy, Opt. Comm. 37, 183 (1981).
13. C. Bordé, G. Camy, B. Decomps, L. Pottier, Cont. Rend. 277, 381 (1973).
14. F. V. Kowalski, W. T. Hill, A. L. Schawlow, Opt. Letters 2, 112 (1978).
15. B. Couillaud, A. Ducasse, in *Laser Spectroscopy*, S.Haroche et al., eds. Lecture Notes in Physics, Vol 43, Springer-Verlag, Berlin, Heidelberg, New York 1975, p. 476.
16. G. C. Bjorklund, M. D. Levenson, Phys. Rev. A24, 166 (1981).
17. J. L. Hall, T. Baer, L. Hollberg, H. G. Robinson, in *Laser Spectroscopy V*, A.R.W.McKellar, T.Oka, B.P.Stoicheff, eds., Springer Series in Optical Sciences Vol 30, Springer-Verlag, Berlin, Heidelberg, New York (1981) p. 15.
18. C. Wieman, T. W. Hänsch, Phys. Rev. Letters 36, 1170 (1976).
19. P. W. Smith, T. W. Hänsch, Phys. Rev. Letters 26, 740 (1971).
20. C. Delsart, J. C. Keller, "Laser Spectroscopy III," J.L.Hall, J.L.Carlsten, eds., Springer in *Optical Sciences, Vol. 7*, Springer-Verlag, Berlin, Heidelberg, New York (1977), p. 154.
21. M. S. Sorem, A. L. Schawlow, Opt. Comm. 5, 148 (1972).
22. J. E. Lawler, A. I. Ferguson, J. E. M. Goldsmith, D. J. Jackson, A. L. Schawlow, Phys. Rev. Letters 42, 1046 (1979).
25. E. E. Marinero, M. Stuke, Opt. Comm. 5, 148 (1972).
26. D. J. Jackson, E. Arimondo, J. E. Lawler, T. W. Hänsch, Opt. Comm. 33, 51 (1980).
27. C. Stanciulesco, R. C. Bobulesco, A. Surmeian, D. Popescu, C. B. Collins, Appl. Phys. Letters 37, 888 (1980).
28. D. R. Lyons, A. L. Schawlow, G.-Y. Yan, Opt. Comm. 38, 35 (1981).
29. A. Siegel, J. E. Lawler, B. Couillaud, T. W. Hänsch, Phys. Rev. A23, 2457 (1981).
30. J. E. Lawler, A. Siegel, B. Couillaud, T. W. Hänsch, J. Appl. Phys. 52, 4375 (1981).
31. T. W. Hänsch, D. R. Lyons, A. L. Schawlow, A. Siegel, Z.-Y. Wang, G.-Y. Yan, Opt. Comm. 37, 87 (1981).
32. Ph. Dabkiewicz, T. W. Hänsch, Opt. Comm. 38, 351 (1981).
33. R. L. Barger, M. S. Sorem, J. L. Hall, Appl. Phys. Letters 22, 573 (1973).
34. J. L. Hall, private communication.
35. R. V. Pound, Rev. Sci. Instr. 17, 490 (1946).
36. T. W. Hänsch, B. Couillaud, Opt. Comm. 35, 441 (1980)
37. T. W. Hänsch, G. W. Series, A. L. Schawlow, Sci. Am. 240, 94 (1979).
38. T. W. Hänsch, I. S. Shahin, A. L. Schawlow, Nature 235, 63 (1972).
39. T. W. Hänsch, M. H. Nayfeh, S. A. Lee, S. M. Curry, I. S. Shahin, Phys. Rev. Letters 32, 1336 (1974).

40. E. R. Cohen, B. N. Taylor, J. Phys. and Chem. Ref. Data 2, 663 (1973).
41. J. E. M. Goldsmith, E. W. Weber, and T. W. Hänsch, Phys. Rev. Letters 41, 940 (1978).
42. J. E. M. Goldsmith, E. W. Weber, F. V. Kowalski, A. L. Schawlow, Appl. Opt. 18, 1983 (1979).
43. E. W. Weber, J. E. M. Goldsmith, Phys. Letters 70A, 95 (1979).
44. E. W. Weber, Phys. Rev. A20, 2278 (1979).
45. Comite Consultatif pour la Definition du Metre, 5é Session-1973, Bureaue International des Poids et Mesures, p. M23.
46. B. W. Petley, K. Morris, R. E. Shawyer, J. Phys. B: Atom. Molec. Phys. 13, 3099 (1980).
47. S. R. Amin, C. D. Caldwell, W. Lichten, to be published.
48. D. E. Roberts, E. N. Fortson, Phys. Rev. Letters 31, 1539 (1973).
49. E. W. Weber, J. E. M. Goldsmith, Phys. Rev. Letters 41, 940 (1978).
50. C. E. Wieman, private communication.
51. L. S. Vasilenko, V. P. Chebotaev, A. V. Shishaev, JETP Letters 12, 113 (1971).
52. N. Bloembergen, M. D. Levenson, in *High Resolution Laser Spectroscopy*, K.Shimoda, ed. (Topics in Applied Physics, Vol. 13,), Springer-Verlag, Berlin, Heidelberg, New York (1976) pp. 315.
53. T. W. Hänsch, S. A. Lee, R. Wallenstein, C. E. Wieman, Phys. Rev. Letters 34, 307 (1975).
54. S. A. Lee, R. Wallenstein, T. W. Hänsch, Phys. Rev. Letters 25, 1262 (1975).
55. C. E. Wieman, T. W. Hänsch, Phys. Rev. A22, 192 (1980).
56. S. R. Lundeen, F. M. Pipkins, Phys. Rev. Letters (1980).
57. G. W. Erickson, Phys. Rev. Letters 27, 780 (1971).
58. P. J. Mohr, Phys. Rev. Letters 34, 1050 (1975).
59. A. I. Ferguson, J. E. M. Goldsmith, T. W. Hänsch, E. W. Weber, in *Laser Spectroscopy IV*, H.Walther and K.W. Rothe, eds. (Springer Series in Optical Sciences, Vol. 21) Springer-Verlag, Berlin, Heidelberg, New York (1979), pp. 31.
60. S. B. Crampton, T. J. Greytag, D. Kleppner, W. D. Phillips, D. A. Smith, and A. Weinrib, Phys. Rev. Letters 42 (1979).
61. R. S. Van Dyck, Jr., P. B. Schwinberg, Phys. Rev. Letters 47, 395 (1981).

STORED-ION SPECTROSCOPY

Hans Dehmelt

University of Washington
Seattle, WA 98195 USA

1. AND 2. LECTURE: SINGLE ELEMENTARY/ATOMIC PARTICLE AT REST IN SPACE

An individual elementary/atomic particle kept at rest in free space for extended periods, is an ideal object for high resolution spectroscopy. All external causes for line broadening or shifts such as 1. and 2. order Doppler and transit time effects as well as Zeeman or Stark effects are eliminated for such a system. This ideal has been approximated most closely so far in experiments on an individual Ba^+ ion localized in a Paul (rf) quadrupole trap to ~2000Å by optical side band cooling and made visible, all accomplished by means of laser

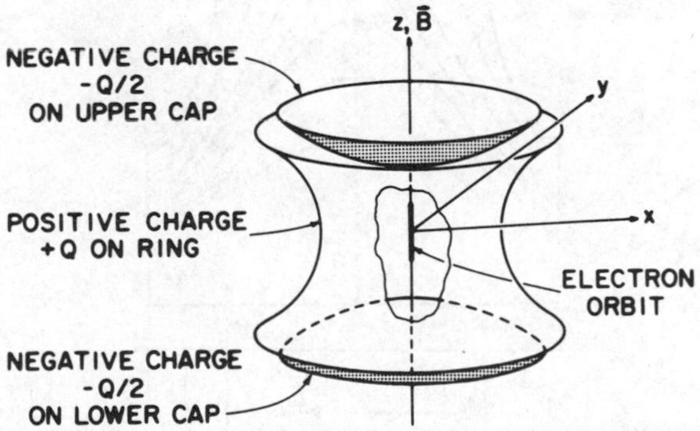

Fig. 1. Mono-electron oscillator mode of electron in Penning trap, the Geonium "atom." The electron moves only parallel to the magnetic field $\vec{B}$ and along the symmetry axis of the electrode structure. Each time it gets too close to one of the negatively charged caps it is turned around and an oscillatory motion results.

beams. High resolution spectroscopy on an individual electron/positron localized to <200 μm by rf side band cooling in a Penning trap employing a 50 kG field has already yielded the most precise data on the magnetic moments of these particles and also provided a severe test of the CPT theorem for charged elementary particles. Furthermore, localization of an elementary particle in space is one of the most fundamental problems in physics and worthy of study on its own merit.

SINGLE ELEMENTARY PARTICLE SPECTRA

Monoelectron Oscillator

In 1955, I became intrigued by the trigger techniques developed for the rf spectroscopy of atoms, especially by the optical double resonance techniques of Kastler and Robert Dicke. Also, in my student days the orthodoxy whose purpose appears to be to invite challenge, had been that it was impossible to measure spin and magnetic moment of a free electron. I took up the challenge, and proposed rf spectroscopy of stored ions in 1956. Extending Kastler's techniques I

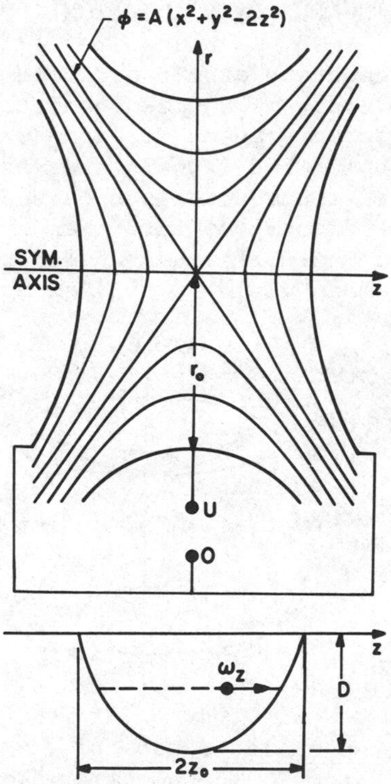

Fig. 2. Hyperbolic electrode configuration employed in ion storage devices useful in spectroscopy. Application of a DC voltage $U=U_0$ at the terminals creates a well of depth $D=U_0/2$ for the choice $r_0^2=2z_0^2$. From (Dehmelt, 1967).

Fig. 3. My anharmonicity-compensated 7-electrode low magnetic field 1959 sealed-off Penning trap tube used in axial and cyclotron resonance experiments in electron clouds (Dehmelt, 1961; Dehmelt and Major, 1962).

developed spin-exchange optical pumping, and by this means succeeded in the first spin magnetic resonance experiment on free electrons in December, 1956. These experiments, eventually yielded a magnetic moment value μ_s = 1.001 116(40) Bohr magnetons.

Attempts to eliminate the inert buffer gas used in these early experiments led to lengthy experiments with various traps. Finally, in 1959, I began to concentrate on the high vacuum Penning trap (magnetic field plus axial electric field, see Figs. 1 and 2) after I realized that the electric shift it induced in the electron cyclo-

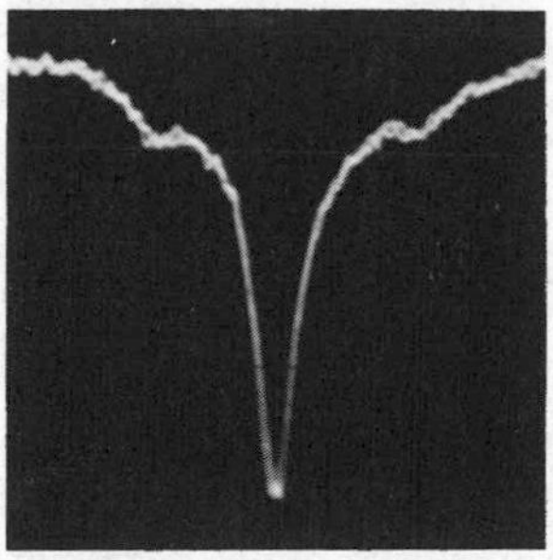

Fig. 4. Axial resonance obtained in 1960 with my 1959 Penning trap tube. The central peak is at ν_z=2.75 MHz. The two satellites are at $\nu_z \pm \nu_m$, where ν_m=100 KHz denotes the magnetron or drift frequency. The cyclotron frequency was ν_c'=37.9 MHz (Dehmelt, 1961).

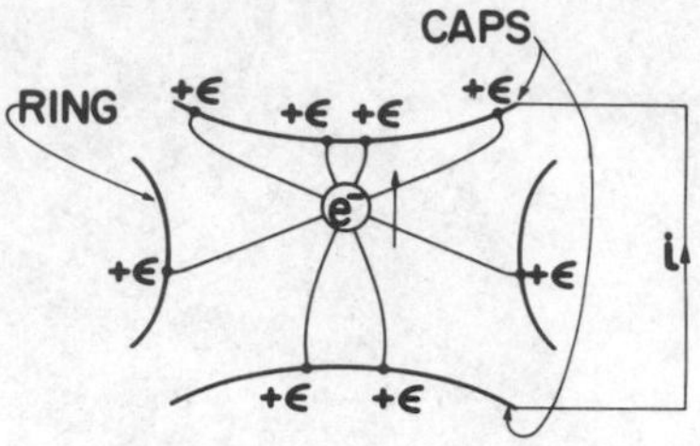

Fig. 5. Displacement currents. When the electron moves upward more
         field lines end on upper cap and additional termination
         charges ε flow through the external circuit.

tron frequency would be constant throughout the trap volume. The
first successful trap tube built by me in 1959 is shown in Fig. 3,
and an axial resonance obtained with it in Fig. 4. (Dehmelt 1961;
Dehmelt and Major, 1962). This resonance was detected via the displacement currents which the oscillating electron cloud induces in
the external circuit, see Fig. 5. While developing this trap in
electron cloud experiments with my graduate student, Fred Walls, and
my postdoctoral associate, Talbert Stein, I also gained further experience in the successful work with Paul traps on the hfs of $^3He^+$
and H_2^+ in my laboratory. Finally, the conditions were ripe to
attempt the realization of an old goal of mine that dated back to the
day I saw my teacher, Richard Becker, in his Electricity and Magnetism
lecture, draw a dot on the blackboard saying, "Here is an electron..."
The goal was the isolation, permanent confinement, and continuous
observation of an individual electron (almost) at rest in space. This
"mono-electron oscillator," whose feasibility I had pointed out in
1962, compare Fig. 6, was demonstrated working together with my postdoctoral associates David Wineland and Philip Ekstrom in 1973, see
Fig. 7.

THERMALIZATION OF ION

$$W_I = kT_0 + (W_{IO} - kT_0)\exp(-t/\tau_{IT})$$

$$\boxed{\tau_{IT} = (4M\,z_0^2)/(e^2 R_s)}$$

OPTIMUM SIGNAL TO NOISE RATIO

INITIAL ENERGY OF ION, W_{IO}, FLOWS
SLOWLY INTO TANK, FAST INTO
BATH, $\tau_{IT} \gg \tau_{TB}$. RETAINED IN TANK
FOR INTERVAL $\approx \tau_{TB}$, $W_T \approx (\tau_{TB}/\tau_{IT})\,W_{IO}$.
THERMAL FLUCTUATIONS OF TANK
ENERGY FOR OBSERVATION TIME $\approx$
τ_{IT} AVERAGE OUT TO $\Delta W_T \approx$
$(\tau_{TB}/\tau_{IT})\,kT_0$, S/N $= W_T/\Delta W_T$;

$$\boxed{S/N \approx W_{IO}/kT_0}$$

NUMERICAL EXAMPLE

$M = 100\,M_H$; $2z_0 = 0.5\,cm$

$C \approx 10^{-11}\,F$; $Q = 100$

$\omega \approx 5 \times 10^5$ CPS; $R_s \approx 2 \times 10^7\,\Omega$

$\tau_{IT} \approx 13$ sec ; $W_{IO} \approx 3\,eV$
S/N ≈ 100, $kT_0 \approx$ $0.03\,eV$

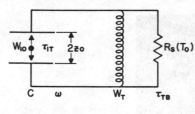

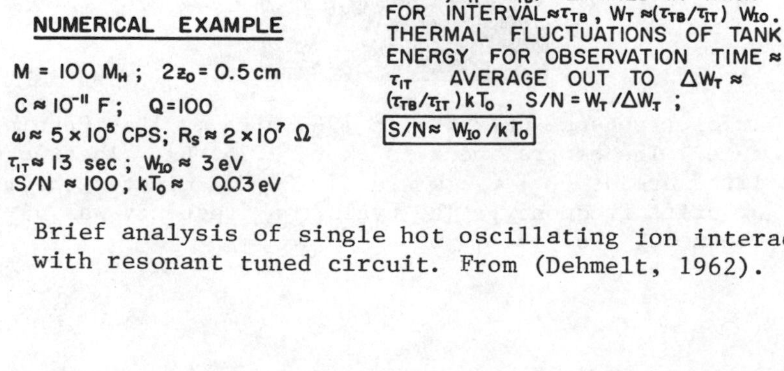

Fig. 6. Brief analysis of single hot oscillating ion interacting
         with resonant tuned circuit. From (Dehmelt, 1962).

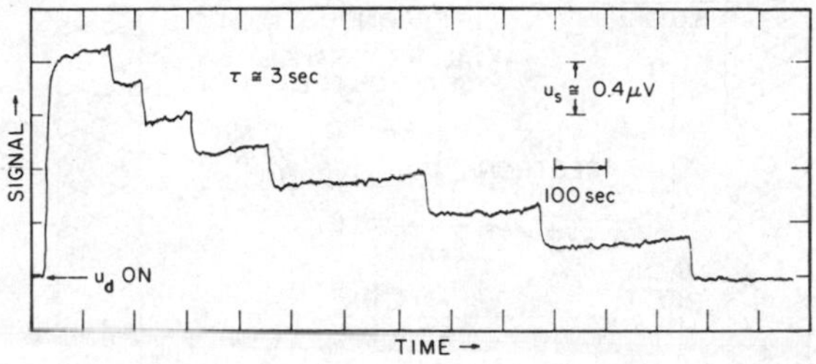

Fig. 7. Recorder trace of forced-oscillation signal versus time. The signal at $\nu_{z0} \approx 55.7$ MHz for an initially injected bunch of electrons decreases discontinuously as the electrons are successively boiled out of the trap by the drive. The last plateau corresponds to a single electron. From (Wineland et al., 1973).

The Non-Destructive Axial Stern-Gerlach Effect

Together with Ekstrom (1973), I worked out a scheme to monitor the spin state of the trapped electron by means of small axial frequency shifts induced by an auxiliary shallow magnetic bottle, the Non-Destructive Axial Stern-Gerlach Effect (to communicate with the electron on the axial frequency via FM radio), See Fig. 8. My scheme was stimulated by rumors about an axial Stern-Gerlach effect for charged particles circulating during my student days in the corridors of Kopfermann's Institute where Wolfgang Paul and Helmut Friedburg invented the magnetic hexapole lens for atomic beams in 1951 (now used in the H-maser). I was able to trace these rumors back to a 1928 proposal by Brillouin of a destructive Stern-Gerlach effect for

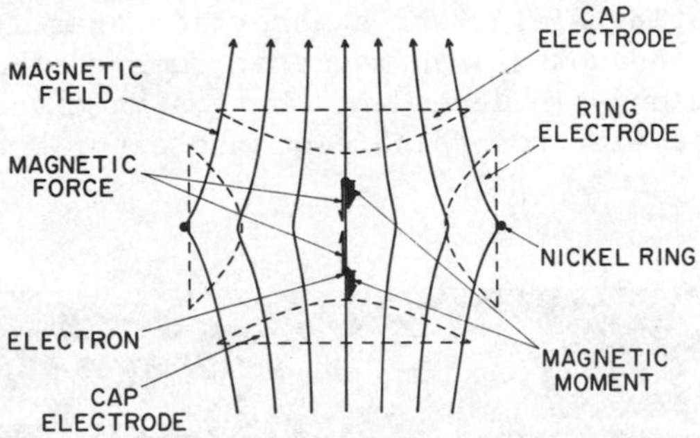

Fig. 8. The Axial Stern-Gerlach Effect (schematic): An electron slowly moving along a field line in an inhomogeneous magnetic field with its magnetic moment parallel/anti-parallel to the field is driven towards stronger/weaker fields. We show here the minute magnetic forces which add to the strong axial electric forces in our experiment and slightly modify the parabolic trapping potential. From (Dehmelt, 1981).

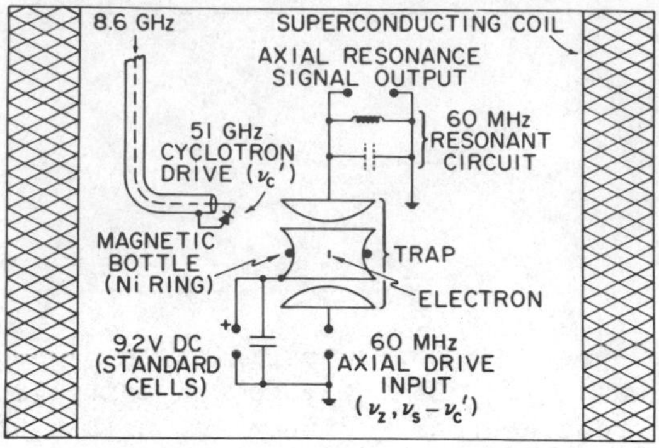

Fig. 9. Geonium spectroscopy experiment (schematic). This apparatus allows the measurement of the cyclotron frequency, ν_c', and the spin-cyclotron-beat (or anomaly) frequency, $\nu_a' = \nu_s - \nu_c'$, on a single electron stored in a Penning trap at $\simeq 4^oK$ ambient. Detection is via Rabi-Landau level-dependent shifts in the continuously monitored axial resonance frequency, ν_z, induced by a weak magnetic bottle. From (Van Dyck, et al., 1978).

electrons: slow electrons move along a field line of a rapidly increasing magnetic field parallel to the z-direction. Those with one spin direction are accelerated by the gradient, collected by an electrode and counted. The other spin direction is repelled. In either case the electrons are rapidly lost from the experiment. This effect was briefly discussed by Wolfgang Pauli (1946) at the 1930 Solvay Congress (Mehra, 1975), and immediately discarded, supposedly on the basis of Heisenberg's uncertainty principle. Pauli was then illustrating Niels Bohr's assertion that it is impossible to measure the spin and associated magnetic moment of a free electron in experiments based on spin dependent changes in classical trajectories. Nevertheless, free electron spin resonance experiments based on de-

Fig. 10. Photograph of 5-electrode 1974 tube. The left half houses the trap, the right ion-getter and cryosorption pumps capable of $<10^{-14}$ torr vacua. From (Van Dyck, et al., 1976).

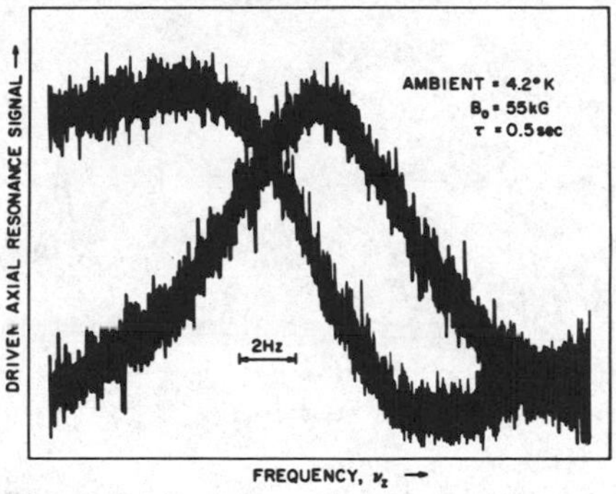

Fig. 11. Axial resonance signals at ≈ 60 MHz, note 8 Hz width. Absorption and dispersion modes are shown. From (Van Dyck, et al., 1978).

structive versions have been attempted by Dicke in 1947-49 and proposed in the past (Bloch, 1953). No results have been reported on this continuing work so far (Dehmelt, 1976). Our novel, nondestructive scheme and subsequent successful experiment resurrects the axial Stern-Gerlach effect for the free electron. Obviously, after our experiment Bohr's assertions cannot claim general validity.

Apparatus

The magnetic bottle used in our apparatus for observation of Geonium Spectra, see Fig. 9 (Van Dyck, et al., 1978), is realized by a nickel wire wound around the ring electrode of the Penning trap which the 18 - 52 kG applied magnetic field magnetized to saturation. The trap tube which my associates, Ekstrom and Robert Van Dyck, Jr.,

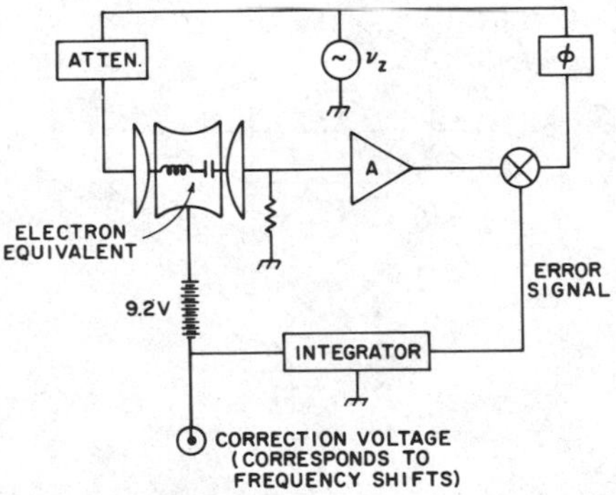

Fig. 12. Detection circuit for the axial resonance at ν_z. The electron acts effectively like an LC series resonant circuit connecting the cap electrodes. The circuit also locks the electron frequency to that of the very stable generator. From (Van Dyck, et al., 1978).

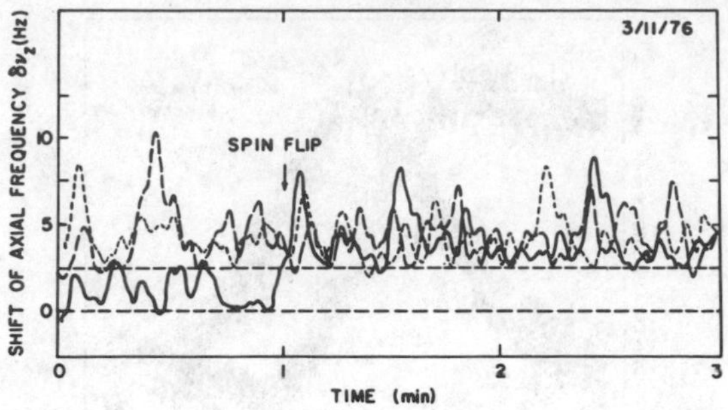

Fig. 13. First spin flip seen in the monoelectron oscillator (geonium). Because of the random fluctuations in the thermally excited cyclotron motion, the axial frequency shift $\delta\nu_z$ associated with the magnetic bottle shows a corresponding unsymmetric fluctuation always staying above a fixed floor for a given spin direction. This floor, however, suddenly changes by 2.5 Hz when the spin is flipped, which occurs occasionally near resonance, $\nu_z - \nu_a \simeq 2$ kHz. No auxiliary drive was used. Three superimposed traces are seen, only the heavy one showing the spin flip. From (Van Dyck, et al., 1976).

built along lines suggested by me, may justifiably be called a little masterpiece of the experimental art, see Fig. 10. In operation it is submerged in a liquid He bath. Fig. 11 shows a ~8 Hz wide axial resonance line at $\nu_z \simeq 60$ MHz obtained with our apparatus. Important contributions to the development of this apparatus were also made by my student, Paul Schwinberg. We observed the first spin flip signal in 1976, see Fig. 13, with the detection circuit shown in Fig. 12.

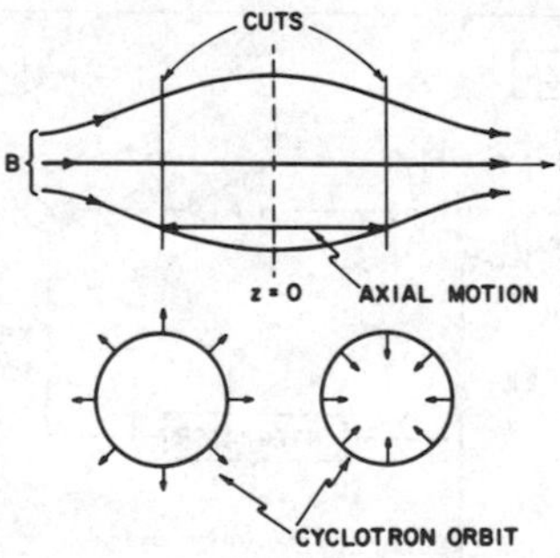

Fig. 14. Mechanism for inducing spin flips by the electron's axial motion in the magnetic bottle. From the electron's frame of reference, a magnetic field component is seen rotating at ν_c', but modulated by the axial motion at ν_a', yielding sidebands at $\nu_c' \pm \nu_a'$ with $\nu_s = \nu_c' + \nu_a'$. From (Van Dyck, et al., 1978).

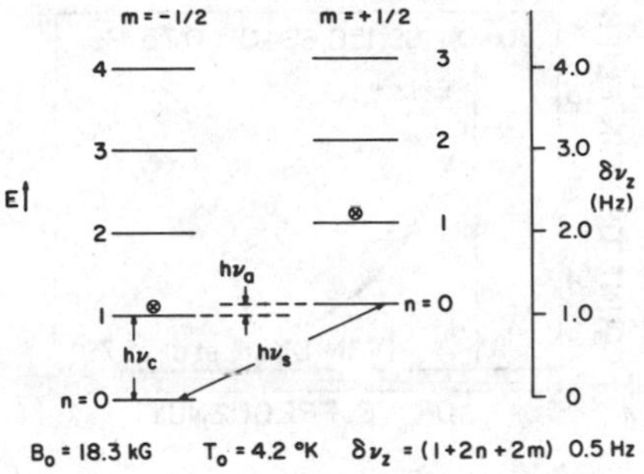

Fig. 15. Electron Rabi-Landau levels and associated magnetic bottle shifts $\delta\nu_z$. From (Van Dyck, et al., 1978).

This experiment marks the advent of serious mono-particle spectroscopy. The trace shows the continuously recorded axial resonance frequency of the electron, the recorder sweeping back and forth until on the third sweep (backward) a spin flip becomes apparent indicating a frequency difference of 2.5 Hz between $m = +\frac{1}{2}$ and $m = -\frac{1}{2}$. Already this constitutes a crude measurement of the magnetic moment. The prominent upwards random fluctuations reflect the dependence of the bottle induced frequency shift $\delta\nu_z = (m + n + \frac{1}{2})$ 2.5 Hz on the thermally excited (4 K) cyclotron motion (quantum number n). The spin flip is an "assisted" Majorana flop (Dehmelt and Walls, 1968) caused by the driven axial motion, at $\nu_z \simeq \nu_s - \nu_c$, and the thermal cyclotron motion at ν_c, through the inhomogenous field of the magnetic bottle. The bottle here serves a distinct, different, second function, see Fig. 14.

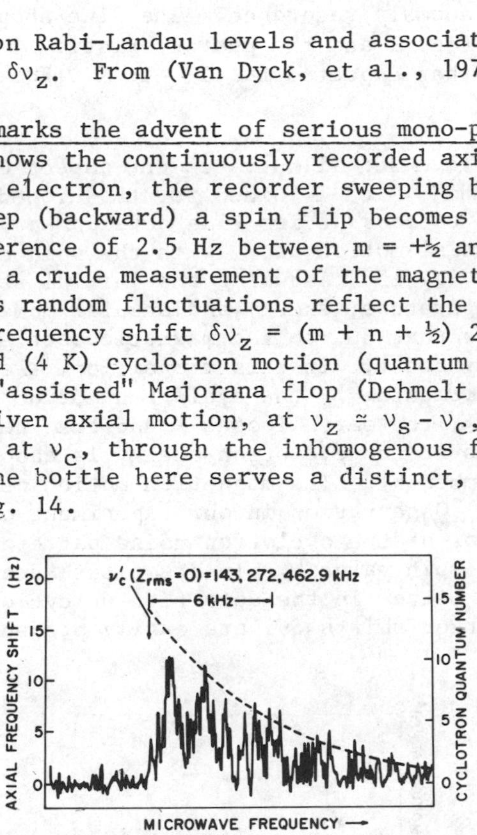

Fig. 16. Geonium cyclotron resonance. The vertical rise-exponential decay line shape exhibiting a signal strength decline to 1/e for a 6 kHz displacement reflects the proportionality of the average magnetic bottle field seen by the electron to the instantaneous thermally excited axial energy. At an axial temperature $T_z \simeq 16$ K the nearly vertical edge allows determination of ν_c' when the electron is at the bottom of the magnetic well ($z = 0$) to ∿500 Hz. From (Van Dyck, et al., 1979).

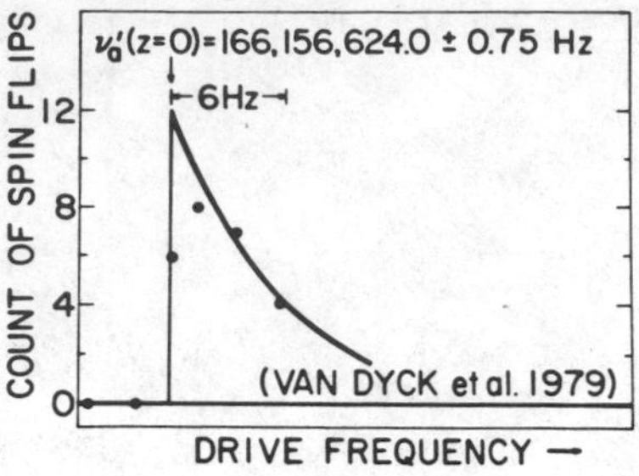

Fig. 17. Geonium anomaly resonance. The line shape is similar to that in Fig. 1 and the characteristic ∿6 Hz displacement is also that expected for $T_z \simeq 16$ K. From (Van Dyck, et al, 1979).

It is instructive and illustrates one aspect of the significance of the term "Geonium" which I coined for the closed system (electron, trap-magnet-earth) to look at its time development due to intentional resonance of the axial motion at ν_z with the spin-cyclotron beat frequency ν_a, see Fig. 15. Assuming the initial state mnkq = ½000 and OK ambient temperature, energy and angular moment conservation considerations show that due to the magnetic bottle a transition to the state -½110 may occur. In this transition, the spin looses an amount of angular momentum $\hbar$, the cyclotron motion gains the kinetic angular momentum $2\hbar$, the earth looses $\hbar$ and the energy excess in the spin state energy over the cyclotron energy is absorbed by the axial motion (quantum number k). The magnetron motion quantum number q remains unchanged. Observation in our experiment of the two different values for the floor of the cyclotron noise patterns corresponding to n = 0 confirms the spin value ½. In its first function the magnetic bottle serves even better in the detection of cyclotron excitation, see Fig. 15, as larger shifts $\delta\nu_z$ are easily produced.

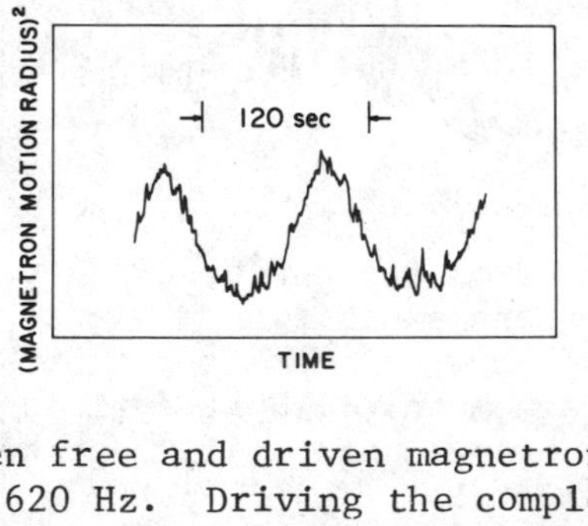

Fig. 18. Beat between free and driven magnetron motions at 35,052.628 and 35,052.620 Hz. Driving the completely undamped magnetron motion only ≃ 10 mHz away from the resonance creates a slow beat from which the resonance frequency may be determined to ≃ 1 mHz. From (Van Dyck, et al., 1978).

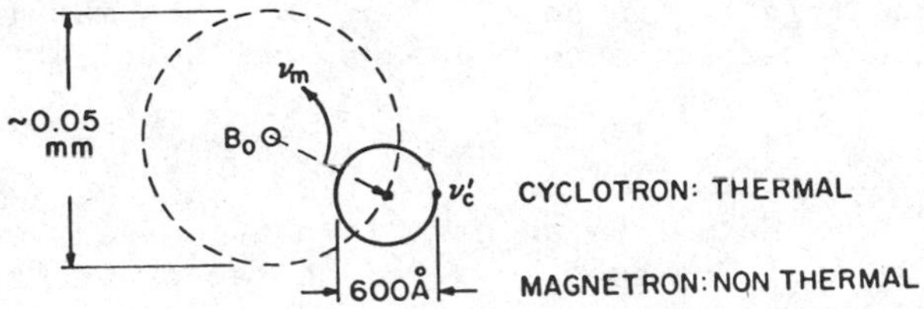

Fig. 19. Geonium orbits. Motion in Penning trap under thermal excitation at ≃ 4 K. From (Van Dyck, et al., 1978).

Geonium Spectra

The most recently obtained cyclotron resonance (Van Dyck, et al., 1979a and b) is shown in Fig. 16. The characteristic vertical rise-exponential decay shape of the line is due to the thermal axial motion at about 20 K through the magnetic bottle. The nearly vertical edge allows determination of the cyclotron frequency to ∿500 Hz or 3 parts per billion for the centered electron. A similarly shaped resonance at the spin-cyclotron beat frequency $\nu_a \equiv \nu_s - \nu_c$ is shown in Fig. 17. Actually, both anomaly frequency and cyclotron frequency measured in our apparatus, now designated ν_a', ν_c', are shifted by the radial electric field in the trap by an amount δ_e with respect to the zero electric field values ν_a, ν_c so that $\nu_c = \nu_c' + \delta_e$ and $\nu_a = \nu_a' - \delta_e$. For an ideal trap one has $\delta_e = \nu_m$ where ν_m denotes the magnetron or drift frequency in the trap which also may be measured to milli-Hz by our magnetit bottle technique, see Fig. 18.

Side Band Cooling

The presence of the magnetic bottle makes it clearly desirable to always localize the electron, whose orbit $\perp z$ is shown in Fig. 19,

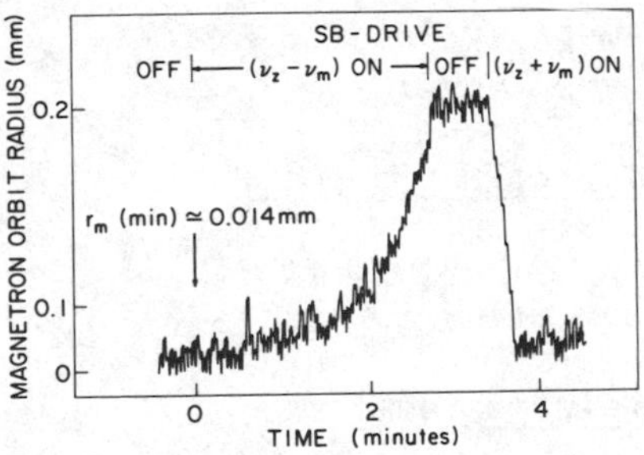

Fig. 20. First demonstration of sideband cooling searched for by Wineland and Dehmelt in 1974. By driving the axial motion on the side bands $\nu_z \pm \nu_m$ it is possible to force the magnetron motion at ν_m to absorb/provide the energy balance $h\nu_m$ and thereby shrink/expand the magnetron orbit. From (Van Dyck, et al., 1978).

as close as possible to the bottle center. This may indeed be achieved by a new technique developed by us, see Fig. 20.

Data and Results

Table 1 shows data from an earlier phase of the experiment, performed at 18 kG. The anomaly a, the g-factor and the frequencies ν_a, ν_c are related by

$$g/2 \equiv \nu_s/\nu_c = 1 + a .$$

Comparing ν_m, and δ_e calculated from the measured ν_c', ν_z values, provides a sensitive check of the axial symmetry of the trap. In the meantime, in about 40 additional runs at 18.6, 32.0 and 51.1 kG, we have collected more precise data and are now quoting (Van Dyck, et al., 1979),

$$\nu_s/\nu_c \equiv g/2 = 1.001\ 159\ 652\ 200\ (40) .$$

This also equals the electron spin magnetic moment value in Bohr magnetons and is currently the most accurately determined parameter of an elementary particle.

Table 1. SAMPLE DATA AND RESULTS (From Van Dyck, et al., 1978)

Measure (1 Run)

$\nu_z = 59\ 336\ 170.14 \pm .10$ Hz
$\nu_m = 34\ 471.9 \pm .1$ Hz
$\nu_c' = 51\ 072\ 915 \pm 10$ kHz
$\nu_a' = 59\ 261\ 337.5 \pm 4.5$ Hz

Calculate

From $2\delta_e\nu_c = \nu_z^2$
$\delta_e = 34\ 469.18 \pm .1$ Hz

Measure Trap Imperfections

$\nu_m - \delta_e = 3.7$ Hz, $(\nu_m - \delta_e)/\delta_e \simeq 10^{-4}$
(Correction to δ_e) << 3.7 Hz

Obtain (8 runs)

With $\nu_c = \nu_c' + \delta_e$, $\nu_s - \nu_c = \nu_a' - \delta_e$,
$a \equiv (\nu_s - \nu_c)/\nu_c$

$a = 1\ 159\ 652\ 410\ (200) \times 10^{-12}$

200×10^{-12} IN a $\rightarrow$ 10 Hz IN ν_a'

Kinoshita (1979) in a private communication from which I quote has compared our result in the form $a_e^{exp} = g/2 - 1$ to a theoretical a-value

$$a_e^{th} = 1\ 159\ 652\ 569\ (150) \times 10^{-12}$$

calculated by him and also obtained the most accurate α-value from it. A detailed account of our work will be published in Physical Review. An updated survey (Dehmelt, 1981) has been published, and a recent review of lepton magnetic moments and their significance has been given by Field, Picasso and Combley (1979). Also, Ekstrom and Wineland (1980) have published a popular article on Geonium physics.

Positron Spectra

Soon after the geonium spectroscopy techniques had been demonstrated on an individual electron Dehmelt et al. (1978) proposed to trap small numbers of positrons by radiation damping in a vacuum for use in spectroscopic experiments (see Fig. 21). Positron g-factor experiments by Schwinberg et al. (1981) using this technique have now yielded the value

$$g\ (e^+)/g(e^-) = 1+(22\pm64)10^{-12}$$

representing the most precise confirmation of the CPT symmetry theorem for a charged particle/antiparticle pair.

SINGLE ATOMIC ION SPECTRA

The 1965 experiment of Fortson, Major and Dehmelt, in which the 0-0 hyperfine transition in the ground state of $^3He^+$ near 8 GHz was observed with a linewidth of ∿10 Hz or a resolution of about 1 part in 10^9 marked the advent of high resolution microwave spectroscopy of stored ions. Samples of ∿10^7 ions were used in these experiments.

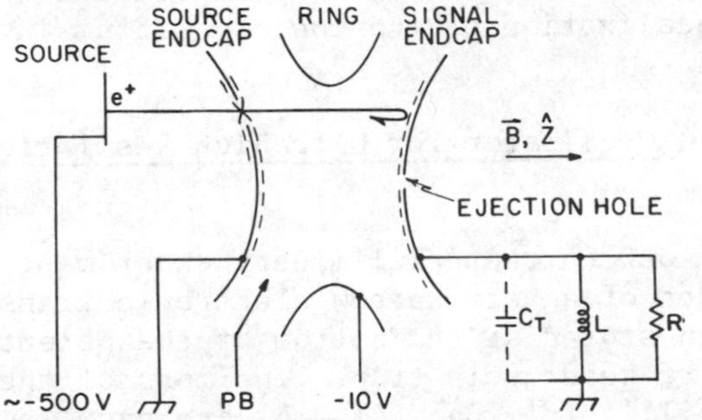

Fig. 21. Continuous catching of β^+ rays by radiation damping. A ∿ 1 mCi source sealed into the trap tube yields ∿ 20 cold (1 meV), trapped positrons/hr. From (Dehmelt et al., 1978).

Already then in our 1966 publication we pointed out the potential usefulness of ion storage for laser spectroscopy. In 1973, Major and Werth now with NASA, reported the record resolution of 2 parts in 10^{10} in an optical pumping experiment on the hyperfine structure transitions near 40 GHz on $\sim 10^5$ stored $^{199}Hg^+$ ions. Also, in 1973, I proposed laser fluorescence spectroscopy on a highly refrigerated, individual Tl^+ ion stored in an rf quadrupole trap, "The Tl^+ Mono-Ion Oscillator," with a projected resolution of 1 part in 10^{14}. Now an isolated atom at rest in free space is a spectroscopists dream. Also, it may make experiments on some of the most fundamental problems in physics possible. However, it was impossible to obtain support and no work could be started in Seattle. In 1975, I added a laser double resonance scheme for effectively amplifying about million-fold the fluorescence intensity used for the detection of the forbidden transition of interest. This scheme also made the study of the similar ions In^+, Ga^+, Al^+ attractive. Incorporation into the Tl^+ experiment of the side band cooling mechanism, searched for by Wineland and Dehmelt in 1974 in experiments on stored electron clouds and found in 1976 by Van Dyck, Ekstrom and Dehmelt in their geonium experiment, was proposed by Wineland and Dehmelt in 1975. Also, in 1974 the award of a Humboldt prize had enabled me to initiate a Ba^+ mono-ion oscillator experiment at the Universitaet Heidelberg in collaboration with P. Toschek and his group to prepare the gound for the more difficult Tl^+ work. At the same time, I proposed a Sr^+ mono-ion oscillator experiment together with H. Walther, then at the Universitaet Köln (1975).

In 1978, the Heidelberg experiment by Neuhauser et al. led to the first isolation and uniform confinement, initially to $< 5\ \mu m$ and eventually (1980) to ~ 2000 Å, which is considerably smaller than the wavelength of the laser light, via laser side band cooling to ~ 6 K and eventually to ~ 10 mK as well as continuous visual and photoelectric observation of an individual Ba^+ ion in a small Paul rf quadrupole trap. Recently, Wineland and Itano (1981), have also succeeded in a Mg^+ mono-ion oscillator experiment and they have established a localization of the ion to $\lesssim 15\ \mu m$ in their Penning trap.

Proposed Mono-Ion Oscillator for Ultrahigh Resolution Laser Spectroscopy

The goal of our current preliminary experiment under preparation is the observation of a very narrow electronic transition at ω_0 on an individual ion stored at the bottom of the potential well formed by a small Paul rf quadrupole trap. The ions of the Group IIIA elements, namely Tl^+, In^+, Ga^+, Al^+, B^+ are excellent candidates for such work. Their metastable lowest $^3P_0$ levels have extraordinary long life times; e.g., the corresponding natural line width of the $3\ ^3P_0$ level of Al^+ is estimated as $\sim 10\ \mu Hz$.

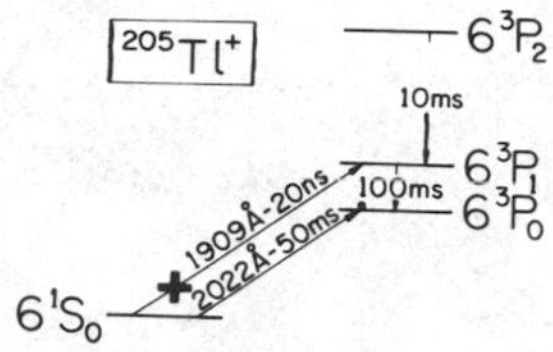

Fig. 22. The four lowest electronic levels of the $^{205}Tl^+$ ion. In the proposed double resonance scheme the forbidden ω_o transition at 2022 Å of interest to the $^3P_0$ level of 50 ms natural life time is observed indirectly: during the temporary shelving of the electron in the metastable $^3P_0$ level it is obviously impossible to observe laser scattering at $\lambda_2 = 1909$ Å. From (Dehmelt, 1975).

"Shelved Optical Electron" Atomic Amplification Scheme

However, in a mono-ion oscillator such a narrow width can only be exploited when a powerful amplification mechanism for the correspondingly low scattering is devised to overcome the large losses in the collection and counting of the scattered photons. Monitoring of the absence of the optical electron from the $^1S_0$ ground state via laser fluorescence at the strong $^1S_0$-$^3P_1$ intercombination line of frequency ω_2 or at the $^1S_0$-$^1P_1$ resonance line provides such an atomic amplification mechanism. For Tl^+, see Fig. 22, where the estimated lifetime of the $^3P_0$ level is $\sim$50 ms, absorption of a single photon at the forbidden $^1S_0$-$^3P_0$ ω_o-transition at 2022 Å suppresses ω_2-laser fluorescence at 1909 Å for about 50 ms, during which interval otherwise $\sim 10^6$ photons might have been scattered. Frequency shifts of the sharp 2022 Å line due to the 1909 Å excitation are obviated by appropriate pulsing schemes. Centering of the ion at the bottom of the trap is achieved via side band cooling by tuning the 1909 Å laser a few megahertz below the $^1S_0$-$^3P_1$ resonance at ω_2 thus that its Doppler side band at $\omega_2 - \omega_v$ is preferentially excited. The vibration frequency in the parabolic trap is denoted by ω_v. Three-dimensional cooling to < 1 mK is achieved by means of a single strategically directed laser beam. Directing the laser beam thus and therefore along an equipotential surface of the trapping rf field also practically eliminates the 1. order Doppler effect due to the forced Ω-micro-motion.

Apparatus

The heart of the apparatus is the small Paul rf trap containing the ion. The design used in the Heidelberg Ba^+ experiments is well suited for the future work on Tl^+, etc. and will therefore be

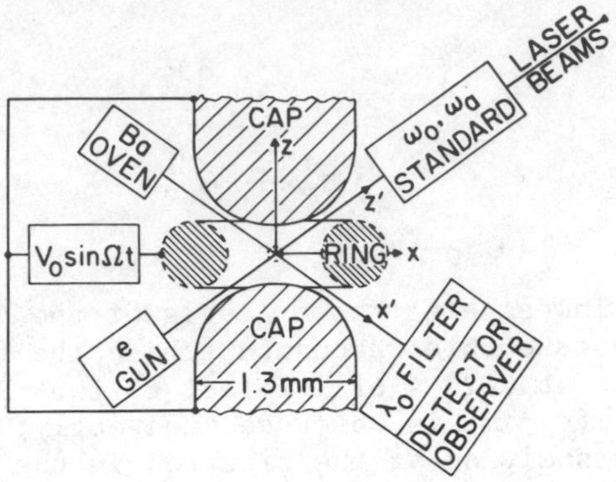

Fig. 23. Apparatus for trapping in ultra-high vacuum, cooling, and visually observing an individual Ba^+ ion (schematic). Ba oven and electron gun were actually not in the xz plane. Vacuum envelope, pump, etc. are not shown. From (Neuhauser, et al., 1978).

sketched here. The trap, see Fig. 23, is formed by spherically ground wire stubs, the "caps" and a wire "ring". Application of an rf voltage $V_0 \simeq 200$ Volt at $\Omega \simeq 2\pi \cdot 18$ MHz created a 3-dimensional well of axial depth ~ 10 Volt (for Ba^+) in which the ion oscillated at an axial frequency $\omega_{vz} \simeq 2\pi \cdot 2.4$ Mhz and perpendicular frequencies $\omega_{vx*}, \omega_{vy*} \simeq 2\pi \cdot 1.2$ MHz with $|\omega_{vx*} - \omega_{vy*}| \simeq 2\pi \cdot 10$ KHz. Ideally the cooling laser beam is directed along the body diagonal of the $x^*, y^*, z^* \simeq z$ coordinate system formed by the principal axes of the effective trapping potential. It suffices to approximate these conditions by slightly deforming the ring electrode eliptically and appropriately directing the laser beam through the gaps between ring and cap electrodes. The resonance fluorescence is likewise observed through the ring-cap gap. A vacuum of $\sim 10^{-11}$ Torr is maintained by an ion getter pump in the sealed off, baked glass envelope. The trap was filled by crossing and ionizing a very weak Ba atomic beam with a very weak electron beam in the center of the trap.

Shifts and Broadening of the Resonance Line

The crucial estimates of such shifts were made in 1973 for Tl^+ in a small rf trap (Dehmelt, 1973). For an estimated temperature ~ 0.1 mK one finds for δ_D, the second order Doppler shift, $\delta_D/\nu_0 = 3kT/Mc^2 \lesssim 10^{-18}$. The estimated quadratic Stark shift $\delta_S \simeq 2\,[\text{mHz}/(\text{V/cm})^2] \cdot E^2$ amounts to 10^{-4} Hz or $\delta_S/\nu_0 \lesssim 10^{-19}$ at 0.1 mK. For small, deep traps it is also necessary to discuss an electric "trap shift" proportional to the applied trap voltage which should occur when ground or excited states have an appreciable electronic electric quadrupole moment eQ. For heavy atoms in a P state with HFS quantum number $F \geq 1$, eQ may be as large as 3 eÅ^2. The order of magnitude of the corresponding m_F-dependent shift δ_Q is $eQ\phi_{zz}$, ϕ_{zz} referring to the electric trapping field. For the Ba^+ rf trap one finds $|eQ\phi_{zz}| \lesssim 30$ KHz. However, since it is an AC shift, it averages out to zero. A neighboring ion at the equilibrium distance in the well (Neuhauser et al., 1980) of $\sim$5 μm would cause an analogous quadrupole shift of $\sim$200 Hz which would not be averaged out by the motion of the ions. For the Penning trap of (Wineland and Itano, 1981) one finds a DC shift $\delta_Q \simeq 2$ Hz. Our Group IIIA mono-ion oscillator experiments are immune to such shifts.

One might think the $\Delta m_F=0$ transitions for our ω_0 line would show no (nuclear) Zeeman effect. However, for the isoelectronic ^{199}Hg isotope a "nuclear" g-factor for the $^3P_0$ state about twice as large as for the $^1S_0$ ground has been measured. This is explained by a small admixture of $^3P_1$ to the $^3P_0$ state via the strong HFS interaction. Estimated Zeeman splittings for Tl^+ amount to $\sim$2 KHz/Gauss requiring fairly elaborate shielding to bring this δ_z shift down to the level of δ_S and δ_D. Outside the Lamb-Dicke dominant carrier regime the amplitude of the carrier is strongly dependent on the

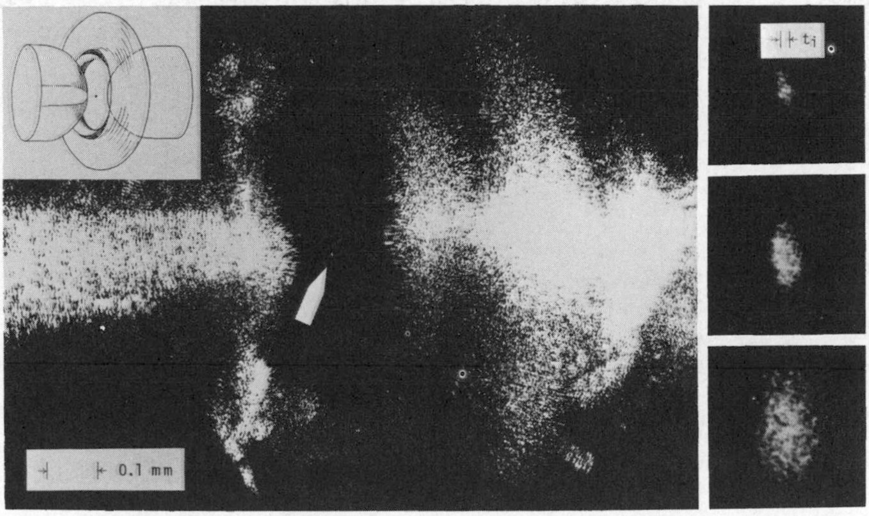

Fig. 24. Micro-photographic images of 1, 2 and 3 trapped Ba^+ ions. The large photograph shows the $\sim$2 μm thick image (white arrow) of a single ion inside the rf quadrupole trap as viewed through the gap between the ring and the left cap-electrodes (trap structure illuminated by scattered laser light). A sketch of the whole trap structure seen from the same angle is inserted. The three small photos, going from top to bottom, show, 10-fold enlarged, the central trap region, containing 1, 2 and 3 ions. From (Neuhauser, et al. 1980).

vibration amplitude. Thus, variations in the vibration amplitude can broaden the carrier (Dehmelt, 1973). So far it has been tacitly assumed that laser sources of arbitrarily high sharpness and short-term stability are available. However, the record in minimal laser spectral width still appears to be a few Hertz. This is still far above fundamental limitations which lie around $\sim 10^{-4}$ Hz.

Results of Preparatory Experiments

The most important result so far obtained in our mono-ion oscillator work (Neuhauser et al., 1980) is the 3-dimensional localization in free space to a region of ~2000 Å diameter of an individual Ba^+ ion, see Fig. 24. The photographic image of the ion has a diffraction limited diameter of 2 μm. It is possible to deconvolute the much smaller localization range by comparing the 1-ion image with a 3-ion image which, obtained under otherwise identical conditions, has a ~20 times larger area and a ~30 times lower peak brightness. This implies $z_0 \simeq \lambda_0/2\pi$ and that the boundary of the Lamb-Dicke regime of the dominant carrier has been reached. The best spectral resolution obtained so far by us (Neuhauser et al., 1981) is a line-width of ~40 MHz for the $6^2S_{1/2} - 6^2P_{1/2} - 5^2D_{3/2}$ two-photon transition (Dehmelt and Toschek, 1975) to the highly metastable $5^2D_{3/2}$ state of Ba^+ which has a lifetime of 17 sec, see Fig. 25. This two-photon transition is obviously still strongly power-broadened.

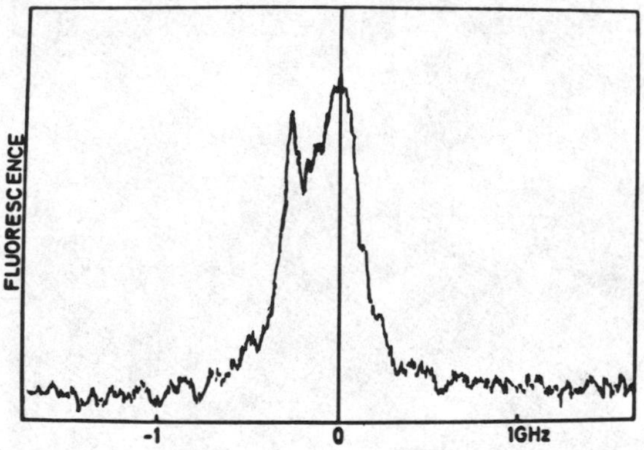

Fig. 25. Resonance fluorescence near 493 nm from an individual barium ion vs. scanned frequency ω_a of 650 nm laser. The 493 nm laser frequency ω was adjusted to $\omega_{sp} - \omega = 2\pi \cdot 280$ MHz to effect side band cooling. The frequency coordinate calibration is given for $\omega_a - \omega_{pd}/2\pi$. The sharp two-photon transition $6^2S_{1/2} - 6^2P_{1/2} - 5^2D_{3/2}$ to the metastable $5^2D_{3/2}$ level occuring for $\omega - \omega_a = \omega_{sp} - \omega_{pd}$ is visible at $\omega_a - \omega_{pd} = -2\pi \cdot 280$ MHz. Here ω_{sp}, ω_{pd} are the respective single photon resonance frequencies. From (Neuhauser, et al., 1981).

More recently, making use of an extension of the shelved optical electron atomic amplification scheme, Wineland and Itano (1981), at Boulder have observed the $[3^2S_{1/2}-\frac{1}{2}]$ - $[3^2P_{3/2}-\frac{1}{2}]$ electronic transition on an individual $^{24}Mg^+$ ion in the ultraviolet practically free of saturation with a width only ∿30 MHz wider than the natural, see Fig. 26. Note decrease of signal on resonance! The shelving level was the well-resolved metastable $[3^2S_{1/2}+\frac{1}{2}]$ Zeeman level while $\Delta m = -1$ optical pumping with the cooling laser pumped the ion into the $[3^2S_{1/2}-\frac{1}{2}]$ level. Weak $\Delta m = 0$ excitation from the latter level with the primary high-resolution laser then caused the ion to fall back into the shelving level, reducing the intense scattering of pumping photons.

Conclusion

After a fairly difficult start, mono-ion oscillator spectroscopy now appears to be off and running, with at least five groups all over the world actively participating in such studies. The demonstration of electronic linewidths and long-term frequency stabilities in such passive devices of 1 Hz and less is likely to act as a strong stimulus for the future development of laser sources of improved short-term stability, much closer to the fundamental limit near ∿10^{-4} Hz and also of new types of laser harmonic generators. Thus, the current promise of an atomic line spectral resolution of about 1 part in 10^{18} may be realized in the not too far future.

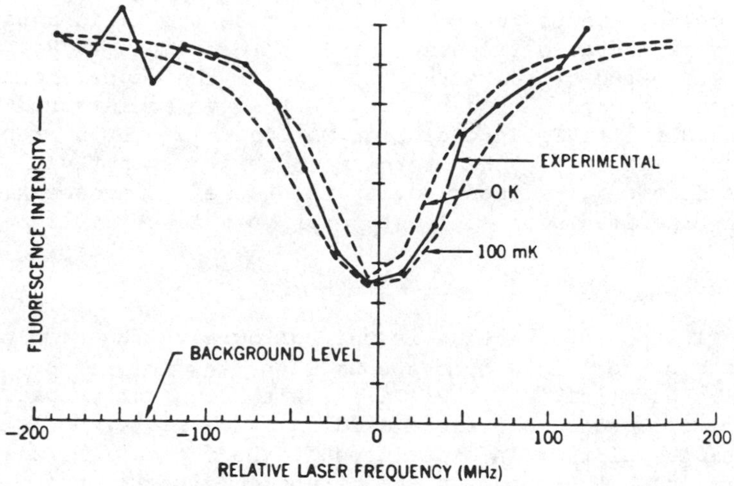

Fig. 26. Ultraviolet transition on individual $^{24}Mg^+$ ion observed by means of new version of shelved optical electron atomic amplification scheme. From (Wineland and Itano, 1981).

3. LECTURE: ION TRAPS

The obvious criteria for usefulness in spectroscopy are minimal disburbance of atomic systems under study, and adequate storage capacity. Long storage times, low kinetic energy, minimal heating processes and simple orbits are desirable. This, more or less eliminates space charge and buffer gas trapping. The oldest high-vacuum device appears to be the Kingdon (1923) trap using a DC electrostatic potential only. Its potential for rf spectroscopy was recognized early (Dehmelt, 1956; Prior et al., 1978). The Lawrence trap or magnetic bottle, well known from cyclotron, accelerators and fusion devices relies solely on an inhomogeneous magnetic field. In contrast to the two previous devices the Penning trap is distinguished by a cyclotron frequency constant throughout its volume. Based on the Penning discharge geometry, it uses a large homogeneous magnetic field plus weak parabolic electric potential. Many spectroscopic experiments have employed the Paul rf electric quadrupole trap (Fischer, 1979). This is the only trap providing a genuine three-dimensional parabolic well. In the following, we shall focus only on the last two trap types.

THE PENNING TRAP

The Penning trap warrants a detailed discussion. Once, extraordinarily narrow linewidths (10 mHz) have been reported by Wineland and coworkers (1981) in hyperfine optical pumping experiments on Mg^+ ions using laser-excitation. Twice, it may be possible to excite with a laser an optical harmonic of the cyclotron resonance of stored electrons which, due to the relativistic mass-shift, is anharmonic, realizing a large-ratio frequency divider (Wineland, 1979). Similarly, as suggested by the author in 1976 at the Copper Mountain Symposium on Frequency Standards and Metrology, an infrared or optical harmonic of a relativistic cyclotron motion driven by a precisely frequency stabilized ~10 GHz microwave generator might yield sufficient excitation of a laser cavity to be useful in some high-resolution experiments or at least to drive a laser amplifier.

Particle Orbits

The principle of the trap is the following. The simplest motion of an electron (ion) in a homogeneous magnetic field is one along a field line parallel, say the z-axis, with constant velocity. If a weak electric quadrupole field axially symmetric around the z-axis of appropriate polarity is superimposed, the motion is still along a straight line but because of the parabolic electric potential well of depth D now present it is sinusoidally oscillatory at frequency ω_z, see Fig. 1 and 2. The motion in the xy plane is easily analyzed for the special case of circular orbits (radius r) centered on the trap center (the origin). Because of the validity of Laplace's equation the radial electric force F_e must have the form $F_e = m\ (\omega_z^2/2)r$, m the electron mass. The force F_e opposes the centripetal magnetic force $e(v/c)B$, v the velocity. Newton's second law then takes the form

$$e(v/c)B - m(\omega_z^2/2)r = mv^2/r \,. \tag{1}$$

Introducing the circular frequency $\omega \equiv v/r$ and the cyclotron frequency $\omega_c \equiv eB/mc >> \omega_z$

$$2\omega(\omega_c - \omega) = \omega_z^2 \text{ follows.} \tag{2}$$

One root of this equation is $\omega = \omega_c'$, the cyclotron frequency slightly shifted by the electric field. Now inspection of (2) shows that it is also satisfied by $\omega = \omega_c - \omega_c'$ which must be the other root. Defining $\omega_c - \omega_c' \equiv \omega_m$ we have

$$\omega_m = \omega_z^2/2\omega_c' \tag{3}$$

and

$$\omega_c = \omega_c' + \omega_z^2/2\omega_c' \,. \tag{4}$$

This shows that circular motion is also possible at a frequency much lower than ω_c namely at the "magnetron frequency" ω_m and that the shift in ω_c may be easily corrected when ω_z is known.

The most general orbit may be obtained by super-posing all three previously described simple orbits. The most amazing feature of the Penning trap is that all three frequencies ν_c', ν_z', ν_m are constants of the trap, i.e., they do not depend on the location of the ion or electron in the trap. For an electron and the parameter values $\nu_c' \simeq 51$ GHz, $\nu_z \simeq 60$ MHz, $\nu_m \simeq 35$ KHz and for thermal (or Brownian motion) excitation of cyclotron and axial motions at liquid He temperature the general orbit consists of a very fast cyclotron motion of

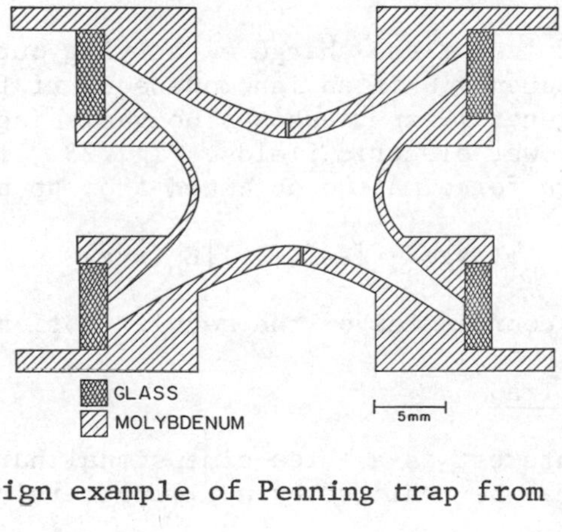

Fig. 27. Design example of Penning trap from (Walls, 1970).

Table 2. Operating Parameters of Penning Trap for Containment of Thermal Electrons

Electrode Dimensions	$2z_0=1cm, r_0/z_0=\sqrt{2}$
Magnetic Field	$H_0=7.8$ kG
DC Operating Voltage	$\|U_0\|=2D=12.8$ Volts
Cyclotron Frequency	$\omega_0/(2\pi)=22.4$ GHz
Axial Oscillation Frequency	$\omega_z/(2\pi)=58$ MHz
Magnetron (Drift) Frequency	$\omega_m/(2\pi)=75$ KHz
Storage Time	Weeks

about 600Å diameter around a more slowly moving guiding center. This guiding center in turn executes a fast axial oscillation of about 0.1 mm peak to peak amplitude and simultaneously carries out a very slow circular motion around the origin of whose radius it has so far not been possible to reduce below about 30 μm diameter. The general equations of motion may be written

$$m\ddot{x} - m\omega_z^2 x/2 + m\omega_c \dot{y} = 0$$

$$m\ddot{y} - m\omega_z^2 y/2 - m\omega_c \dot{x} = 0 \qquad (5)$$

$$m\ddot{z} + m\omega_z^2 z = 0 \ .$$

Their linearity justifies the superposition of simple orbits and their essential content is summarized in (3) and (4).

A basic design example of a Penning trap used in many spectroscopic experiments on electrons and atomic ions is shown in Fig. 27, and Table 2 gives an example of operating parameters for this trap.

PAUL RF TRAPS

A particle of mass m and charge e carrying out a forced fast oscillation at frequency Ω in an inhomogeneous rf field $E_0 \cos \Omega t$, $E_0 = E_0(xyz)$, in general experiences a nonvanishing average force directed towards lower electric fields, Fig. 28. In the limit of large Ω the average force may be obtained from an effective potential

$$\psi(\overline{xyz}) = [e/(4m\Omega^2)]E_0^2(\overline{xyz}) \qquad (6)$$

where $\overline{xyz}$ are the coordinates of the secular motion.

Axially Symmetric Trap

Of special interest is a three-dimensional harmonic potential. Such a potential may be realized by applying merely an AC potential

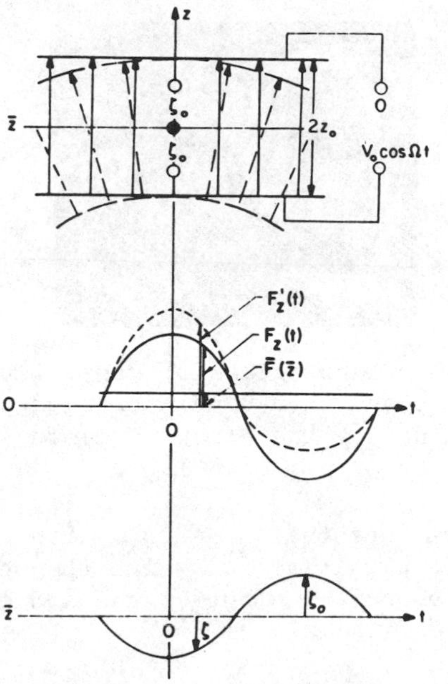

Fig. 28. Principle of rf electric quadrupole ion cage. While the force $F_z(t)$ causing the micromotion displacement ζ of an ion from its guiding center $\overline{z}$ has a vanishing time average for a homogeneous rf field (full lines) this is no longer the case in an inhomogeneous field of the same intensity (broken lines). Here the averaging of the modified instantaneous force $F_z'(t)$ over the essentially unaltered micromotion $\zeta(t)$ leads to a small, nonvanishing average force $\overline{F}(\overline{z})$, which always points toward the region of weaker rf fields independent of the sign of the ionic charge. From (Dehmelt, 1967).

$V = V_o \cos \Omega t$ to the terminals of the trap shown in Fig. 2. The resulting effective potential has the form

$$\psi(\overline{xyz}) = \overline{D}(\overline{x}^2+\overline{y}^2+4\overline{z}^2)/(4z_o^2) \tag{7}$$

independent of the sign of the charge. The effective axial depth $\overline{D}$, $[\overline{D}]$ = volt, is related to the applied rf amplitude V_o by

$$V_o = \sqrt{2}[1+\tfrac{1}{2}(r_o/z_o)^2](\Omega/\overline{\omega}_z)\overline{D} \tag{8}$$

where $\overline{\omega}_z$ is the frequency of the free secular axial oscillation in the effective potential well. Values of the stability parameter $\Omega/\overline{\omega} > 10$ are desirable. Fig. 29 shows a photograph of the motion of a single charged metal particle in the xz plane. One recognizes the fast Ω micromotion and the Lissajous pattern of the xz secular

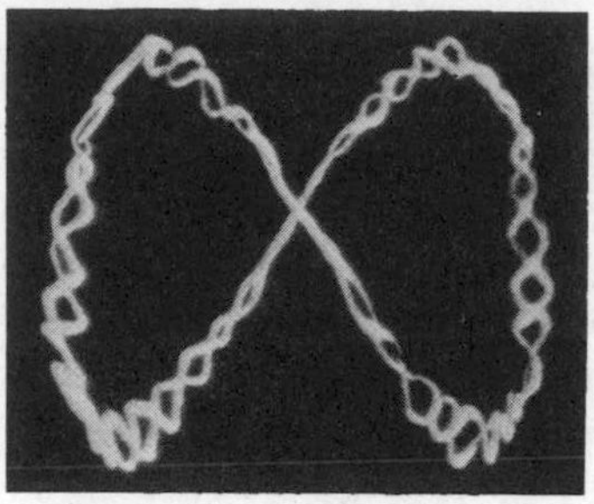

Fig. 29. Retouched microphotograph of the 2:1 Lissajous trajectory in the x-z plane of a single charged particle of aluminum dust contained by the electrodynamic suspension system. From (Wuerker er al., 1959).

motion reflecting the fact that $\bar{\omega}_z = 2\,\bar{\omega}_x$. Fig. 30 shows the xz motion for a cloud of particles for which the secular motion has been completely damped out by a buffer gas and only the micromotion is visible.

<u>Small Trap Design Example</u>

A design example for a small ion trap is given in Fig. 31 and in Table 3 (Church, 1969). For ∿ 7,000 protons initial characteristic storage times of ∿2 minutes were realized without cooling (Fig. 32). When the axial ion motion was coupled to a tuned circuit of resonance frequency $\sim\bar{\omega}_z$ and at room temperature, see Fig. 6, the characteristic storage time could be stretched to ∿45 minutes on this trap. For fixed Ω systematic studies of ion lifetimes and ion cloud temperatures have been undertaken when V_0 and therefore $\bar{D}$ are varied. The data are summarized in Table 4. They confirm the concept of an ion cloud in temperature equilibrium with itself. Even without tuned circuit cooling the ion temperature T_m is much lower than the well depth, $e\bar{D}/kT_m \approx 15$. The temperatures T_m are

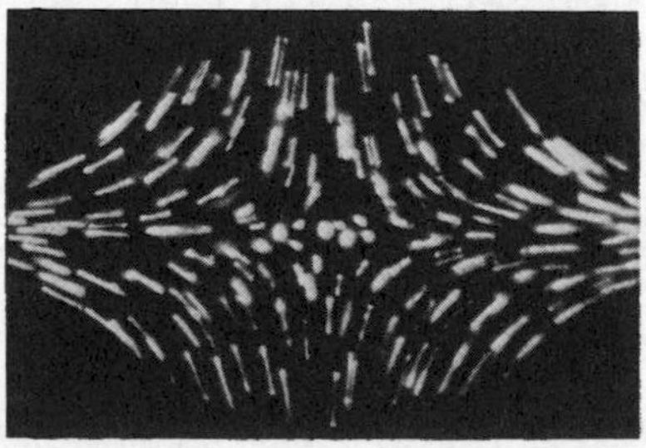

Fig. 30. Cloud of charged metal particles suspended in symmetrized potential well as viewed in the xz plane. From (Wuerker et al., 1959).

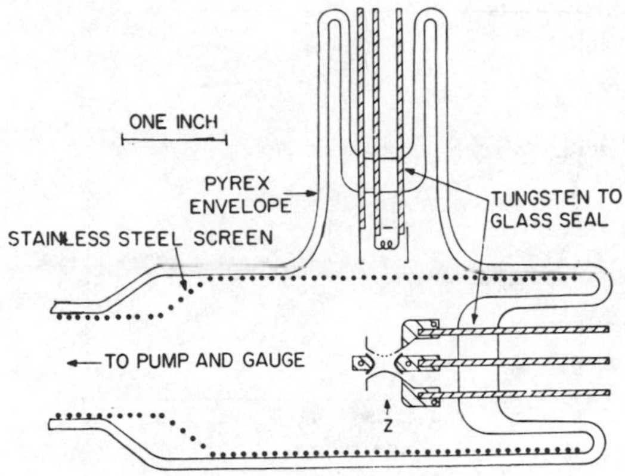

Fig. 31. A cross section of the ion trap and glass portion of the vacuum system. The holes in the ion trap electrode for the passage of electrons have been exaggerated for clarity. From (Church and Dehmelt, 1969).

assumed to be maintained by evaporation cooling even in the presence of slight heating effects associated with the micromotion: the slow loss of ions in the high energy Maxwell-Boltzmann tail, each removing the large amount of heat energy $\sim e\bar{D}$, suffices to stabilize the cloud. For about 2,000 ions via LC resonant cooling it was possible to reduce T_m further by nearly another 70%, to T_e for which a minimum value of $\sim$800 K for $\bar{D}$=2.6 V was found.

Race Track Trap

RF ion storage devices are not restricted to the geometry of Fig. 1. Fig. 33 shows a "race track" design (Church, 1969) derived from the widely used Paul-Raether mass filter. In Fig. 34 a plot of proton numbers versus storage time is given for a similar, improved circular device. At the base pressure of $\sim 2 \times 10^{-10}$ Torr of residual gas (helium) the respectable characteristic proton storage time of $\sim$5 minutes was found, also inverse proportionality to the He pressure.

Table 3. Small Symmetrized Ion Trap Parameters

Field dimension $z_o = 2^{-\frac{1}{2}} r_o$	0.16 cm
Trapping rf frequency $\Omega/2\pi$	146 MHz
rf Amplitude V_o	485 V
DC Bias U_o	5.5 V
Ion oscillation frequency $\bar{\omega}_z/2\pi$	4.4 MHz
Well depth $\bar{D}$ (axial and radial)	$\sim$10 V
Calculated maximum ion number N_{max}	2×10^7
Maximum measured ion number N	7×10^3
Estimated base pressure	7×10^{-12} Torr

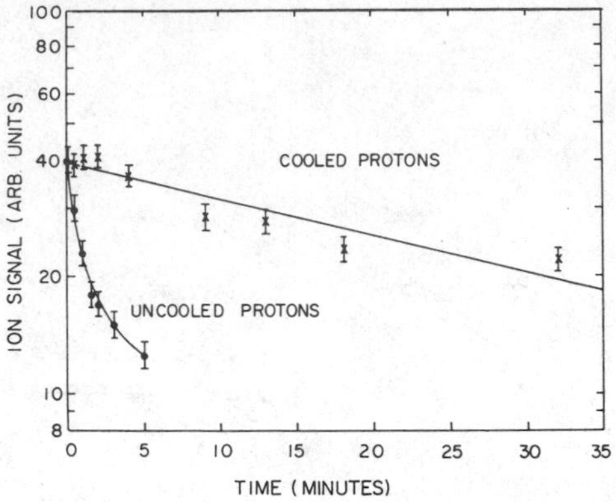

Fig. 32. Decay of the stored proton population with time for the two cases of cooled and uncooled protons. The well depth was D = 10.4 V. From (Church and Dehmelt, 1969).

ION CLOUD IN TEMPERATURE EQUILIBRIUM

The fact that the maximum observed ion density n observed by Church in the LC-cooled clouds remains 10-100 times lower than the space charge limit n_{max} is in agreement with the hypothesized Ω-rf heating in proton-proton collisions. While in the Heidelberg experiments 2 presumably isotopically identical Ba^+ ions could be compressed to the theoretical minimum separation (5μ) via intense laser cooling, clouds of ∿50 ions of natural isotopic composition could only be compressed to 50 μm diameter corresponding to a density about

Table 4. Small Ion Trap Data

Well Depth $\bar{D}$ (V)	10.4	4.6	2.6	3.9
$\bar{\omega}_z/2\pi$ (MHz)	4.4	2.95	2.2	2.7
V_o (Volts)	485	215	123	183
C (pF)	7	10	23	7
$3\tau_{zo}$ (sec)	13	13	17	7.3
T_B (T_m) (sec)	100	100	100	120
T_B (T_e) (sec)	2600	--	2400	2400
kT_e/e (V)	0.33	0.16	0.07	0.08
kT_m/e (V)	0.93	0.30	0.16	0.21
N_c (typical)	3000	1600	2000	1600
N_t (typical)	3000	2300	2000	2000

V_o Peak trapping rf voltage.
C Tank circuit capacitance.
$3\tau_{zo}$ Ion cooling time constant.
$T_B(T_m)$ Ion loss time constant at the temperature T_m.
$T_B(T_e)$ Ion loss time constant at the temperature T_e.
T_m Evaporation cooled temperature.
T_e LC-cooled ion temperature.
N_c Ion number measured by exciting a coherent ion signal.
N_t Ion number from tank circuit temperature measurements.

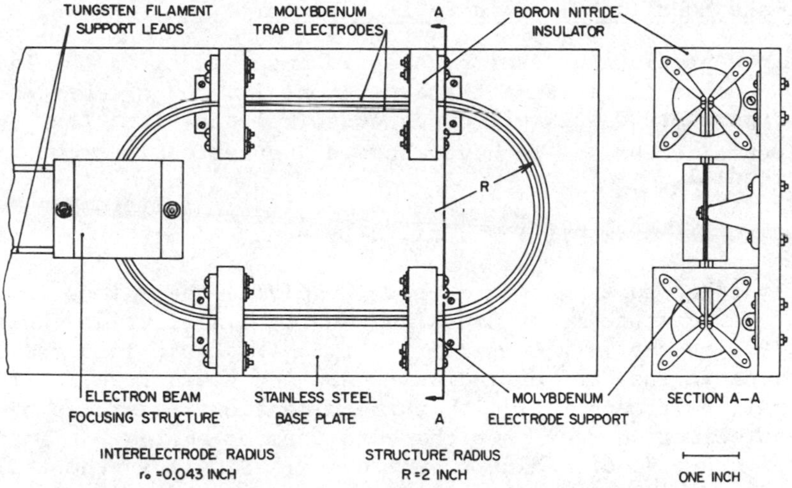

Fig. 33. "Racetrack" ion trap. From (Church, 1969).

$n_{max}/100$ (Neuhauser et al., 1978 and 1980). Here Ω rf heating in collisions involving different Ba^+ isotopes is the likely cause. This process also explains the rapid loss of such clouds in the absence of laser cooling.

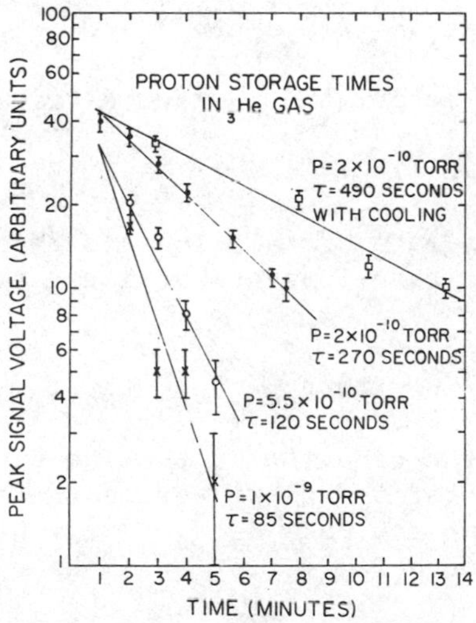

Fig. 34. Decay of the stored proton population in the trap with time for long times at several pressures. The background gas was helium in all cases. From (Church, 1969).

Approximate Model of Cloud in Perfect Vacuum

We now attempt to develop an approximate model in the limit $kT_i \ll e\bar{D}$ capable of yielding the experimentally found dependence of ion storage times T_B on well depth $e\bar{D} \equiv e\psi(r_o)$ and other trap parameters (Dehmelt, 1967). We have assumed a spherically symmetric pseudpotendial

$$\psi(r) = d \cdot r^2, \; d \cdot r_0^2 = e\bar{D} \tag{9}$$

initailly taken to extend to $r \gg r_o$. We divide the N ions in the cloud at temperature T_i of Gaussian density distribution (Knight and Prior, 1979) $\propto \exp[-(x^2+y^2+z^2)/r_c^2]$, $r_c^2 \equiv r_0^2(kT_i/e\bar{D})$, into two groups: the $\sim$N ions in the core of radius r_c and the small fraction $f N$ in the Maxwell-Boltzmann tail with three-dimensional energies $\bar{W} > \bar{W}_B \gg kT_i$. In an approximation we assume the core ions of energy $\sim 3kT_i$ to fill the core volume V_c of radius r_c uniformly. Similarly the tail ions of energy $\gtrsim \bar{W}_B$ are assumed to fill the volume $V_t = (4\pi/3)r_t{}^3$, of radius r_t, $d \cdot r_t^2 = \bar{W}_B/3$ uniformly. The assumption of an ion cloud at thermal equilibrium at temperature T_i is justified, because the observed storage times are much longer than estimated ion-ion collision times. Most of the tail ions have energies only slightly larger than $\bar{W}_B$. These tail ions return to the core of volume $V_c = (4\pi/3)r_c^3$, $dr_c^2 = kT_i$, when they move through V_c with a kinetic energy $E_k \simeq \bar{W}_B$ due to collisions with the slower core ions. Assuming that the tail ions spend the fraction V_c/V_t of their time inside the core and using Spitzer's (1956) formula for the slowing down time in a plasma

$$t_s = m^{1/2}(2E_k)^{3/2}/(8\pi e^4 n \ln\Lambda) \equiv C \cdot (2E_k)^{3/2} n^{-1} \tag{10}$$

one finds for the characteristic tail-core transition time

$$t_{tc} \simeq (V_t/V_c) \cdot C \cdot (2W_B)^{3/2}(V_c/N) = C\,(2W_B)^{3/2}(V_t/N) \; . \tag{11}$$

The corresponding loss rate of tail ions is fN/t_{tc}. In equilibrium the rate N/T_B at which core ions are promoted to the tail must equal the loss rate:

$$N/T_B = fN/t_{tc}, \; T_B = t_{tc}/f \; . \tag{12}$$

In order to describe the behavior of the ion cloud in a finite-sized trap with cap-cap separation $2z_o$ we assume that any ion with sufficient energy which we now take as $\bar{W}_B$ to reach the spherical equipotential surface through $00z_o$, $z_o = r_o$ will be quickly lost (at the caps),

$$dz_0^2 = \bar{W}_B \; . \tag{13}$$

This allows us to identify T_B with the ion lifetime in the trap due to evaporation cooling. With

$$1/f = 2(kT_i/\bar{W}_B)^2 \exp(\bar{W}_B/kT_i) \tag{14}$$

we may now write with (12), $r_t^2/z_o^2 = 1/3$ and with the abbreviation $\bar{W}_B/kT_i \equiv b$

$$T_B = (C/N)\,\frac{8\pi}{3}\,(2/3)^{3/2}\bar{W}_B{}^{3/2} z_o^3 b^{-2} \exp b. \tag{15}$$

Finally, with $\ell n\,\Lambda \simeq 20$ we have

$$T_B \simeq 5\cdot 10^5 A^{1/2} N^{-1} z_o^3\, \bar{W}_B{}^{3/2} b^{-2} \exp b$$

$$[z_o] = \text{cm},\; [\bar{W}_B] = \text{eV}\,. \tag{16}$$

In order to compare this result with Church's data we interpret $\bar{W}_B = d\cdot z_o^2$ as an effective well-depth and z_o as an effective axial well-dimension. To be determined from the experimental parameters A, N, d, T_B, kT_i by solving (16) for z_o. Example (I), A=1, N=3000, T_B=100 sec, kT_i=0.93 eV, d=706 eV/cm^2 yields z_o (eff.)=0.126 cm to be compared with the geometrical value z_o=0.16 cm and $\bar{W}_B$=6.5 eV to be compared with $e\bar{D}$=10.4 eV. Example (II), A=1, N=1500, T_B=100 sec, kT_i=0.3 eV, d=180 eV/cm^2 gives z_o (eff.)=0.123 cm and $\bar{W}_B$=2.73 eV vs. $e\bar{D}$=4.6 eV. That the effective trap dimension z_o (eff.) for ion loss must be smaller than the geometrical one, z_o, is obvious for large micromotion amplitudes associated with small Ω/ω. Another possible contributing cause is contact potential effects which may be expected to increase for low $e\bar{D}$ values.

Cloud With Ion-Ion RF Heating

The ion loss process just modeled is associated with evaporation cooling or heat loss from the cloud at the rate $\sim N\,\bar{W}_B/T_B$ which compensates some unspecified heat input. Even in a perfect vacuum there may be a heat input due to ion-ion collisions in the electric rf field. This is certainly the case when the ion cloud consists of an isotopic mixture. We attempt to describe this case by assuming an average heat input $a_i 3kT_i$, $o<a_i<1$ per ion collision. The corresponding heat flux is $N\cdot a_i 3kT_i/t_c$ where t_c is the Spitzer (1956) self collision time for the core ions,

$$t_c = (C/0.7)(3kT_i)^{3/2}(V_c/N)\,. \tag{17}$$

In equilibrium holds

$$N\, a_i 3kT_i/t_c = N\,\bar{W}_B/T_B \tag{18}$$

from which follows

$$\bar{W}_B/kT_i = -\ell n\, a_i + 0.81\,. \tag{19}$$

Even for a heating coefficient a_i as low as 0.01 this shows the rather high ion temperature $kT_i/\bar{W}_B \simeq 1/4$ independent of ion number N.

Cloud With Ion-Atom RF Heating

Taking the heat input in ion collisions with heavier atoms (at rest) of mass m_r as $a_r \cdot 3kT_i$, $a_r = a_r(m/m_r)$, per collision, m the ion mass (Dehmelt, 1967), one has in equilibrium of heating and cooling rates

$$N \cdot a_r 3kT_i/T_r = N\,\bar{W}_B/T_B \text{ or } T_B/T_r = b/3a_r \tag{20}$$

T_r the ion-residual atom collision time. The equations (16) and (20) may now be solved for b and T_B. With abbreviation $C_r = 1.5 \cdot 10^6 A^{1/2} N^{-1} z_0^3\, \bar{W}_B^{3/2} a_r/T_r$ one finds

$$b = 3\ell nb - \ell nC_r\ . \tag{21}$$

With the expansion $\ell nb \simeq 1.3+0.1b$ valid near $b \simeq 10$ this yields

$$b \equiv \bar{W}_B/kT_i = 1.43[\ell nN \cdot T_r - 0.5\ell nA - 3\ell nz_0 - 1.5\ell n\bar{W}_B - \ell na_r - 10.3]\ . \tag{22}$$

For the storage time T_B now follows

$$T_B = b \cdot T_r/3a_r\ . \tag{23}$$

In the numerical example A=25, z_0=0.08 cm, $\bar{W}_B$=2 eV, N=10^3, T_r=10^3 sec, a_r=1/3, we find b=13.7, kT_i=0.15 eV, $T_B \simeq 13\ 000$ sec, while N=10, T_r=100 sec, yields b=3.9, $kT_i \simeq 0.5$ eV, and $T_B \simeq 400$ sec. Equation (23) is in rough agreement with the data of Fig. 34 as b is approximately constant.

4. LECTURE: ION COOLING

The following methods have been used in the past, compare (Dehmelt, 1967 and 1969): (A) Cooling by light buffer gas, see Fig. 30. (B) Coupling the ion motion to a cold resonant circuit, see Figs. 6, 9, 16, 21 and Table 4, or the free space radiation field, Figs. 13, 15, 19 and 21. (C) Evaporation cooling, see Table 4. (D) Adiabatic reduction of well depth. (E) Side band cooling, see Figs. 20 and 24. Here we only have space to discuss (E), compare (Neuhauser et al., 1980 and 1981).

SIDE BAND COOLING

Thomson Atom in Doppler Regime

Without much loss of generality we discuss the cooling process for a (non-saturable) Thomson atom, here defined as an electron elastically bound with characteristic frequency ω_0 to a much heavier homogeneous positive spherical rigid charge distribution. The atom at z is restrained to a harmonic vibration at ω_v along the z axis, $z(t)=z_o \cos \omega_v t$ and irradiated with a plane wave

$$E_y \propto \cos(\omega t - z/\lambda\!\!\!^{-}),\ \lambda\!\!\!^{-} \simeq \lambda\!\!\!^{-}_o = c/\omega_o\ . \tag{24}$$

The electric field seen by the moving atom then has the form

$$E_y \propto \cos[\omega t - (z_o/\lambda\!\!\!^{-}) \cos \omega_v t]\ . \tag{25}$$

The usual expansion of (25) yields spectral components at $\omega + n\omega_v$, n=0, ±1,±2,±3,... of relative power $J_n^2(z_o/\lambda\!\!\!^{-})$. These spectral components now excite the atom independently according to their position with respect to the Lorentzian response function

$$g(\omega) = \{1+[2(\omega-\omega_o)/\gamma]^2\}^{-1} \tag{26}$$

with the $\omega+n\omega_v \simeq \omega_o$ components predominating. Each of the exciting spectral components $\omega+n\omega_v \simeq \omega_o$ is then re-radiated in the atoms rest frame. Again, through the Doppler effect when viewed from the lab-frame each re-emitted component $\omega+n\omega_v$ develops its own Doppler side-bands $(\omega+n\omega_v)+m\omega_v$, m=±1,±2,±3 symmetric around $\omega+n\omega_v \simeq \omega_o$. A simple lab-frame energy balance shows that while photons of energy $\hbar\omega$ are absorbed, photons of average energy $\sim\hbar\omega_o$ are re-emitted. For $\omega<\omega_o$ this implies cooling of the oscillatory motion by an amount $\sim\hbar(\omega_o-\omega)$ per photon. For the case $\omega_v \ll \gamma \ll \omega_o - \omega \leq \omega_v\beta$, $\beta \equiv z_o/\lambda\!\!\!^{-} \gg 1$ a simple expression for the cooling rate may be obtained. The power spectrum seen by the ion may be approximated by

$$P(\omega') \propto \{1-[(\omega'-\omega)/\beta\omega_v]^2\}^{-\frac{1}{2}}\ . \tag{27}$$

Of the spectral components ω' only those in a γ-wide band around ω_o are scattered appreciably. We now further approximate $g(\omega)$ by a square band of height 1 and width γ and $P(\omega')$ by a square band of height $(2\beta\omega_v)^{-1}$ and width $2\beta\omega_v$. Introducing, S, the resonant scattering rate for the stationary Thomson atom we obtain for the heat loss from the vibrational energy W or the cooling power

$$\dot{W} = -S\cdot\gamma\cdot\hbar(\omega_o-\omega)/2\beta\omega_v\ . \tag{28}$$

In an alternative model one may begin with the observation that the atom in its rest frame sees an excitation at a modulated frequency $\omega'(t)$. For brief intervals Δt resonance occurs when $\omega'(t)$ falls into the γ-wide band around ω_o and then the atom emits bursts of photons which, in the laboratory frame, have an average energy $\hbar\omega_o$. Approximating the time variation of $\omega'(t)$ by a periodic linear rise from $\omega-\beta\omega_v$ to $\omega+\beta\omega_v$ in the half-period π/ω_v and a linear fall back to $\omega-\beta\omega_v$ in again a time π/ω_v the fractional time the atom spends scattering photons is obviously $(\gamma/2\beta\omega_v)$. This again leads to formula (28). This way of looking at the problem also shows that the saturation problem arising in a real two-level atom for $S>\gamma$ is not relieved by the oscillatory motion in the trap as the spreading-out of the power in the strictly monochromatic ω-radiation over the many (2β) spectral components in the side-band picture would suggest if their phases were random. For the maximum unsaturated ($S\simeq\gamma/3$) heat loss rate one has with the tuning $(\omega_o-\omega)=\beta\omega_v$

$$(\dot{W})^*\simeq\gamma^2\hbar/3 \ . \tag{29}$$

A numerical example, Mg^+, $\lambda_o=0.28$ μm, $\gamma\simeq2.7\times10^8$ Hz well-depth $e\bar{D}_z=10$ eV, well-width $=1.5$ mm, $\omega_v=2\pi\cdot1.9$ MHz yields for $kT_i\simeq2.5$ eV, $z_o=0.38$ mm, $(\dot{W})^*\simeq16$ eV/sec, for $\omega_o-\omega\simeq2\pi\cdot16$ GHz. For a cloud in a spherical well one may expect cooling rates somewhat smaller than those given by (28) with $\beta\simeq\langle z_o\rangle/\lambda\!\!\!^{-}_o$ now.

Thomson Atom in Lamb-Dicke Regime

For $(z_o/\lambda\!\!\!^{-}_o)\equiv\beta\ll1$, the Lamb-Dicke regime, the expansion of formula (26) becomes particularly simple:

$$E_y \simeq \cos\omega t-\beta\cos\omega_v t\sin\omega t \ . \tag{30}$$

This shows that the spectrum now consists only of the dominant carrier plus two weak side bands at $\omega\pm\omega_v$ of fractional power $f=\beta^2/4$. In this limit a fairly general expression for the cooling power may easily be obtained. The re-emission rate of $\omega_-\equiv\omega-\omega_v$, ω, $\omega_+\equiv\omega+\omega_v$ photons in the ion rest frame is given by

$$Sfg(\omega_-),\ \sim S\cdot1\cdot g(\omega),\ Sfg(\omega_+) \ . \tag{31}$$

Again, through the Doppler effect these re-emmitted ω_-, ω, ω_+ components in the lab-frame develop Doppler components of their own. As this does not affect the heating-cooling balance very much a rough approximation is justified: We replace the continuous $\cos^2\theta_y$ re-radiation pattern by a simpler discrete one, in which the power is re-radiated in equal parts along the +x, -x, +z, -z axes. Only the parts re-radiated along ±z, i.e., ½ of the total power, are affected by the Doppler effect which gives rise to two weak side bands each $f/2$ of the power in the respective ion rest-frame parent component. The relative powers of the re-emitted lab-frame components $\omega_--\omega_v$, ω_-, $\omega_-+\omega_v$; $\omega-\omega_v$, ω, $\omega+\omega_v$; $\omega_+-\omega_v$, ω_+, $\omega_++\omega_v$ are proportional to

$$\tfrac12f^2g_-,\ fg_-,\ \tfrac12f^2g_-;\ \tfrac12fg_\omega,\ g_\omega,\ \tfrac12fg_\omega;\ \tfrac12f^2g_+,\ fg_+,\ \tfrac12f^2g_+$$
$$g_-\equiv g(\omega_-),\ g_\omega\equiv g(\omega),\ g_+\equiv g(\omega_+) \ . \tag{32}$$

For most important application g_-, g_ω, $g_+<1$ and $f\ll1$ hold which justifies neglecting all terms containing f^2, which will be important later on. In the lab-frame one then has finite re-emission rates for $(\hbar\omega_-)$-, (average energy $\hbar\omega$)-, and $(\hbar\omega_+)$-photons with corresponding respective heating, neutral and cooling rates $Sfg_-\hbar\omega_v$, $Sg_\omega\hbar\omega_v$, and $Sfg_+\hbar\omega_v$. Various lab-frame spectral components overlap and the corresponding interference terms have been neglected in a further approximation. For the net cooling power follows now

$$\dot{W} = -Sf\hbar\omega_v(g_+ - g_-) \tag{33}$$

and with $W=(M/2)z_0^2\omega_v^2$ a characteristic cooling time

$$\tau_c \equiv -W/\dot{W} = \text{const} \tag{34}$$

may be defined.

Cooling Limit

A cooling limit now is established by quantum effects. Taking the ion to be initially in the vibrational energy level v, lab-frame re-emission of ω_+, ω_- photons must be accompanied by transitions $v \rightarrow v-1$ and $v \rightarrow v+1$. Recalling $f=z_0^2/4\lambda\!\!\!^-{}_0^2$ we expect $f(v \rightarrow v-1)$ to differ from $f(v \rightarrow v+1)$ as, so to speak, the vibrational amplitude shrinks during one transition and in the other grows. This is born out as the quantum-mechanical analog of $f=z_0^2/4\lambda\!\!\!^-{}_0^2=W/2M\omega_v^2\lambda\!\!\!^-{}_0^2$ is $f_\mp = f(v \rightarrow v\pm 1)$ with

$$f(v \rightarrow v\pm 1) = (W_{v+\frac{1}{2}\pm\frac{1}{2}} - W_0)/2M\omega_v^2\lambda\!\!\!^-{}_0^2 = (v+\tfrac{1}{2}\pm\tfrac{1}{2})/v_L . \tag{35}$$

Here

$$v_L \equiv 2(M\omega_v^2\lambda\!\!\!^-{}_0^2/2)/(\hbar\omega_v/2) >> 1 \tag{36}$$

marks the boundary of the Lamb-Dicke regime and W_v are the vibratory energy eigenvalues (Dehmelt, 1979). This modifies the relative (finite) re-emission rates for ω_-; $\omega-\omega_v$, ω, $\omega+\omega_v$; ω_+ photons of formula (32) to f_-g_-; $\frac{1}{2}f-g_\omega$, g_ω, $\frac{1}{2}f+g_\omega$; f_+g_+ and now follows

$$\dot{W} = -S\hbar\omega_v[-f_-g_- - \tfrac{1}{2}(f_- - f_+)g_\omega + f_+g_+] \text{ or}$$
$$\dot{W} = -S\hbar\omega_v[-(v+1)g_- - \tfrac{1}{2}g_\omega + vg_+] . \tag{37}$$

Setting $\dot{W}=0$ yields the desired minimal average vibrational level

$$\langle v\rangle_{min} = [g(\omega_-)+\tfrac{1}{2}g(\omega)]/[g(\omega_+)-g(\omega_-)] . \tag{38}$$

To illustrate we set $\omega_0-\omega=\gamma/2$ and $\omega_v<<\gamma$ and obtain $\langle v\rangle_{min} \simeq \gamma/2\omega_v$ and the minimum average vibrational energy $\langle W\rangle_{min} \simeq \hbar\gamma/2$. More elaborate treatments yielding essentially the same results are available (Wineland and Itano, 1979; Javanainen, 1980).

It is a pleasure to acknowledge discussions with J.C. Bergquist, Rainer Blatt, Gary Janik, Warren Nagourney, Robert Puff, T. Sauter, G. Werth, and David Wineland. Mitzie Johnson typed the manuscript and David Azose did the photographic work.

REFERENCES

Bloch, F., 1953, Physica, 19, 821.
Brillouin, L., 1928, Proc. Nat. Acad. Sciences, 14, 756.
Church, D., 1969, Thesis, University of Washington; Journal of Applied Physics, 40, 3127, and 3421.
Dehmelt, H. G., 1956, Phys. Rev., 103, 1125.
Dehmelt, H., 1961, Progress Report NSF-G5955, May 1961, "Spin Resonance of Free Electrons."
Dehmelt, H., 1962, Bull. Am. Phys. Soc., 7, 470.
Dehmelt, H., and Major, F. G., 1962, Phys. Rev. Lett., 8, 213.
Dehmelt, H. G., 1967 and 1969, Advan. At. Mol. Phys., 3, 53, and 5, 109.
Dehmelt, H., and Walls, F., 1968, Phys. Rev. Lett., 21, 127.
Dehmelt, H., and Ekstrom, P., 1973, Bull. Am. Phys. Soc., 18, 727.
Dehmelt, H. G., 1973, Bull. Am. Phys. Soc., 18, 1571.
Dehmelt, H. G., 1975, Bull. Am. Phys. Soc., 20, 60.
Dehmelt, H., and Toschek, P., 1975, Bull. Am. Phys. Soc., 20, 61.
Dehmelt, H., and Walther, H., 1975, Bull. Am. Phys. Soc., 20, 61.
Dehmelt, H. G., 1976, in: "Atomic Masses and Fundamental Constants" Vol. 5, Edited by J. H. Sanders and A. H. Wapstra, Plenum.
Dehmelt, H., Schwinberg , P., and Van Dyck, Jr., R. S., 1978, Int. J. Mass. Spec. and Ion Physics, 26, 107.
Dehmelt, H., 1976, Nature, 262, 777, and 1979, Bull. Am. Phys. Soc., 24, 634.
Dehmelt, H., 1981, in: "Atomic Physics 7," edited by Daniel Kleppner and Frank Pipkin, Plenum.
Ekstrom, P., and Wineland, D., 1980, Scientific American, 243, No. 2, p. 104.
Field, J., Picasso, E., and Combley, F., 1979, Soviet Physics Uspekhi, 22(4), 199.
Fischer, E., 1959, Z. Physik, 156, 1.
Fortson, E. N., Major, F. G., and Dehmelt, H. G., 1966, Phys. Rev. Lett., 16, 221, and Schuessler, H. A., Fortson, E. N., and Dehmelt, H. G., 1969, Phys. Rev., 187, 5.
Friedburg, H., and Paul, W., 1951, Naturwiss., 38, 159.
Itano, Wayne M., and Wineland, D. J., 1981, Phys. Rev. A., to be published.
Javanainen, J., 1980, Appl. Phys., 23, 175.
Kinoshita, T., 1979, Private Communication.
Major, F. G., and Werth, G., 1973, Phys. Rev. Lett., 30, 1155.
Mehra, Jagdish, 1975, "The Solvay Conferences on Physics," D. Reidel Co., Dordrecht, Boston, p. 184 and 198.
Neuhauser, W., Hohenstatt, M., Toschek, P., and Dehmelt, H., 1978, Phys. Rev. Lett., 41, 233.

Neuhauser, W., Hohenstatt, M., Toschek, P. E., and Dehmelt, H., 1980, Phys, Rev., A22, 1137.

Neuhauser, W., Hohenstatt, M., Toschek, P. E., and Dehmelt, H. G., 1981, in: "Spectral Line Shapes," B. Wende, editor. Walter de Gruyter & Co., Berlin, New York.

Pauli, W., 1946, in "Prix Nobel," p. 134. Eds. du Griffon, Neuchâtel.

Prior, M. H., and Wang, 1975, Phys. Rev. Lett., 35, 29.

Schwinberg, P. B., Van Dyck, Jr., R. S., and Dehmelt, H. G., 1981, Phys. Rev. Lett., (in press).

Van Dyck, Jr., R. Ekstrom, P., and Dehmelt, H., 1976, Nature, 262, 776.

Van Dyck, Jr., R. S., Schwinberg, P., and Dehmelt, H., 1978, in: "New Frontiers in High-Energy Physics," B. Kursunoglu, A. Perlmutter, and L. Scott, editors, Plenum.

Van Dyck, Jr., R. S., Schwinberg, P., and Dehmelt, H., 1979a, Bull. Am. Phys. Soc., 24, 758.

Van Dyck, Jr., R. S., Schwinberg, P. B., and Bailey, S. H., 1979b, in: "Atomic Masses and Fundamental Constants," J. A. Nolen, Jr. and W. Beneson, editors, Plenum.

Walls, F., 1970, Thesis, University of Washington.

Wineland, D., Ekstrom, P., and Dehmelt, H., 1973, Phys. Rev. Lett., 31, 1279.

Wineland, D., and Dehmelt, H., 1975, Bull. Am. Phys. Soc., 20, 637.

Wineland, D. J., and Itano, W. M., 1979, Phys. Rev., A20, 1521.

Wineland, D. J., and Itano, W. M., 1981, Physics Letters, 82A, 75.

Wuerker, R. F., Shelton, H., and Langmuir, R. V., 1959, J. Appl. Phys., 30, 342.

PROGRESS IN TUNABLE LASERS

H. Welling

Universität Hannover, Institut für Quantenoptik

Welfengarten 1
D - 3000 Hannover 1
West Germany

INTODUCTION

No doubt, tunable lasers play an important role in scientific investigations and in technical applications. Therefore we find a strong interest to improve tunable lasers and to extend their frequency range: Dye lasers today have excellent properties as tunable systems but their operation area is limited to the visible region and the close vicinity. Within the last four to five years there was a strong development of color center lasers and they became the near infrared counterpart to the dye lasers. The active material in dye and color center lasers - dyes and color center crystals - are no natural products. However dyes were already available in large numbers, whereas each new color center crystal suitable as laser material had to be developed, making the progress in color center lasers so tedious. Nevertheless the state of the art of color center lasers was continuously improved. In the first part of this lecture recent improvements that have been achieved by my coworker G. Litfin will be reported.

Having in mind that developments of new dyes and new color center crystals need a tremendous effort we investigated many natural diatomic molecules, if they are suited as active materials in tunable lasers. My coworker B. Wellegehausen worked for several years in this field. Results of his dimer laser investigations and new concepts for tunable lasers with dimer molecules will be discussed in the second part of the lecture.

If we continue our way to find simple systems or simple principles for tunable lasers, we should reverse the process of two photon absorption, where the sum energy of the two absorbed photons should be equal to difference energy of two levels. This leaves one degree of freedom for the energy of one absorbed photon. The reverse of the two photon absorption is the stimulated two photon emission. Using the amplification of this stimulated two photon emission in a coupled resonator, the two photon laser may become a tunable system. Preliminary results of this subject achieved by my coworker D. Frölich will be reported in the third part of the lecture.

NEW DEVELOPMENTS IN COLOR CENTER LASERS

In recent years color center lasers (CCLs) became an important part in the group of lasers which are continuously tunable over wide spectral ranges. These lasers are optically pumped solid state devices using color centers as the active species. The operation conditions and characteristics as well as the capability of this laser system are comparable to dye lasers. The emission range (0.8 - 3.33 μm), however, offers many new possibilities; mainly in the field of molecular and solid state physics.

Cw operation has been obtained in $F_A(II)$[1,2], $F_B(II)$[3], F_2^+ [4,5] and F_2^+ like centers [6,7,8]. $F_A(II)$ and $F_B(II)$ CCLs do not show any bleaching or fading effects so that operation over periods of several months is possible with one crystal. Therefore spectroscopic investigations with CCLs have mainly been carried out with these lasers[9,10,11]. In contrast most of the F_2^+ and F_2^+ like centers show a temporal bleaching effect so that applications of these lasers are more difficult. In the following table we compare the performance data of cw CCLs and dye lasers. It can be seen that the power output and the relative tuning ranges are of same magnitude. CCLs show a better mode behaviour and have for the free running system an essentially smaller emission linewidth. Looking for broader applications of CCLs in the future, further research on the material aspect is necessary. By all means the thermal stability of F_2^+ lasers has to be improved and the fading caused by laser radiation should be reduced. There is a strong interest to find color centers that allow laser action

Table I Comparison of performance data of CCLs and dye lasers.

	F_A/F_B	F_2^+	Dye
Output power	<250 mW	< 1 W	< 2 W
Slope efficiency	< 10 %	< 60 %	< 30 %
Tuning range	< 15 %	< 23 %	< 15 %
Total tuning range	2.2 - 3.33 μm	0.8 - 1.78 μm	0.4 - 1.1 μm
Frequency behaviour (linear cavity)			
Modes (free running)	2 - 3	100 - 1000	100 - 1000
P_{SM}/P_{MM}	75 %	10 - 20 %	10 - 20 %
Frequency behaviour (ring cavity)			
Frequency selecting elements in unidirectional operation	——	Birefringent filter + Faraday Rotator	Birefringent filter + unidirectional device + 1 FPE

up to 4μm. Cw CCL systems have already today a high degree of technical performance, but for the optical excitation expensive ion lasers are necessary. Because of technical difficulties there was a lack of well developed and simple pulsed CCLs. Many applications, however, require high peak power that is only achievable with pulsed pump sources. In particular photochemical processes need simpler and more efficient pump schemes resulting in low cost of photon production. In this first part of the lecture we will discuss new results on material research and report on new excitation schemes for pulsed tunable operation.

New centers for stable operation in the 1 to 2 μm regime

Although color center lasers cover the 0.8 to 3.33μm range so far the operation of F_2^+ and F_2^+ like centers (0.8 - 2.0μm) is still not satisfying due to slow bleaching effects and both difficult center production and storage. Recently progress has been achieved by OH doping of the crystals yielding an improved behaviour of F_2^+ center lasers[5,8]. Anyway operation of F_2^+ center lasers is still not comparable to F_A(II) and F_B(II) center lasers. Therefore the search for new centers allowing stable and rigid laser operation in the 0.8 to 2μm range is necessary. A new center which can possibly replace the F_2^+ center has recently been found in Tl^+ doped alkali halide crystals[12].

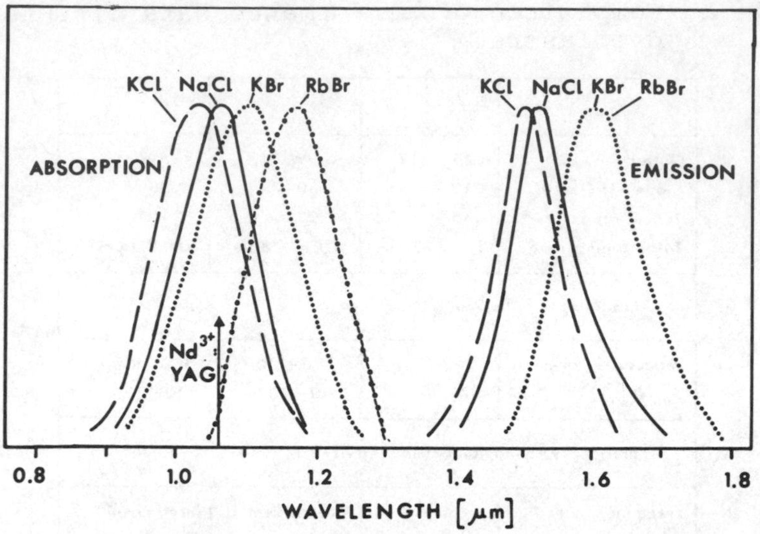

Fig. 1 Absorption spectra of different Tl^+ doped crystals at 77 K after electron beam coloration and exposure to white light.

The structure of this new center is not yet determined, although the spectroscopic behaviour is similar to that of F_2^+ centers including small stokes shift and high efficiency for the emission. Considering the high stability, however, the new center is similar to an $F_A(II)$ or $F_B(II)$ center. Fig. 1 shows the absorption spectra of these Tl^+ centers in different alkali halide crystals. Unfortunately the emission bands do not vary strongly with the host crystal resulting only in a small overall tuning range. However, doping with similar ions (Ga^+, In^+) can probably extend the emission range of this center.

First results of laser experiments are shown in Fig. 2. A KCl : Tl^+ crystal which was e-beam irradiated at - 40 C and exposed to white light was pumped with λ = 1.06μm using a Nd^{3+} YAG laser. Tunable laser operation has been achieved in the 1.42 to 1.58μm range with an output power of ~ 140 mW.

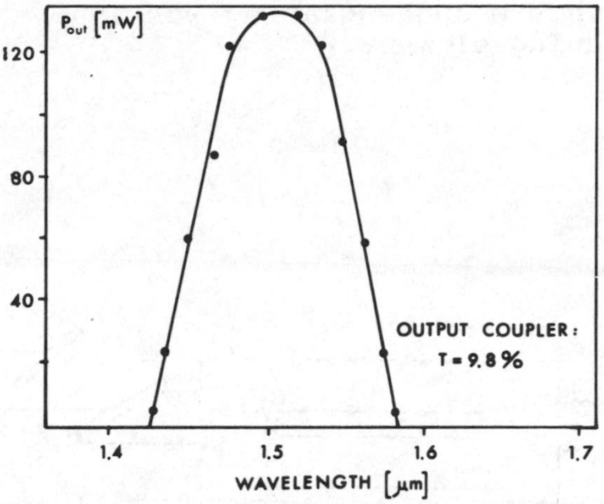

Fig. 2 Tuning curve of KCl : Tl^+ excited at 1.06μm

<u>Tunable laser operation with flashlamp pumped CCLs</u>

Color center lasers are particularly suited for operation with broad band incoherent light sources. The threshold pump intensities are low (I_{thr} ~ 200 W/cm^2),the absorption bands are broad (Δλ ~ 300 nm) and the alkali halide crystals used as the host material have a fairly high thermal conductivity (0.4 W/cm K). Flashlamp pumped operation of a CCL has already been demonstrated in 1965 by F. Fritz and E. Menke[13]. However at that time the concept of broadly tunable lasers has not yet been considered and the authors did not utilize the tuning potential of this laser. They only demonstrated laser action and obtained a few mW of output power near the band center of an F_A(II) and KCl:Li crystal at 2.71μm. In the work reported here tunable laser operation with flashlamp pumping has been achieved with three different F_A(II) and F_B(II) centers in KCl and RbCl hosts. Fig. 3 shows the schematical set-up of the flashlamp pumped CCL. Cylindrical color center rods containing ~ 6 $\cdot 10^{16}$ F(II) centers per cm^3 are mounted on a copper cold finger in one focal line of a cylindrical ellipse. The rods are excited by light from a small Xenon flashtube mounted in the second focus of the ellipse. In order to allow for the required low temperature operation of the crystal the entire elliptical chamber is evacuated and the copper cold finger is connected to a dewar vessel. The cavity consists of a gold coated mirror with long radius of curvature and a grating working as output coupler and tuning element.

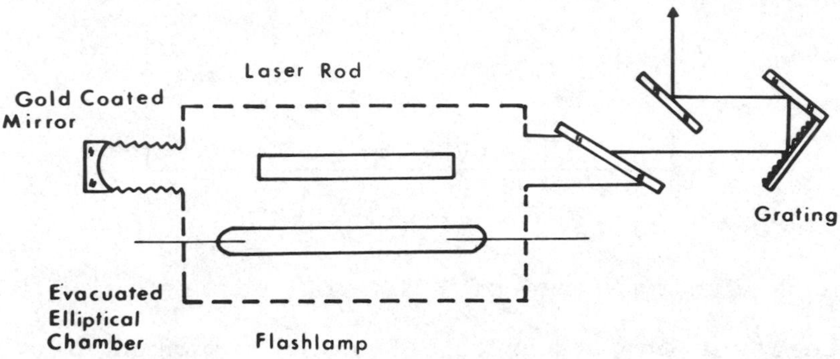

Fig. 3 Schematics of the flashlamp pumped CCL system

As the excited state lifetime of F (II) centers is only on the order of 80 - 250 ns energy storage is limited in these active media. Therefore the laser threshold depends on the momentary power and not on the input energy.

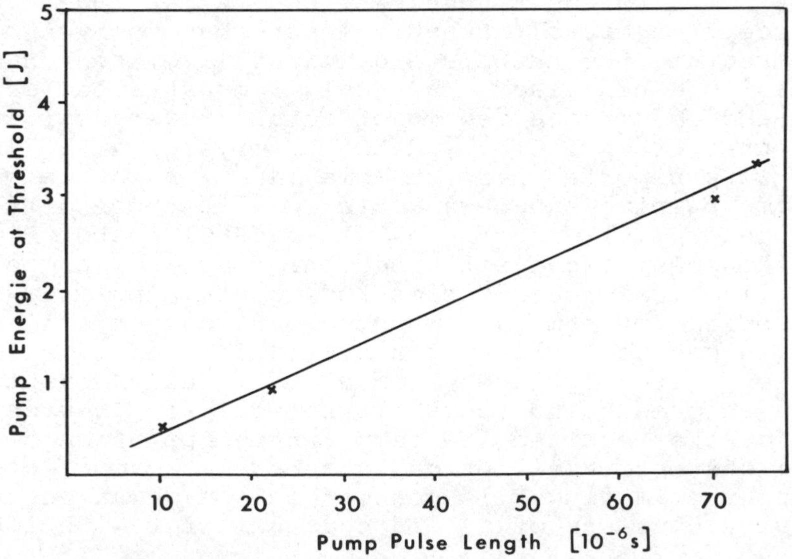

Fig. 4 Pump energy at threshold as a function of the pump pulse duration

Fig. 4 demonstrates this behaviour showing the dependence of the threshold pump energy on the pulse duration. From the linear slope an electrical input power of 3 kW to the lamp can be determined as the threshold pump power for an F_A(II) KCl:Li laser operating at the band center at 2.7µm. The lowest threshold pump energy observed with this laser was 0.3J with 10µs pulse duration.

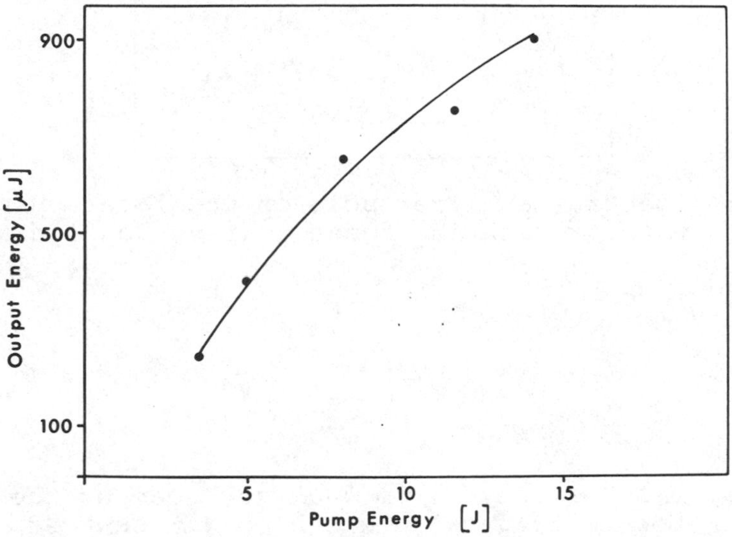

Fig. 5 Output performance of the flashlamp pumped CCL at 2.7µm

In fig. 5 the output characteristic of the laser is shown. With 15 J input, which is the maximum energy the flashlamp can handle at a pulse duration of 25µs, an output energy of nearly 1 mJ was achieved. Considering the rod size of only 3 mm diameter and 20 mm length and the simple flashlamp (i.d. = 2 mm, arc length 32 mm) it is obvious that the system is scalable, so that an output energy of 100 mJ per pulse is feasible, resulting in 1 W average at 10 Hz repetition rate. Such experiments are currently underway with laser rods of Ø 5 mm and 70 mm length. The emission linewidth measured with a Littrow mounted grating as shown in fig. 3 is on the order of 70 GHz in agreement with the passive bandwidth of the grating. A tremendous narrowing of the emission line can be achieved with a grazing incidence grating configuration similar to that described by Littman[5].

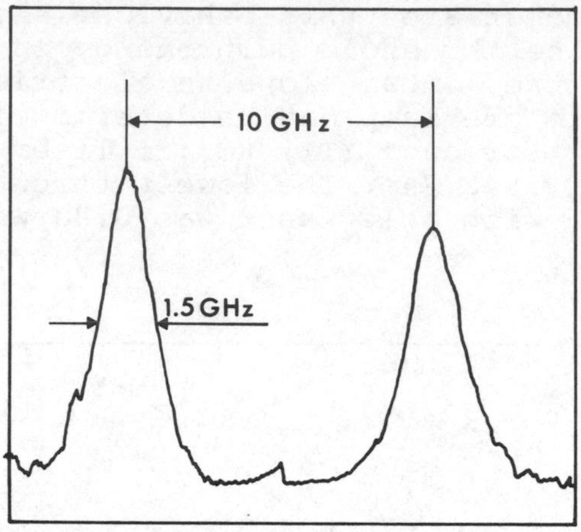

Fig. 6 Spectrum analyzer scan of the laser output with grazing incidence grating configuration

Fig. 6 shows a spectrum analyzer scan of the laser output, demonstrating a linewidth of 1.5 GHz. Probably the linewidth can be decreased even further by more accurate adjustment of the grazing incidence grating.

L-band excitation of CCLs

Nitrogen and excimer lasers are among the most reliable and powerful pump sources for excitation of pulsed tunable lasers. These lasers are powerful and in the case of excimer lasers even provide high output energies with short pulse length. Therefore in many experiments using tunable dye lasers these pump sources have been used especially for the investigation of non-linear processes or time resolved studies. So far F_A(II) and F_B(II) center lasers have been pumped by excitation of the visible absorption bands between 500 and 750 nm using tandem pump sources e.q. nitrogen laser pumped dye lasers[14] or frequency doubled Nd^{3+} YAG lasers[15,16]. However the F_A(II) and F_B(II) centers show also transitions in the uv range, called the L-bands[17]. Excitation of these bands also results in infrared fluorescence so that direct excitation of the color center crystal with nitrogen or excimer laser lines is possible. This excitation scheme results in much simpler set-ups including the advantage of higher output power and energy. The origin of the L-bands still remains unclear although it is very likely that the absorption of a photon excites the electron to a bound transition in the valence band and after relaxation to the upper laser level immediately (within < 10 ns) produces the infrared fluorescence. The only problem occuring with L-band excitation is that these absorption bands are fairly weak (on the order of 1/10 of the visible transition). This problem can be solved by increasing the center concentration compared to crystals which are optimized to be pumped in the visible range. In this way efficient laser operation has been achieved with nitrogen and excimer laser pumped CCLs.

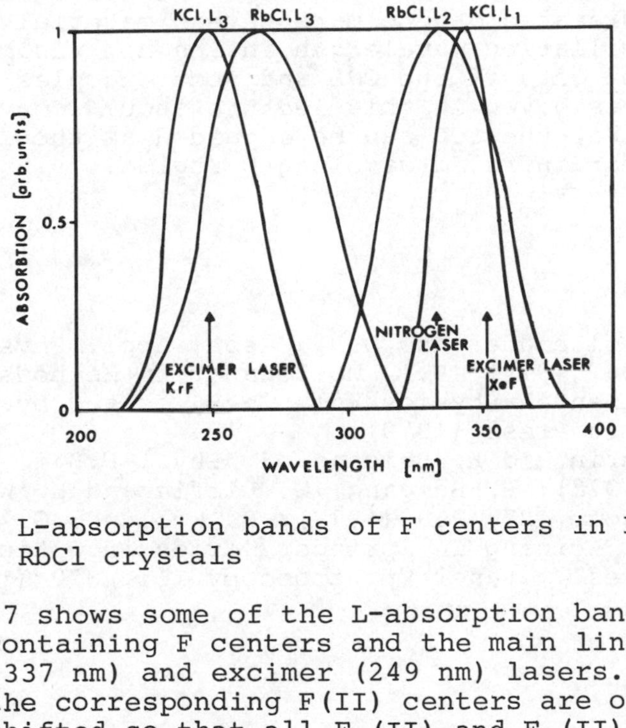

Fig. 7 L-absorption bands of F centers in KCl and RbCl crystals

Fig. 7 shows some of the L-absorption bands of KCl and RbCl containing F centers and the main lines of the nitrogen (337 nm) and excimer (249 nm) lasers. The L-bands of the corresponding F(II) centers are only slightly shifted so that all F_A(II) and F_B(II) center lasers which have been developed so far show absorption lines at both excitation wavelengths.

The laser experiments have been performed using a transversal pumping scheme. The cavity used in these experiments is the same as described in the previous chapter on flashlamp pumped CCLs. In first experiments

the output power was usually less than 1 kW but conversion efficiencies of 1% seem to be feasible, so that output energies of 10 mJ may be obtainable with excimer laser pumping. The tuning range of the lasers was not increased so far due to problems of laser oscillation between the end surfaces of the crystal.
In conclusion a tunable infrared system seems to be feasible providing 10 mJ pulses in the 2 - 3.5μm regime by excitation with a KrF laser.

Conclusions

In the development of CCLs there is still a large potential of new possibilities for increased output power, new emission ranges and improved operation conditions. Compared with dye lasers the speed of progress was less dramatic due to more severe material problems and an oscillation wavelength in the non visible region. Performance data of the CCL and some examples of new developments given in this lecture should convince however, that the CCL can be regarded as the dye laser for the near infrared wavelength regime.

References

1. L.F. Mollenauer and D.H. Olson, Appl. Phys. Lett. 24, 386 (1974); L.F. Mollenauer in Methods of Experimental Physics Vol. 15 B, edited by C.L. Tang, Academic Press (1979).
2. G. Litfin and R. Beigang, J. Phys. E:Sci. Instr. 11, 984 (1978); R. Beigang, G. Litfin and H. Welling, Opt. Comm. 22, 269 (1977); H. Welling, G. Litfin and R. Beigang in Springer Series in Optical Sciences 7, Laser Spectroscopy III, 370 (1977)

3. G. Litfin, R. Beigang and H. Welling, Appl. Phys. Lett. 31, 381 (1977); K.P. Koch, G. Litfin and H. Welling, Opt. Lett. 4, 387 (1979)
4. L.F. Mollenauer, Opt. Lett. 1, 164 (1977); L.F. Mollenauer, D.M. Bloom and A.M. Del Gaudio, Opt. Lett. 3, 48 (1978)
5. W. Gellermann, F. Lüty, K.P. Koch and G. Litfin, Phys. Stat. Sol. (A) 57, 411 (1980)
6. L.F. Mollenauer, Opt. Lett. 5, 188 (1980) and 6, 342 (1981)
7. I. Schneider and M.J. Maronne, Opt. Lett. 4, 390 (1979); I. Schneider and A. Marquardt, Opt. Lett. 5, 214 (1980)
8. W. Gellermann, F. Lüty, K.P. Koch and H. Welling, Opt. Comm. 35, 430 (1980)
9. R. Beigang, G. Litfin and R. Schneider, Phys. Rev.A 20, 229 (1979)
10. G. Litfin, C.R. Pollock, R.F. Curl and F.K. Tittel, J. Chem. Phys. 72, 6602 (1980); G. Litfin, C.R. Pollock, J.V.V. Kasper, R.F. Curl and F.K. Tittel, IEEE J. Quant. Electr. 11, 1154 (1980)
11. D.J. Jackson, E. Arimondo, J.E. Lawler and T.W. Haensch, Optics Comm. 29, 357 (1979)
12. W. Gellermann, F. Lüty and C.R. Pollock, to be published. Many thanks to Gellermann, Lüty and Pollock for permission to quote their unpublished results here.
13. B. Fritz and E. Menke, Solid State Comm. 3, 61, (1974)
14. D.J. Jackson, J.E. Lawler and T.W. Haensch, Opt. Comm. 29, 116 (1979)
15. G.C. Bjorklund, L.F. Mollenauer and W.J. Tomlinson, Appl. Phys. Lett. 29, 116 (1976)
16. A.S. Sudbo, M.M.T. Loy, P.A. Roland and R. Beigang, Optics. Comm. 37, 417 (1981)
17. F. Lüty in Physics of Color Center Lasers edited by W.B. Fowler, Academic Press (1968)

OPTICALLY PUMPED LASERS WITH DIATOMIC MOLECULES

The simplest molecular species, homonuclear or heteronuclear diatomics, offer a variety of possibilities for interesting new laser systems in the visible and near UV spectral region. Bound-bound electronic transitions in these molecules may yield efficient multiline lasers, while bound-free electronic transitions may result in continuously tunable coherent radiation. Examples for the bound-bound type are the long known discharge pumped nitrogen and hydrogen lasers, the dissociatively pumped HgBr laser or the coherently excited continuously operating dimer lasers [1-5] as Li_2, Na_2, K_2, Bi_2, S_2, Te_2 and I_2, which have been developed in the past few years. Examples for lasers operating between bound-free transitions are the excimer lasers [6], which are being investigated intensively due to their high power capabilities. The excimer lasers use molecules which only exist in an excited electronic state, while the ground state is more or less repulsive. For laser purposes, bound-free transitions in stable molecules can, however, also be regarded. An interesting system, discussed already in the early days of laser research, is the bound-free $^3\Sigma_g^+ \rightarrow {}^3\Sigma_u^+$ transition in molecular hydrogen [7,8] which, in principle, could yield a tunable coherent source in the 160 nm to 500 nm range. Recently corresponding transitions in Na_2 [9] or NaK [10] have been proposed and are presently investigated [11]. In this article we will briefly review principles and features of coherently excited lasers with homonuclear diatomic molecules and outline attempts to realize new tunable molecular lasers by optical pumping techniques.

Cw dimer lasers

Homonuclear diatomic molecules can be found in the saturated vapors of practically all elements. At a temperature of 800 K the sodium vapor, for example, contains about 5% Na_2 molecules at a partial pressure of 0.4 mbar. For many elements the absorption bands of these dimers lie in the visible or near uv spectral region; these molecules can therefore be excited with existing powerful cw lasers as argon, krypton or even dye lasers. A typical excitation-emission cycle within the energy level diagram of a dimer molecule is shown in Fig. 1 in case of Na_2. The pump laser populates a certain rotational-vibrational level in an excited electronic state and creates population inversion with respect to the much less populated higher lying rotational-vibrational levels of the electronic ground state. As dipole transitions between rotational-vibrational levels of the same electronic state are strongly forbidden, fluorescence or laser emission will directly start from the upper pump level. Neighboured levels can only be populated by collisions; consequently the lower laser levels must be depopulated by collisions too, in order to prevent bottlenecking. To achieve a steady state population inversion, the relaxation rate S_2 of the lower laser level 2 must therefore exceed the spontaneous transition rate A_{32}. This condition can usually

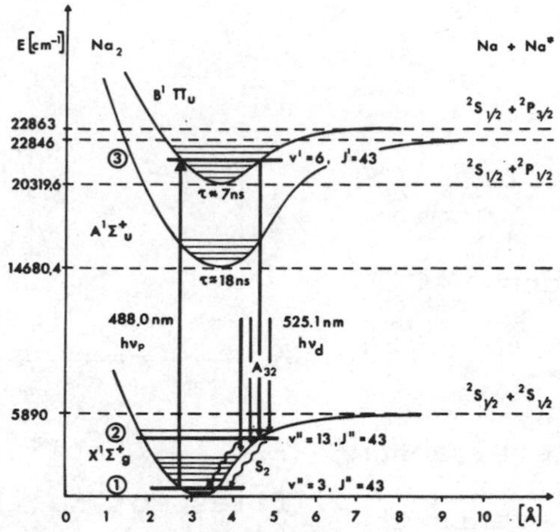

Fig. 1 Energy level diagram of Na_2 with lowest bound electronic states (after Ref. 23) and dimer laser excitation-emission cycle.

be fulfilled by a suitable choice of temperature and pressure[5]. Figure 1 shows, that optically pumped dimer lasers are three level laser systems, where two coherent light fields are coupled by a common level. In such a system, amplification can not only occur due to population inversion but also due to a two photon or stimulated Raman-type scattering process, which is resonantly enhanced by an intermediate level. By this process a new behaviour of the system is generated, that differs appreciably from normal inversion lasers and makes this system attractive both for principal laser studies as well as for special applications.

A narrow band (single frequency) pump laser with

frequency ω_p only excites a velocity subgroup u_p of the molecular ensemble, which is then responsible for the laser emission (Fig. 2). For $\omega_p \neq \omega_{13}$, the center emission frequencies in the direction of the pump radiation or opposite to this direction are given by $\omega_d^{\pm} = \omega_{32}\ (1 \pm u_p/c)$ with ω_{13} and ω_{32} being the molecular transition frequencies.

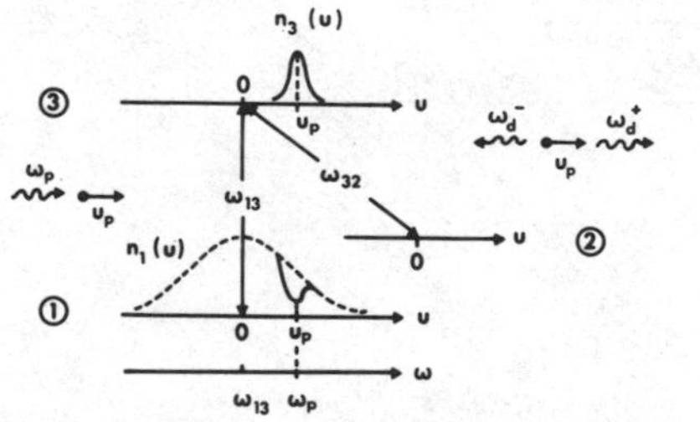

Fig. 2 Three level scheme with sub Doppler excitation. n(u) represents the velocity dependent population density.

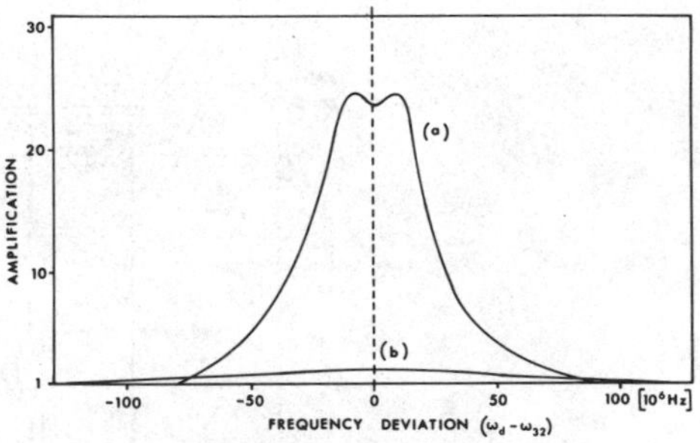

Fig. 3 Amplification profiles for Na_2 $B^1\Pi_u$ (6,43)→ $X^1\Sigma_g^+$ (13,43) laser transition (525 nm). Pump: 488 nm argon laser, temperature 780 K, length of vapor zone 10 cm.
(a) Forward profile; pump intensity 85 Wcm^{-2}. The slight splitting of the profile is due to the dynamic Stark effect.
(b) Backward profile; pump intensity 860 Wcm^{-2}.

The material therefore has an unisotropic emission (amplification) frequency characteristic, which alone results from the Doppler effect and the narrowband (sub Doppler) excitation. In addition, the resonance condition for the absorption of a pump photon ω_p and emission of a laser photon $\omega_d^{\pm}$ in a two photon process introduces a strong asymmetry in the amplification lineshape, with a high and narrow forward profile and a low and broad backward profile (Fig. 3). The difference in the line profiles is comparable to the line shape behaviour in two photon spectroscopy with co or counterrunning laser beams. For a more detailed discussion of the forward-backward amplification asymmetry, the line shape at high pump intensities (dynamic Stark effect) and gain calculations in dimer laser systems see Ref. 5. A direct consequence of the strong preferation of the forward direction is, that in a dimer ring laser system laser oscillation will only occur in the direction of the pump beam. Furtheron, due to the narrow amplification profile in case of single frequency excitation, only a single mode of the dimer laser will oscillate. Thus, unidirectional single frequency oscillation is inherent in ring laser systems. As the center emission frequency is determined by the selected velocity group, the dimer laser can be tuned within the Doppler profile by tuning the pump laser. For a Na_2 ring laser system a tuning range of about 4 GHz could be achieved[5]. According to the Raman scheme, it should also be possible to tune the laser much further away from resonance. Pure Raman laser lines in Li_2 have recently been observed[12]. The most important properties of cw dimer lasers are its amplification asymmetry and easy to achieve single frequency oscillation, extremely low threshold pump intensities, good conversion efficiencies and possible oscillation on many lines spanning broad spectral ranges. Fig. 4 shows the spectral emission ranges, approximate number of observed laser lines, transitions and pump laser lines for the presently operating cw dimer lasers. Some important data and operation conditions are summarized in Table I. More details can be found in Ref. 1 - 5. Recently continuous laser oscillation in Se_2 on many lines in the range of 380 nm to 700 nm has been obtained by optical excitation with an argon ion laser at 351.1 nm[13]. Considering applications, dimer lasers are obviously well suited for investigations on the dimer molecules itself, and have been used for measurements of transition moments and relaxation rates[5], precise determination of hyperfine splittings[14] and various kinetic measurements[15]. Recently laser oscillation has also been achieved using a molecular sodium beam as laser medium[16]. By this way, molecular beams with high concentrations of molecules in selected vibrational-rotational states can be generated for systematic kinetic studies.

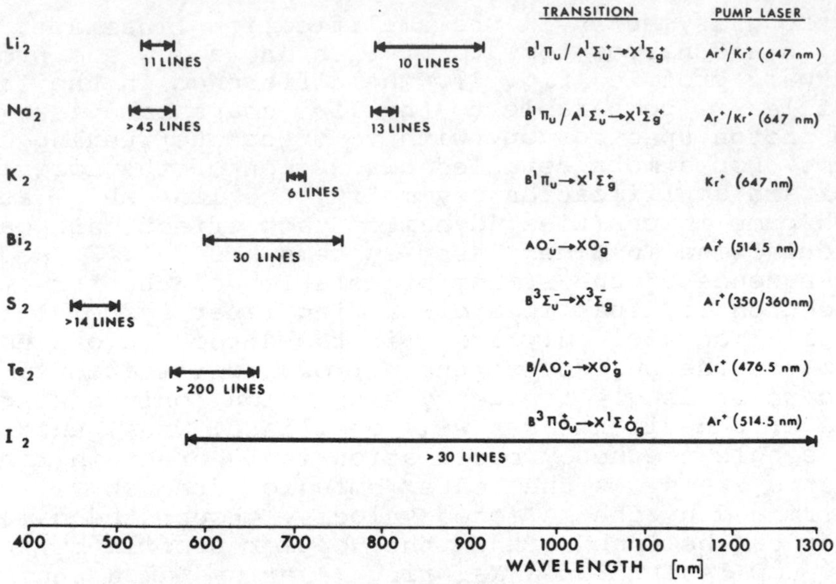

Fig. 4 Spectral ranges of cw dimer laser lines

Table I Data of cw dimer lasers

Molecule	Pump Line [nm]	Pump Power [W]	Output Power [mW] Multiline	Observed Gain [cm^{-1}]	Typical Threshold [mW]	Operation Conditions Remarks
Li_2	476.5	1	15	0.05	< 100	~ 1000 °C, vapor zone 10 cm
	647	1	30			system not optimized
Na_2	472.7	1	70	0.2	< 20	~ 500 - 600 °C
	488	2.5	250			vapor zone 3 - 10 cm
						lowest thresholds < 1 mW
	647	1	20			buffergas pressures up to 150 mbar
K_2	647	1	5		< 200	not optimized
Bi_2	514.5	3.5	350	0.01	< 200	~ 900 - 1000 °C, vapor zone 20 cm buffergas pressures up to 150 mbar
S_2	351	2	5		< 800	Cell temperature 1000 °C reservoir temperature 250 °C vapor zone 18 cm system not optimized
Te_2	476.5	1	20	0.008	< 20	~ 600 - 700 °C, vapor zone 18 cm high density of lines > 300 (550 - 660 nm)
I_2	514.5	3.5	420	0.002	< 100	~ 20 °C, vapor zone 50 - 100 cm. strong quenching by buffergases (< 2 m bar)

Quite general, dimer lasers are simple to operate optically pumped systems, which can efficiently convert the radiation of fixed frequency cw pump lasers into many laser lines in other spectral regions. Operation and laser set-up of these dimer lasers are very similar to cw dye lasers, however, the laser material is stable and inexpensive and can be handled in long living closed cells or heatpipes. It it possible that dimer lasers may replace dye lasers in all those cases where only fixed laser lines are needed.

Concepts for new tunable lasers with diatomic molecules

The optically pumped dimer lasers use bound-bound electronic transitions which, at low pressure, yield discrete spectra and therefore fixed laser lines. To achieve continuous tunability, the Raman aspect may be considered or application of high pressures. For a typical pressure broadening coefficient of about 10 MHz/mbar, buffer gas pressures of more than 1 bar are necessary to achieve overlap between rotational-vibrational levels of the Na_2 molecule. It is however very questionable whether laser oscillation will be possible under these conditions, as the dimer laser emission is already quenched at much lower pressures[5]. Therefore it seems to be more promising to consider bound-free transitions in the triplet manifold of these molecules. Fig. 5 shows a more detailed Na_2 level scheme. Of interest are those transitions that terminate at the lowest repulsive state, yielding continuous spectra, for example the recently discussed emission on the $b^3\Sigma_g^+ \rightarrow X^3\Sigma_u^+$ transition[9]. Diffuse spectra from these molecules have first been observed in discharges[17-19] and recently upon uv or two photon excitation[20-22]. In case of optical pumping, collisional energy transfer processes and molecular disturbances must be involved, as direct optical excitation of triplet levels is not possible. By excitation of Na/Na_2-vapor with 351 nm radiation of an XeF excimer laser, we could achieve laser oscillation at molecular lines around 915 nm (Fig. 6), probably on the $C^1\Pi_u \rightarrow 2^1\Sigma_g^+$ transition. In connection with this emission new diffuse fluorescence bands around 850 nm and 960 nm are obtained. The intensity especially of the 850 nm band strongly increases with increasing laser emission on the molecular lines. Considerung the molecular data of Konowalow[9,23] and the observed emission behaviour, the diffuse band around 850 nm might correspond to the $b^3\Sigma_g^+ \rightarrow X^3\Sigma_u^+$ transition, where the upper state is populated from the $2^1\Sigma_g^+$ state via internal molecular processes (disturbances, curve crossings). More detailed investigations to clarify the

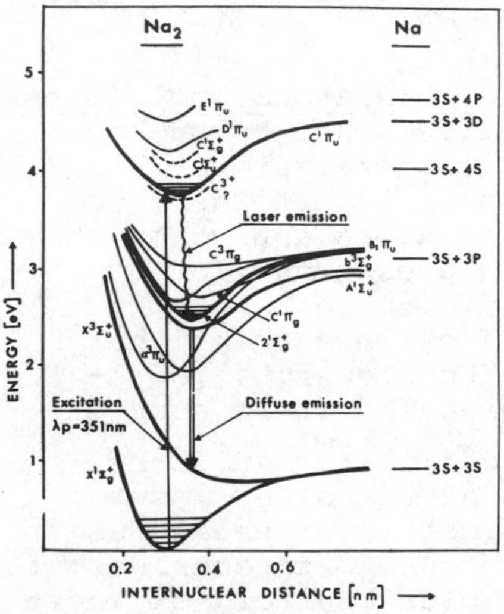

Fig. 5 Energy level diagram of Na_2 (after Refs. 9,23) with proposed transitions for observed laser lines and diffuse emission spectrum.

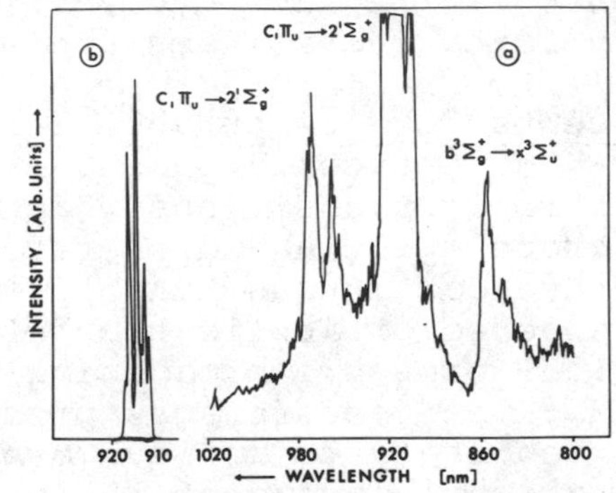

Fig. 6 Na_2 emission spectrum upon excitation with 351 nm XeF excimer laser. Temperature 760 K, pump power 50 mJ.
(a) Spectrum, showing diffuse emissions around 860 nm and 960 nm and laser oscillation around 915 nm. The spectrum is not corrected with regard to the multiplier response (RCA 8852)
(b) Laser oscillation spectrum around 915 nm with 10^8 attenuator.

the nature of the diffuse emissions and its possible use for a tunable laser are presently underway.

A more conventional method to achieve tunable lasers is to regard bound-free molecular transitions in excimer molecules. Already long time proposed candidates especially for the visible part of the spectrum are metal vapor-noble gas (or mercury) excimers like Na-Xe,

Cs-Xe or Tl-Hg. A new way to produce these excimer molecules has recently been demonstrated[24]. It produces the necessary excited atoms Na*, Cs* or Tl* by photodissociation of metal halides. By a suitable choice of the photodissociation wavelength, great amounts of excited atoms are produced, without having ground state atoms. The population inversion thus initially achieved in the atomic system (atomic photodissociation laser) is then transfered into the excimer system by collisions. Calculations indicate, that good chances exist to realize with this technique optically pumped new excimer lasers.

References

1. B. Wellegehausen, S. Shahdin, D. Friede, H. Welling; Appl. Phys. 13, 97 (1977).
2. B. Wellegehausen, K.H. Stephan, D. Friede, H.Welling; Opt. Commun. 23, 157 (1977).
3. J.B. Koffend, R.W. Field; J. Appl. Phys. 48, 4468 (1977).
4. B. Wellegehausen, D. Friede, G. Steger; Opt. Commun. 26, 391 (1978).
5. B. Wellegehausen; IEEE J. Quantum Electr. 15, 1108 (1979).
6. Ch.K. Rhodes, Ed.; Excimer Lasers, Topics in Applied Physics 30, Springer Verlag, New York 1979.
7. F.G. Houtermans; Helv. Phys. Acta 33, 933 (1960).
8. C.V. Heer; J. Appl. Phys. 41, 1875 (1970).
9. D.D. Konowalow, P.S. Julienne; J. Chem. Phys. 72, 5815 (1980).
10. E.J. Breford, F. Engelke; Chem. Phys. Lett. 53, 282 (1978)
11. S. Shahdin, Z. Ma, B. Wellegehausen; to be published.
12. C.N. Man-Pichot, A. Brillet; IEEE J. Quantum Electr. 16, 1103 (1980).
13. B. Wellegehausen, A. Topouzkhanian, C. Effantin, J. d'Incan; to be published.
14. J.B. Koffend, S. Goldstein, R. Bacis, R.W. Field, S. Ezekiel; Phys. Rev. Lett. 41, 1040 (1978).
15. J.B. Koffend, R.W. Field, D.R. Guyer, S.R. Leone; Laser Spectroscopy III, 382, Springer Series in Optical Sciences 7, New York 1977.
16. K. Bergmann; private communication
17. H. Bartels; Z. Physik 73, 203 (1932).
18. M.M. Rebbeck, J.M. Vaughan; J. Phys. B:Atom. Molec. Phys. 4, 258 (1971).
19. J.P. Woerdmann, J.J. de Groot; Chem. Phys. Lett. 80 220 (1981).
20. J.P. Woerdman; Opt. Commun. 26, 216 (1978).
21. A. Kopystynską,P. Kowalczyk; Opt.Com. 28, 78 (1979).

22. M. Allegrini, L. Moi; Opt. Commun. 32, 91 (1980).
23. D.D. Konowalow, M.E. Rosenkrantz, M.L. Olson; J. Chem. Phys. 72, 2612 (1980).
24. S. Shahdin, K. Ludewigt, B. Wellegehausen; IEEE J. Quantum Electr. 17, 1276 (1981).

TWO PHOTON AMPLIFICATION

Progress in the solution of the material problems connected with most tunable lasers is to be expected if tunability of the laser can be obtained irrespective of the level widths of the laser material under consideration, in principle, this can be achieved by the use of stimulated 2-photon emission. In 1964 P. Sorokin et.al. suggested[1] to use this type of emission as the basic process in a laser. They mainly concentrated on the development of a pulsed 2-photon amplifier with the ability to store large amounts of energy[2]. This aspect of 2-photon stimulated emission was also investigated by other authors[3-6]; the level configuration considered in this context is shown in fig. 1a. However, the 2-photon transition rate in a level system where the intermediate level is a purely virtual level (all real levels "2" connected to the upper laser level "3" by electric dipole transitions are situated above "3" or below the lower laser level "1") is rather small. Therefore, neither 2-photon amplification nor 2-photon oscillation was ever observed in such a configuration except as transient effect with the technique of adiabatic rapid passage[7].

2-photon coherence effects are most easily observed in a 3-level configuration with the intermediate level on or close to resonance, where they produce asymmetry of the lineshape and anisotropy of the absorption or amplification process[8-10]. In most 3-level experiments with a cascade configuration as in fig. 1b. either one of the optical fields was a weak probe field such as spontaneous emission[11,12] or they dealt with 2-photon absorption with the exception of a very recent investigation[13] where true 2-photon stimulated emission was achieved using strong pulsed fields. We observed useful 2-photon amplification from a continuous incoherently produced population inversion in a gas laser. A 3-level system interacting with 2 strong optical fields is most properly described with the density matrix formalism; this treatment leads to the following set of "generalized rate equations"[14] which include single and multi-photon processes as well:

$$\frac{dN_3}{dt} = \gamma_3 N_3^0 - R_{31}(N_3-N_1)n_P n_L - R_{32}(N_3-N_2)n_P - \gamma_3 N_3 \qquad (1a)$$

$$\frac{dN_2}{dt} = \gamma_2 N_2^0 + R_{32}(N_3-N_2)n_P - R_{21}(N_2-N_1)n_L - \gamma_2 N_2 \qquad (1b)$$

$$\frac{dN_1}{dt} = \gamma_1 N_1^0 + R_{21}(N_2-N_1)n_L + R_{31}(N_3-N_1)n_P n_L - \gamma_1 N_1 \qquad (1c)$$

Here N_i is the population density of the i´th level, $n_P = |\mu_{32}E_P|^2/(2h)$ and $n_L = |\mu_{21}E_L|^2/(2h)$ are the intensities of the optical fields with μ_{ij} as the electric dipole matrix element between states i and j, E_P and E_L as the corresponding field amplitudes. N_i^0 are the velocity dependent zero-field population

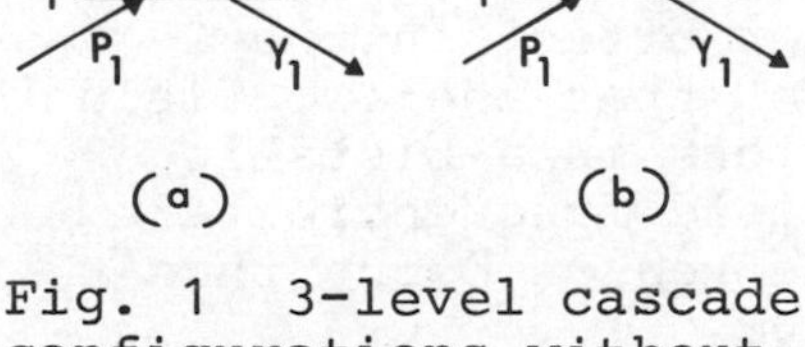

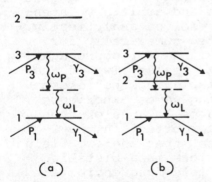

Fig. 1 3-level cascade configurations without (a) and with (b) real intermediate level.

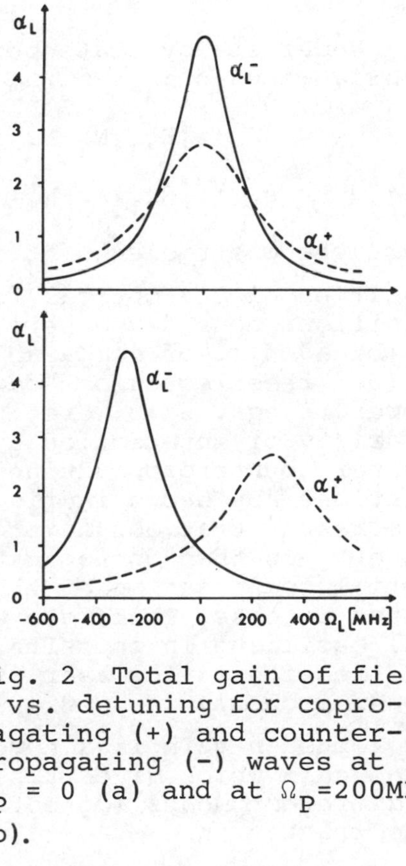

Fig. 2 Total gain of field L vs. detuning for copropagating (+) and counterpropagating (-) waves at $\Omega_P = 0$ (a) and at Ω_P=200MHz (b).

densities, and γ_i are relaxation constants. Therefore, the first term in eqs. 1a-c is an incoherent pump rate, the last term is a relaxation rate. Expressions for R_{ij} are given in [14]; it should be noted however, that all these coefficients are functions of the field intensities and of the detunings of the optical fields from resonance, whereas the relative direction of propagation is contained essentially in R_{31}. For this reason we call those terms in eqs. 1a-c containing R_{31} 2-photon transition rates; this nomenclature is also justified by the asymptotic behaviour of R_{31} for vanishing fields and large detunings, yielding the well known second order perturbation theory 2-photon transition rate. Correspondingly the terms containing R_{32} or R_{21} are called single-photon transition rates, although at arbitrary intensities or close to resonance an absolute distinction between these two types of transitions cannot be made.

Under steady state conditions, the fields P and L experience gain $\alpha_J = \dot{n}_J/n_J$ (J=P,L) given by[15]

$$F_P \cdot \alpha_P = n_L \cdot \int R_{31}(N_3-N_1)\cdot dv + \int R_{32}(N_3-N_2)\cdot dv \qquad (2a)$$

$$F_L \cdot \alpha_L = n_P \cdot \int R_{31}(N_3-N_1)\cdot dv + \int R_{21}(N_2-N_1)\cdot dv \qquad (2b)$$

where the coefficients F_J relate the n_J to the photon densities of the optical fields. Under the general conditions considered here (homogeneous and inhomogeneous broadening of comparable size and arbitrary intensities) these equations cannot be solved analytically. Therefore eqs. 2a,b were integrated numerically for a variety of intensities and detunings for copropagating and for counterpropagating optical fields. Two of these solutions are shown in fig. 2, displaying the foreseeable result that counterpropagating waves experience higher gain than copropagating waves because of the partial compensation of the Doppler effect in the 2-photon process. Therefore we expect that the two fields will oscillate in opposite directions in a bistable unidirectional ringlaser if the field frequencies are chosen properly ($\Omega_L \approx -\Omega_P\omega_L/\omega_P$) the reason being that the 2-photon gain is a homogeneous contribution to the total gain which gives rise to strong cross-saturation of different modes (directions of propagation) on the same transition.

This cross-saturation of different directions of propagation has to be considered carefully if the cascade laser is a standing wave laser. If spatial modulations of the population densities can be neglected ($\Omega_p > \gamma_3 + \gamma_2$ and $\Omega_L > \gamma_2 + \gamma_1$), then correct results are obtained from eqs. 1 and 2 by decomposing the standing waves into two travelling waves each [16] and summing over all possible contributions to the transition rates. Some typical evaluations of this type are given in fig. 3, which show a marked asymmetry in the amplification lineshape of one transition when the optical field on the other transition is off resonance. Also shown in fig. 3 are the corresponding output power lineshapes which should be observed in a standing wave laser. These power profiles were calculated as follows: Eq. 2 (generalized for standing waves as described above) gives the gain on each transition as a function of the intracavity intensities n_J; both optical fields suffer some constant loss V_J per round trip. By equating the saturated gain α_J to the loss V_J (J=P,L) the intensities are obtained implicitly. The numerical solutions to this set of equations show (fig. 3) that asymmetry and splitting are more pronounced than in the gain line; the detection of 2-photon amplification in a standing wave laser is based on these features.

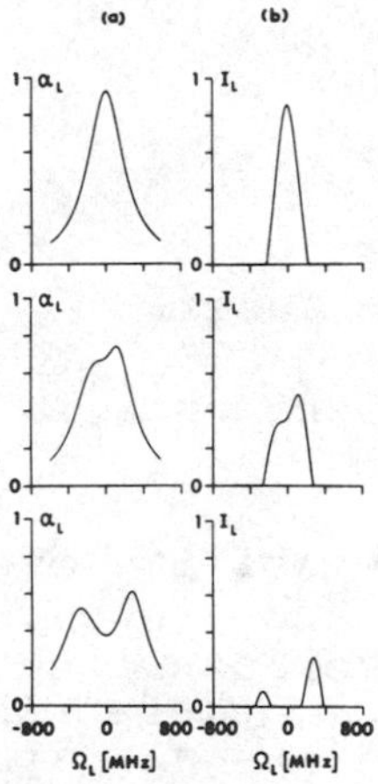

Fig. 3 Gain (a) and outputpower (b) vs. detuning for standing waves at different values of Ω_P.

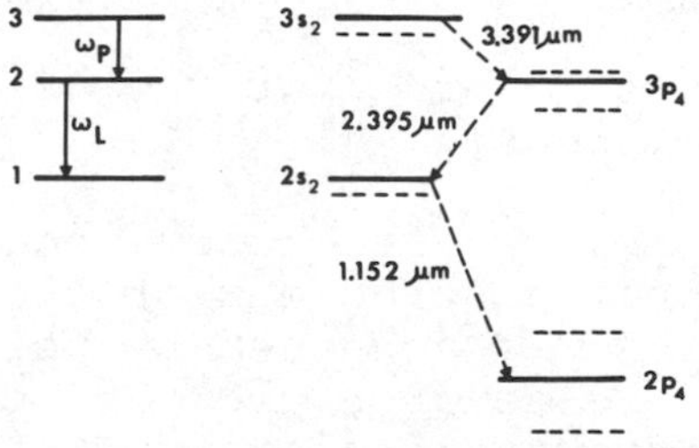

Fig. 4 Correspondence between levels 1, 2 and 3 and the levels of the Ne atom.

Our experiments were performed in a He-Ne discharge; the correspondence between levels 1, 2 and 3 and the levels of the Ne atom is shown in fig. 4. As in the discharge not only the 3s level but also the 2s level is populated, the $2s_2 - 2p_4$ transition had to oscillate broadband (several longitudinal and transversal modes) at a high power level in order to depopulate the $2s_2$ level uniformly.

In the travelling wave experiment the transitions at 2.4 μm and 3.4 μm oscillate at a single frequency in the ringlaser shown in fig. 5 without being forced into either direction. If the detunings were chosen properly spontaneous unidirectional oscillation occured on both transitions, and the directions of propagation were strictly opposite, indicating the importance of the 2-photon gain for the overall amplification of the laser system.

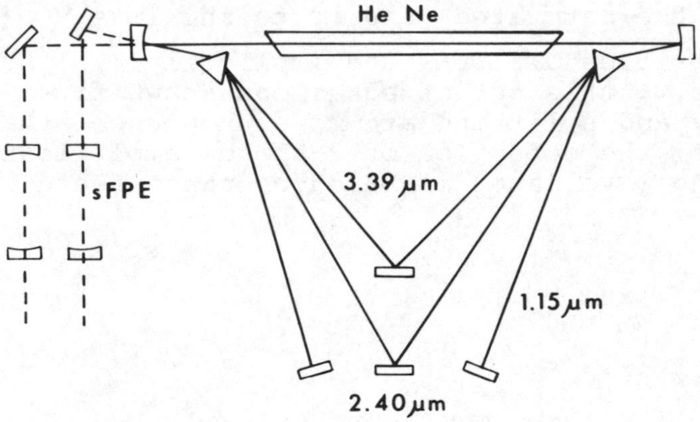

Fig. 5 Travelling wave ring laser.

The setup for the standing wave experiments is shown in fig. 6.

Single frequency oscillation and tunability over several hundred MHz of the transitions under investigation was achieved by using high finesse Fox-Smith mode selectors which are electronically regulated in the conventional way. In the experiment, the detuning of the 3.4 μm line is set to some fixed value Ω_p. Then the 2.4 μm line is tuned slowly and its output power is detected through a fast scanning Fabry-Perot interferometer with 750 MHz free spectral range. The output power lineshape is then obtained as the envelope of a multimode of such interferometer fringes on a storage oscilloscope.

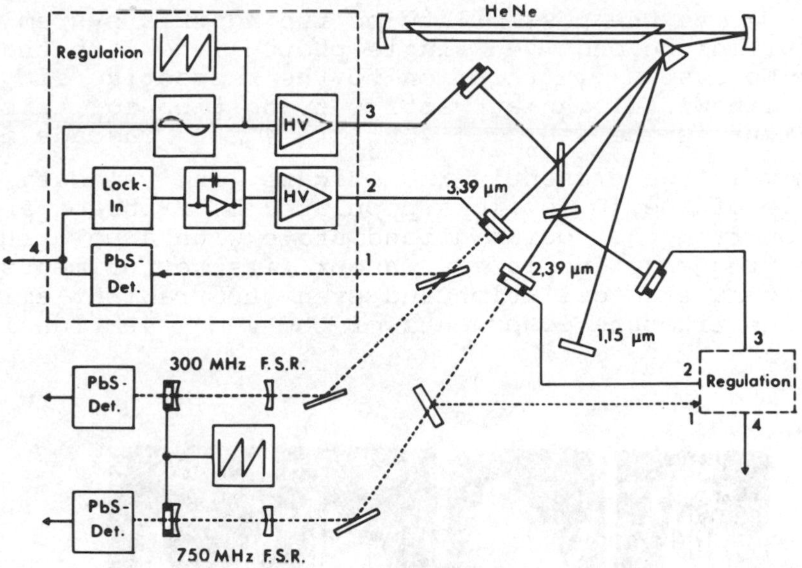

Fig. 6 Standing wave laser.

Some results are shown in fig. 7 for increasing values of Ω_p. The observed profiles agree qualitatively with the theoretically predicted lineshapes. For sufficient detunings of the 3.4 μm transition, two relative maxima can be resolved clearly, where the higher peak contains the stronger 2-photon contribution. Particularly interesting is fig. 7g where only the 2-photon enhanced peak oscillates, demonstrating the ability of the 2-photon process to increase the tuning range of cascade lasers beyond the value predicted by single photon theory.

Using modified frequency selectors and an improved stabilization system, combined tuning ranges in excess of 4.5 Doppler widths were achieved. The limitation arises from the steep decrease in 2-photon gain in Doppler broadened media at detunings of a few Doppler widths which could not overcome the rather high cavity losses. This problem can be solved by using a gain medium with a narrower lineshape and a larger electric dipole matrix element like an optically pumped alkali vapour. In such a system single frequency oscillation occurs spontaneously and there is no need for lossy frequency selectors. As the gain on resonance can be several orders of magnitude higher as compared to the He-Ne discharge, we expect that also the 2-photon gain off resonance is correspondingly higher. Both effects contribute to the possibility of tuning this system much further beyond its single photon linewidth than the He-Ne system (in relation to the respective single-photon linewidth). An attractive transition for this experiment is the $4^2D_{5/2} \rightarrow 4^2P_{3/2} \rightarrow 4^2S_{1/2}$ cascade in sodium which we presently investigate. One velocity ensemble of the 4D level is populated by 2-photon absorption from the sodium ground state with a powerful single frequency dye laser. In our first experiments we obtained laser oscillation and even superradiant emission on 2 finestructure components of the 4D → 4P transition.

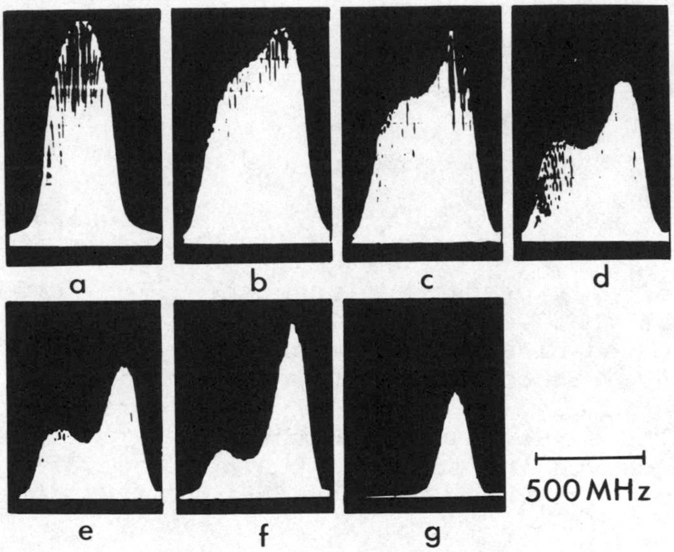

Fig. 7 Output power profiles of the 2.4μm transition for various detunings of the 3.4 μm transition: a: 0MHz; b: 80MHz; c: 100MHz; d: 120 MHz; e: 130MHz; f: 140MHz; g: 160MHz.

References

1. P.P. Sorokin and N. Braslau, IBM J. of Res. and Dev. 8, 177 (1964)
2. D.S. Bethune, J.R. Lankard and P.P. Sorokin, J. Chem. Phys. 69, 2076 (1978)

3. L.M. Narducci, W.W. Eidson, P. Furcinitti and D.C. Eteson, Phys. Rev. A 16, 1665 (1977)
4. H.P. Yuen, Appl. Phys. Lett. 26, 505 (1975)
5. V.S. Letokhov, ZhETF Pis'ma 7, 284 (1968)
6. R.L. Carman, Phys. Rev. A 12, 1048 (1975)
7. M.T. Loy,Phys. Rev. Lett. 41, 473 (1978)
8. T. Hänsch and P. Toschek, Z. Phys. 236, 213 (1970)
9. M.S. Feld and A. Javan, Phys. Rev. 177,540 (1969)
10. B.J. Feldman and M.S. Feld, Phys. Rev. A 5,899 (1972)
11. H.K. Holt, Phys. Rev. Lett. 20, 410 (1968)
12. T.W. Ducas, M.S. Feld, L. W. Ryan, Jr., N. Skribanowitz and A. Javan, Phys. Rev. A 5, 1036 (1972)
13. B. Nikolaus, D.Z. Zhang and P.E. Toschek, Phys. Rev. Lett. 47, 171 (1981)
14. R. Salomaa, J. Phys. B 10, 3005 (1977)
15. H.H. Schlemmer, D. Frölich and H. Welling, Opt. Comm., (1980)
16. D. Frölich and H.H. Schlemmer, Proc. Int. Conf. on Lasers 79, 817 (1979)

UV - GENERATION IN CW DYE LASERS

L. Wöste

Institut de Physique Expérimentale
Ecole Polytechnique Fédérale de Lausanne
PHB-Ecublens, 1015 Lausanne, Suisse

I. INTRODUCTION

Molecules whose electronic transitions lie in the UV portion of the electromagnetic spectrum far outnumber those in the visible. Consequently, it is of considerable chemical interest to extend tunable coherent light sources to shorter wavelengths. Potential applications are numerous including photo-chemical, kinetic, analytical and spectroscopic studies. Because dye lasers are limited to wavelengths above 330 nm, the production of deeper UV light requires the use of non linear optics such as sum frequency mixing and harmonic generation[1].

Under well chosen conditions two light sources at the frequencies ω_1 and ω_2 produce radiation at the sum frequency $\omega_3 = \omega_1 + \omega_2$. This is only possible, when the phasematching condition

$$\omega_3 \, n_{\omega_3} = \omega_1 \, n_{\omega_1} + \omega_2 \, n_{\omega_2}$$

is satisfied, where n_ω is the refractive index of the non linear medium at the corresponding frequency ω. Second harmonic generation (SHG) is a special case with $\omega_1 = \omega_2 = \omega$ and $\omega_3 = 2\,\omega$.

Phasematching can be obtained by tuning the direction of propagation of the input beam - referred to as angle tuning - or by tuning the temperature of the nonlinear medium - referred to as temperature tuning (see also fig. 1).

When using focussed Gaussian input beams the output power can be calculated by use of a formula given by G.D. Boyd and D.A. Kleimann[2], where in a first approximation the SHG - power P_{ω_3} is

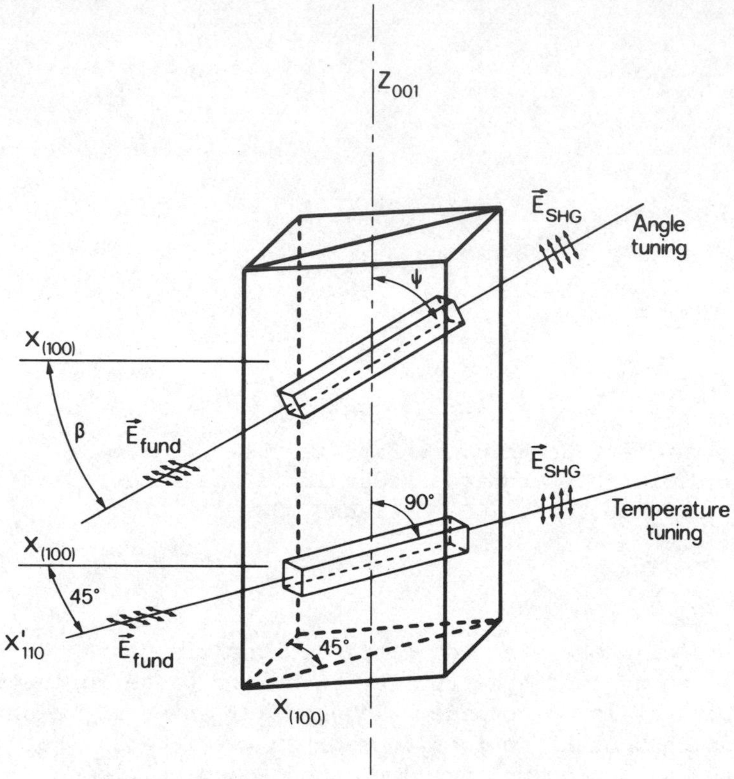

Fig.1 : Orientations of the crystal for angle and temperature tuning.

proportional to the square of the input power P_{ω}. Various parameters determine the conversion efficiency like the orientation and homogeneity of the crystal, its nonlinear coefficient, its length, the linewidth and focussing parameter of the input light, etc. In pulsed laser systems conversion efficiencies of $\sim$ 40 % can be obtained, using angle tuning outside the laser cavity[3]. However, the bandwidth of pulsed lasers is limited by the Fourier transform of the pulse duration. For narrower bandwidths, therefore, cw operation is necessary.

The low output signal of cw systems, however, is very unfavourable for an efficient frequency conversion. This makes it advisable to use the high circulating power inside the laser cavity. Since, however, angle tuning would sensitively disturb the operation of the laser resonator, temperature tuning is advised.

In 1976, Frölich et al. first doubled the frequency of a Rh 6 G laser inside the cavity using an ADA crystal[4]. Wagstaff and Dunn performed a similar experiment inside the cavity of a single mode ring laser, reaching UV - output in the range 292 - 302 nm[5]. Mariella reached wavelengths near 247 nm by doubling a Coumarin 480 dye laser[6], while Clough and Johnston reported intracavity doubling of a Coumarin 535 dye laser to produce UV output in the range 257 - 260 nm[7]. Fig. 2 indicates the wavelength range, that can be covered for SHG by using various crystals at various temperatures.

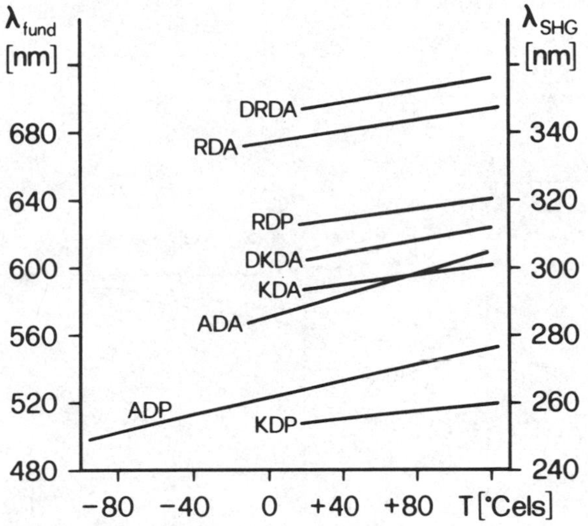

Fig. 2 : Temperature tuning ranges of various crystals[11].

Spectroscopic applications, however, are still rare : With the crystal inside the cavity the complexity of the system makes it hard to align and to operate. Furthermore, the temperature accuracy that is required for phasematching the crystal is about ± 0.05 °C. This makes it difficult to actively tune the laser. We therefore propose an improved resonator design that allows easy operation, involving electrooptical tuning of the crystal for servo controlled single frequency scans ranging over 30 GHz.

II. THE RING RESONATOR

The basic resonator design is shown in fig. 3. The two mirrors M_2 and M_3 are focussed on the fluorescent spot, which is formed by the focussed ion laser pump beam. The mirror curvatures are between 7,5 and 2,5 cm, accordingly if high of low gain dyes are being used. The reflective surface of the mirrors is placed into the center of gimbal mounts on top of linear translation stages. This allows independently to optimize the focal and angular adjustments. From the fluorescent spot M_2 and M_3 direct to the flat mirrors M_4 and M_5, which initially are aligned to 180° reflexion in order to form a linear resonator over M_2 and M_3. The insertion of an optical element at Brewsters angle causes a parallel beam offset. This would shift the overlap point on the flat mirror surface and disturb the alignment in an existing ring configuration. In a linear laser configuration, however, this has no influence. All elements like the birefringent filter, the optical diode, the galvo plates and the etalon, therefore, are inserted into the linear configuration. Once this is established and optimized, M_4 and M_5 are slightly tilted around to form a ring.

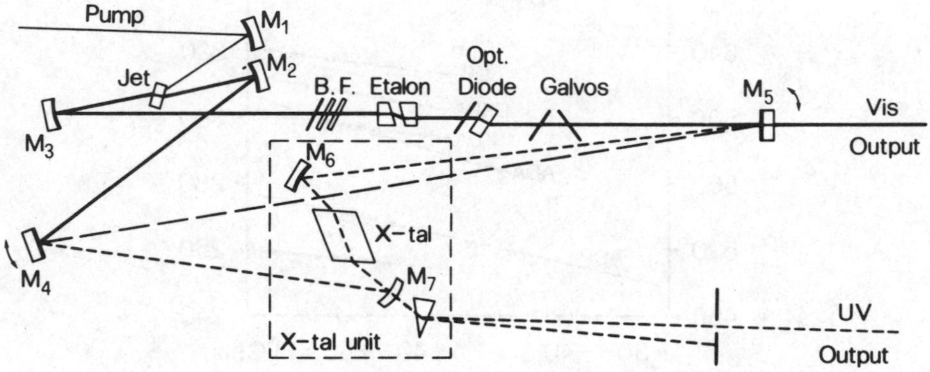

Fig. 3 : Dye laser configuration for successively operating the resonator as a linear laser, a ring laser and an intracavity SHG ring laser.

The optical diode consists of an optically active quartz plate and a Faraday rotator glass. Since glasses with a high Verdet constant tend to absorb, the plate thickness should be kept as low as possible by applying strong magnets and rotating the polarization just as much as necessary. A rotation of 2,5° proved to be sufficient. Since optical activity and the Faraday effect are strongly wavelength dependent, different plate-thicknesses were tailored to obtain this value for various wavelength ranges with a stack of Cobalt-Samarium magnets (∿ 3.2 KGauss).

Single mode operation was obtained with a tunable uncoated 1 cm Brewster etalon[8]. The etalon was slaved to a longitudinal mode, cavity tuning was obtained by galvo-scanning the Brewster plates. With a 12 % output coupler the ring provided 3.2 Watts single frequency output power, when pumping an aqueous solution[9] of Rh 6 G with 17 Watts of an Ar^+ - laser (all lines).

III. FREQUENCY DOUBLING

For SHG - operation, the output coupler is replaced by a high reflector, and a UV - unit is inserted in the laser resonator. The UV - unit consists of a 2 cm crystal, cut at Brewster's angle, and two focussing mirrors M_6 and M_7. The latter are coated to reflect the visible light and to transmit the UV. Their inner radius of curvature is 12 cm, the outer curvature is designed to recollimate the divergent UV - ouptut. The angles of incidence on M_6 and M_7 are chosen to compensate for astigmatism due to the formula given by Kogelnik et al.[10]. The whole unit is prealigned to provide parallel in - and output beam characteristics. The crystal is placed inside a vacuum chamber which is schematically shown in fig. 4.

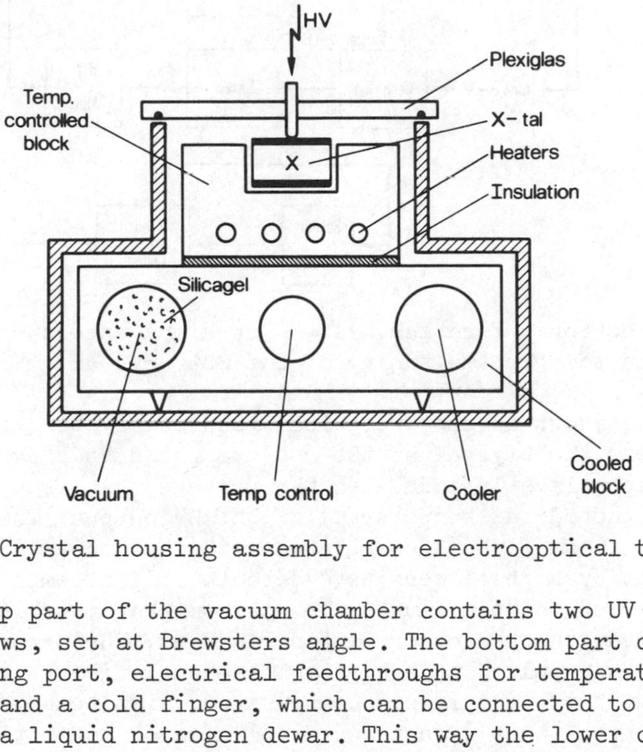

Fig. 4 : Crystal housing assembly for electrooptical tuning.

The top part of the vacuum chamber contains two UV - grade quartz windows, set at Brewsters angle. The bottom part contains a vacuum pumping port, electrical feedthroughs for temperature stabilization, and a cold finger, which can be connected to either a cryostat or a liquid nitrogen dewar. This way the lower copper block inside the vacuum chamber is permanently being kept cold. It is stuffed with molecular sieve to provide efficient cryogenic pumping. The upper cold block contains the crystal. It is kept at the desired temperature by being cooled from the bottom while being heated against with cartridge heaters. The crystal is plated on top and bottom with a gold layer, in order to connect it to a high voltage source. Due to the magnitude of the electrooptical tuning coefficients of various crystal types[3], shifts corresponding to changes of the phase match temperature of several degrees centigrade can be induced by the electrical field, before voltage breakdown inside the crystal occurs. This highly reduces the temperature stability requirement.

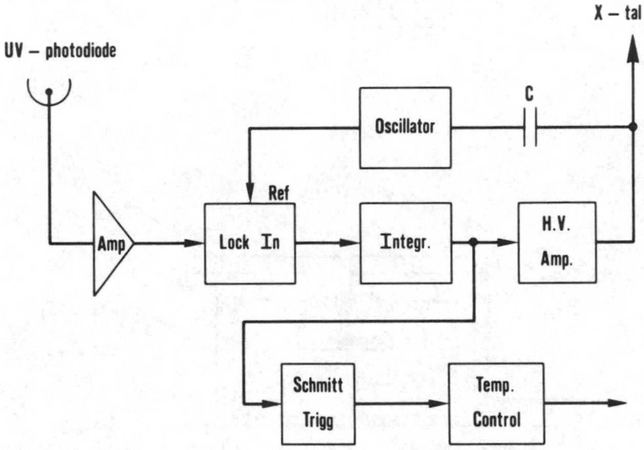

Fig. 5 : Electronic feedback system for simultaneous electrooptical and temperature tuning of the SHG crystal.

Fig. 5 shows an electrooptical feedback system, that automatically keeps the crystal at the optimum phasematching condition : An oscillator slightly modulates the voltage that is applied to the crystal. A photodiode keeps track of the UV output. The resulting power modulation is phase correlated with the modulation signal, and recorded by a phase sensitive detector. If a temperature variation or wavelength scan causes the optimum phasematching position to be detuned, the phase correlation changes. The error signal is integrated and applied to a high voltage amplifier. This way the phase mismatch is immediately compensated. Simultaneously the temperature control unit is activated to raise or lower the crystal temperature according to the voltage which is presently applied to the crystal. This temperature change will follow slowly. Simultaneously the voltage is reduced to a low level, in order to prevent thermal runaway breakdown, which may occur, when high voltage is applied to a crystal too long[11].

IV. RESULTS

So far Rh 6 G and Coumarin 515 (Exciton) dye lasers have been doubled, using ADA and ADP crystals. The Rh 6 G laser was pumped with a 10 Watts (all lines) Ar^{+} - laser. Single frequency powers of 20 mWatts were obtained. The latter, required a second thin etalon (0,1 mm) to be inserted, because the laser showed a tendency to modejump away from the optimum phasematch. A tuning range from 285 to 310 nm was obtained. The Coumarin 515 laser was pumped with 2,5 Watts of the violet line of a Kr^{+} - laser. Single frequency powers of ∿ 100 μWatts were obtained, when operating around 254 nm. Spectroscopic applications of this laser are presented elsewhere[12].

ACKNOWLEDGMENTS

This work has been partially performed in the laboratories of Prof. R.N. Zare at Stanford University, and at the Lexel Corp. Palo Alto. The author wishes to thank them for the support and many stimulating discussions.

REFERENCES

1. F. Zernicke and J.E. Midwinter, Applied Nonlinear Optics, A.A. Ballard ed. (Wiley, New York, 1973).
2. G.D. Boyd and D.A. Kleimann, J. Appl. Phys. 39, 3597 (1968).
3. D. Hon, High Average Power, Efficient Second Harmonic Generation, in: Laser Handbook, vol. 3 ed. M.L. Stitch (North-Holland Pub. Co., Amsterdam, New York, 1979).
4. D. Frölich, L. Stein, H.W. Schröder and H. Welling, Appl. Phys. 11, 97 (1976).
5. C.E. Wagstaff and M.H. Dunn, J. Phys. D : Appl. Phys. 12, 355 (1979).
6. R.P. Mariella, J. Chem. Phys. 71, 94 (1979).
7. P.N. Clough and J. Johnston, Chem. Phys. Letters 71, 253 (1980).
8. A.D. Berg, U.S. Patent 4.081.760 (1978).
9. S. Leutwyler, E. Schumacher and L. Wöste, Opt. Comm. 19, 197 (1976).
10. H.W. Kogelnik, E.P. Ippen, A. Dienes and C.V. Shank, IEEE, J. Quant. Electron. QE - 8, 373 (1972).
11. Quantum Technologie Inc. Data Sheet 702 and 707 (1979).
12. C.R. Webster, L. Wöste and R.N. Zare, Opt. Comm. 35, 3 (1980).

APPLICATIONS OF LASER SPECTROSCOPY

Herbert Walther

Sektion Physik, Universität München and
Max-Planck-Institut für Quantenoptik
D 8046 Garching/München, Fed. Rep. Germany

1. INVESTIGATION OF CHEMICAL REACTIONS WITH LASERS

The laser has become an important tool for studying elementary chemical processes and for analytical applications, while the use of lasers to initiate chemical reactions by selective excitation of reactants is still very much in its infancy and the awaited breakthrough has not yet occurred. It may be that recent developments in laser technology, e.g. excimer lasers will change the whole picture, and that applications in chemistry such as laser-induced chemical reactions or synthesis will come close to reality.

Laser photochemistry is characterized by high selectivity and intensity of the radiation source. The three most important types of reaction in laser-induced chemistry are summarized in the following:

(a) Electronically or vibrationally excited molecules are produced in a bath of reactants; the activation barrier for a reaction is thus reached or surmounted.
(b) Molecules are ionized by single or multi-photon absorption and either undergo further reactions with other molecules or unimolecularly decay into fragments.
(c) Molecules are dissociated and a high concentration of reactive free radicals is produced, reacting with other molecules.

The laser photon is expensive, and therefore the application of lasers to synthesis is only profitable in the case of small-scale production of expensive compounds, e.g. in isotope separation[1] or when many product molecules are created for every photon

Table1. Laser applications in chemistry. Examples of studies of elementary processes

Field of application	Method	Aim and applications
Laser-induced chemistry	selective electronic or vibrational excitation, dissociation, ionization	acceleration of bimolecular reactions (see Table 2), free radical chemistry initiation of chain reactions [2] practical applications (realized): isotope separation [1] product purification [7] synthesis [7] laser photolytic deposition of structures (microchemistry) [3]
Analytical application	selective excitation or ionization, laser-induced fluorescence	investigation of internal product state distribution (reaction dynamics, reaction kinetics, see also Secs. 2 and 3 of this paper) selective ion source [4,6] single atom and molecule detection (spectral analysis microprobing of materials) [4] detection of trace constituents over large distances in the atmosphere (see Sec. 4 of this paper) time-resolved measurements (time resolution 10^{-13}s) (see Ref. [5] and Sec. 3 of this paper).

Table 2. Examples for laser-induced chemical processes

Reaction	Change in rate constant	Experimental technique
Vibrational excitation $K + HCl^{+}(v = 1) \rightarrow KCl + H$	$k^{+}/k \sim 10^{2}$	Crossed molecular beams, analysis by mass spectrometers. Brooks et al. [8]
$HCl^{+}(v = 1) + O \xrightarrow{k_r^{+}} OH + Cl$ $\xrightarrow{k_d^{+}} HCl(v = 0) + O$	$(k_r^{+} + k_d^{+})/k_r \sim 4 \cdot 10^{3}$ $k_r^{+} \gg k_d$	Quenching of HF^{+} fluorescence Wolfrum et al. [9]
$HCl^{+}(v = 1) + H \xrightarrow{k_r^{+}} H_2 + Cl$ $\xrightarrow{k_d^{+}} HCl(v = 0) + H$	$(k_r^{+} + k_d^{+})/k_r \sim 60$ $k_r^{+} \ll k_d^{+}$	
$HCl(v = 2) + Br \rightarrow HBr + Cl$	$k^{+}/k \sim 10^{11}$	
$HF^{+}(v = 1) + Ba \rightarrow BaF^{+} + H$ $HF^{+}(v = 1) + Ca \rightarrow CaF^{+} + H$ $HF^{+}(v = 1) + Sr \rightarrow SrF^{+} + H$	Vibrational excitation of the product increased	Product analysis by laser-induced fluorescence Zare et al. [10]
Electronic excitation $NO_2^{*} + CO \rightarrow NO + CO_2$	$k^{*}/k \sim 10$	Herman et al. [19]

absorbed, i.e. when the photochemical quantum yield is larger than unity. A good example of the latter case is laser initiation of radical chain reactions. In this way Wolfrum et al. obtained extremely pure vinyl chloride from $C_2H_4Cl_2$.[2] It was demonstrated in laboratory experiments that a substantial increase in the yield can be achieved if already existing production equipment for vinyl chloride is fitted out with lasers. However, the lasers with sufficiently high power needed for large-scale production are not yet available.

A rough survey of laser applications in chemistry is given in Table 1 with a few pertinent references. Some examples of laser-induced processes are compiled in Table 2. The table shows that despite the fact that industrial applications of laser-induced chemistry are still lacking, useful information has been obtained from the study of elementary processes.

Besides its usefulness in selective preparation of reactants, the laser has also proved its worth in the analysis of reacting molecules and reaction products. Especially in measurements of the nascent internal state distribution many interesting results on reaction dynamics and kinetics have been obtained.

Table 3. Analysis of reaction products by laser-induced fluorescence (selection of papers)

Product molecule		References
BaX	X = F, Cl, Br, I	[10, 13 - 18]
CaF, SrF		[10]
BaO		[11, 12]
MO	M = Sc, Y, La, Ac, Al	[20, 21]
OH, OD		[22 - 26]
IF		[27, 29]

The method usually applied for this purpose is laser-induced fluorescence. To analyze the internal state distribution, the laser is tuned to electronic transitions, starting from groups of different rotational-vibrational levels of the electronic ground state and ending on the same level of an excited state; the total fluorescence is observed. The internal state distribution of the lower levels is then evaluated by means of the fluorescence intensity and the Franck-Condon factors. In order to obtain the nascent state distribution the experiment has to be performed under collision-free conditions; a crossed molecular beam setup therefore usually has to be used in these experiments.

Table 3 shows a few examples of experiments performed by the laser-induced fluorescence method, developed by Zare et al.[11] The method has been extensively applied so far to the oxides and halides of the group II and group III elements and the OH radical. In the following, the extension of this technique to the measurement of the internal product state distribution of interhalogen molecules performed at our laboratory will be discussed as an example of this technique.

Experiments have been carried out for several elementary reactions forming iodine monofluoride, which is, from the spectroscopic point of view, the most suitable interhalogen molecule to be studied by laser-induced fluorescence. The investigated reactions are:

$$F + CH_3I \rightarrow IF + CH_3$$
$$F + CF_3I \rightarrow IF + CF_3$$
$$F + IX \rightarrow IF + Cl \qquad \text{for } X = Cl, Br, I$$

The vacuum chamber is evacuated with a cryopump. The reactant beams strike the 14 K surface connected to the closed cycle cooler and condense. In this way even reactive atoms or molecules can be studied and no damage to the pumping systems can occur, as would be the case if conventional oil diffusion pumps were used (Fig. 1). For details of the experiments see Refs.[27,29,30]

The reagent beams are uncollimated and have thermal energies corresponding to 300 K. The F atoms are produced by a microwave discharge either in CF_4 or in F_2. In contrast to previous investigations, the fluorescence of the product molecules in this experiment is induced by a cw dye laser with a spectral output narrowed by a birefringent filter.

Compared with measurements with pulsed lasers, there is considerable gain in the signal-to-noise ratio. The minimum detectable number density of molecules in a given internal quantum state under the conditions of this experiment is only about $10^2/cm^3$.

A further advantage of cw excitation is the straightforward application of phase-sensitive detection, e.g. the fluorescence of reagent molecules can be suppressed by chopping the F atom beam source. This advantage has been fully utilized in the study of F + X reactions.

Figure 2 shows the excitation spectrum of the $X^1\Sigma^+-B^3\pi(0^+)$ transition of IF which is formed by the reaction $F+CF_3I \rightarrow IF+CF_3$.

From the fast scan (upper trace) relative population densities of the individual vibrational states of the product molecules were derived by using the integrated band intensities, the relative laser power, and the Franck-Condon factors determined by Clyne and McDermid.[28] At a higher resolution (lower trace) the rotational structure of individual bands is sufficiently resolved to allow determination of the fraction of energy entering product rotation.

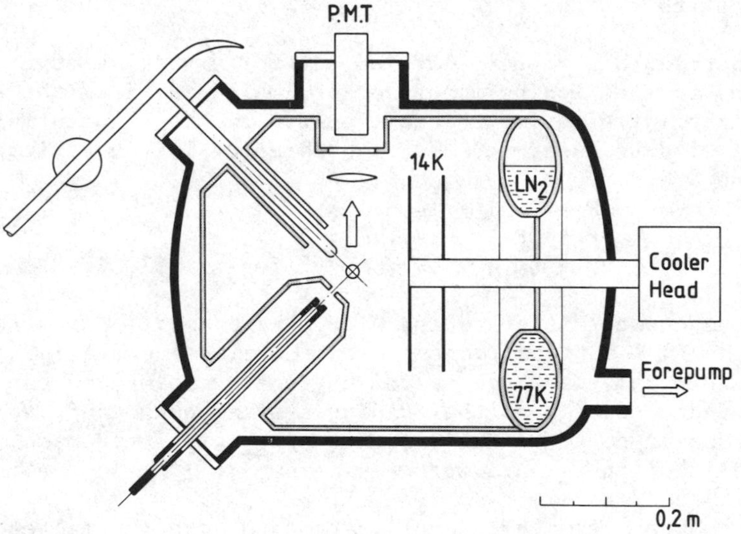

Fig. 1. Cross-section of the cryopumped vacuum vessel. The liquid-N_2-cooled surfaces surround the entire reaction region. The laser beam intersects the crossing region of the two molecular beams perpendicularly to the plane of the drawing. Upper left: the F atoms are produced by a microwave discharge. Lower left: heated nozzle for the iodides.

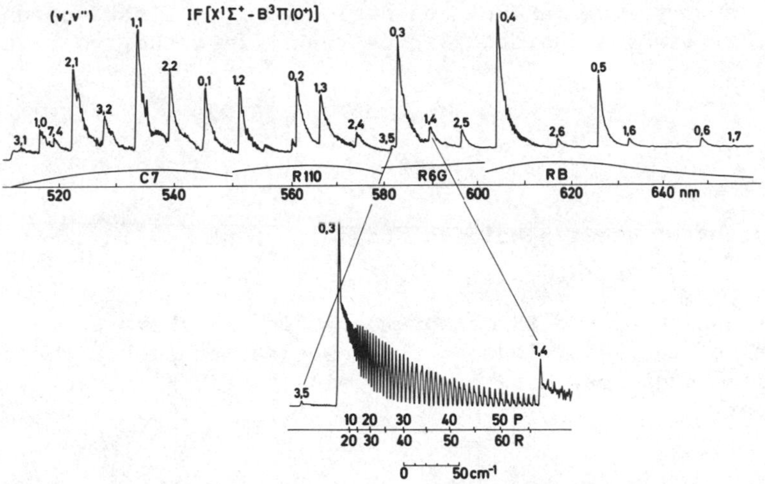

Fig. 2. Excitation spectrum of IF formed in the reaction of F + CF_3I. The signal-to-noise ratio for the strongest bands is of the order of 10^3.

The internal product excitation for the reactions studied is summarized in the following table:

Reaction	$E_{tot}[kJ\ mol^{-1}]$	$\langle f_V \rangle$	$\langle f_R \rangle$
F + CH_3I	45.5	0.15 ∓ 0.02	0.14 ∓ 0.03
F + CF_3I	54.4	0.11 ∓ 0.02	0.12 ∓ 0.02
F + ICl	70.0	0.59 ∓ 0.09	0.10 ∓ 0.02

$\langle f_V \rangle$, $\langle f_R \rangle$ denote the mean fraction of the total energy available E_{tot} which is channeled into product vibration and rotation, respectively. The listed data together with the results of a more detailed investigation[27,30] lead to the conclusion that the first two reactions proceed via a long-lived complex. This is in agreement with an angular distribution measurement by Farrar and Lee.[31] The third reaction shows, as also the other reactions of the group F + IX, a bimodal characteristic. The numbers given for f_V and f_R refer to the part of the IF molecules in excited vibrational levels (for F+ICl peaking at V=6). The second maximum in the population distribution of the vibrational levels is at V=0[30]. The fractional vibrational energy of this lower maximum is very small and can be neglected in a first approximation. The bimodal characteristics result from the two reaction paths

$$F + IX \begin{cases} \rightarrow IF^{+} + X\,(^{2}P_{3/2}) \\ \rightarrow IF + X\,(^{2}P_{1/2}) \end{cases}$$

The fraction of excited halogen atoms is about 0.9 in the cases of X = I.

Discussion of the IF measurements clearly shows the enormous potential of the laser-induced fluorescence method for the study of chemical dynamics. More examples will be given in Sec. 3 of this paper.

2. VIBRATIONAL-RELAXATION STUDIES

The knowledge of vibrational relaxation of molecules is very important in connection with laser-induced chemistry. Only when the vibrational energy is stored long enough in the reactant is a laser-initiated process possible. In the following, two experiments in the gas phase were discussed. Relaxation measurements in liquids and solids will be covered in the talk presented by A. Laubereau.[5] The relaxation rates in the gas phase are of the order of μs, and so they can be measured with laser pulses with durations in the ns region.

The energy accumulated in a specific vibrational mode of a polyatomic molecule is in general transferred rather rapidly to other vibrational modes by collisions (V-V transfer). For a number of molecules it has been shown that this process may occur through a variety of vibrational channels on a time scale that is short compared with the transfer times into translational and/or rotational degrees of freedom (V-T/R transfer).

Much information about V-V transfer processes has been collected in experiments using laser-induced fluorescence and laser-double-resonance techniques. With a few exceptions, fixed-frequency lasers have been applied; only a restricted number of molecules could therefore be investigated. The use of tunable infrared lasers greatly increases the possibilities, so that more systematic studies have now become feasible.

With the setup described here, V-V transfer in ethylene and acetylene has been measured by means of laser-induced fluorescence. The excitation source is a tunable parametric oscillator which covers the infrared spectral region up to 3.5 μm.

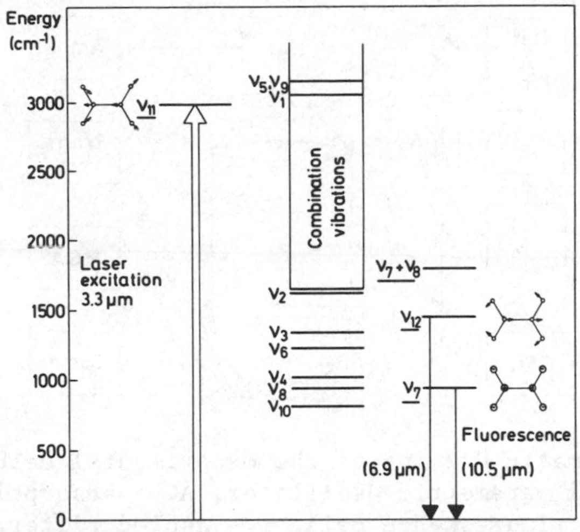

Fig. 3. Simplified energy level diagram of the vibrational levels of ethylene. Strong infrared active modes are underlined.

In the experiment on ethylene the Q branch of the ν_{11} vibrational mode of ethylene at 3.3 μm was excited. The rise time of the ν_7 fluorescence was measured with a detection system with a response time of 600 ns, which was short enough to determine the V-V transfer rates from ν_{11} to ν_7. These transfer rates were also measured with admixtures of rare gases (Fig. 3).

For the measurements on acetylene the molecules were excited into the ν_3 and $(\nu_2+\nu_4+\nu_5)$ vibrations at 3300 cm^{-1} (see Fig. 7). Time-dependent fluorescence signals of two transitions were observed: from ν_5 to the ground state (13.7 μm) and from ν_2 to ν_5 (8.0 μm).

The experimental setup is shown in Fig. 4. Most of the details have been described elsewhere.[32,33] The laser pulses had an energy of 4 μJ, a pulse length of 8o ns and a spectral width of 0.5 cm^{-1}. With a small part of the laser energy an absorption measurement was made in order to adjust the frequency of the parametric oscillator to a strong absorption line in the 3 μm region. The time-dependent signals obtained from the Ge:Hg detector were amplified, recorded with a transient digitizer, averaged in a multichannel analyzer, and subsequently stored in a computer for further evaluation.

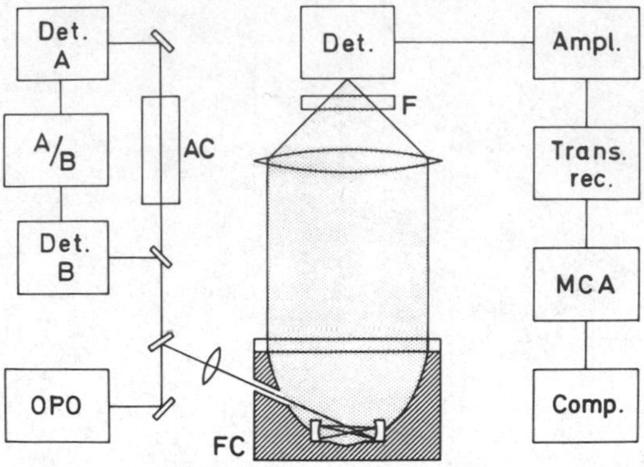

Fig. 4. Schematic drawing of the experimental setup (OPO = optical parametric oscillator, AC = absorption cell, FC = fluorescence cell, F = cooled filter, MCA = multichannel analyzer).

Most of the measurements were done with a fluorescence cell giving a fluorescence yield of 50 %. The cell was the shape of a paraboloid (see Fig. 4). A multipass system was used for the exciting laser beam. This consisted of two concave mirrors in a near-concentric setup. With the same absorbed energy the parabolic cell increased the fluorescence yield by a factor of about 5 compared with standard cells.

Figure 5 shows a typical signal for ethylene. It shows the fluorescence from the ν_7 level. Immediately after the laser pulse, the fluorescence intensity rises very sharply and then decays exponentially with two different time constants. This behaviour is interpreted as follows: The initial sharp rise of the fluorescence intensity describes the growth of the population of the ν_7 mode by vibrational transfer from ν_{11}. The fast decay is caused by the transfer of vibrational energy into translation and/or rotation, a process which occurs mainly through the lowest vibrational states. The populations in these lower states are believed to equilibrate rapidly.

Finally, the slow decay of the fluorescence signal in Fig. 5 is attributed to the temperature equilibration between the irradiated gas volume and the cell walls. This was proved by the fact that the slow decay vanished when the heat capacity of the system was increased by adding large amounts of a rare gas. In addition, the decay rate was found to be inversely proportional to the pressure.

The ratio of the slow thermal decay amplitude to the peak intensity was measured to be 15 % - 20 % (see Fig. 5(b)). (In Fig. 5(c), the peak of the fluorescence intensity is not reproduced in scale owing to the slower detector rise time.) The theoretical value of this quantity is determined by the ratio of the vibrational heat capacity to the total heat capacity. For ethylene, this ratio is calculated to be 27 %, including all levels which give a considerable contribution to the vibrational heat capacity (eight fundamentals below 1600 cm^{-1}).

The details of the results on the vibrational relaxation of ethylene and acetylene are described in [32] and [33] respectively. In the following only the main features are briefly reviewed.

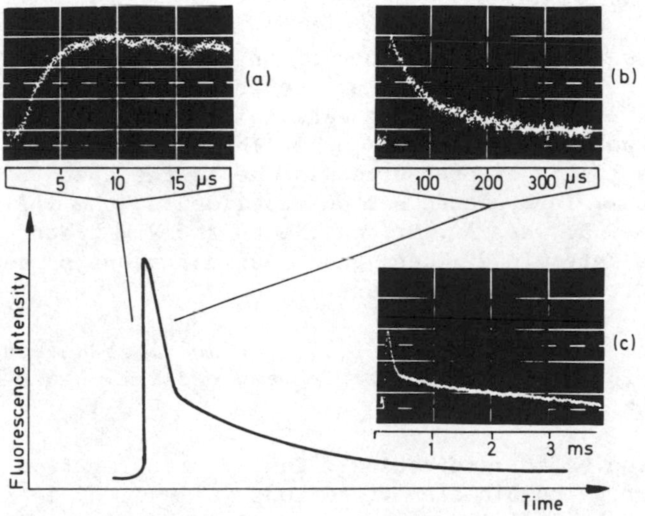

Fig. 5. Time dependence of the 10.5 μm fluorescence signal at 1 torr ethylene (ν_7 mode). (a), (b), (c): actual signals with appropriate time resolution. [In (c), a slower detector rise time was used. The peak of the fluorescence intensity is therefore not reproduced in scale.]

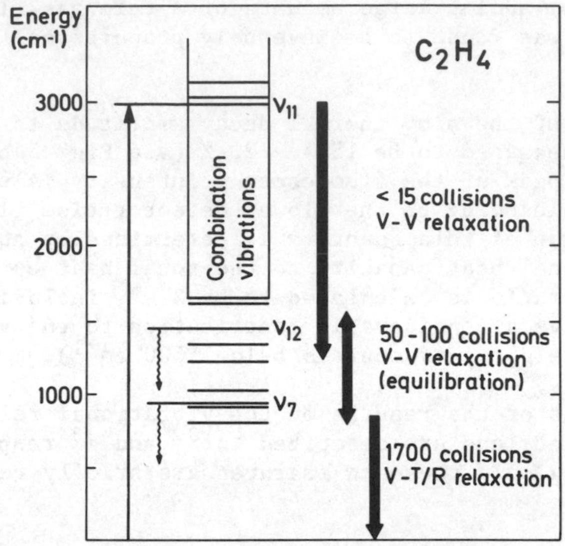

Fig. 6. Simplified vibrational-relaxation scheme of ethylene. The relaxation paths are characterized by collision numbers.

Ethylene: The pressure dependence of the V-V transfer process was determined in pure ethylene at pressures between 0.8 and 3.5 torr. Measurements at lower pressures were not successful owing to the vanishing absorption in the gas cell. The limitation of the measurements to higher pressures was due to the speed of the detection system. In a second set of experiments, the influence of the rare gases He, Ne, Ar, Kr, and Xe on the V-V transfer process $\nu_{11} \rightarrow \nu_7$ was determined. These gases, at different pressures, were mixed with 1.5 torr ethylene.

From the measurements the following model is deduced for the energy transfer in ethylene after a weak excitation of the ν_{11} mode (see Fig. 6).

(a) First step (after excitation): Energy transfer from ν_{11} to neighbouring combination vibrations followed by decay into the fundamentals with lower energy or direct decay into these fundamentals. This proceeds in a time shorter than 15 collisions (corresponding to the ν_{12} activation rate).

(b) Second step: Thermal equilibration of the lower levels. The time scale of this process is given by the fast decay rate of ν_{12}.

(c) Third step: Relaxation of these equilibrated vibrational levels into rotational/translational energy in about 1700 collisions.

Acetylene: Laser-induced acetylene fluorescence was observed in two different spectral regions at 13.7 and 8.0 μm. The signals were separated by means of cooled interference filters in front of the detector.

The fluorescence signals at 13.7 μm were detected through a filter with a maximum transmission of 60 % and a half-width of 2.9 μm centred at 13.7 μm. Possible transitions in this wavelength region are as follows: $\nu_5 \rightarrow$ ground state, $2\nu_5 \rightarrow \nu_4$. An absorption cell 5 cm in length filled with 100 torr acetylene placed between the fluorescence cell and the infrared detector blocked about 90 % of the fluorescence light. The fluoresence at 13.7 μm is therefore mainly due to the transition $\nu_5 \rightarrow$ ground state, i.e. its time dependence gives information about the population of ν_5.

For the fluorescence measurements at 8.0 μm a cooled filter with a maximum transmission of 62 % and a half-width of 2.5 μm centred at 6.9 μm was used. The measured fluorescence signal is attributed to the transition $\nu_2 \rightarrow \nu_5$ at 8.0 μm.

The measurements with rare gases as collision partners yield collision numbers for the deactivation of ν_2 and ν_5 which are between 13 and 25 times larger, respectively, than for self-deactivation. These factors are quite usual for collision partners that cannot take up internal energy. The ν_2 deactivation rate for each rare gas is about 1.3 times larger than the corresponding rate of ν_5. This may be an indication that in rare gas collisions ν_2 is deactivated by a process involving an energy gap of a magnitude comparable with that of V-T/R relaxation.

The relaxation model for acetylene is summed up in Fig. 7. For simplicity, only the main relaxation pathways are indicated by arrows. Possible intermediate steps would be the activation of sum vibrations, which, however, decay very fast, so that their omission does not change the general picture.

After laser excitation and rapid V-V activation of the lower fundamental modes the most prominent feature of the relaxation in acetylene is the very slow V-V deactivation of ν_2 whose rate is comparable with the V-T/R relaxation rate.

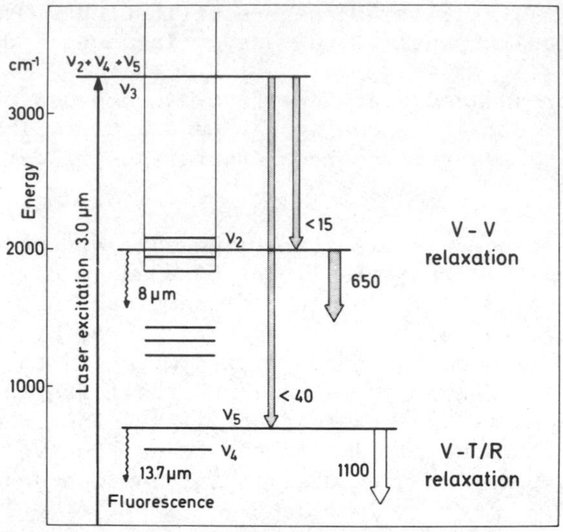

Fig. 7. Simplified vibrational-relaxation scheme of ethylene. The relaxation paths are characterized by collision numbers.

Following activation, the stretching vibration ν_2 is depopulated by a very slow process in about 650 collisions. This cannot be interpreted as a simple V-T/R relaxation process. In such a case the energy deficit of nearly 2000 cm^{-1} which had to be transferred into translation would lead, according to the Lambert-Salter plot[34], to a collision number severeal orders of magnitude larger than the experimental value. A contribution of V-T/R relaxation from ν_2 to the ground state can therefore be ignored and the measured deactivation rate of ν_2 can be interpreted as a pure V-V process transferring energy into the neighbouring and lower combination vibrations and fundamentals. From the measured data, however, it is not possible to determine the exact relaxation path out of ν_2.

This "bottlenecking" in ν_2 is very surprising because the energy of the three combination modes $(\nu_4+2\nu_5)$, $(2\nu_4+\nu_5)$, and $(3\nu_4)$ is within kT around ν_2, and energy transfer to these modes should be very fast. The energy deficit, however, is not the only parameter that influences vibrational relaxation. Two factors will tend to decrease the probability for collisional transfer to the neighbouring states:

(a) The number of quantum jumps required is rather large in this case.

(b) The types of vibrational modes are different: ν_2 is a stretching vibration, and the neighbouring modes are combinations and harmonics of the bending vibrations.

The results demonstrate that for acetylene the usual model of vibrational relaxation in polyatomic molecules cannot be applied (see, for example, results for ethylene). It is obvious that in the case of acetylene the CC stretching vibration is effectively isolated from the other modes during V-V relaxation, i.e. it can store energy for a very long time. This, in principle, might open up possibilities for laser-driven chemical reactions.

3. STUDY OF MOLECULE SURFACE INTERACTION BY LASERS

A detailed understanding of surface reactions is of major technical and scientific interest for several reasons:

(a) Large-scale industrial processes are realized using catalytic reactions; important examples are ammonia synthesis, oil cracking, methanol production, catalytic combustion etc.
(b) In materials science surface reactions are essential in connection with corrosion, steel hardening, or hydrogen embrittlement.
(c) Reactions between surfaces and electrolytes play a role in electrochemistry, e.g. in fuel cells or in hydrogen production by photolysis.
(d) In biology and medicine surface reactions are of importance, e.g. for bioadhesion or biomembranes.

A large number of methods have been developed to study surfaces. At the beginning the solid state properties of well defined surfaces and the adsorbed layers were investigated and a detailed picture of the atomic positions and electron structure of many adsorption and coadsorption systems was obtained. This knowledge about the static system was the prerequisite for the study of the dynamics of the interaction and surface reactions.

The characterization of a reaction is obtained by studying the reaction kinetics and the reaction dynamics. The first one is described by the rate constants and the microscopic details of the reaction, whereas the reaction dynamics is described by the reaction path, and the dependence of the reaction cross-section on internal and translational energies of the reactants and on the orientation of the reactants. The experimental studies of reaction dynamics therefore requires molecular beam type experiments where no secondary processes such as collisions influence the results. This is not necessarily the case for studies of reaction kinetics.

Especially in connection with studies of reaction dynamics, the laser is a useful diagnostic tool. In the following a few experiments of this type will be discussed.

As long as only atoms are involved in the collision process the measurement of the angular distribution and the momentum change provides sufficient insight into the scattering dynamics. However, when molecules are scattered, additional information on the change of the internal energy is necessary. Recently, the laser-induced-fluorescence method has been successfully used in several experiments to determine the influence of the surface interaction on the molecular rotational distribution.[35-43] It has been shown that for the case of a carbon-covered Pt(111) surface the rotational degree of freedom of the scattered NO molecules is only partly accommodated to the surface temperature.[35,37] A similar effect has also been observed for molecules which experience only weak inelastic interaction with a solid surface, such as CO/LiF (001)[38] and NO/Ag(111),[39,40] as well as for NO molecules desorbing from a Ru(001) surface.[41] In the following, our own investigations of the angular and rotational distributions of NO molecules scattered from different surfaces will be described in more detail.

In the experiments the laser-induced fluorescence of the scattered molecules is measured in the $^2\Sigma \leftarrow {}^2\pi$ (0-0) transition of NO before and after the scattering process. The details of the experimental setup are described in Refs.[35,37,43] The UHV scattering chamber (10^{-10} torr) contained a rotatable quadrupole mass filter for determining the angular distribution of the scattered molecules. A supersonic NO beam with a particle flux of about 2×10^{14} molecules $cm^{-2}s^{-1}$ was scattered from the surface of a Moore-type pyrographite crystal, from a carbon-covered Pt surface, or from a Pt(111) surface. The graphite crystal consisted of microcrystals most of whose c-axes pointed in the direction perpendicular to the surface, and whose a-axes were randomly oriented. The surfaces could be cooled to 130 K with liquid nitrogen and heated via the tantalum support leads. In the following, the experiments with the graphite crystal will be described in more detail as an example. (The experimental setup is shown in Fig. 8.)

The properties of the graphite surface were probed by scattering a supersonic He beam from the crystal.[44] The full width at half maximum of the specular lobe was twice the width of the incident beam. The peak intensity of the specularly scattered He atoms was only about 9 per cent of the incident beam intensity, which is small compared with measurements on graphite single crystals.[45] According to these experiments a relatively high density of surface defects had to be assumed, which were due mainly to grain boundaries and twin crystals with slightly misoriented surface normals.

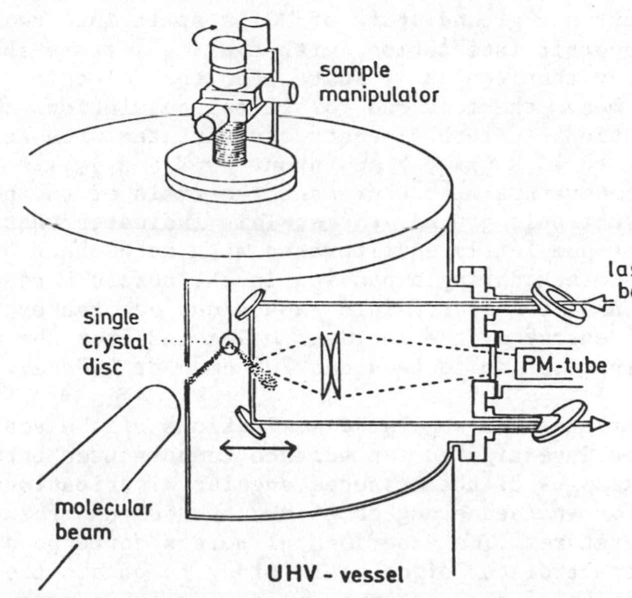

Fig. 8. Experimental arrangement (schematic). The laser beam enters and leaves the vacuum chamber through Brewster angle quartz windows as shown. Inside the chamber it is deflected by aluminium-coated mirrors which are displaceable to the left and right as shown, so that the incoming NO beam or the scattered molecules leaving the surface could be excited and observed.

For temperatures below 700 K chemical reactions between the surface atoms and impurity molecules in the beam (1 %, with main constituents N_2O, NO_2 and N_2) could be excluded. An estimate can be given for the NO coverage of the graphite surface during the scatttering experiment. From measurements with a chopped NO beam at a surface temperature of 140 K an upper limit for the residence time of 4×10^{-5}s can be deduced. With the given particle flux and a sticking probability of unity, the upper limit of the NO coverage was estimated to be about 10^{-6} monolayers.

The frequency-doubled radiation of an excimer-pumped dye laser was used to probe the NO molecules (5 ns pulses at a rate of 5 Hz, pulse energy 10µJ in the desired wavelength range around 226 nm). Rotational state distributions in the electronic ground state could be derived from the measured fluorescence spectra as described previously.[35]

The electronic ground state of NO is split into two states owing to spin-orbit interaction, with the $^2\pi_{3/2}$ state about 120 cm^{-1} higher than the $^2\pi_{1/2}$ state. For the molecules of the incident beam, the measured rotational populations show Boltzmann distributions in both fine-structure states with temperatures corresponding to 40 K ($^2\pi_{3/2}$) and about 70 K ($^2\pi_{1/2}$). The second temperature also expresses the ratio of the populations of the two electronic ground states. This indicates that these states are not completely equilibrated with each other by the collision processes during expansion in the nozzle kept at room temperature. Assuming isenthalpic expansion, one can estimate the translational energy of the incident molecules from the measured rotational distribution to be about 700 cm^{-1} or 0.08 eV.

The rotational and angular distributions of the scattered NO molecules were investigated for surface temperatures between 130 K and 780 K. Examples of the measured angular distributions are shown in Figure 9 for incidence angles of 30^0 and 60^0 and for different surface temperatures. The experimental points correspond directly to the mass spectrometer signals. In order to obtain the angular flux distribution of the scattered molecules it is necessary to correct for the velocity of the molecules, which may be different for the isotropically and specularly scattered molecules. In the present experiment the velocities of the scattered molecules were not analyzed, and therefore the corresponding corrections could not be considered.

In the figure one can distinguish broad scattering lobes in the direction close to specular reflection, and underlying cosine-like distributions caused by diffusive scattering. The specular scattering lobe is interpreted as due to weakly inelastic scattering processes. The half-width of the lobe is about 40^0 and is independent of the scattering angle. As the surface temperature is raised, the direction of the lobular maximum approaches an angle to the surface normal, corresponding qualitatively to the predictions of the hard cube model.[46,47]

Figure 9 shows that as the surface temperature increases, the fraction of molecules scattered in a cosine distribution decreases, while the specularly scattered fraction grows. A considerable fraction of the observed diffusively scattered particles, however, is thought to be due to the surface roughness of the crystal used. This is shown by the remaining isotropic part obtained at the highest temperatures investigated.

The rotational distributions were measured for both diffusively scattered and specularly reflected molecules. In both cases the rotational temperatures of both electronic ground states were de-

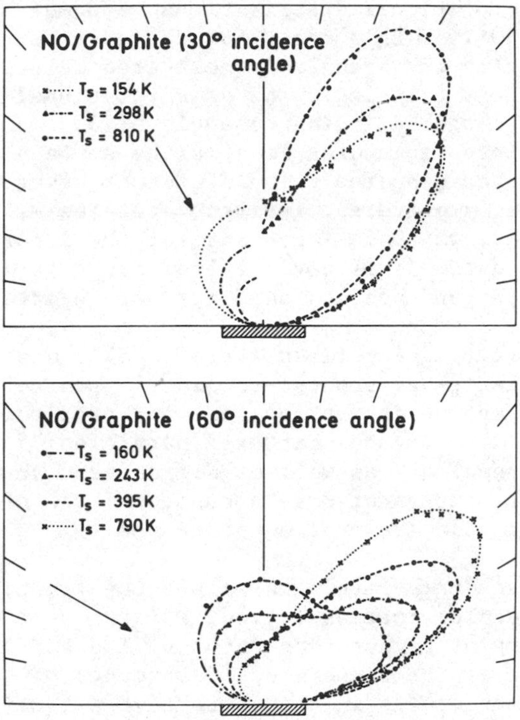

Fig. 9. Angular distributions of NO molecules scattered from a graphite surface at different temperatures with incidence angles of 30^0 (upper graph) and 60^0 (lower graph).

termined. The result is a Boltzmann distribution between the rotational levels with the same temperature for both fine-structure states. This means that the rotational levels and the two fine structure states are in thermal equilibrium after the scattering process. This was the case for all surface temperatures between 130 K and 800 K, and for both the diffusively scattered and the specularly scattered molecules.

The dependence of the rotational temperature on the surface temperature is shown in Fig. 10. Each point and the corresponding error bar is obtained by a least squares fit to the measured population distribution. The solid line corresponds to complete accommodation of the rotational degree of freedom to the surface temperature. The experimental points follow this line up to a surface temperature of 170 K, they deviate at higher temperatures and from 350 K upwards the rotational temperature converges to a value of about 250 K.

Most of the rotational distributions were measured for angles of incidence of 30^0. In this case, for the given experimental geometry, predominantly the specularly scattered molecules were probed by the laser beam (circles). The same rotational temperatures were obtained at a second incidence angle of 60^0, where mainly diffusively scattered molecules leaving the surface perpendicularly were investigated (squares). The different experimental geometries, however, imply different incident energies with respect to the surface normal, which is important for the interpretation of the experimental data. As is obvious from the data in Fig. 10, the same temperature dependence is found for both scattering geometries.

Surface scattering experiments are usually discussed in terms of three basic scattering processes: elastic scattering, inelastic scattering and trapping/desorption, as characterized by typical angular distributions of the scattered particles.[48,49] The observation of specularly as well as diffusively scattered NO molecules may thus be interpreted as a superposition of inelastic scattering and trapping/desorption processes.

As the surface temperature is raised the fraction of specularly scattered molecules increases, indicating a predominance of inelastic scattering at higher temperatures. A possible explanation of this behaviour is the temperature dependence of the residence time τ of the NO molecules in the attractive potential well of the graphite surface:

$$\tau = \nu^{-1} \exp(E_d/kT_s),$$

where T_s is the surface temperature and E_d is the desorption energy of the NO molecules (estimated to be 0.12 eV for low coverage[50]). The frequency factor ν is assumed to be $10^{13}s^{-1}$. This equation yields residence times of between $\sim 10^{-9}$ s for surface temperatures of 150 K and ~1 ps for T_s = 700 K. As the transit time of the molecules through the potential well of the surface is estimated to be about 1 ps, the molecules are assumed to undergo many hundred surface collisions in the case of lower surface temperatures, leading to a higher fraction of diffusively scattered particles. For higher temperatures the number of surface collisions decreases, resulting in a growing specularly scattered fraction.

The essential result we find in the present experiment is that within the range of rotational energies investigated the rotational distributions of the scattered molecules correspond to Boltzmann distributions of the same temperature for both electronic ground states, and that the ratio of the overall populations of both fine-structure levels exhibits the same temperature. This only agrees with the temperature of the graphite surface below 170 K. At higher surface temperatures there is an increasing discrepancy (Fig. 10).

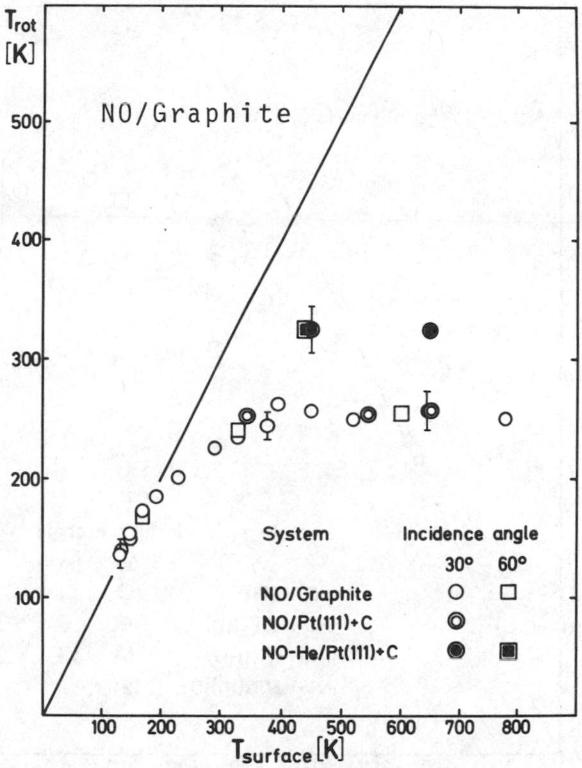

Fig. 10. Plot of the measured rotational temperature of the scattered NO molecules versus the surface temperature for different surfaces and incidence angles (see text). The straight line symbolizes full rotational accommodation

A similar deviation was obtained in a desorption experiment where a rotational temperature of 235 K was measured for NO molecules thermally desorbing from a 450 K Ru(001) surface, in spite of a very long residence time.[41] Assuming that there is no reactive interaction of the NO/Ru system, the result of the experiment shows that the final rotational distribution is predominantly determined by the exit channel of the interaction; the decisive processes leading to this final rotational distribution seem to take place when the molecules are receding from the surface. For the adsorbed molecules, rotation is hindered by the binding to the surface; thus, rotational energy is taken up when the molecules desorb from the surface. Experimental results for NO scattered from a Pt(111) surface support this interpretation[51]: in spite of residence times of the order of ms and the observation of primarily diffusive scattering indicating a trapping/desorption process, rotational temperatures much lower than the surface temperatures are found for $T_s > 400$ K (Fig. 11).

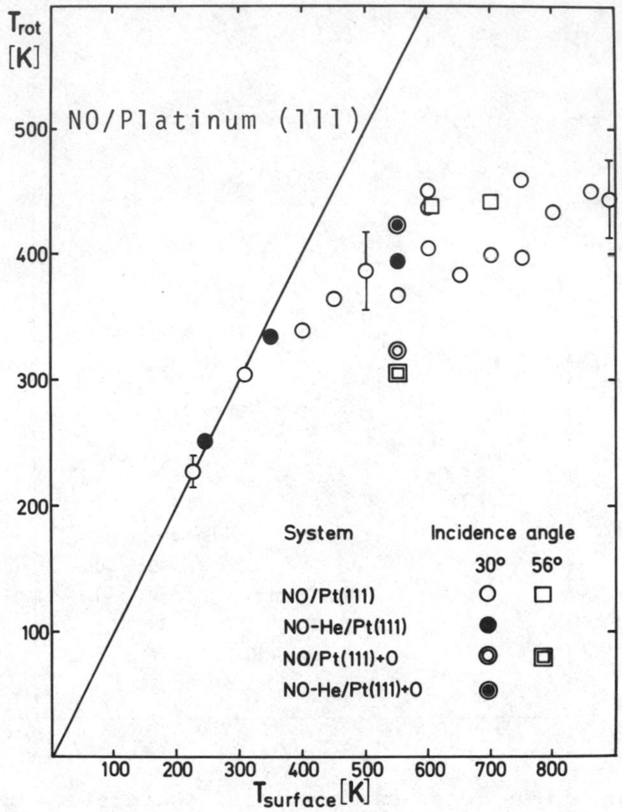

Fig. 11. Plot of the measured rotational temperature of the scattered NO molecules versus the surface temperature for different surfaces and incidence angles (see text). The straight line symbolizes full rotational accommodation

In the experiment with the graphite surface we do not have a pure adsorption/desorption process since a rather large fraction of the molecules is specularly scattered. The result presented in Fig. 10 must therefore be due primarily to other processes.

Even at very low surface temperatures the angular distribution shows a specular contribution (Fig. 9) indicating molecular residence times comparable with those at high surface temperatures. This means that the measured rotational accommodation at lower temperatures (Fig. 10) cannot be strongly influenced by the residence time of the molecules on the surface. It is most likely that in the region of surface temperatures higher than 400 K the rotational temperature of the scattered molecules is influenced by the kinetic energy of the incoming molecules, and that a redistribution of the energy into the different degrees of freedom of the molecules takes place during the interaction with the surface. This is supported by an experiment in which a seeded NO beam is scattered from a graphite surface.[51] In this case, for surface temperatures larger than 400 K, a higher rotational temperature has been observed (Fig. 10); a corresponding result has not been obtained in the scattering of NO from Pt(111), as shown in Fig. 11. It is quite clear that the description of the processes as given here must be rough and incomplete, and that a perfect understanding of the dynamics of the molecule-surface interaction can only be obtained by trajectory calculations.

4. LASER INVESTIGATIONS IN THE ATMOSPHERE

The laser is an ideal instrument for determining properties of the atmosphere by absorption and scattering processes: its high spectral density and low divergence are especially useful for such measurements. Lasers with continuous and discrete tunability can be used to measure specific components selectively. A survey of methods used to date is given in Ref.[52]

Scattering of laser light in the low atmosphere is dominated by Mie scattering caused by aerosols, clouds, dust and other particulates. Rayleigh scattering from the molecular constituents of the atmosphere is two orders of magnitude lower. It becomes dominant above heights of 30 km, where virtually no aerosols exist. It is therefore possible to derive from the scattering intensity at these heights information about the gas density. One can thus determine the pressure and temperature distributions; even seasonal variations of these parameters were studied in RADAR-like experiments.[53]

For the analysis of gases in the lower atmosphere, fluorescence or Raman scattering can be measured. Absorption measurements yield the highest sensitivity and are very simple, but give concentrations only integrated along the path of the light. To cover a larger area by the laser beam, mirrors can be used. A very simple setup uses topographic targets for reflection of the light back to its source for detection. By the use of mirror arrays, absorption measurements, such as tomography, can yield the spatial distribution of gas constituents.[54]

Normally, absorption measurements can be performed with low-power lasers (e.g. diode lasers). Even arc lamps with spectral filtering can be applied. This setup was very successful in detecting SO_2, N_2O, CH_2O, O_3, NO_2 and NO_3 with very high sensitivity.[55]

A remarkable improvement of the detectivity in absorption measurements becomes possible by a heterodyne technique, where a tunable laser is used as local oscillator and a photodetector as mixer.[56] The signal at the intermediate frequency is amplified by a suitable amplifier of narrow bandwidth. This technique (most advantageous in the infrared spectral range) yields an increase in sensitivity of several orders of magnitude, allowing appreciable lengthening of the absorption paths. With heterodyne detection, fast detectors used at room temperature (e.g. pyroelectric detectors) are equally sensitive as cooled infrared detectors.

As mentioned above, the observation of absorption excludes measurement of the density distribution of a gas. This disadvantage can be overcome by using Mie scattering (which is very strong in the lower atmosphere) as a "mirror". The position of the "mirror" can be determined from the time elapsing between the emission of the laser pulse and the detection of the backscattered signal. The dependence on wavelength of the Mie scattering is low. It is therefore possible to eliminate the local variation of the scattering by measurement at two different wavelengths. Only one of these has to be absorbed by the gas to be detected. Both wavelengths must be sufficiently close together to ensure that their Mie scattering is in fact equal.

This method, called "differential absorption", was first applied to pollution detection by Rothe et al.[58] Theoretical studies[57] had shown before that it is the most sensitive of all LIDAR methods.

In the first measurements[58,59] the apparatus was equipped with a tunable dye laser. The concentration of NO_2 near a chemical factory was measured. With only 1 mJ of pulse energy, concentrations as low as 0.2 ppm could be detected at distances of 4 km. The distribution of the NO_2 concentration above the plant was determined by varying the direction of the laser beam. Connecting points of equal concentration measured in five different directions, a map of isolines was obtained. It becomes evident from this picture which building is the source of the pollutant. By means of a dye laser the differential absorption method can also be applied to SO_2 and O_3. SO_2 measurements were carried out by, for example, Svanberg et al.[60]

When the differential absorption method is extended to a larger number of different pollutants, measurements have to be carried out in the infrared range of the spectrum. The most universal setup would be obtained by using a continuously tunable laser. As such lasers are at present unsuitable for use in the field, molecular lasers (e.g. DF, HF, CO, CO_2 and N_2O lasers) must be employed and measurement must rely on accidental coincidence of laser emission lines with absorption lines of the pollutants.

Several pollutants can even be detected simultaneously by a multi-line measurement or sequential measurements using different emission lines.

In the following, some results obtained from a setup equipped with a multi-gas laser are reported. Details of the apparatus are published elsewhere.[61,62]

The first experiments to be discussed here were conducted at the cooling tower of the power station in Meppen, in cooperation with Electricité de France. The objective of these measurements was to determine the concentration of water vapour around the cooling tower in order to obtain data comparable with computational results of the distribution.

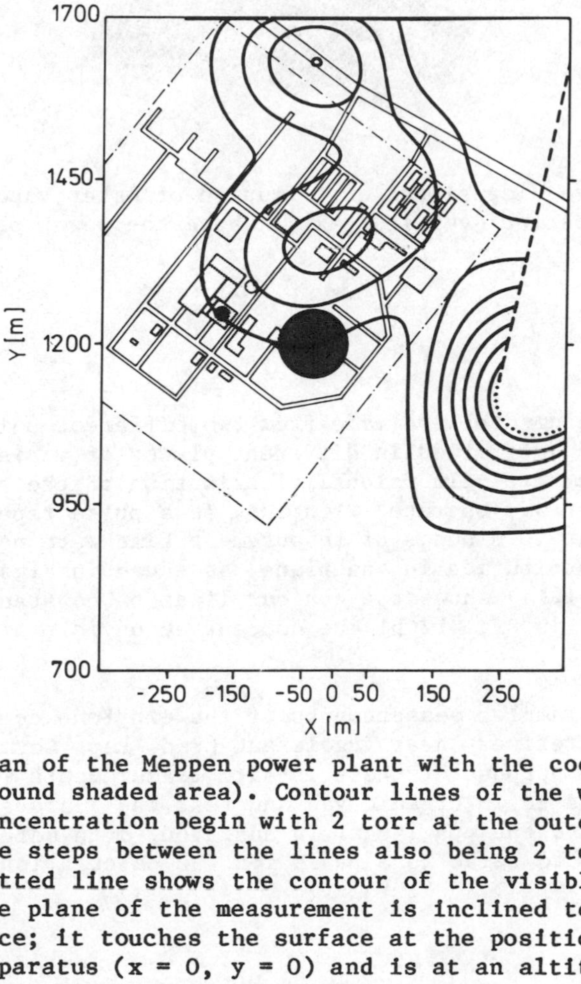

Fig. 12(a). Plan of the Meppen power plant with the cooling tower (round shaded area). Contour lines of the water vapour concentration begin with 2 torr at the outermost line, the steps between the lines also being 2 torr. The dotted line shows the contour of the visible plume. The plane of the measurement is inclined to the surface; it touches the surface at the position of the apparatus (x = 0, y = 0) and is at an altitude of about 100 m above the cooling tower.

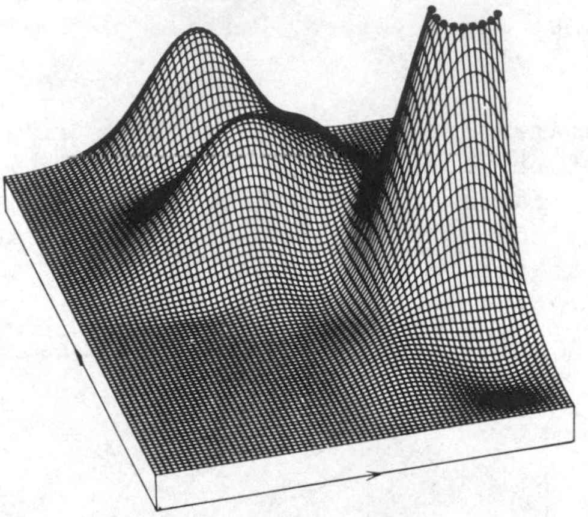

Fig. 12(b). Two-dimensional distribution of water vapour concentration (vertical axis) above the power plant.

The measurements were made from two different points; the concentration was determined in different planes of varied elevation. This yielded a three-dimensional distribution of the concentration. The laser was operated with CO_2. At a pulse repetition rate of 70 Hz, about 10 minutes of measurement time were necessary to obtain the distribution in one plane, as shown in Figs. 12(a) and 12(b). Figure 12(a) shows the contour lines of constant concentration, while in Fig. 12(b) the concentration is plotted on the vertical axis.

As next example, measurements of the ethylene concentration around an oil refinery near Ingolstadt (Fed. Rep. Germany) are discussed. The purpose of these in-situ measurements was to extend the experiments to an organic gas and test the improvement of the detectivity into the ppb range. In addition, data were to be collected that could serve as a basis for the calculation of the distribution.

The ethylene detected leaks out of the distillation plant. These measurements were also made with a CO_2 laser placed at a distance of about 500 m from the refinery. An example is shown in Figs. 13(a) and 13(b).

The measurements on air pollution described here demonstrate the present technical knowledge of laser investigations in the atmosphere. The differential absorption method is now at the stage where it can be enlisted for routine measurements.[53]

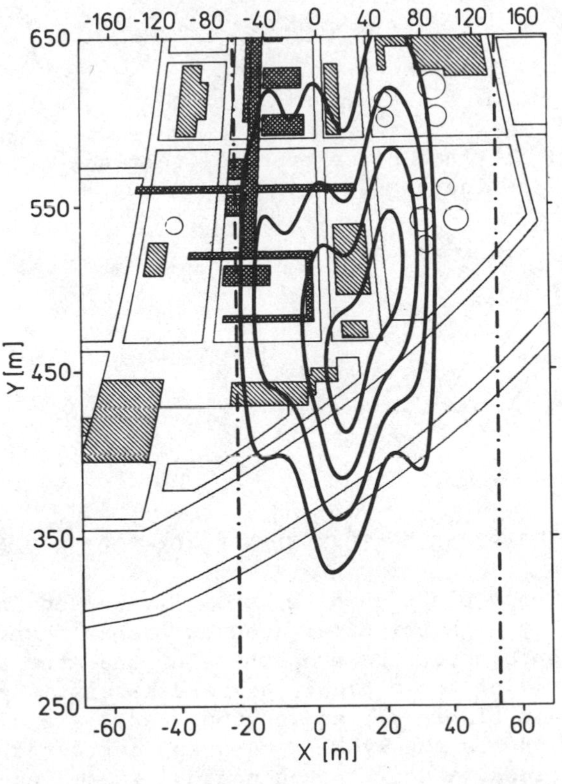

Fig. 13(a): Map of the refinery and contour lines of ethylene concentration. The lines start with 20 ppb, the inward increase being 20 ppb.

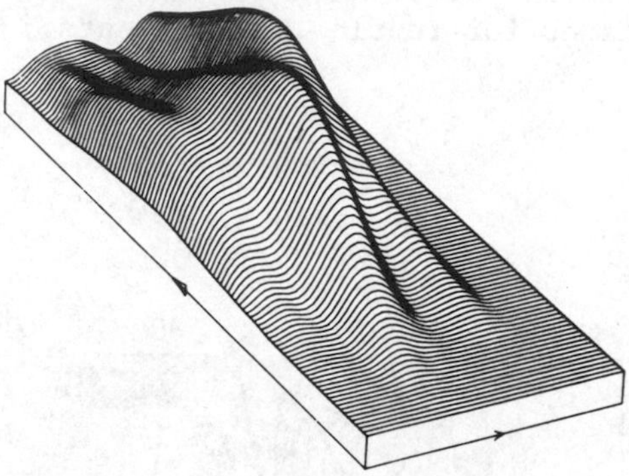

Fig. 13(b). Ethylene concentration (vertical axis) above the refinery.

5. INTERACTION OF RYDBERG ATOMS WITH BLACK-BODY RADIATION

When a valence electron of an atom is excited into an orbit with sufficiently high principal quantum number n and therefore far from the ionic core, the properties of the atom appear hydrogenic. The energy of these highly excited levels is given by Rydberg's formula (Table 4), and so the states are also called Rydberg states. R^* is the Rydberg constant corrected for the mass of the nucleus, and δ_l the phenomenological quantum defect of the states of angular momentum l. For states of low l, where the orbits of the classical Bohr-Sommerfeld theory are ellipses of high eccentricity, the penetration and polarization of the electron core by the valence electron lead to large quantum defects and strong departures from the hydrogenic behaviour. As l increases, the orbits become more circular and the atom becomes more hydrogenic, δ_l changing with l^{-5}.

Table 5. Properties of Rydberg atoms

Energy: $E_n = -R/(n-\delta_\ell)^2 = R/n^{*2}$

n^* effective quantum number

Radius: $\langle r \rangle = a_o n^{*2}$

a_o: Bohr radius

Lifetimes: $\tau \sim n^{+3}$

spontaneous transition probability $n \rightarrow n'$: $A_{n \rightarrow n'}$

$$A_{n\rightarrow n'} = \frac{e^2}{c^3}\omega^3 |r_{nm}|^2$$

case a: $n' < n$ then $|r_{nm}| \ll 1$ due to small overlap of wavefunctions

$\underline{A_{n\rightarrow n'} \sim n^{-3}}$

case b: $n' \approx n$ $\omega = En - E_{n'} \sim n^{-3}$ and $|r_{nn'}| \sim n^{+4}$

$\underline{A_{n\rightarrow n'} \sim n^{-5}}$

Table 4. Properties of hydrogen Rydberg states

	n=2	n=10	n=30	n=100	Units
Energy	3.4	0.14	0.015	$1.4 \cdot 10^{-3}$	eV
Radius	4	100	900	10 000	a_o
Area	16	10^4	$8.1 \cdot 10^5$	10^8	Πa_0^2
Frequency	$2.5 \cdot 10^6$	6600	240	6.6	GHz
Lifetime	$2 \cdot 10^{-3}$	0.51	23	1 600	µs
Ioniz. Field	$5 \cdot 10^7$	$3 \cdot 10^4$	420	3	V/cm

Further properties of the Rydberg atoms are also listed in Tables 4 and 5.[63] Remarkable is the large spontaneous lifetime of the states, scaling as n^{*3} when n' is very much different from n, or as n^{*5} when n' and n are close. The size of $A_{n\to n'}$ still depends on the angular momentum l. This can be understood by simple classical arguments as explained in the following.

For low angular momentum states (core penetrating) the lifetime τ can be deduced from the third Kepler law. Accordingly, the electron period T is given by $T \approx (n^2 a_0)^{3/2} \approx n^3$ (in the classical picture T must be proportional to τ since transition to a lower orbit is always more probable when the electron approaches the core).

For high angular momentum orbitals in the classical picture the electron radiates continuously and decreases its radius. The acceleration of the electron is inversely proportional to the square of the radius of the orbit, and therefore the power of the emitted radiation scales as n^{-8}. The distance between neighbouring Rydberg levels changes as n^{-3}; this gives a characteristic time needed for each step of $n^{-3}/n^{-8} = n^{-5}$, corresponding to the lifetime.[64]

The square of the matrix element $r_{nn'}$ $(n \approx n')$ scales as n^4, showing a rather high transition probability for induced transitions. Rydberg atoms therefore strongly absorb microwave or far infrared radiation. A consequence is that blackbody radiation may cause a strong mixing of the states. This is especially the case for states with high angular momentum since the spontaneous life-times of these are much larger, and therefore the induced transitions can be saturated much more easily than for the lower l states.

The induced transition rate due to black-body radiation is proportional to $|r_{nn'}|^2\ d\phi/d\upsilon$, where $d\phi/d\upsilon$ is the energy flux of the black-body radiation per unit bandwidth and unit surface area. At low frequencies (Rayleigh-Jeans limit) $d\phi/d\upsilon$ changes as υ^2, which, because the distances between the Rydberg states scale as n^{-3}, means that $d\phi/d\upsilon \approx n^{-6}$. With $r_{nn'}{}^2 \sim n^4$ one gets an induced transition rate varying as n^{-2}. The ratio between the induced transition rate to the spontaneous rate therefore changes as n (for low l) and as n^3 (for high l). This means that for a given atom and a given temperature there exists an n above which the black-body induced rate overcomes the spontaneous rate.

The influence of black-body radiation on Rydberg atoms was first demonstrated in lifetime measurements. For instance, Gallagher et al.[66] observed that the measured lifetimes of the 16p and 17p states of Na are three times shorter than expected;

the shorter lifetime was supposed to be due to black-body interaction. Haroche et al.[67] found a population transfer to nearby levels which could not be explained by spontaneous decay. More direct evidence of interaction with black-body radiation was observed later.[68,69,70]

The influence of black-body radiation is demonstrated in Fig. 14. For this measurement the 5s23f state of the Sr atom was excited by a pulsed dye laser.[65] The Rydberg atoms were detected by field ionization. For this purpose a field ramp was used so that the different Rydberg states were successively ionized starting with the levels closest to the ionization limit.[64] The field ramp was started for the first measurement 1μs after the laser excitation. The measurements performed with larger time delays 2,6 and 12 μs clearly show the increasing population change due to the strong interaction with black-body radiation (see also Ref.[68]).

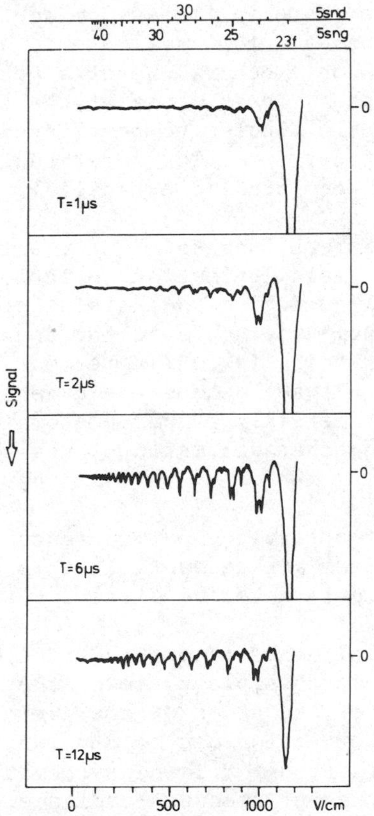

Fig. 14. Influence of black-body radiation on Sr Rydberg atoms. For details see text.

A consequence of the long radiative decay time of Rydberg levels and the very large value of the electric dipole matrix elements is that the saturating power for transitions between closely lying Rydberg levels is very small. The corresponding saturating power fluxes are proportional to n^{-10} for low and to n^{-14} for high angular momentum states. A very vivid way of describing the behaviour is to express the saturating power flux in terms of number of photons per surface of the size λ^2 and per lifetime. (The size λ^2 corresponds to the resonant cross-section.) For $n \approx 30$ one obtains 10^2 and 1 for low and high angular momentum states, respectively. This means that for high angular momentum states a single photon is required (in the chosen units) to saturate the transition to a neighbouring Rydberg level.[64]

In most of the experiments, the Rydberg atom is prepared in a particular initial state by a short laser pulse. As discussed, the spontaneous transition rate to closely lying levels is very small, and therefore it is easy to achieve the conditions for collective emission even for small atomic densities. This super-radiance phenomenon[71] occurs when the collective field radiated in phase by the Rydberg atoms at the location of any of them causes the atomic dipoles to be flipped in a time shorter than the spontaneous decay time. The super-radiance phenomenon for Rydberg atoms has been studied extensively by several authors.

Haroche and co-workers (see Ref.[64] for a review) performed the experiments using field ionization detection. The ion current is observed time resolved during the raising of the ionizing field ramp. The threshold was achieved for densities of the order of 10^8 atoms/cm^3, this being in fair agreement with theory. Evidence of fluctuations between successive experiments corresponding to the same number of initially prepared atoms have also been observed, this also being characteristic of the phenomenon of super-radiance.

Liberman et al.[72] observed super-radiance of Rydberg atoms in a multiple atomic beam arrangement. In this experiment the emitted radiation could be detected direct.

Surrounding the volume of laser-excited Rydberg atoms with a resonant cavity, it is possible to lower the super-radiance threshold by a factor of the order of the cavity finesse. In this way transient maser-action can be observed with a very small number of Rydberg atoms.[73] These systems, working in a transient regime reminiscent of super-radiance, generate a few hundred ns pulses of millimeter-wave radiation. They are characterized by extremely low inversion density thresholds, as mentioned above ($N \approx 4 \times 10^3$ atoms typically), and by very small peak power outputs (in the 10^{-11} to 10^{-12} watt range). This study is interesting in the context of new tests of superradiance and electrodynamic theory. It also opens new possibilities in the important domain of millimeter amplification and detection. In this connection Rydberg atoms are useful in two types of devices. The first are the coherent amplifiers just mentioned which amplify the field to be detected and are thus also sensitive to its phase.[74] The second are sensitive to just the radiated energy. This type will be described in more detail in the following.

There are many applications of wide ranging importance for detectors in the submillimeter region, e.g. infrared and radio astronomy, diagnostics of plasmas for nuclear fusion, stratospheric monitoring and materials research. The investigation of new principles for detectors is therefore as important as the further development and improvement of known detector principles. The ultimate sensitivity obtainable for detection of any radiation is of course reached when single photons can be monitored with a high probability and when the noise of the signal is only determined by the quantum noise of the radiation. The quantum noise limit of the radiation can be reached in the visible or near infrared spectral range since there the available photomultipliers allow the photocurrent of a single photon to be amplified to a value exceeding the noise of the dark current and the amplifier.

Ducas et al.[75] demonstrated the very sensitive detection of low-power far infrared laser radiation at 600 GHz by inducing transitions between Rydberg states of sodium atoms. To check the ultimate sensitivity of the Rydberg detector to microwave or far infrared radiation, two improvments of the previous experiments have to be effected. First, the population of the Rydberg atoms must be performed by cw lasers in order to increase the duty cycle and, in addition, the walls of the surrounding chamber have to be cooled to a low temperature, so that the influence of the thermal background radiation is minimized. In the following an experiment of this type will be described.[69]

Sodium atoms of an atomic beam were excited to high-lying ^2D-states in two steps via the $3^2P_{3/2}$ intermediate state. The $3^2S_{1/2}$, F=2 → $3^2P_{3/2}$, F=3 hyperfine transition (589 nm) is saturated with circularly polarized light by means of a single mode cw dye laser stabilized to this transition. A fraction of the atoms in the $3^2P_{3/2}$, F=3 state is then excited to the n^2D state by means of a multimode dye laser whose cavity length is wobbled in order to obtain a more homogeneous intensity distribution over the laser line width of 100 GHz at a wavelength of about 4150 Å.

The interaction region is surrounded by a box cooled down to 14 K in order to keep the background of black-body radiation as low as possible (Fig. 15). The slit where the atomic beam enters the cooled box is covered with a wire mesh to reduce microwave and far infrared radiation emitted by the atomic beam oven. The two dye laser beams (DL1 and DL2) intersect the atomic beam at right angles. After a path of 20 mm the excited atoms leave the box through a second wire mesh which also acts as one plate of the capacitor used for field ionization. If an atom is field-ionized, the ion is accelerated and leaves the capacitor through the mesh in the negative plate and is accelerated to the entrance of a channeltron multiplier.

A small flap is mounted at one side of the cooled box. By opening this flap, the Rydberg atoms can be exposed to the radiation of a heated wire. The microwave transitions induced in the Rydberg atoms by the thermal radiation of the wire are then monitored via field ionization. For the experiments described here the atoms were excited to the 22^2D state. The strength of the electric field was adjusted, so that atoms in the 22^2D state are not ionized. However, any transition induced to a higher state is detected through the ion signal.

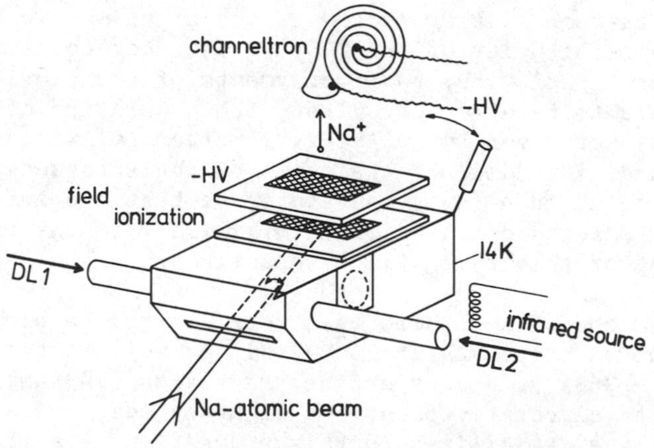

Fig. 15. Experimental setup. The flap at the right side of the box allows the thermal radiation of the infrared source to enter the box.

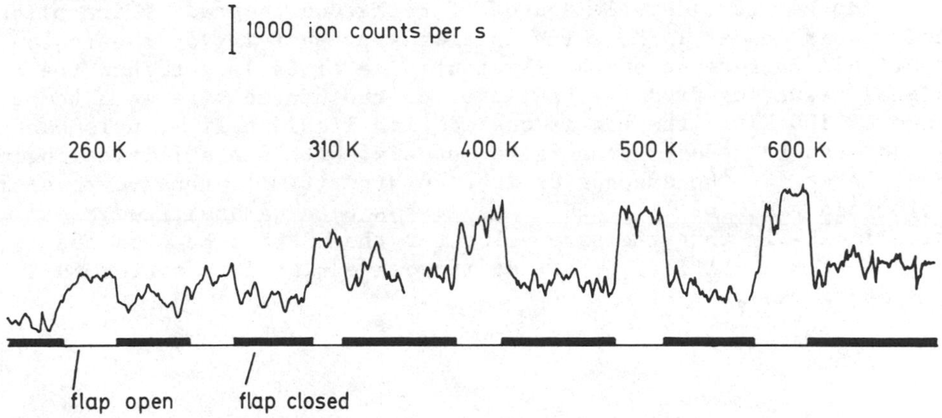

Fig. 16. By opening and closing the flap the ion signal is changed. The background (not shown in the figure) is about five times the signal induced by the infrared source at 310 K and is due to the 14 K blackbody radiation emitted by the walls of the box.

By opening and closing the flap of the box the influence of the radiation from the heated wire could be investigated. Figure 16 shows the change of the ion count rate correlated with the opening and closing of the flap for different temperatures of the heated wire.

The lifetime of the 22^2D state is 10 μs. Only about 5 % of the initially excited Rydberg atoms therefore reach the field ionization region. The $\Delta n=1$ transition is most likely to be induced (Fig. 17). The 23^2P state has a lifetime of about 100 μs. All atoms excited by the microwave therefore reach the field ionization region. The ratio between the rate of the detected ions and the rate of absorbed photons gives the quantum efficiency of the device of 3×10^{-3}.

As a result, the noise equivalent power (NEP) of the detector is 10^{-17} W/Hz$^{1/2}$, using an output bandwidth of 1 kHz. The NEP of this preliminary setup favourably compares with the NEP of other detectors, which is at least one order of magnitude larger. The NEP was calculated by assuming a background noise which is equal to the signal at 310 K. Figure 16 shows that the noise of the signal is not statistical, a major contribution coming from slow intensity and frequency fluctuations of the second laser. If the signal is only observed for a few seconds, it is much more stable, indicating that a better NEP can be achieved by stabilizing the second laser.

Another considerable source of background signal is the blackbody radiation at 14 K, which is always present inside the cooled box. This background signal is about five times larger than the signal resulting from the radiation of the heated wire at a temperature of 300 K. If the box is cooled with liquid helium, this background can be reduced by a factor of six. With a stabilized single mode laser for the seconds excitation step it will thus be possible to obtain a NEP of 10^{-19} $WHZ^{-1/2}$. It has been demonstrated experimentally that the same value for the NEP can be obtained if coherent amplification of the microwave signal is performed in a Rydberg maser.[76]

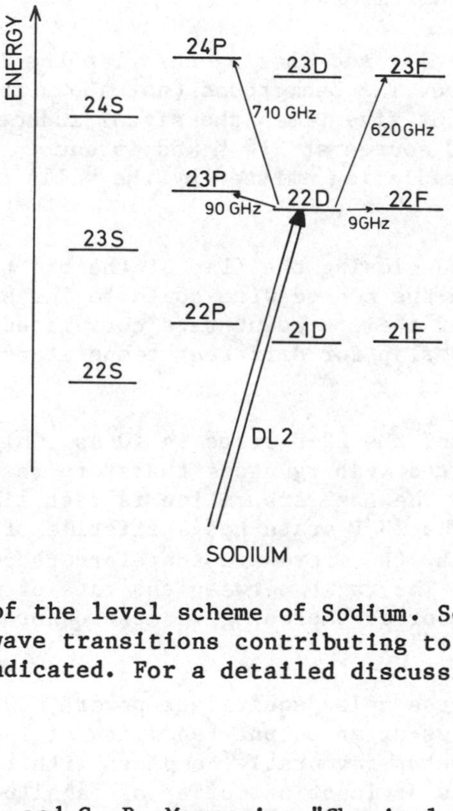

Fig. 17. Part of the level scheme of Sodium. Some of the microwave transitions contributing to the signal are indicated. For a detailed discussion see Ref.[76]

REFERENCES

1. V. S. Lethokov, and C. B. Moore in: "Chemical and Biochemical Applications of Lasers" Vol. III, ed. C. B. Moore (Academic Press, New York, 1977, p. 1.
2. J. Wolfrum, to be published Ber. Bunsenges. physik. Chem. and Report No 21 of the MPI für Strömungsforschung, 1981
3. T. F. Deutsch, D. J. Ehrlich, and R. M. Osgood Jr., Appl. Phys. Lett. 35:175 (1979)
4. V. S. Letokhov, Contribution in this volume.

5. A. Laubereau, Contribution in this volume.
6. P. V. Ambartzumian, and V. S. Letokhov, Appl. Opt. 11:354 (1972)
7. A. Ben-Shoul, Y. Haas, K. L. Kompa, and R. D. Levine in: "Lasers and Chemical Change", Springer Series in Chemical Physics 10 (Springer Verlag, Berlin, Heidelberg, New York) 1981
and "Photoselective Chemistry", Part 1 and 2, eds. J. Jortner, R. D. Levine and S. A. Rice (Wiley, Interscience, New York) 1981
8. T. J. Odiorne, P. R. Brooks, J. V. V. Kasper, J. Chem. Phys. 55:1980 (1971)
J. C. Pruett, F. R. Grabinger, P. R. Brooks, J. Chem. Phys. 60:3335 (1974)
9. J. Wolfrum, Ber. Bunsenges. physik. Chem. 81:114 (1977)
and M. Kueba, J. Wolfrum, Ann. Rev. Chem. 31:47 (1980)
10. J. G. Pruett, R. N. Zare, J. Chem. Phys. 64:1774 (1976) and Z. Karny, R. N. Zare, J. Chem. Phys. 68:3360 (1978) and Z. Karny, R. C. Estler, R. N. Zare, J. Chem. Phys. 69:5199 (1978)
11. A. Schultz, H. W. Cruse, and R. N. Zare, J. Chem. Phys. 57:1354 (1972)
12. P. J. Dagdigian, H. W. Cruse, A. Schultz, R. N. Zare, J. Chem. Phys. 61:4450 (1974)
13. H. W. Cruse, P. J. Dagdigian, R. N. Zare, Foraday Discussion 55:277 (1973)
14. G. P. Smith, R. N. Zare, Chem. Phys. 64:2632 (1976)
15. P. J. Dagdigian, H. W. Cruse, R. N. Zare, Chem. Phys. 15:249 (1976)
16. C. R. Dickson, J. B. Kinney, R. N. Zare, Chem. Phys. 15:243 (1976)
17. W. Schmidt, A. Sigel, A. Schultz, Chem.Phys. 16:161 (1976)
18. M. Rommel, A. Schultz, Ber. Bunsenges. physik. Chem. 81:139 (1977)
19. I. P. Herman, R. P. Mariella, Jr., A. Javan, J. Chem. Phys. 68:1070 (1978)
20. K. Liu, J. M. Parson, J. Chem. Phys. 67:1813 (1977) and J. Chem. Phys. 68:1794 (1978)
21. L. Pasternack, P.J. Dagdigian, J. Chem. Phys. 67:3854 (1977)
22. J. A. Silver, W. L. Dimpfl, J. H. Brophy, J. L. Kinsey, J. Chem. Phys. 65:1811 (1976)
23. R. P. Mariella, Jr., B. Lantzsch, V. T. Maxson, A. C. Luntz, J. Chem. Phys. 69:5411 (1978)
24. J. H. Brophy, J. A. Silver, J. L. Kinsey, J. Chem. Phys. 62:3820 (1975)
25. Ch. T. Rettner, J. F. Cordova, J. L. Kinsey, J. Chem. Phys. 72:5280 (1980)
26. E. J. Murphy, J. H. Brophy, G. S. Arnold, W. L. Dimpfl, J. L. Kinsey, J. Chem. Phys. 74:324 (1981)
27. L. Stein, J. Wanner, H. Walther, J. Chem. Phys. 72:1128 (1980).

28. M. A. A. Clyne, I. S. McDermid, J. Chem. Soc. Feraday Trans. II, 73:1094 (1977)
29. L. Stein, J. Wanner, H. Figger, and H. Walther, in: "Laser-Induced Processes in Molecules", edited by K.L. Kompa and S. D. Smith, Springer Series in Chemical Physics, Vol. 6, Springer Verlag, Berlin, Heidelberg, New York, 1979, p.232
30. T. Trickl, PhD Thesis, Universität München 1982
31. J. M. Ferrar, Y. T. Lee, J. Chem. Phys. 63:3639 (1975)
32. J. Häger, W. Krieger, T. Rüegg H. Walther, J. Chem. Phys. 70:2859 (1979)
33. J. Häger, W. Krieger, T. Rüegg, H. Walther, J. Chem. Phys. 72:4286 (1980)
34. J. D. Lambert, R. Salter, Proc. R. Soc. (London) Ser. A 253:277 (1959)
35. F. Frenkel, J. Häger, W. Krieger, H. Walther, C. T. Campbell, G. Ertl, H. Kuipers, J. Segner, Phys. Rev. Lett. 46:152 (1981)
36. G. M. McClelland, G. D. Kubiak, H. G. Rennagel, R. N. Zare, Phys. Rev. Lett. 46:831 (1981)
37. F. Frenkel, J. Häger, W. Krieger, H. Walther, C. T. Campbell, G. Ertl, H. Kuipers , J. Segner in: "Laser Spectroscopy V", ed. by A. R. W. McKellar, T. Oka, B. P. Stoicheff, Springer, Berlin, Heidelberg, New York 1981, p. 425
38. J. W. Hepburn, F. J. Nothrup, G. L. Ogram, J. C. Polanyi, and J. M. Williamson, Chem. Phys. Lett. 85:127 (1982)
39. A. W. Kleyn, A. C. Luntz, and D. J. Auerbach, Phys. Rev. Lett. 47:1169 (1981)
40. A. C. Luntz, A. W. Kleyn, and D. J. Auerbach, J. Chem. Phys. 76:737 (1982)
41. R. R. Cavanagh, and D. S. King, Phys. Rev. Lett. 47:1829 (1981)
42. L. D. Talley, W. A. Sanders, D. J. Bogan, and M. C. Lin, Chem. Phys. Lett. 66:500 (1981)
43. F. Frenkel, J. Häger, W. Krieger, H. Walther, G. Ertl, J. Segner, W. Kielhuber, Chem. Phys. Lett. 90:225 (1982)
44. D. L. Smith, and R. P. Merrill, J. Chem. Phys. 52:5861 (1970)
45. G. Boato, P. Cantini, C. Guidi, R. Tatarek, Phys. Rev. B 20:3957 (1979)
46. R. M. Logan, and R. E. Stickney, J. Chem. Phys. 44:195 (1966)
47. W. L. Nichols, and J. H. Weare, J. Chem. Phys. 63:379 (1975)
48. W. H. Weinberg, and R. P. Merrill, J. Chem. Phys. 56:2881 (1971)
49. W. H. Weinberg, and R. P. Merrill, J. Vac. Sci. Technol. 8:718 (1971)
50. C. E. Brown, and D. G. Hall, J. Colloid Interface Sci. 42:334 (1973)
51. F. Frenkel, J. Häger, W. Krieger, H. Walther, G. Ertl, H. Robota, J. Segner, and W. Vielhaber, to be published
52. H. Walther, Laser Investigations in the Atmosphere, in: "Festkörperprobleme (Advances in Solid State Physics)", Vol. XX, p. 327, ed. J. Treusch, Vieweg, Braunschweig 1980

53. D. K. Killinger, A. Mooradian, eds., Workshop on Optical and Laser Remote Sensing, Proceedings of the Workshop Springer Verlag, Berlin, Heidelberg, New York, to be published 1982
54. R. L. Byer, L. A. Shepp, Opt. Lett. 4:75 (1979)
55. U. Platt, D. Perner, H. W. Patz, Journ. Geophys. Rev. 84:6329 (1979); D. Perner, U. Platt, Geophys. Lett. 6:917 (1979)
56. R. T. Menzies, M. S. Shumate, Science 184:570 (1974)
57. R. L. Byer, M. Garbuny, Appl. Optics, 12:1496 (1973)
58. K. W. Rothe, U. Brinkmann, H. Walther, Appl. Phys. 3:114 (1974)
59. K. W. Rothe, U. Brinkmann, H. Walther, Appl. Phys. 4:181 (1974)
60. S. Svanberg in: "Surveillance of Environmental Pollution and Resources by Electromagnetic Waves - Principles and Applications", T. Lund, ed., Nato Advanced Study Institute Series, D. Reichel Publishing Company, Dordrecht, Holland 1978
61. K. W. Rothe, H. Walther in: "Tunable Lasers and Applications", A. Mooradian, T. Jaeger, P. Stokseth, eds., Springer Series in Optical Sciences, Vol. 3, Springer Verlag Berlin, Heidelberg, New York 1976
62. W. Baumer, PhD Thesis, Ludwig-Maximilians Universität, München, July 1979
63. H. A. Bethe, and E. A. Salpeter, "Quantum mechanics of one and two electron atoms", Springer-Verlag, Berlin-Göttingen-Heidelberg 1957
64. S. Haroche in: "Atomic Physics 7", D. Kleppner, and F. M. Pipkin, eds., Plenum Press, New York, London 1981, p.141
65. G. Rempe, Diplomarbeit, Ludwig-Maximilians Universität München, 1982
66. T. F. Gallagher, and W. E. Cooke, Phys. Rev. Lett. 42:835 (1979)
67. S. Haroche, C. Fabre, P. Goy, M. Gross, and J. M. Raymond in: "Laser Spectroscopy IV", H. Walther, and K. W. Rothe, eds., Springer Series in Optical Sciences, Vol. 21, Springer Verlag, Berlin, Heidelberg, New York 1979
68. E. J. Beiting, G. F. Hildebrandt, F. G. Kellert, G. W. Foltz, K. A. Smith, F. B. Dunning, and R. F. Stebbing, J. Chem. Phys. 70:3551 (1971)
69. H. Figger, G. Leuchs, R. Straubinger, H. Walther, Opt. Comm. 33:37 (1980)
70. P. R. Koch, H. Hieronymus, A. F. J. Van Raan, W. Raith, Physics Letters 75 A:273 (1980)
71. R. H. Dicke, Phys. Rev. 93:99 (1954)
72. A. Crubelier, S. Liberman, P. Pillet, Phys. Rev. Letters 41:1237 (1978)
73. M. Gross, P. Goy, C. Fabre, S. Haroche, J. M. Raimond, Phys. Rev. Letters 43:343 (1979)
74. L. Moi, C. Fabre, P. Goy, M. Gross, S. Haroche, P. Encrenaz, G. Beaudin, B. Lazareff, Opt. Comm. 33:47 (1980)

75. T. W. Ducas, W. P. Spencer, A. G. Vaidyanathan, W. H. Hamilton, and D. Kleppner, Appl. Phys. Lett. 35:382 (1979)
76. S. Haroche, private communication.

HIGH RESOLUTION LASER SPECTROSCOPY OF RADIOACTIVE ATOMS AND RYDBERG ATOMS

S. Liberman

Laboratoire Aimé Cotton - CNRS

Orsay, France

I   <u>High resolution laser spectroscopy of radioactive atoms</u>

- Purpose of the measurement
- Principle of the experiments - the magnetic detection
- Production of the atoms
- Excitations of the atoms - the laser set-up
- Results

II   <u>High resolution laser spectroscopy of Rydberg atoms</u>

II.1   Purpose of the experiments
- The laser set-up
- Principle of the experiments
- Photoionization of Rb atoms in the presence of an electric field
- Interpretation

II.2   Photoionization of Na atoms in the presence of an electric field
- Polarization characteristics
- Influence of a small magnetic field
- Level crossing experiment - Lifetime modifications
- Lightshift of Rydberg levels

III   <u>Superradiance</u>

- Introduction

III.1 Polarization characteristics
- Observations
- Interpretation
- Statistical study

III.2 Superradiance with long duration pumping pulses
- Relaxation oscillations
- Interpretation

III.3 Collision induced superradiance

HIGH-RESOLUTION LASER SPECTROSCOPY IN THE FAR INFRARED

F. Strumia

Istituto di Fisica, Università di Pisa

Piazza Torricelli, 2 I-56100 Pisa, Italy

INTRODUCTION

The far-infrared (FIR) part of the electromagnetic spectrum (1 mm - 30 μm) has been investigated far less than other regions, mainly because of the lack of suitable photon sources. Only in the past decade has the situation changed dramatically as a consequence of the discovery of the optically pumped FIR laser. Both spectral sources and fluorescent detection in the FIR cannot be realized in the laboratory since in spontaneous emission the transition rate is proportional to ν^3. Only the black-body was available as a source before the discovery of the laser. Also, classical linear absorption spectroscopy leads to unsatisfactory results in the FIR. In fact, the relative Doppler linewidth $\Delta\nu_D/\nu \simeq 10^{-6}$ is independent of frequency. But on the other hand, any spectroscope based either on diffraction or interference (λ measurement) is scaled proportional to λ. Its size for the FIR must be about 100 times larger than for the visible in order to reach the same resolving power. A very high resolution with laboratory-size equipment can be obtained only in a frequency measuring apparatus. It has been practically realized by using the MIM diode and heterodyne techniques [1]. Again, this method works properly only with such highly monochromatic and coherent sources as the laser. Thus, high-resolution spectroscopy could not be extended to the FIR region before the discovery of the laser. In 1964 the first FIR laser lines were observed in gaseous media excited by an electric discharge: Ne (85 and 133 μm), H_2O (78 μm), D_2O (72 μm), and HCN (337 μm) [2]. The output power from a molecular laser was stronger and at the mW level in cw operation. The importance of this discovery was immediately recognized and extensive researches for new molecules and new lines were carried out. However, only three more molecules (DCN, H_2D, SO_2) were found to

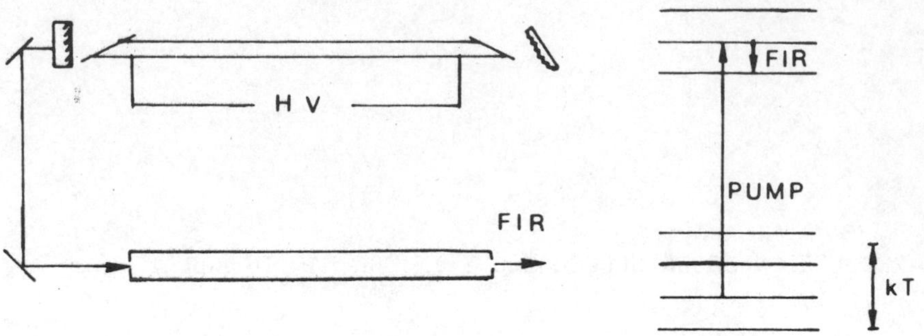

Fig. 1 - Basic scheme of an OP FIR laser.

give a few FIR laser lines. This poor result was found to be related to the unfavorable inversion mechanism [2].

The key step toward a more general FIR laser source was made by T.Y. Chang and T.J. Bridges in 1970 with the discovery of the optically pumped molecular FIR laser [3]. They observed that the 9-P(20) line of the CO_2 laser was strongly absorbed by the CH_3F molecules as a consequence of a near perfect coincidence with the $\nu_3\, {}^qQ_2(12)$ vibrational line. By putting CH_3F gas in a mirror resonator a strong laser emission at 496 μm was observed when the CO_2 radiation was also introduced into the resonator (Fig. 1).

The laser action is a consequence of pumping the CH_3F molecules into the rotational level of the excited vibrational state by the CO_2 radiation. A population inversion is possible because the energy of the pump radiation is larger than KT so that the other levels of the excited vibrational state are much less populated. It is obvious that this scheme is quite general and laser action is expected every time a molecular line is in resonance with a pumping radiation of sufficient power and with $h\nu_p > kT$. Since the FIR lasers are generally associated to rotational transitions, molecules with a permanent electric dipole moment are also necessary. As for the pump laser, a power of a few watts is sufficient for cw operation of the FIR laser with an output power at the mW level. However the frequency of the pump laser must be kept not too much larger than kT in order to have a favorable quantum conversion efficiency $\frac{1}{2}\, \nu_{FIR}/\nu_p$. The CO_2 laser is the ideal pump source for FIR lasers since $h\nu \simeq 4kT$ (the similar N_2O laser has also been used successfully). Laser emission on about one hundred lines in the 9-11 μm region can be easily obtained with a cw output power of 4 - 25 W by using an intracavity grating in a CO_2 laser of conventional design. The average separation of the laser lines is 40 - 50 GHz while the tuning for each of them is 60 - 80 MHz in conventional lasers with a spectral coverage of about 0.2%. The average number of lines in a roto-vibrational molecular band is 10^3 - 10^4. As a consequence, 2 to 20 coincidences are espected for each molecular band in the 9-11 μm region. By changing the FIR resonator length for each pump transition one or more FIR laser lines can be obtained according to the selection rules. Up to the present, 88 molecules (Table I) have been found to lase, giving more than 1600 lines in the 12 μm - 2 mm range. Most of this lines were found to lase in cw mode. Each of these lines can be frequency tuned by an amount that was found to be of the order of the Doppler linewidth. In fact, a common feature of the cw FIR laser is a working pressure that leads to $\Delta\nu_H \simeq \Delta\nu_D$, where $\Delta\nu_H$ is the pressure-dependent homogeneous linewidth. As a consequence, the FIR region is covered only by a discrete set of frequencies in spite of the large number of discovered laser lines. Nevertheless, high-resolution laser spectroscopy in the FIR is now possible. In describing the various methods we will distinguish between experiments performed on the active medium of the FIR laser itself and experiments performed by irradiating an independent sample with the FIR laser output. The first kind of spectroscopy is peculiar to the optically pumped FIR laser since the absence of an electric discharge on the active medium leaves it open to any kind of probing perturbation.

OPTICALLY PUMPED FIR LASER

Let us now consider in more detail the relevant parameters involved in an O.P. FIR laser. In Fig. 2 the levels involved in the laser scheme are shown. N and M are the total populations of the rotational levels of the ground and excited vibrational states respectively. The radiation of a pump laser at frequency ν_p excites molecules from the level 1 to the level 2 and a population inversion between levels 2 and 3 is obtained. If f_1, f_2, and f_3 are the occupational fractions of the levels such that $\sum_M f_i = 1$ and $\sum_N f = 1$, in the absence of the pump laser we have

$$n_1 = f_1 N \quad ; \quad m_2 = f_2 M \quad ; \quad m_3 = f_3 M \qquad (1)$$

$$M \ll N \quad ; \qquad f_2 < f_3$$

TABLE I

Optically pumped FIR lasing molecules. The molecules in the frames are of particular interest either for the number of laser lines or for the high output power.

CH_3F	CDF_3	**CH_3Br**	**CH_3I**	**CH_3CN**
$^{13}CH_3F$	**CH_3Cl**	CD_3Br	CD_3I	CD_3CN
CD_3F	CD_3Cl	CF_3Br	$^{13}CD_3I$	CH_3NC
$^{13}CD_3F$		CF_3I		CH_3NO_2
CH_3OH	CD_2HOH	CH_3NH_2	**CH_2F_2**	**$HCOOH$**
CD_3OH	CH_2DOH	CH_3ND_2	**CD_2F_2**	**$HCOOD$**
CD_3OD	CH_2DOD	CH_3NHD	CH_2Cl_2	**$DCOOH$**
CH_3OD	CH_3SH	CF_2Cl_2	CD_2Cl_2	**$DCOOD$**
$^{13}CH_3OH$			$CHClF_2$	$H^{13}COOH$
CH_3CH_2F	**CH_2CF_2**	CH_3CCH	$C_2H_4(OH)_2$	
CH_3CHF_2	CH_2CHBr	CH_3CH_2OH	H_2NNH_2	
CH_3CF_3	CH_3CH_2I	$(CH_3)_2O$	$CHFCHF$	
CH_3CH_2Cl	CH_2CHCN	CH_3CHO		
CH_2CHCl	CH_2CHF	CH_3COOD		
CF_4	$^{13}CF_4$	SiF_4	SiH_4	SF_6
OCS	**D_2O**	**NH_3**	$HCCF$	$HDCO$
$O^{13}CS$	D_2S	**$^{15}NH_3$**	$FClO_3$	D_2CO
ClO_2	DHS	NH_2D	HF	$HFCO$
O_3	FNC	ND_3	SO_2	$(H_2CO)_3$
$NOCl$	NSF	PH_3	$S^{18}O_2$	

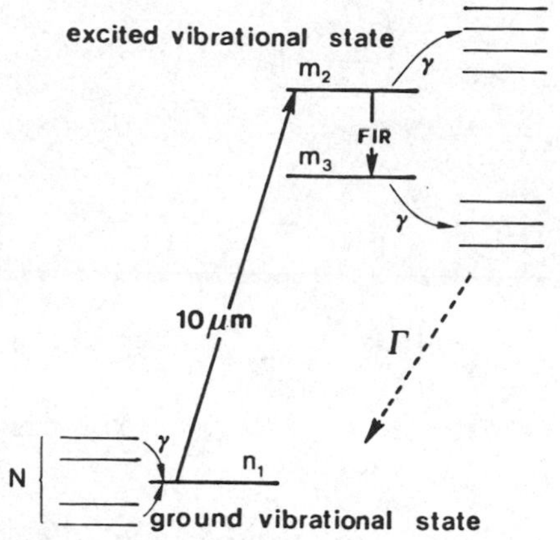

Fig. 2 - Schematic energy level diagram of an optically pumped FIR laser.

The relaxation between levels is a consequence of collisions. The rotational relaxation γ within a given vibrational level is very fast, $\gamma \simeq 10^8$ Torr^{-1}sec^{-1}, while the vibrational relaxation Γ is much slower, $\Gamma \simeq 10^3$ Torr^{-1}sec^{-1}. As a consequence, the wall collisions proportional to $p^{-1}\, d^{-2}$ (p, gas pressure; d, diameter of the laser resonator) also play an important role in avoiding a bottleneck. On the other hand, the spontaneous emission is negligible, and we have

$$\Gamma_{coll} + \Gamma_{diff} \simeq 10^3 \text{ Torr sec}^{-1} + 10^3 \text{ Torr}^{-1}\text{sec}^{-1} \qquad (2)$$

If $\nu_p \simeq 10$ μm, the pump transition is inhomogeneously broadened ($\Delta\nu_D \simeq 10\text{-}20\ \Delta\nu_H$) and only a fraction of the molecules in level 1 can be simultaneously excited. We have a velocity-dependent excitation rate

$$W(v) \propto \frac{I(\nu_p)}{\left[\nu - \nu_p(1+\frac{v}{c})\right]^2 + (\frac{\Delta\nu_H}{2})^2}\, e^{-(\frac{v}{\Delta v})^2} \qquad (3)$$

where $I(\nu_p)$ is the intensity of the pumping beam. At a steady state

$$f_1\, N\, W = \Gamma M$$

$$\left(\frac{d\,m_2}{dt}\right)_p = f_1\ N\ W = -\left(\frac{d\,m_2}{dt}\right)_c = \gamma\left[m_2 - f_2\ M\right] \tag{4}$$

and

$$m_2 = f_1\ N\ \frac{W}{\gamma}\left[1 + f_2\frac{\gamma}{\Gamma}\right];\ m_3 = f_3 M = f_1 N\ \frac{W}{\Gamma}\ f_3$$

$$m_2 - m_3 = f_1\ N\ \frac{W}{\gamma}\left[1 - \frac{\gamma}{\Gamma}\ (f_3 - f_2)\right] \tag{5}$$

In order to have a positive gain, the quantity in the brackets must be positive.

Since $h\nu_F < KT$ we can write $f_3 \simeq f_2\ (1 + h\nu_F/KT)$ and

$$\frac{h\nu_F}{KT}\ f_2\frac{\gamma}{\Gamma} < 1 \tag{6}$$

is the threshold condition for a cw FIR laser. Let us consider as an example a hypothetical FIR line at 500 µm, $h\ \nu \simeq 0.1$ KT, and $f_2 \simeq 10^{-3}$. The value of the left side of eq. (6) is ≃ 10 for a pressure of 1 Torr and no steady state laser emission is possible. On the other hand, the value is 0.1 for a pressure of 0.1 Torr. In fact, no cw FIR laser has been observed for pressures larger than 0.5 Torr. From a rate equation modeling the conversion efficiency of infrared pump power P_p into the FIR emission P_F is predicted to be [4]

$$\frac{P_F}{V} = \frac{P_p}{V}\ \frac{1}{1+g_2/g_3}\ \frac{\nu_F}{\nu_p}\ \frac{T}{A+T}\ \frac{\alpha L}{\alpha L+a_p}\left[1-\frac{h\nu_F}{KT}\ f_2\frac{\gamma}{\Gamma}\right] \tag{7}$$

where g_2 and g_3 are the level degeneracies, T the FIR mirror transmission, A the cavity losses at FIR, α the absorption coefficient for the pump, L the cavity length, and a_p the cavity losses at ν_p. From eq. (7) is evident the importance, for a good conversion, of molecules with a high vibrational relaxation rate Γ and a large absorption coefficient α for the pump radiation. The value of α may be modified by saturation effects of spectroscopic interest, such as the Lamb dip and the nonlinear Hanle effect, to be discussed later. Three kinds of FIR laser cavity are used. They are shown in Fig.3. The open mirror cavity was introduced early, in Chang and Bridges's original work [3]. It has a high Q factor and no constraints

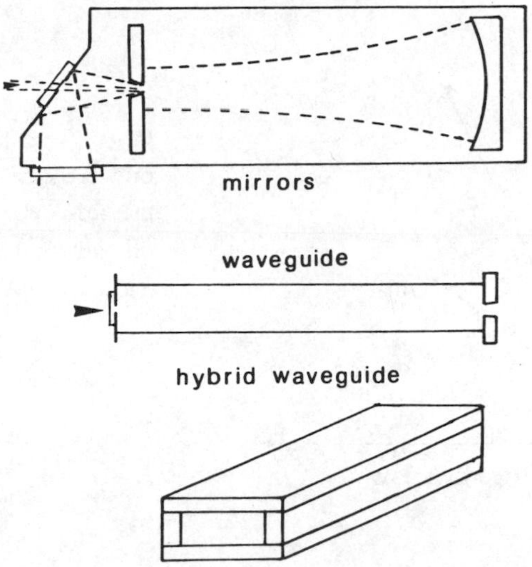

Fig. 3 - Resonator schemes used for the optically pumped FIR laser.

for the field polarization. Both properties are important for spectroscopy. However, two principal drawbacks limit the conversion efficiency of the mirror resonator: i) the mode overlap between the pump field and the FIR field is poor, ii) the radius of the mirror must be large in order to limit the diffraction losses and as a consequence the diffusion contribution to vibrational relaxation is small. To overcome these problems the oversized circular waveguide resonator was introduced with a diameter between 1 and 4 cm. The waveguide material may be either metallic or dielectric. In the first case the principal mode is the TE_{01} with a minimum along the axis and the electric field along concentric circles in the cross section [5]. In the second case the principal mode is the EH_{11} with the E field linearly polarized in an arbitrary direction as for the mirror cavity [5]. This is an important feature since a linearly polarized pump field induces a FIR gain preferentially polarized either parallel or orthogonal to the pump beam. A third useful waveguide resonator employs a hybrid waveguide of rectangular cross section with the longer side made of metal and the shorter one of a dielectric. In this case the E field at FIR is forced to be parallel to metal plates [6]. Standard sizes are 0.5-1 x 3-4 cm. The hybrid waveguide, being a plane parallel condenser, is very im-

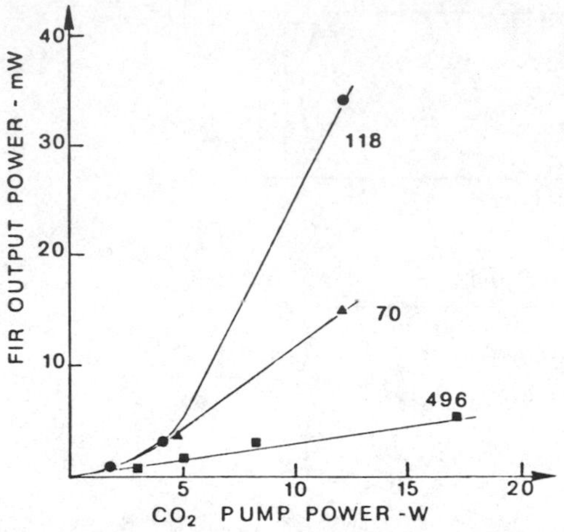

Fig. 4 - FIR laser output power for the 118 µm and 70 µm lines of CH_3OH and for the 496 µm line of CH_3F.

portant for studying the Stark effect on the FIR lasers [7]. In Fig. 4 are shown the output powers obtained for three of the most popular FIR laser lines in the case of a waveguide laser consisting of a pyrex tube of 34 mm i.d. and 1 m in length. By increasing the tube length the FIR power was proportionally increased. For I_p = 12 W and L = 154 cm a maximum power of 60 mW was obtained for the 118.8 µm line. Both powers were measured at the laser output and are not corrected for window reflection and absorption losses.
For each absorption transition in coincidence with a pump laser line one or more FIR laser lines are observed provided that I_p is sufficiently high and that the associated rotational transitions are in the FIR range. Low J levels are normally associated to microwave transitions that cannot lase within the FIR resonator. The FIR transitions directly connected with the pumped level are one for linear or symmetric top molecules, two in the case of inversion splitting (NH_3) or lambda doubling, two or three or more in the case of asymmetric top molecules. Other FIR lines can be obtained as cascade pumping from the strongest lines. A general scheme is shown in Fig. 5. Of relevant spectroscopic interest is the relative polarization between the pump and FIR E field. The CO_2 pump radiation is linearly polarized and induces a preferential orientation in the excited molecules that can be either parallel or orthogonal depending on the ΔJ selection rule. Therefore also the FIR radiation is linearly polarized and, if the FIR resonator does not impose a constraint, it is found to follow Chang's rule [8].

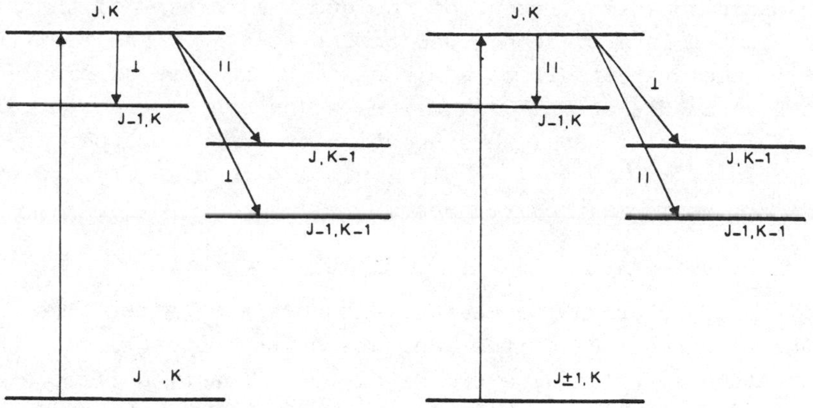

Fig. 5 - Selection rules and relative polarization for FIR laser lines .

$$\begin{aligned} \Delta J_p + \Delta J_F &= \text{even} \qquad E_F \,//\, E_p \\ \Delta J_p + \Delta J_F &= \text{odd} \qquad E_F \perp E_p \end{aligned} \tag{8}$$

FIR SPECTROSCOPY ON THE LASER MEDIUM

The general scheme of the possible FIR laser spectroscopy experiments is shown in Fig. 6. It is worth noting that the Stark or Zeeman absorption spectroscopy is more sensitive when using intracavity cells whenever possible instead of external cells as in Fig. 6.

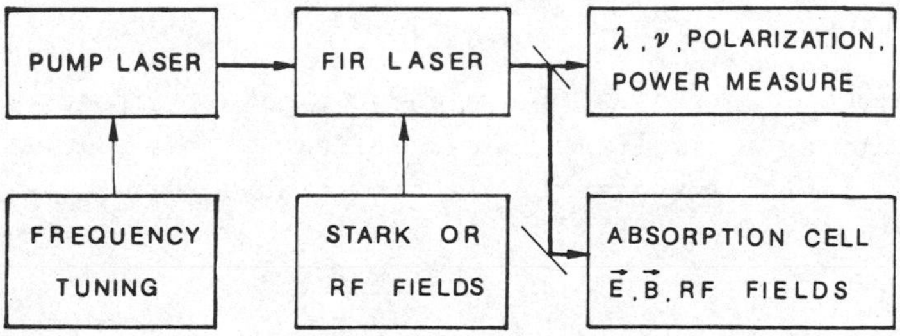

Fig. 6 - Block diagram of the FIR laser spectroscopy

An unambiguous assignment of the quantum numbers of the levels involved in the laser action based on the frequency and relative polarization measurements is possible only in the case of simple molecules such as NH_3 or CH_3F. A unique advantage of the optically pumped FIR lasers is the interconnection through the upper level of the pump line with one or more FIR lines. This is a very important check of the assignment correctness.

Power, polarization, ν and λ measurement

Power and polarization measurements are easily performed on FIR lasers as a consequence of the adequate output power level. Wavelength measurements are always limited by diffraction effects and mechanical imperfections. The FIR cavity itself can be used as a scanning interferometer with a resolution of the order of a MHz. This is useful for resolving very close lines like Stark multiplets. Absolute λ measurements performed by measuring the longitudinal spacing of the fundamental mode are limited to about one part in 10^4 even when the cavity length is interferometrically monitored [9]. Better precision is obtained in frequency measurements by using an MIM diode [1] and heterodyne techniques. The unknown FIR frequency can be directly related either to harmonics of a microwave frequency standard or to another laser of known frequency [10]. The resolution is limited only by the homogeneous linewidth of the FIR laser and the uncertainty in finding the line center. The resulting relative precision is of the order of one part in 10^7 and comparable to that obtained in the microwave spectroscopy. An example is shown in Fig. 7.

In the case of complex or less known molecules, other spectroscopic data are needed for a successful assignment of the quantum numbers. In optically pumped lasers they can be obtained by adding a further electromagnetic field on the laser medium.

Stark effect

Polar molecules possess a permanent electric dipole moment and therefore display a significant Stark effect on the allowed rotational transitions. The energy splitting of the levels is given for a symmetric top rotor by [13]

$$W_S = \mu E \frac{K\ M}{J(J+1)} \tag{9}$$

where μ is the electric dipole moment in debyes (1 D = 5.0344 MHz/V cm^{-1}). A further term quadratic in the field is very small for the

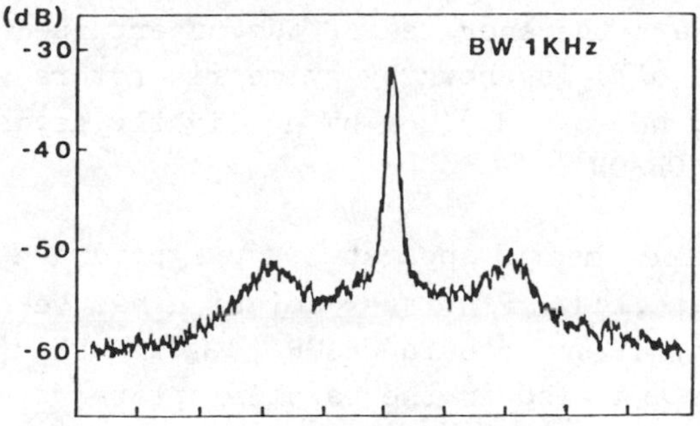

Fig. 7 - Beat note obtained by mixing a CH_3F FIR laser with the 8th harmonic of a 75 GHz klystron in a point contact MIM diode. The full width of the signal at −3db points is about 2 KHz. The klystron is phase-locked to a 5 MHz low noise crystal oscillator. Total scanning time is 10 seconds and the FIR frequency 604297,46 MHz (from ref. 11-12).

fields that can be applied to the active medium of the FIR laser and can be neglected. The Stark effect is more complex in the case of the asymmetric rotors since K is no longer a good quantum number. As a general rule a perturbation treatment is given and the first term is quadratic in the field

$$W_S = (A' + B' M^2) E^2 \qquad (10)$$

where A' and B' are functions of the molecular constants. A linear Stark effect is observed also in asymmetric rotors when the K splitting is very small or when the K degeneracy is restored by an additional effect. The Stark energy splitting in this case is given again by eq. (9), but the effective value of μ may now change from one state to another. An important case of linear Stark effect in asymmetric rotors is that of molecules (as CH_3OH) with an internal rotation which restores the K degeneracy for most of the levels. In other cases the K splitting is comparable to the Stark energy and the level splitting is given by

$$W_S = \frac{W^+ + W^-}{2} \pm \left[\left(\frac{W^+ - W^-}{2}\right)^2 + \left(\frac{\mu EKM}{J(J+1)}\right)^2\right]^{\frac{1}{2}} \qquad (11)$$

where W^+ and W^- are the energies of the unperturbed levels. A Stark effect as in eq. (11) is shown by symmetric rotors with inversion splitting (e.g., NH_3 or PH_3), or by a slightly asymmetric top (e.g., the A states of CH_3OH).

As for the experimental apparatus, the hybrid waveguide cavity is the most convenient [7,14]. FIR laser emission has been observed for fields up to many KV/cm both in CH_3OH [7] and NH_3 [15]. In the hybrid waveguide and also in the parallel plate waveguide the E field of the FIR radiation is forced to be parallel to the metal plates. The selection rule for the M quantum number is as a consequence forced to be $\Delta M = \pm 1$. This polarization constraint does not exist for the pump radiation [14] and two orthogonal cases can be investigated

$$\begin{aligned} &\text{i)} \quad \begin{cases} E_p \;/\!/\; E_s & : \quad \Delta M = 0 \\ E_F \perp E_s & : \quad \Delta M = \pm 1 \end{cases} \\ &\text{ii)} \quad \begin{cases} E_p \perp E_s & : \quad \Delta M = \pm 1 \\ E_F \perp E_s & : \quad \Delta M = \pm 1 \end{cases} \end{aligned} \tag{12}$$

The intensity of each M component is proportional to the Clebsch-Gordan coefficients. In FIR laser emission not all the M components can be observed since the gain for the weaker ones may be below the threshold. As an example the first resolved Stark pattern in laser emission [11,16] is shown in Fig. 8. The relative intensity pattern for the two orthogonal cases of eq. (12) is quite different. This effect can be understood by assuming that the collisional mixing between the M sublevels of a given state is negligible. As a consequence the other sublevels of the excited state do not participate in the laser emission when the cavity setting allows laser action only for a given resolved M component. This is a first important experimental result. In fact, the relative intensity of the M pump transitions must also be taken into account in computing the relative small signal gain of the lines of a Stark multiplet [16]. One or more FIR laser lines are expected for each pumped upper level, depending on the selection rule. In suitable molecules, like CH_3OH, a triad is expected as shown in fig. 5. By assuming a linear Stark effect (eq. (9)) the relative frequency of the Stark components can be computed, to a first approximation, to be

$$\begin{cases} \Delta J = -1 \;;\; \Delta K = 0 \;;\; \Delta M = \pm 1 \\ \Delta\nu_S = \dfrac{\mu E\, K}{J(J^2-1)}\left[2\,M - (J+1)\;\;\Delta M\right] \end{cases} \qquad (13)$$

$$\begin{cases} \Delta J = 0 \;;\; \Delta K = -1 \;;\; \Delta M = \pm 1 \\ \Delta\nu_S = -\dfrac{\mu E}{J(J+1)}\left[M + (K-1)\;\;\Delta M\right] \end{cases} \qquad (14)$$

$$\begin{cases} \Delta J = -1 \;;\; \Delta K = -1 \;;\; \Delta M = \pm 1 \\ \Delta\nu_S = \dfrac{\mu E}{J(J-1)}\left[\dfrac{2K-1-J}{J+1}\,M - (K-1)\,\Delta M\right] \end{cases} \qquad (15)$$

where μ was assumed to be the same for both upper and lower laser levels.

The case of eq. (15) is of particular interest since for $K \simeq J/2$ the first term in the bracket becomes negligible and all the M components collapse into only two close packets corresponding to $\Delta M=\pm 1$.

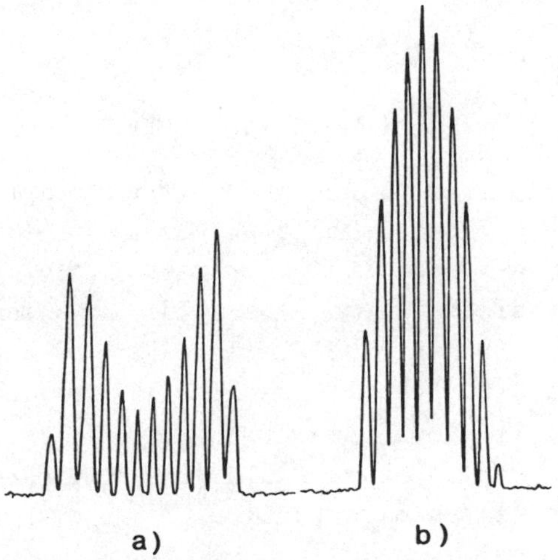

Fig. 8 - FIR laser oscillation on the individual Stark components of the CH_3F 496 µm line. In a) the pump selection rule was $\Delta M=0$ and in b) $\Delta M=\pm 1$. the electric field intensity was 1.9 KV/cm in both cases.

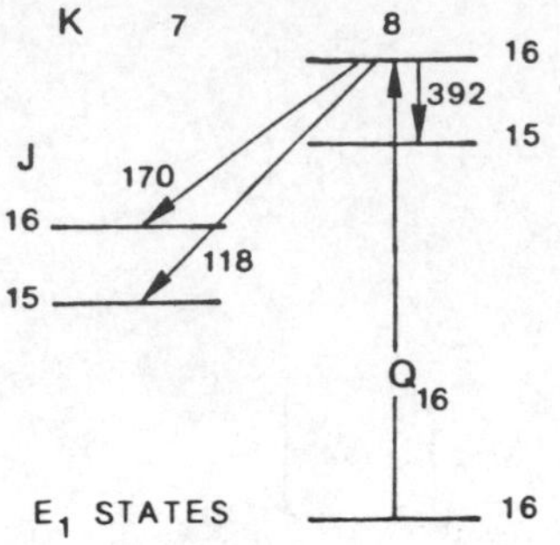

Fig. 9 - Scheme of the CH_3OH levels involved in the 118.8 μm laser emission.

In this case the FIR lasing lines is seen to split into only two components whose intensity is of the same order as the unperturbed line and whose separation is $\Delta\nu_S \simeq \mu E/J$.

As an example we will consider the popular 118.8 μm line of CH_3OH pumped by the CO_2 - 9P(36) line. The scheme of the levels involved is shown in Fig. 9. The other laser lines of the triad are at 170 μm and 392 μm respectively, the last being also the weakest.

The Stark effect for the 118 and 170 μm lines can be computed to be respectively

$$\Delta\nu_S(118) = (-0.11M \pm 13)\cdot E \quad \text{MHz/KV cm}^{-1} \tag{16}$$

and

$$\Delta\nu_S(170) = 1.66\ (M \pm 7)\cdot E \quad \text{MHz/KV cm}^{-1} \tag{17}$$

As a consequence the first line is expected to split linearly into two components at a rate of about 26 MHz/KV cm^{-1}, while the second one is expected to decrease its power output rapidly as a function of the Stark field since the gain is split into many components so weak as to be below the threshold. In Figs. 10 and 11 are shown the experimental results for the 118 μm line, in good agreement with the theory. In Fig. 12 is shown the output power decrease of the 170 μm line as a function of the Stark field [17] . No frequency shift or splitting was observed, as expected.

A Stark splitting into two components is of relevant practical interest when frequency tunable FIR laser lines are requested and is shown by many of the strongest laser lines of CH_3OH. A more extensive review of the Stark effect on optically pumped FIR lasers can be found in ref. 7 and 18.

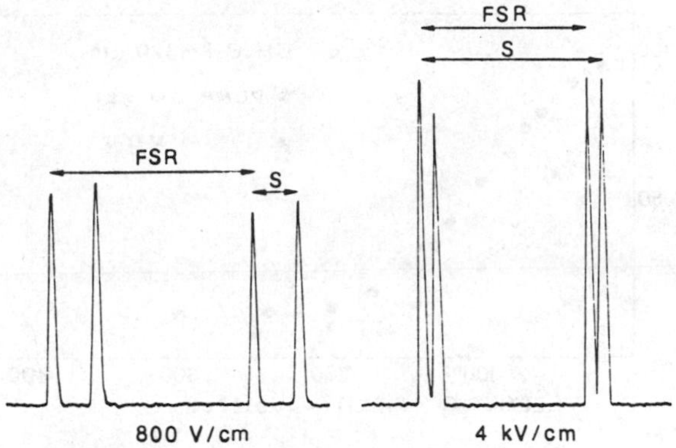

Fig. 10 - Interferogram of the observed Stark splitting (S) of the 118.8 μm line. The free spectral range (FSR) of the laser cavity is also indicated.

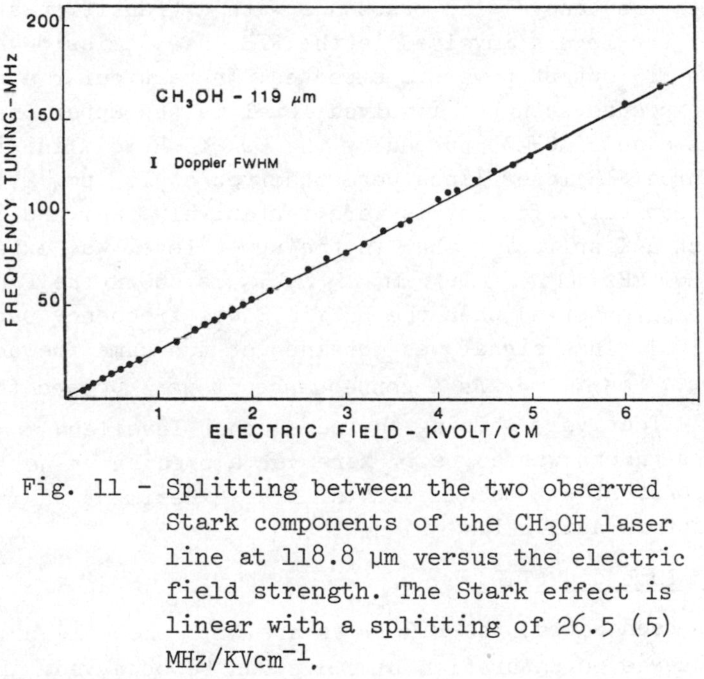

Fig. 11 - Splitting between the two observed Stark components of the CH_3OH laser line at 118.8 μm versus the electric field strength. The Stark effect is linear with a splitting of 26.5 (5) $MHz/KVcm^{-1}$.

Triple resonance

Beside a static field, radiofrequency or microwave radiation can also be applied to the FIR laser active medium. In this case the most useful experimental setup is a dielectric waveguide in-

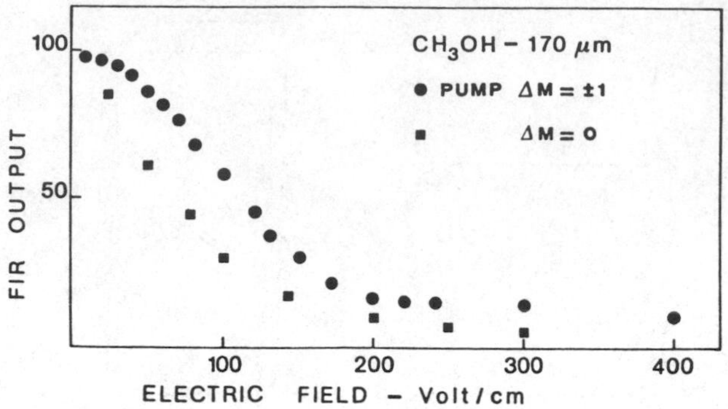

Fig. 12 - Relative output power of the 170 μm laser line as a function of the applied Stark field.

serted in a coaxial or rectangular waveguide fed with the RF field. Whenever its frequency is in resonance with a transition starting from one of the levels involved in the FIR laser emission a variation in the FIR output power is expected. In particular a power decrease is expected when the involved level is the upper one. As an example the case of CH_3OH pumped by the CO_2-10 R(38) line is illustrated. Three FIR laser lines were observed at 163 μm, 251 μm, and 450 μm respectively. The levels were tentatively assigned [19] as A state with a K splitting that in the upper level was expected to be around 300 MHz. (Fig. 13a). In Fig. 13b is shown the FIR laser power decrease observed when the RF field was frequency tuned at 370.8 MHz [20]. This signal was obtained at the same frequency for all the three FIR lines. As a consequence it was proved that the lines were effectively sharing the same upper level and that the proposed assignment was correct. Moreover a precise value for the splitting constant was measured. A similar experiment was also performed on formic acid [21].

Transferred Lamb dip (TLD)

As has been noted, the FIR lines are homogeneously broadened and as a consequence no saturation Lamb dip can be observed. However, the pump transition is Doppler broadened, and a pump saturation dip is present which can be detected in the FIR laser power emission when the pump radiation is frequency tuned (TLD). This effect was observed early [22], but its spectroscopic relevance was demonstrated only recently [23]. In particular, a precise measurement of the

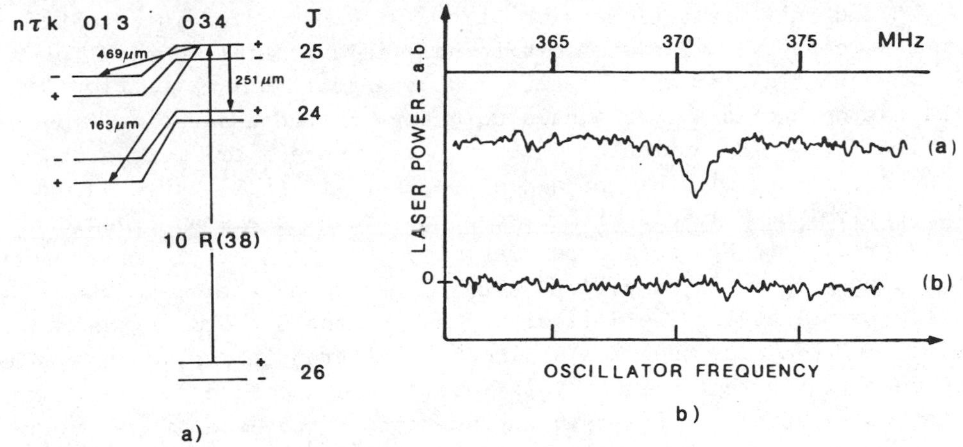

Fig. 13 - a) Schematic of the levels involved in the action of lasers pumped by the CO_2 10R(38) laser lines. b) Resonance signal for the (25,4) excited level observed on the FIR laser output as a function of the radio frequency. (a) CO_2 laser on; (b) CO_2 laser off.

pump detuning can be obtained by using the TLD and a resolution within the homogeneous linewidth of the pump transition can be reached [24]. This result is very important for determining whether different FIR laser lines are sharing the same pump transition.

Nonlinear Hanle effect

The nonlinear version of the Hanle effect was observed in the past in intracavity experiments in the visible and near infrared [25]. Recently it was also introduced to explain the power enhancement observed on optically pumped three-level FIR lasers [17]. As is well known, in the presence of saturation the absorption coefficient of a Doppler broadened line can be written as

$$I(\omega) = \frac{\alpha_0}{\sqrt{1+S}} \exp - \left[\frac{\omega \quad \omega_0}{\Delta_D} \right]^2 \qquad (18)$$

where $S = I/I_s$ is the saturation parameter. In the absence of an external field the natural quantization axis of the system is determined by the polarization and propagation direction of the laser beam. For a linearly polarized beam and an electric dipole interaction the quantization axis is parallel to the E vector of the EM field and only transitions with $\Delta M=0$ are induced. This selection rule is conserved if a static external field is applied parallel to the quantization axis. There is no change in the saturation of the system and as a consequence no change in the absorption coefficient $I(\omega)$. In particular $I_F(\omega) = I(\omega)$, where $I_F(\omega)$ is the absorption coefficient in the presence of a static field large enough to resolve the M levels with respect to the homogeneous linewidth. This prediction has been experimentally confirmed [17]. On the other hand, only $\Delta M = \pm 1$ transitions are induced by a laser beam when the external static field is orthogonal to the E field of the radiation. In this case the $\Delta M = -1$ and $\Delta M = +1$ transitions saturate independently when the degeneracy is removed and a change in the absorption is expected. It is worth noting that in the case of an inhomogeneous line whose M components are resolved with respect to the homogeneous linewidth the $\Delta M = \pm 1$ transitions starting from a given M sublevel involve molecules with different velocity components along the beam direction, i.e. different molecules. As an example for a transition $J=0\to 1$ and a laser beam linearly polarized orthogonal to the static field ($\Delta M=\pm 1$) the absorption increase can easily be calculated to be

$$R(S) = \frac{I_F}{I} = \sqrt{\frac{1+2S}{1+S}} \tag{19}$$

assuming Δ_D much larger than the homogeneous linewidth Δ_H. For intermediate static field intensities the coherence between the M sublevels must be taken into account. A density matrix formalism treatment was performed [26], however lengthy and often meaningless calculations are involved in high J molecular transitions. A simple calculation of the absorption increase in the limiting case of non-degenerate M sublevels is more convenient for understanding the relevance of the effect. R(S), as defined by eq. (19), can be computed as a function of J and ΔJ selection rule by using the Clebsch-Gordan coefficients formalism [27].

By assuming $\Delta_H \ll \Delta_D$, the exponential term in eq. (18) can be neglected and a simple expression for R(S) can be obtained for Q branch transitions

$$R_Q(S,J) = \frac{\sum_{-J}^{+J}\left(\dfrac{J(J+1)-M(M-1)}{\left(1+\frac{3}{4}\,S\,\dfrac{J(J+1)\;M(M-1)}{J(J+1)}\right)^{\frac{1}{2}}} + \dfrac{J(J+1)-M(M+1)}{\left(1+\frac{3}{4}\,S\,\dfrac{J(J+1)-M(M+1)}{J(J+1)}\right)^{\frac{1}{2}}}\right)}{4\sum_{-J}^{+J}\dfrac{M^2}{\left(1+S\dfrac{3M^2}{J(J+1)}\right)^{\frac{1}{2}}}} \tag{20}$$

and for P and R branches transitions

$$R_P(S,J+1) = R_R(S,J) =$$

$$= \frac{\sum_{-J}^{+J}\left(\frac{(J+M)^2+3(J+M)+2}{\left(1+\frac{3}{2}S\frac{(J+M)^2+3(J+M)+2}{(2J+1)(2J+3)}\right)^{\frac{1}{2}}} + \frac{(J-M)^2+3(J-M)+2}{\left(1+\frac{3}{2}S\frac{(J-M)^2+3(J-M)+2}{(2J+1)(2J+3)}\right)^{\frac{1}{2}}}\right)}{4\sum_{-J}^{+J}\frac{(J+1)^2-M^2}{\left(1+6S\frac{(J+1)^2-M^2}{(2J+1)(2J+3)}\right)^{\frac{1}{2}}}} \quad (21)$$

Obviously $R(S) \equiv 1$ for the selection rule $\Delta M=0$.

In Fig. 14 the computed results are shown for several J values and for a saturating beam constant over its entire cross-section.

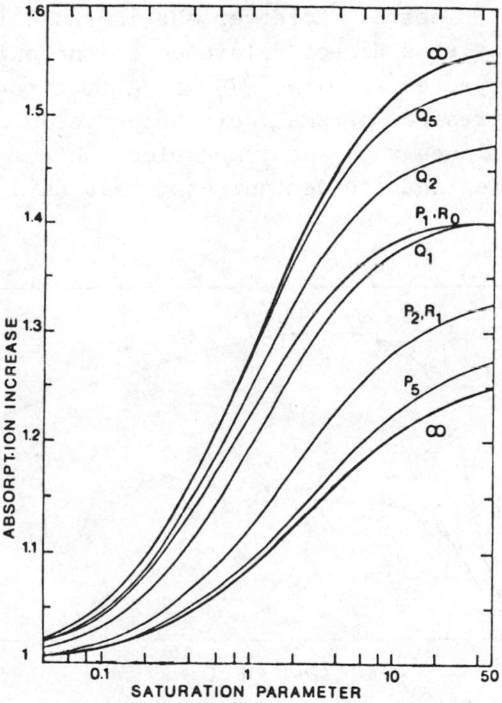

Fig. 14 - Nonlinear Hanle effect relative absorption increase as a function of the saturation parameter S.

We have the same maximum absorption increase ($R = \sqrt{2}$) for Q_1, R_0, and P_1 lines. By increasing the J values, the Q_J and the P_{J+1} or R_J lines show a rapid convergence to two limiting curves. $R_Q(J,S)$ increases with J up to a maximum value of 1.57, while $R_P(J+1) = R_R(J)$ decreases to an asymptotic value of 1.27. In both cases the effect is quite large and can be easily detected. Fig. 15 shows a typical dependence of the absorption increase versus the electric field for two different CH_3OH molecular transitions responsible for the two strong FIR laser lines at 118 μm and 96 μm respectively. The upper curve refers to the Q_{16} line irradiated by the 9P(36) CO_2 laser line (Fig. 9). At low electric fields (up to 0.6 KV/cm) a large absorption increase due to the nonlinear Hanle effect is detected. The signal decrease at higher electric fields shows the Doppler profile of the absorption line. The lower curve refers to the R_{26} line irradiated by the 9R(10) CO_2 laser line. As expected the absorption increase for about the same CO_2 laser power is smaller. The nonlinear Hanle effect is a convenient and simple method for the direct measurement of the saturation parameters of the pump transition of FIR lasers. Moreover the increase in the absorption coefficient has also a direct relevance in the output power of the FIR lasers as expected from eq. (7). An output power increase of FIR lasers in the presence of small electric fields was observed early for some strong lines when the pump selection rule was $\Delta M = \pm 1$ [11,28]. A systematic investigation demonstrated that this effect was a com-

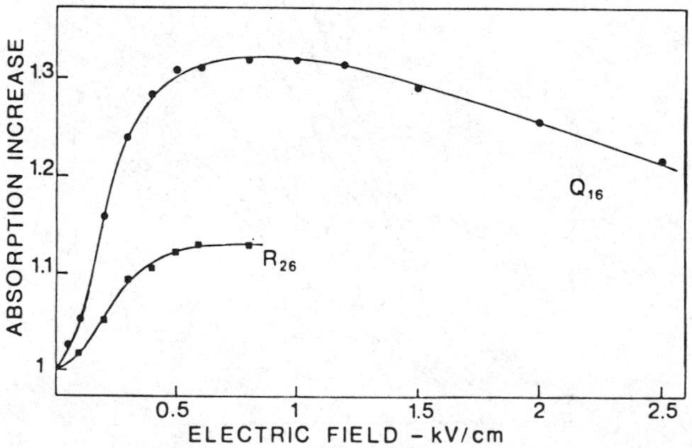

Fig. 15 - Nonlinear Hanle effect as a function of the applied electric field in the case of a Q branch line and an R branch line, respectively.

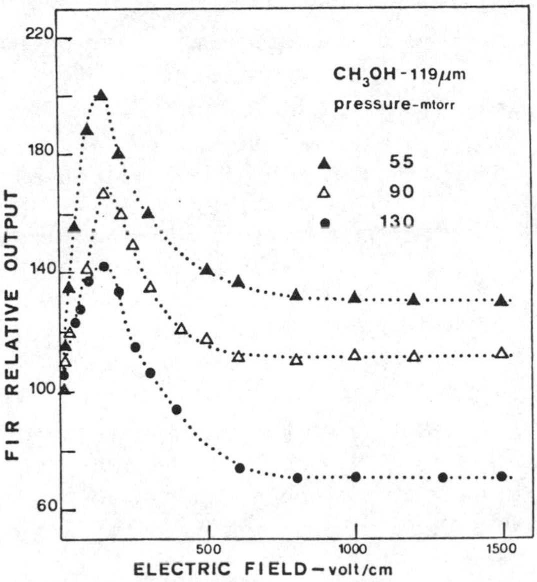

Fig. 16 - Power enhancement effect on the CH_3OH 118.8 μm laser line pumped by the CO_2 9P(36) line as a consequence of the nonlinear Hanle effect.

bined result of the nonlinear Hanle effect and of the absence of collisional mixing of the M sublevels [17]. Fig. 16 shows a typical enhancement effect. This result is to be compared to that of Fig. 12 for the 170 μm line.

The pump line is the same (Fig. 9), and as a consequence the absorption also increases. The key point is again the Stark effect (eqs. (16) and (17)). For the 170 μm line the gain decrease as a consequence of a splitting in many components is so rapid that it overcomes the increase in the absorption coefficient. In conclusion, the nonlinear Hanle effect can give complementary spectroscopic information both on the pump and on the FIR transitions.

<u>Methods for new OP FIR laser lines</u>

From a spectroscopic point of view, there is a minimum number of coincidences and FIR laser lines that are necessary for a reliable fit to the parameters of the molecule under investigation. As a general result the number of the experimentally observed FIR laser lines observed with a conventional CO_2 laser is not enough. For understanding the spectrum the data can be implemented by other techniques such as double resonance, Stark spectroscopy, and diode laser spectra. Nevertheless our present knowledge of molecules of moderate complexity, such as CH_3OH, is unsatisfactory and new experimental data are needed. A straightforward method for obtaining new FIR laser lines, beside the use of new molecules, is a more careful search around the pump lines of well known molecules by using FIR resonators of reduced losses and more pump power. A second approach is to develop new techniques for obtaining new additional pump transitions:

i) Stark effect shift

The Stark effect can be used also for bringing in resonance lines that are close to the pump lines but outside the tuning range of the pump laser. At present this method has been successfully applied to NH_3 and CH_3OH [7]. Other molecules appear to be good candidates but the method is confined only to molecules with a large Stark effect in the pump transition.

ii) Two-photon or cw Raman pumping

More recently the feasibility of these techniques has been demonstrated [29] in NH_3. Several new lines were discovered with a pump offset of few hundred MHz. However the two-photon techniques are limited to cases where both microwave and FIR transitions are connected to the same level, such as NH_3. As a further example let us consider the case of CH_3OH discussed as an example for the triple resonance method. With more pump power, three new lines are obviously expected starting from the level $(034,25)^-$ which is populated by two-photon pumping.

iii) Extended tunability of the pump laser

This method is quite obvious and was demonstrated early by using a TEA CO_2 laser [30]. Since then the method has been extensively used for generating high peak power pulsed lasers and for discovering FIR laser lines in molecules, like D_2O, without any close coincidence with the pump laser [31]. However a TEA laser is a broad band pump laser (typically 1-2 GHz).

As a consequence it is not easy to determine the pump frequency and the experimental data cannot be used unambiguously for spectroscopic purposes. Furthermore it has been observed experimentally that only the strongest absorbing lines within the TEA laser bandwidth yield FIR emission. Other potentially favorable pump lines with weaker

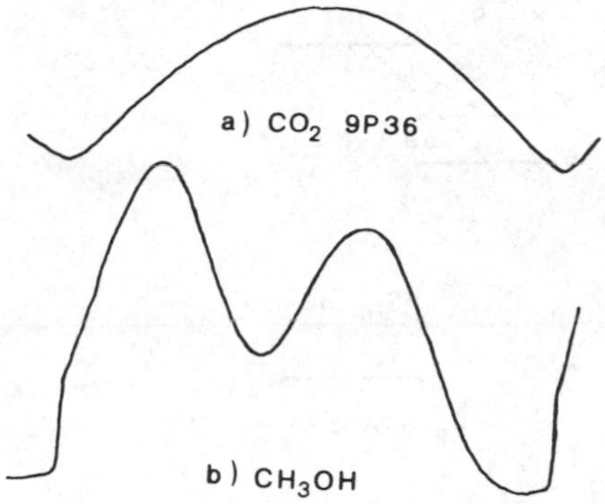

Fig. 17 - Absorption optoacoustic signal of CH_3OH around the CO_2 9P(36) line (b). The CO_2 waveguide laser mode with a tuning of 240 MHz is shown in (a).

absorption are masked out (see eq. (7)). The number of FIR laser lines observed by using TEA lasers was less than expected from the increased tunability. The cw waveguide CO_2 laser is an alternative approach for increasing the frequency tuning range of the pump source. For several years now the cw waveguide lasers have demonstrated single line and single mode operation over a frequency tuning range of 0.1 ~ 1 GHz. However until recently the operation with a grating reflector yielded an average power of about 1 watt, too low to be attractive for pumping FIR lasers. More recently a compromise was reached between tuning range and output power by using pyrex or fused silica waveguides. A single mode and single line output was obtained with an average power of 3 to 8 W cw and a tuning range of 240 MHz, three to four times larger than that of a conventional CO_2 laser [32,33], obtaining FIR laser emission for the first time with a waveguide CO_2 laser as a pump [32]. Since then, many new FIR laser lines have been discovered with a pump offset larger than the tuning range of conventional lasers, yielding precise spectroscopic information that led to level assignments in most cases [30]. As an example we will again consider the case of the CH_3OH 118 μm line pumped by the CO_2 9P(36) line (Fig. 9). Figure 17 shows a CH_3OH absorption optoacoustic signal recorded by tuning the 9P(36) line over the 240 MHz range of the waveguide laser.

Near the center one can observe the strong absorption of the Q(E_1, 16, 8) line responsible for the pump of the 118, 170, 392 μm laser lines. An even stronger absorption can be observed at the low frequency side of the CO_2 line center. By using the CO_2 laser tuned near the maximum of this new absorption four new strong FIR laser lines were discovered. Their pump frequencies were measured with sub-Doppler accuracy by using the transferred Lamb dip. It was found

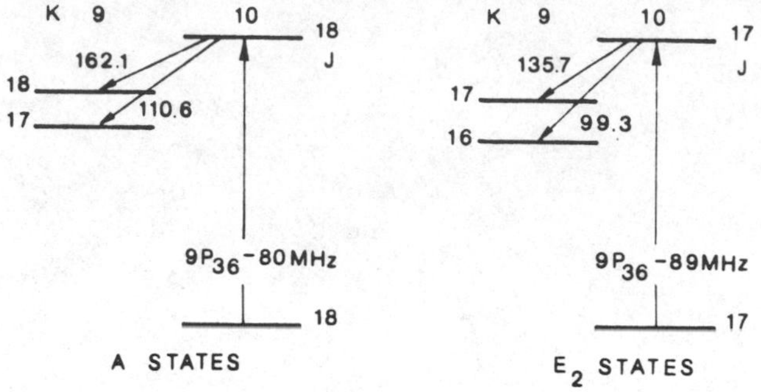

Fig. 18 - Level assignment of the new CH_3OH FIR laser lines discovered by using a CO_2 waveguide laser as a pump around the 9P(36) line.

that the FIR laser was pumped by two different transitions; one, located at -80 MHz, was responsible for the FIR lasers at 162.1 and 110.6 μm, the other, located at -89 MHz, for the FIR lasers at 135.7 and 99.3 μm. Both the lines at 110.6 and 99.3 μm showed also a linear Stark splitting into two components as for the 118.8 μm line. These experimental data were sufficient for the assignment of the associated levels as shown in Fig. 18 [34]. It is worth noting that only the line at 110.6 μm was observed by using a TEA CO_2 laser as a pump.

Spectroscopic use of the FIR laser radiation

The OP FIR lasers cover all the region from 2 mm → 30 μm with a comb of many lines of mW average output power. The relative tuning around each line is limited to a few parts in 10^6 with the exception of few strong CH_3OH lines that can be tuned to about one part in 10^4 by using the Stark effect. The spectroscopic use of these fixed frequency sources is limited to the standard techniques of laser Stark spectroscopy and of magnetospectroscopy. Both were used in the FIR even before the discovery of the OP FIR laser by using the HCN or D_2O discharge laser [35,36].

Intracavity optically pumped magnetospectrometers such as that shown in Fig. 19 have been successfully used in recent years [37]. This technique has a very high sensitivity, and valuable results have been obtained especially in the case of free radicals of interest in astrophysics [37]. A more general application of the FIR laser to gaseous spectroscopy is expected if sum and difference sideband gene-

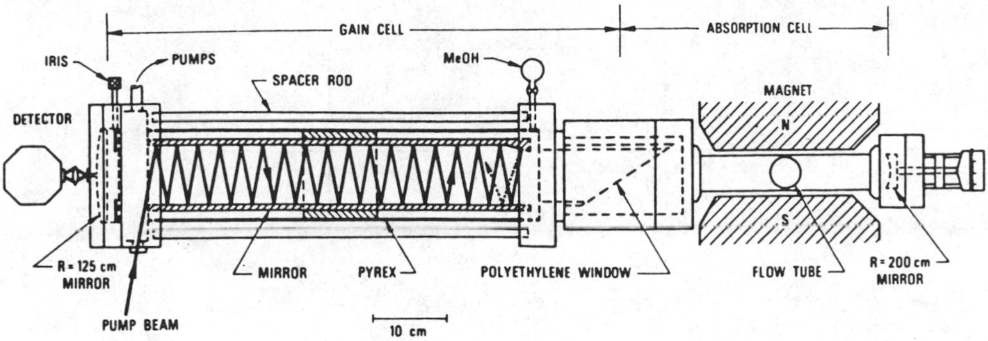

Fig. 19 - Scheme of a OP FIR laser magnetic resonance spectrometer.

ration could be obtained at a resonable power level by a nonlinear mixing of laser radiation with that of tunable microwave sources.

The first practical results with this technique were obtained in 1978 [38,39] by using as a nonlinear device either a Schottky diode or a MIM diode mixer. An output sideband power larger than 10^{-7} W was observed in both experiments with a frequency coverage up to 100 GHz. As a consequence it became possible with this method to cover continuously all the FIR region up to a few THz by using only a few strong laser lines. The resolution in frequency is limited to about 1 MHz as a consequence of the laser frequency jitter. By using a second mixer and a stabilized microwave local oscillator to monitor this jitter the frequency of the generated sidebands could be determined with an accuracy better than 50 KHz [40]. High resolution spectroscopy by using the generated sideband was performed either in excited [41] or ground vibrational states [40]. For the detection a helium-cooled bolometer was used in ref. 40. In ref. 41 a fraction of the laser beam was used to feed a second mixer, placed after the absorption cell, for a heterodyne detection with an excellent signal to noise ratio. The scheme of the experimental apparatus is shown in Fig. 20. The detection scheme of Fig. 20 has also been used for FIR heterodyne astronomy. By using the NASA infrared telescope on Mauna Kea (Hawaii) it was possible to detect the J:6↔5 CO line from Orion [42]. The H_2O vapor in the atmosphere is the major limiting factor for this kind of astronomic spectroscopy. Nevertheless an extensive use of FIR laser heterodyne detection is expected in the future [43].

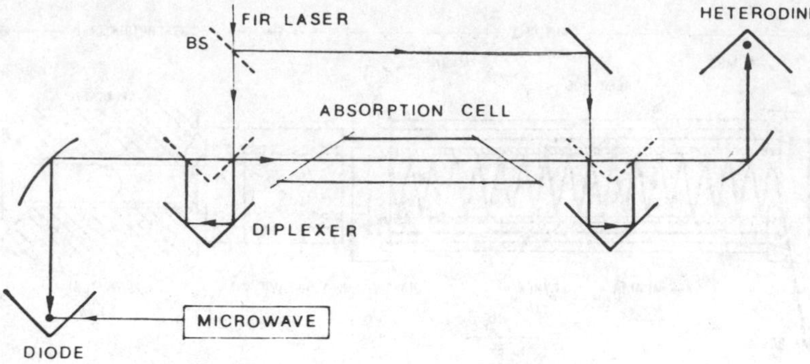

Fig. 20 - Scheme of a sideband double resonance spectrometer.

Conclusion

FIR laser spectroscopy has shown a rapid and successful development in the last ten years. The feasibility of different methods has been demonstrated and new effects discovered. A systematic and extensive use is expected in the near future when the new experimental techniques will become accessible in a larger number of laboratories.

References

1. K.M. Evenson, D.A. Jennings, F.R. Petersen, and J.S. Wells: in "Laser Spectroscopy III" J.L. Hall and J.L. Carlsten Eds. Springer Verlag - Berlin, New York, 1977 pag. 56.
   Jennings D.A., F.R. Petersen, and K.M. Evenson: in "Laser Spectroscopy IV" H. Walther and K.W. Rothe Eds. Springer Verlag Berlin-New York 1979 pag. 39
2. P.D. Coleman:IEEE J. Quantum Electron., QE9, 130 (1973)
3. T.Y. Chang and T.J. Bridges: Opt. Commun., 1, 423 (1970)
4. D.T. Hodges: Infrared Phys. 18, 375 (1978)
5. E.A.J. Marcatili and R.A. Schmeltzer: Bell Syst.Tech.J. 43, 1783 (1964)
6. B. Adam and F. Kneubühl: Appl.Phys. 8, 281 (1975)
7. F. Strumia and M. Inguscio: "Stark Spectroscopy and Frequency Tuning in O.P. FIR Molecular Lasers" in "Infrared and Millimeter Wave" Vol. V K.J. Button ed., Academic Press, New York 1982, pag. 129-213.

8. T.Y. Chang: in "Nonlinear Infrared Generation" Y.R. Shen ed. Springer Verlag, Berlin, New York 1977 pag. 216-272
9. J.M. Lourtioz, J. Pontnau and F. Julien, Infrared Phys. 20, 231 (1980)
10. F.R. Petersen, K.M. Evenson, D.A. Jennings, J.S. Wells, K. Goto, and J.J. Jimenez: IEEE J. Quantum Electron. QE-11,838(1975)
F.R. Petersen, K.M. Evenson, D.A. Jennings and A. Scalabrin: IEEE J.Quantum Electron. QE16, 319 (1980)
11. M. Inguscio, P. Minguzzi, A. Moretti, F. Strumia and M. Tonelli: Appl. Phys. 18, 261 (1979)
12. E. Bava, A. De Marchi, A. Godone, R. Benedetti, M. Inguscio, P. Minguzzi, F. Strumia and M. Tonelli: Optics Commun. 21, 46 (1977)
13. C.H. Townes and A.L. Schawlow: "Microwave Spectroscopy" McGraw-Hill, New York(1955).
M. Gordy and R.L. Cook: "Microwave Molecular Spectroscopy" Interscience Publishers, New York, (1970)
14. G. Bionducci, M. Inguscio, A. Moretti and F. Strumia: Infrared Phys. 19, 297 (1979)
15. M. Redon, C. Gastaud and M. Fourrier, Infrared Phys. 20, 93 (1980)
16. F. Strumia: in "Coherence in Spectroscopy and Modern Physics", F.T. Arecchi, R. Bonifacio and M.O. Scully eds. Pag.381, Plenum Press, New York, (1978)
17. M. Inguscio, A. Moretti, F. Strumia: IEEE J. Quantum Electron. QE-16, 955 (1980)
18. M. Inguscio, A. Moretti, G. Moruzzi and F. Strumia: Int. J. Infr. Millim. Waves 2, 943 (1981)
19. J.O. Henningsen: IEEE J.Quantum Electron. QE-13, 435 (1977)
20. E. Arimondo, M. Inguscio, A. Moretti, M. Pellegrino and F. Strumia: Optics Lett. 5, 496 (1980)
21. D. Dangoisse, A. Deldalle and P. Glorieux: J. Chem. Phys. 69, 5201 (1978);
D. Dangoisse and P. Glorieux: Opt. Commun. 32, 246 (1980)
22. H.R. Fetterman, C.D. Parker and P.E. Tannenwald: Optics Commun. 18, 10 (1976)
23. M. Inguscio, A. Moretti and F. Strumia: Optics Commun. 30, 355 (1979)
M. Inguscio; Contribution to this book
24. M. Inguscio, A. Moretti and F. Strumia: Optics Commun. 35, 64 (1980)

25. A.C. Luntz, R.G. Brewer, K.L. Foster and J.D. Swalen: Phys. Rev. Lett. 23, 951 (1969)
A.C. Luntz and R.G. Brewer: J. Chem. Phys. 53, 3380 (1970)
J.S. Levine, P. Boncyk and A. Javan: Phys. Rev. Lett. 22, 267 (1969)

26. M.S. Feld, A. Sanchez, A. Javan and B.J. Feldman: Publ. n. 217 du CNRS, Paris 1974 pag. 87

27. F. Strumia, M. Inguscio, A. Moretti: in " Laser Spectroscopy V" A.R.W. McKellar, T. Oka and B.P. Stoicheff eds. Springer Verlag, Berlin - New York 1981 pag. 255

28. M.S. Tobin and R.E. Jensen: Appl. Opt. 15, 2023 (1976)

29. G.D. Willenberg: Opt. Lett. 6, 372 (1981) and references therein

30. H.R. Fetterman, H.R. Schlossberg and J. Waldman: Opt. Commun. 6, 156 (1972)

31. T.A. De Temple: "Pulsed optically pumped FIR lasers" in "Infrared and Millimeter Waves - vol. I" K.J. Button ed. Academic Press, New York 1979 pagg. 129-184

32. N. Ioli, G. Moruzzi and F. Strumia: Lett. Nuovo Cimento 28, 257 (1980)

33. N. Ioli, G. Moruzzi, M. Pellegrino and F. Strumia: Conf. Digest Sixth Int. Conf. on Infr. Millim. Waves (1981) K.J. Button Ed. IEEE CAT n° 81 CH1645-1 MTT paper F-1-8

34. M. Inguscio, N. Ioli, A. Moretti, G. Moruzzi and F. Strumia: Opt. Commun. 37, 211 (1981)
J.O. Henningsen, M. Inguscio, A. Moretti and F. Strumia: IEEE J. Quant. Electron QE-18, 1004 (1982)

35. K. Uehara, T. Shimizu and K. Shimoda: IEEE J. Quantum Electron. QE-4, 728 (1968)

36. K.M. Evenson, H.P. Broida, J.S. Wells, R.J. Mahler and M. Mizushima: Phys. Rev. Lett. 21, 1038 (1968)

37. K.M. Evenson, D.A. Jennings, F.R. Petersen, J.A. Mucha, J.J. Jimenez, R.M. Charlton and C.J. Howard: IEEE J.Quantum Electron. QE-13, 442 (1977) and references therein

38. D.D. Bicanic, B.F.J. Zuidberg and A. Dymanus: Appl. Phys. Lett. 32, 367 (1978)

39. H.R. Fetterman, P.E. Tannenwald, B.J. Clifton, C.D. Parker and W.D. Fitzgerald: Appl. Phys. Lett. 33, 151 (1978).

40. F.V. Van den Heuvel, W. Leo Meerts and A. Dymanus: J. Mol. Spectr. 84, 162 (1980)

41. W.A.M. Blumberg, H.R. Fetterman and D.D. Peck: Appl. Phys. Lett. 35, 582 (1979)

42. H.R. Fetterman and W.A.M. Blumberg: in "Laser Spectroscopy V" A.R.M. McKeller, T. Oka, B.P. Stoicheff eds. Springer Verlag Berlin-New York 1981 pag. 76

43. C.B. Cosmovici, Inguscio M., F. Strafella and F. Strumia: Astr. and Space Sc. 60, 475 (1979).

SUB-DOPPLER AND SUBHOMOGENEOUS SPECTROSCOPY IN THREE-LEVEL SYSTEMS: SUPERHIGH RESOLUTION AND INFORMATION+

Massimo Inguscio

Istituto di Fisica, Università di Pisa
piazza Torricelli, 2 I-56100 Pisa, Italy

INTRODUCTION

As is well known, nonlinear laser techniques provide means for high-resolution atomic and molecular spectroscopy. The power of these techniques can be further increased when they are applied to coupled three-level systems. In the present paper new high-resolution techniques developed in optically pumped molecular lasers are described. They refer to both Doppler-Doppler and Doppler-homogeneous folded configurations. Velocity-selective excitation yields Doppler tuning of the coupled transitions; sub-Doppler saturation features generated in the absorption can be observed in the re-emission. In some cases resolutions beyond the limitations imposed by the homogeneous linewidths in two-level spectroscopy are achieved.

These superresolving techniques suggest an interpretation in terms of information capacity of atomic and molecular spectroscopy.

Coupled Three-Level Lasers

Generation of far infrared (FIR) laser radiation by means of optical pumping of molecules has received increasing attention in recent years. Basic performances and applications to spectroscopy can be found in [1]. The scheme of operation is illustrated in Fig. 1a.

+Work jointly supported by the Gruppo Nazionale Struttura della Materia del CNR and by Ministero della Pubblica Istruzione.

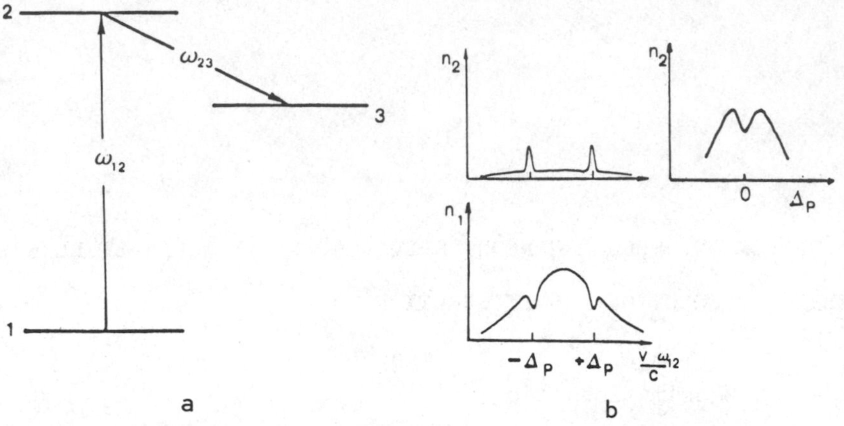

Fig. 1

The absorption of 10 μm pump radiation leads to inversion between rotational levels in an excited vibrational state and to FIR laser action. The analysis of the optical pumping process is usually based on a rate equations (RE) treatment. The importance of multiple quantum processes, neglected in the RE model, was stressed in [2].

The active molecules can be considered as a coupled three-level system interacting with applied radiation fields. A peculiarity of the system is that one of the coupled transitions itself generates a laser emission. Different processes manifest themselves, depending on whether one or both of the coupled transitions are homogeneously or Doppler broadened.

The homogeneous width of the levels is determined by collisions and saturation:

$$\Delta\nu_{hom} = \Delta\nu_{coll} \sqrt{1 + I/I_{sat}} \qquad (1)$$

$\Delta\nu_{coll} \simeq$ 10-40 MHz/Torr for the polar molecules active in the FIR. At the typical cw operation pressures (5-500 mTorr), $\Delta\nu_{hom}$ will be of the order of a few MHz, weakly affected by the actual pressure: at lower pressures the saturation intensity I_{sat} decreases and the saturation broadening compensates the decrease of $\Delta\nu_{coll}$.

The inhomogeneous Doppler width of the transitions is of the order of several tens of MHz in the IR pump, while in the FIR it can range from less than 1 MHz to more than 10 MHz, depending on the wavelength and on the molecular mass. Therefore the coupled three-level system can be Doppler-Doppler or Doppler-homogeneous broadened, depending on the FIR wavelength, the pressure, and the pump intensity and consequent power broadening.

In the case of fully Doppler-broadened systems, it is possible to observe tuning of the FIR emission by tuning the pump radiation across the absorption Doppler profile.

If a detuning Δp is introduced between the pump frequency and the center of the absorption, only the molecules with the proper velocity component along the laser direction, $v = c\Delta p/\omega_{12}$, can be excited. If the homogeneous broadening is negligible, the FIR emission is Doppler shifted, and the gain curve consists of the sum of two Lorentzian curves centered at frequencies

$$\Omega_{+-} = \omega_{23} \pm \omega_{23}\ \Delta p/\omega_{12} \tag{2}$$

corresponding to FIR co- and counterpropagating with the pump beam. (Here we neglect Raman-type processes, which cause an anisotropy in intensity and width of the two curves.) As a consequence, by scanning the FIR cavity when a Δp detuning is present, two distinct peaks will be present in the FIR emission if the splitting

$$\Delta\nu_{FIR} = 2\omega_{23}\ \Delta p/\omega_{12} = 2\Delta p\lambda_{IR}/\lambda_{FIR} \tag{3}$$

exceeds the homogeneous width. That is illustrated in Fig. 2, which shows the emission profile for the new [3] CH_3OH 62.98 μm line, as recorded for different offsets of the waveguide CO_2 laser 10R(34) pump. The FIR shift is in agreement with the theoretical slope $\lambda_{IR}/\lambda_{FIR} = 0.16$. The effect can be helpful in determining the pump offset and is quite similar to that observed in dimer lasers [4].

The effect can also be observed by scanning the pump frequency while the FIR resonator is tuned off the resonance position: the FIR signal will be at a local minimum when the pump laser is at the line center and displays two maxima as the pump offset sweeps the FIR frequency through resonance [3].

A more general effect, observable also in the Doppler-homogeneous broadening case, is the IR-FIR transferred Lamb dip (TLD), illustrated in Fig. 1b.

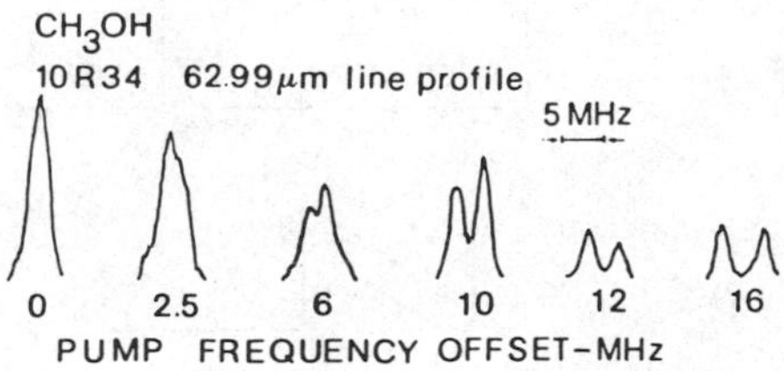

Fig. 2

Transferred Lamb Dip

Since the pump is effected in a resonator configuration, the system interacts with forward- and backward-propagating IR beams and with standing wave FIR radiation. A saturation Lamb dip is generated in the IR Doppler profile; hence the decrease in absorption and in the number of excited molecules (Fig. 1b).

A typical optoacoustic recording of the IR Lamb dip is shown in Fig. 3a [5]. The overall FIR laser intensity depends on the number of excited molecules; therefore zero pump detuning causes a dip also in the FIR laser emission (Fig. 3b).

The 2→3 transition generates a laser emission, and hence is easy to investigate by means of the experimental arrangement shown in Fig. 3c. The IR saturation effect is "amplified" by the non-linearity of detection via FIR laser oscillation (Fig. 3a,b): dip contrasts as large as 100% can be obtained by a proper choice of the laser threshold. TLD is a general technique and has been applied in a wide wavelength range (from 37 to 1200 μm) and to several lasing molecules (CH_3F, CH_3Br, CH_3OH, CF_2Cl_2, NH_3,...).

Since the effect originates in the absorption, the center frequency does not depend on the FIR resonator tuning. This is for instance shown in Fig. 4, where the TLD on the CH_3OH 133.12 μm line is recorded at different pressures. The changes affect the emitted intensity, the tuning of the FIR resonator (via the refractive index), and hence the symmetry of the FIR line, but not the frequency position of the TLD. Also, the width of the TLD is not strongly affected by the pressure, consistent with Eq. (1).

These features have encouraged experimental investigations of the IR transition via coupled FIR emission. The simultaneous recording of TLD and CO_2 tuning curves makes possible the determination of pump offsets: the detection in the coupled FIR emission avoids the ambiguity that exists in two-level spectroscopy when several absorptions are located within the CO_2 tuning range.

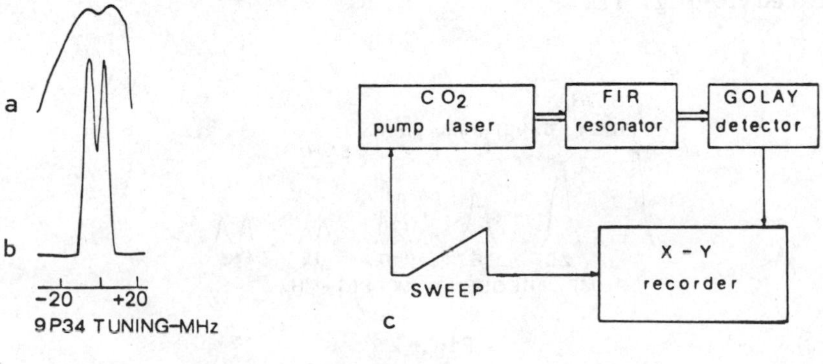

Fig. 3

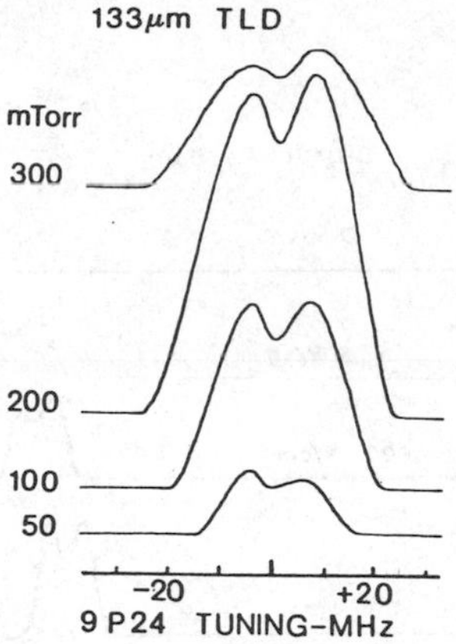

Fig. 4

It is also possible to investigate the Stark effect in the pump transition by monitoring the FIR emission, and so obtain information about the quantum numbers and the permanent dipole moment values of the states involved. A typical example is illustrated in Fig. 5, which refers to the optical pumping of a CH_3OH line at 205.6 μm [6] whose pump transition is close to the high-frequency end of the tuning range of the 9P(34) waveguide CO_2 laser (±100 MHz in the recording). At zero electric field the line center is not accessible and no TLD is detected. When the pump transition is Stark tuned across the CO_2 tuning range, the different offsets of the TLD allow the investigation of the effect. Various examples, also of quadratic effects, are described in [7].

In the following we discuss applications of the TLD technique in which the homogeneous linewidth is also overcome.

Subhomogeneous Spectroscopic Resolution

Generally, nonlinear spectroscopy of the transitions between two energy levels permits the resolution of structures larger than the homogeneous width. However, in Fig. 6a,b cases of failure are shown. In case a the structure W_1 of level 1 is not observable since the homogeneous width of level 2 causes the linewidth $\Gamma_1 + \Gamma_2$ of the associated 1→2 transition to be too large. In case b the energy splittings are larger than the homogeneous width in both levels, but they are nearly equal. Therefore the photons associated with the 1→2 transitions have nearly the same energy and cannot be resolved in the homogeneous linewidth. The three-level configuration can be successful in overcoming the difficulties present in the two-level scheme. In Fig. 6c the three-level version of case a is shown. The system interacts simultaneously with two laser beams at frequencies ω_{12} and ω_{23}: the width of the optical resonances is $\Gamma_1 + \Gamma_3$, and hence independent of the upper coupling level

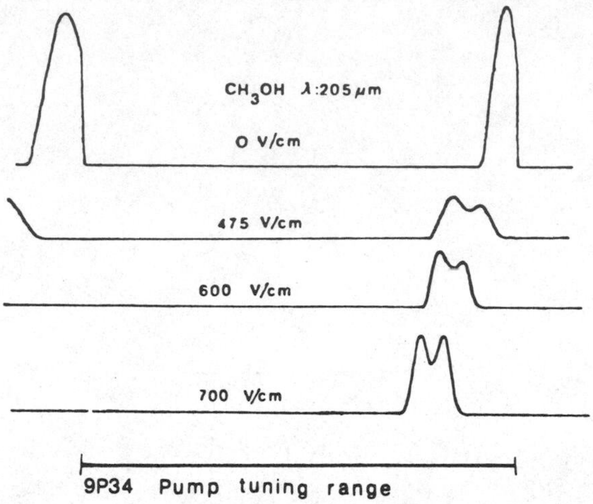

Fig. 5

lifetime. Very narrow resonances were actually reported by coherent two-photon transitions in a three-level optical pumping of sodium [8,9,10] and by two-step resonant scattering in I_2 [11].

The three-level version of case b is illustrated in Fig. 6d. The photons associated with the 1→2 transitions, unresolved within the homogeneous linewidth, are coupled to the photons of the 2→3 transition, where the line structures are resolved. The TLD technique can be used to transfer and resolve the 1→2 transition saturation spectrum in the coupled 2→3 transitions.

Experimental examples of the failure illustrated in Fig. 6b are found in Stark spectroscopy of polar molecules: 1 and 2 can be regarded as two different vibrorotational levels, W_1 and W_2 as the Stark splittings. For a symmetric or slighly asymmetric rotor, to a first-order approximation, the energy separation between contiguous Stark sublevels of a (J,K,M) level changes linearly with the electric field:

$$W = -\mu[KM/J(J+1)]\cdot E \tag{4}$$

where J is the total angular momentum, K and M the components along the symmetry axis of the molecule and the external field direction, and μ is the permanent electric dipole moment of the state (Fig. 7a).

Depending on the radiation polarization, transitions are induced with $\Delta M = \pm 1$ selection rules. The 1→2 line consists of various components equally spaced by a quantity $|W_2-W_1|\cdot E$. In case of a ^qQ branch transition ($\Delta J = 0$, $\Delta K = 0$), the difference between W_1 and W_2 is due only to the μ values in the two states differing only by a few parts in 10^2. Hence the Stark line structure is

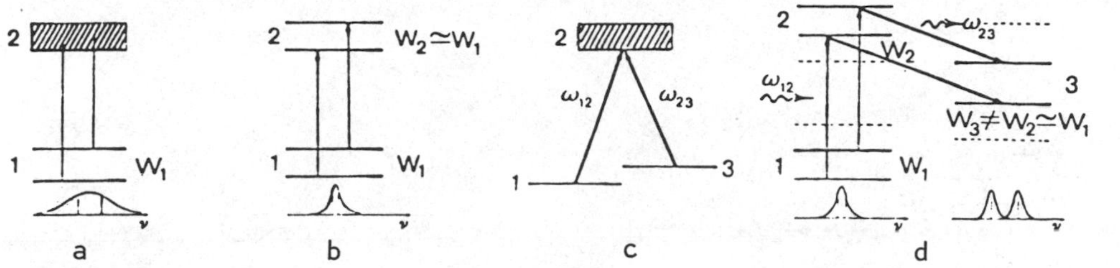

Fig. 6

unresolved in spite of an energy sublevel splitting larger than the level width in both 1 and 2 states. Higher resolution, overcoming the homogeneous linewidth limitation, is obtained in molecular lasers optically pumped in the presence of an electric field (Fig. 7a): the infrared 1→2 transition Stark structure is unresolved, because of the above considerations, but in the rotational FIR transition the changes in J and/or K values cause W_2 to be significantly different from W_3. Provided that M changing collisions can be neglected [12], in the presence of an electric field the 1→2→3 cycle is simply divided into cycles corresponding to the different M sublevels in the 1 state. The saturation dip of a given Stark component of the IR 1→2 transition is transferred and detected on the corresponding Stark component of the coupled FIR emission. The resolution of Stark structures inside the homogeneous linewidth [13] is illustrated here for the three-level cycle yielding FIR action at 496 μm (Fig. 7) from CH_3F.

The three-level quantum numbers are (J,K):(12,2)→(12,2)→(11,2) with Stark splittings W_1 = 11.9 MHz/KVcm^{-1}, W_2 = 12.2 MHz/KVcm^{-1}, W_3 = 14.5 MHz/KVcm^{-1}. The level homogeneous width is $\gamma \simeq$ 6 MHz. Therefore the resolution of the Stark components of the 2→3 transition is obtained for electric field intensities of about 1 KVcm^{-1} and actually shown in Fig. 7b for a field of 1.5 KVcm^{-1}. By setting the FIR cavity length for oscillation on a given component, it is possible to detect the TLD corresponding to the coupled components in the 1→2 transition (Fig. 7c). Note that these components are spaced ~500 KHz, i.e., much less than the homogeneous linewidth. The infrared saturation spectrum has been transferred and investigated in the FIR emission, observing structures unresolved within the homogeneous linewidth of the two-level transition.

Superhigh Resolution and Information

Thanks to the above-described results, the TLD can be included among the few techniques in which subnatural resolution is obtained. These recent developments suggest a reexamination of the concept of resolution for a spectroscopic apparatus. In fact the usual definition that two spectral lines can be resolved if their separation is larger than a given fraction of the linewidth can be subjective and depend on practical limitations of the instrument and not on

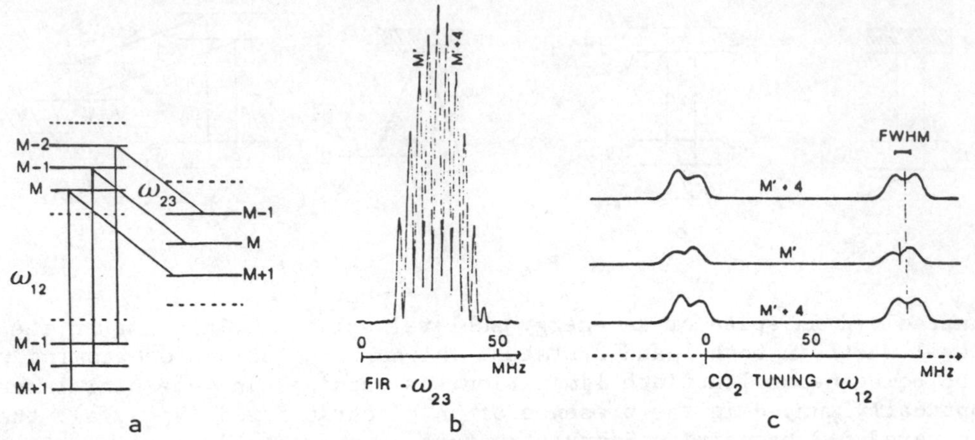

Fig. 7

intrinsic properties of the investigated sample and the experimental techniques. Moreover, it is difficult to take the S/N ratio into account objectively in the definition. For instance, in some applications of time-resolved spectroscopy it is possible to narrow the lines to less than their natural width by proper data handling. But the price paid for this is loss of intensity, and this, together with the wiggles introduced in the curves, makes it hard to predict the identification of weak spectral features in spite of reduced linewidths.

An objective definition of the power of a spectroscopic technique can be given in terms of its information capacity.* It is reasonable to define the capacity of a spectrometer as the greatest number of independent information units (bits) obtainable from a recording, applying to spectroscopic resolution a treatment similar to that by Toraldo di Francia for an exact definition of the resolving power of an optical instrument [14].

Following [15], the information capacity in a spectral range $\Delta\nu$ can be written as

$$P = p_0 \frac{\Delta\nu}{\Delta\nu_0} \quad \text{(bits)} \tag{5}$$

where p_0 is the number of information units (bits) obtained within the resolved spectral width $\Delta\nu_0$. An estimated order of magnitude for p_0 is ten, which corresponds to measuring the intensity within

---

*This possibility was suggested by stimulating discussions with F.T. Arecchi, following the delivery of this lecture by the author.

$\Delta\nu_o$ with an accuracy of 10^{-3}. A decrease of $\Delta\nu_o$, or an increase of the "resolving power" at a frequency ν_o defined as $R = \nu_o/\Delta\nu_o$, in general leads to an increase of the bits obtained in the spectral range $\Delta\nu$, if p_o is constant or decreases more slowly than $\Delta\nu_o$. Now the basic idea is that there is a finite number of independent information units "intrinsic" in the observation of a radiative transition under ideal circumstances. That is a quantitative limit to the power of a single experiment. Superresolving techniques should be treated with this limit in mind. Subnatural spectroscopy can sometimes lead to the observation of new structures, but also at the same time to the loss of some other information.

Of course an objective improvement in the overall information can be obtained by properly using results from independent experiments. Also, new information can be obtained from a given experiment if "a priori" information is used. The latter is actually the case in the experiment described in the present paper. In fact, in the case of photons with nearly the same energy (Fig. 6b), the overlap within the homogeneous width makes the information capacity of the sub-Doppler spectrum much lower than that evaluable for saturation spectroscopy according to Eq. (5). In particular, the information on the frequency separation between the components of the transition is lost. The TLD spectrometer (Figs. 6d, 7) gives these data, using "a priori" information to "label" the 1→2 photons. The observer makes use of the "a priori" information by setting the FIR resonator for oscillation on a given Stark component. That is equivalent to knowing the M sublevels involved in the 1→2→3 cycle, and hence in the 1→2 transition. In the presence of M randomizing collisions the experiment cannot work. The widths of the levels are the same and also the resolution in the IR (1→2) and in the FIR (2→3), but the subhomogeneous TLD resolution is lost because collisions erase the extra information used in the experiment.

Foreseen Applications

Among the various high-resolution spectroscopy applications of this technique, yielding features narrower than the homogeneous linewidth, the precise determination of molecular excited dipoles can be foreseen. In general these measurements are carried out by microwave Stark spectroscopy, involving low J values, or by laser Stark spectroscopy. In both cases the calibration of the electric field is a very critical point. On the other hand, precise meaurements in the FIR region are very useful for understanding the dependence of μ on vibrational, torsional, or rotational quantum numbers.

With the TLD technique, the Stark splittings of the 2→3 FIR transitions (Fig. 7a,b) depend on the electric dipole moment of the excited level, while the frequency separation in the TLD recordings (Fig. 7a,c) depends on the difference of the dipole moments in the ground and excited states. In the limit in which the quadratic contribution can be neglected and the Stark effect considered linear, measurements at the same E values allow the determination of $(\mu_{exc} - \mu_{gs})/\mu_{gs}$, independent of a precise calibration of the field, as ratio of frequency separations in the FIR and in the IR. Both these measurements can be performed with high accuracy by beating with reference lasers.

Acknowledgments

F.T. Arecchi, E. Arimondo, J.O. Henningsen, and G. Moruzzi are thanked for many useful discussions, and F. Strumia for the continuous interest in the author's work.

References

1. F. Strumia: "High resolution FIR laser spectroscopy," this volume.
2. D. Seligson, M. Ducloy, J.R.R. Leite, A. Sanchez, and M.S. Feld: "Quantum mechanical features of optically pumped cw FIR lasers," IEEE J. Quantum Electron. QE-13, 468 (1977).
3. J.O. Henningsen, M. Inguscio, A. Moretti, and F. Strumia: "Observation and assignment of large offset FIR laser lines in CH_3OH optically pumped by a CO_2 waveguide laser," IEEE J. Quantum Electron. QE-18, 1004 (1982).
4. H. Welling, this volume.
5. M. Inguscio, A. Moretti, and F. Strumia: "IR-FIR transferred Lamb-dip spectroscopy in optically pumped molecular lasers," Opt. Commun. 30, 355 (1979).
6. M. Inguscio, A. Moretti, and F. Strumia: "IR-FIR Stark spectroscopy of CH_3OH around the 9P(34) CO_2 laser line," Conf. Dig., 4th Int. Conf. IR-mm Waves, Miami 1979; IEEE Cat. CH 1384-7 MTT p. 205, and to be published.
7. F. Strumia and M. Inguscio: "Infrared and millimetric waves: coherent sources and applications," Vol. 5, pp. 129-213, K. Button, Ed., Academic Press, 1982.
8. G. Alzetta, A. Gozzini, L. Moi, and G. Orriols: "An experimental methods for the observation of RF transitions and laser beat resonances in oriented Na vapour," Nuovo Cimento 36B, 5 (1976).
9. E. Arimondo, and G. Orriols: "Nonabsorbing atomic coherences by coherent two-photon transitions in a three-level optical pumping," Lett. Nuovo Cimento 17, 333 (1976); G. Orriols, Nuovo Cimento 53B, 1 (1979).
10. M.S. Feld, M.M. Burns, T.U. Kühl, G. Pappas, and D.E. Murnick: "Laser-saturation spectroscopy with optical pumping," Opt. Lett. 5, 79 (1980).
11. R.P. Hackel, and S. Ezekiel: "Observation of subnatural linewidths by two-step resonant scattering in I_2 vapor," Phys. Rev. Lett. 42, 1736 (1979).

12. M. Inguscio, F. Strumia, K.M. Evenson, D.A. Jennings, A. Scalabrin, and S.R. Stein: "Far-infrared CH_3F Stark laser," Opt. Lett. 4, 9 (1979).
13. M. Inguscio, A. Moretti, and F. Strumia: "Spectroscopy inside the homogeneous linewidth in three-coupled level system," Opt. Commun. 35, 64 (1980); J. Opt. Soc. Am. 70, 1390 (1980).
14. G. Toraldo di Francia: "Resolving power and information," J. Opt. Soc. Am. 45, 497 (1955); "Capacity of an optical channel in presence of noise," Opt. Acta 2, 5 (1955).
15. V.S. Letokhov, in: "High-Resolution Laser Spectroscopy," K. Shimoda, Ed., p. 142, Topics in Applied Physics, Vol. 13, Springer, 1976.

INFRARED - FAR INFRARED SPECTROSCOPY OF METHANOL BY OPTICALLY PUMPED FAR INFRARED LASER EMISSION*

J.O. Henningsen

Physics Laboratory I
H.C. Ørsted Institute
University of Copenhagen
Denmark

INTRODUCTION

The classical techniques for studying the vibration-rotation spectra of molecules are broadband, based either on monochromators or on the Fourier transform principle. Their advantage is that the entire spectral region of interest can be covered, and the general view obtained in this way greatly facilitates the interpretation of a spectrum. Their main disadvantage is that they offer a resolution which is rarely better than 1 GHz, while the Doppler width is typically about 100 MHz. Even for fairly simple molecules there are often spectral regions where the density of lines is very high, and where there are no recognizable patterns. In such cases, a conventional unresolved spectrum will provide only rudimentary insight into the molecular vibration-rotation states.

With the advent of infrared gas lasers, the situation changed insofar as the Doppler limit could be reached and, with saturation techniques, even surpassed. However, being line sources with very limited tunability, gas lasers can be used only if they are in accidental coincidence with molecular absorption lines. This shortcoming was remedied when continuously tunable diode lasers, covering a large part of the infrared spectral region, became available. With such lasers an entire band can be mapped with Doppler limited resolution, and the only remaining problem then is the interpretation. This, however, is not trivial, and if the molecule has internal degrees of freedom, or if several vibrational modes are

*Work supported by the Danish Natural Science Research Council under grants No. 11-9148 and 11-9344.

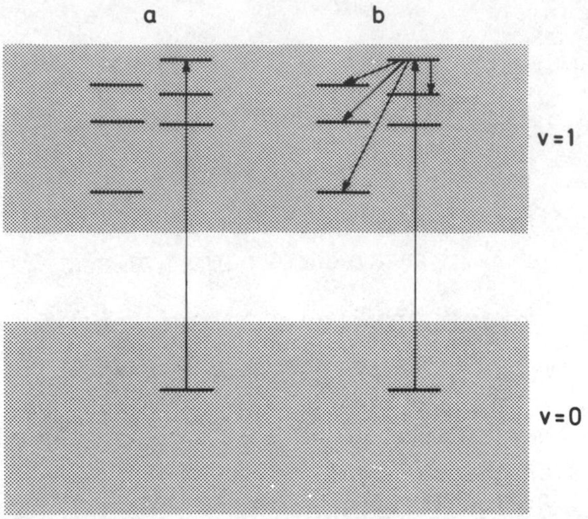

Fig. 1. Infrared linear and non-linear spectroscopy (a) provides information about the frequencies of allowed transitions to vibrationally excited states. Optically pumped far infrared lasers (b) provide simultaneous information about the infrared transition and one or several coupled far infrared transitions within the vibrationally excited state.

interacting, it may be insurmountable. In such cases, a fruitful interplay can be established between diode laser spectroscopy and a technique based on optically pumped far infrared laser emission. In the diode laser spectrum one can frequently recognize families of lines by taking advantage of the systematic shift in line positions as a single quantum number is changed. If one member of such a family is in near coincidence with a CO_2 laser line, it can be pumped, and in this way population inversion may be created between adjacent rotational levels of the excited vibrational state. By confining the gas inside a properly tuned resonator, this may result in laser emission on one or several far infrared frequencies.[1] While a saturation spectroscopy experiment, which can of course be performed using the same coincidence, provides information about an isolated point of the vibrationally excited manifold of states, the optical pumping technique informs not only about this point, but also about its environments (Fig.1), and this significantly alleviates the assignment problem.

Experimental Arrangement

The key components of the spectrometer are the pump laser and the far infrared resonator. In order to allow pumping of relatively weak lines and to relax the coincidence requirement, the pump laser should deliver at least about 50 Watt. A good solution is to use a short laser for obtaining maximum tunability, and operate it in a long pulse mode to obtain a peak power of about five times the cw power. For maximum versatility, several types of far infrared resonators can be used. Weak lines require an open resonator with spherical mirrors in order to minimize the losses, as well for the trapped pump radiation as for the far infrared radiation. Stronger lines can be investigated with either hollow dielectric or metal waveguides, the former being preferable owing to the simple mode structure. Stark effect investigations call for a hybrid waveguide where two plane parallel metal plates at the same time guide the radiation and act as Stark electrodes, while lateral confinement is provided either by dielectric walls, or by using an optically stable mirror configuration. The pump radiation is usually coupled into the far infrared resonator through a hole in one mirror, and the far infrared radiation is coupled out either through the same or a different hole. Hole coupling works at all wavelengths, and is preferable under circumstances where output power optimization and output beam quality are of minor importance. The far infrared signal can be detected by any reasonably sensitive detector since the power level, once the laser is above threshold, is typically between a few microwatt and a few milliwatt. By making one mirror movable, the resonator itself can be used as a scanning Fabry-Perot interferometer to determine the wavelengths to a few parts in 10^4. Subsequently, the frequencies can be measured to one part in 10^7 by heterodyning with a suitably chosen difference frequency between two CO_2 lasers.[2] A far infrared polarizer is needed for analyzing the polarization of the laser emission, and a selection of high and low pass filters is useful for simplifying the resonator scans if several far infrared lines are simultaneously present.

The Methanol Spectrum

The potential of the technique is best illustrated by way of an example. Methanol is chosen because the good overlap between the strong C-O stretch band, centered at 1034 cm^{-1}, and the CO_2 laser spectrum, makes it well suited for optical pumping, and also because its spectrum in this region turns out to be too complicated to be analyzed solely by conventional spectroscopy. The molecule is a nearly symmetric top with hindered internal rotation of the OH group relative to the CH_3 group (Fig.2). A large permanent dipole moment with components both parallel and orthogonal to the quasi symmetry axis leads to rotational selection rules $\Delta J = 0,\pm1$ and $\Delta K = 0,\pm1$, where J and K are angular momentum quantum numbers. The quantization of the internal rotation introduces a torsional quantum number $n = 0,1,2,...$ and an additional label $\tau = 1,2,3$ for each n, owing to the tunnelling through the three fold hindering barrier. Thus, a particular state can be specified as $(n\tau K,J)^v$, where the superscript labels the vibrational state. Selection rules for n are not subject to restrictions. However,

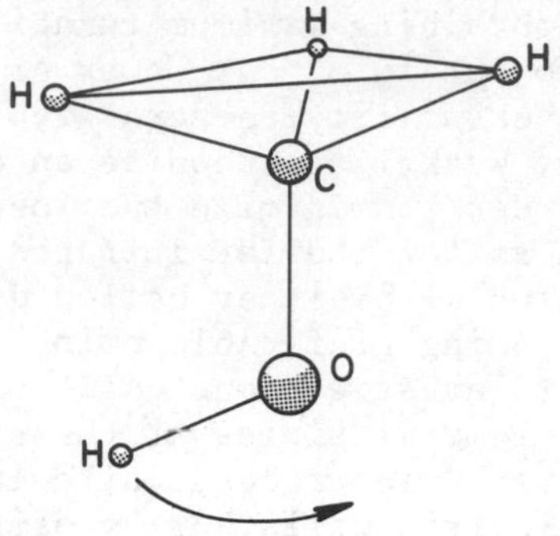

Fig. 2. Structure of the methanol molecule.

for each given K the three τ correspond to distinctive internal rotation symmetries, with transitions between states of different symmetry being strictly forbidden. The asymmetry causes shifts of all levels, and in addition a splitting of certain low K levels. Both effects can be treated by perturbation theory.

The energy of a particular state can be expressed as an expansion in powers of J(J+1) with coefficients that depend parametrically on the remaining quantum numbers:

$$E(n\tau K,J)^{v} = E^{v}_{vib} + W^{v}(n\tau K) + B^{v}(n\tau K)J(J+1)$$
$$-D^{v}(n\tau K)J^{2}(J+1)^{2} - H^{v}(n\tau K)J^{3}(J+1)^{3}\ldots$$

For the vibrational ground state the coefficients can be evaluated from a 20-parameter model whose input parameters are the various moments of inertia, expansion coefficients for the three fold barrier, and a set of constants accounting for centrifugal effects and for external-internal rotation interaction.[3,4] Using this model, ground state energies can be evaluated to better than 0.001 cm^{-1} for $n = 0$, $K \lesssim 6$ and $J \lesssim 15$, and the immediate aim of a vibrational state analysis is to find suitably modified parameters which provide similar accuracy for the excited molecule.

Assignment of Observed Far Infrared Lines

Having obtained far infrared emission from a particular coincidence, the next task is to assign the transitions with a minimum of a priori assumptions about the spectrum. A powerful aid in this respect is the existence of quasi combination relations among emission lines. Frequently, three lines are observed according to the scheme of Fig.3, leading to the relation $a \simeq c-b$. The ambiguity associated with the interchangeability of a and b is

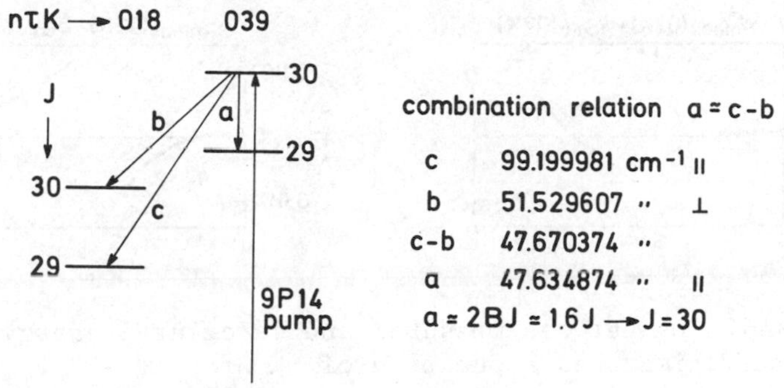

Fig. 3. Combination relation among emission lines. Polarizations identify b as the J→J transition, and the pump transition as P or R type.

removed by a polarization analysis. Linearly polarized pump radiation will lead to an anisotropy in the far infrared gain, since the angular momenta of the excited molecules will be oriented preferentially parallel to the pump polarization if the absorption line is Q-type, and orthogonal if it is P- or R type. This, in turn, leads to linearly polarized far infrared emission, parallel to the preferred direction for Q-type emission, and orthogonal for P- or R type emission. Consequently, the b transition always has a polarization different from that of a and c. Having identified a, one can usually determine J with sufficient confidence from a knowledge of the ground state rotational constant.

Owing to the strong coupling between internal and external rotation, the energy for states of a given internal rotation symmetry depends in a non-systematic way on K. Comparison with the ground state provides an important clue to the symmetry species and the K-assignment, but if this proves insufficient, one may resort to Stark effect experiments. This is not always feasible, since the progressive lack of cooperation between different M components of the far infrared transition with increasing electric field, may reduce the gain below threshold before a distinctive pattern has developed. However, if $K \simeq \frac{1}{2}J$, it can be shown that the far infrared signal will split into two components corresponding to $\Delta M = +1$ and $\Delta M = -1$, with all M components moving at essentially the same rate with the electric field.[5] In such cases K can be determined if J is known. Far infrared emission corresponding to $\Delta M = 0$ transitions is never observed, since the guiding action of the Stark plates constrain the far infrared polarization to be always orthogonal to the static electric field.

The C-O Stretch Torsional Ground State

Data for methanol has been gathered by many different groups, and complete references can be found elsewhere.[6] By using the

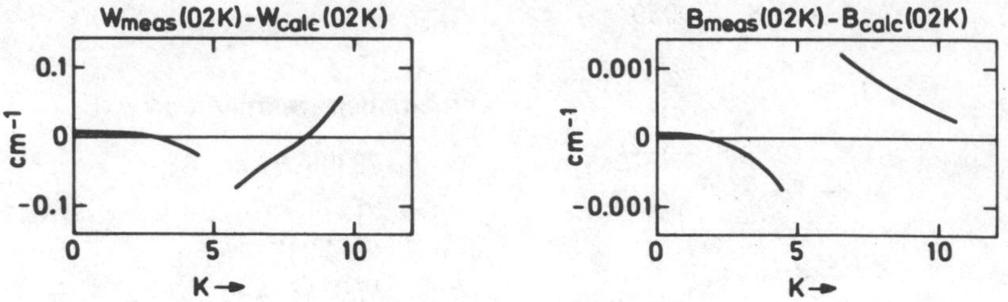

Fig. 4. Deviation between measured and calculated energy expansion coefficients as a function of K for $\tau = 2$.

techniques described in the preceding section, a number of pump transitions involving K-values up to 10 were identified, and a diode laser spectrum of the C-O stretch R-branch region[7] was subsequently used for determining the energy expansion coefficients. Despite extensive efforts it proved impossible to reproduce these coefficients by using the ground state model with modified input parameters. However, for a suitable choice, the deviations between calculated and measured coefficients was found to follow a suggestive pattern[8] indicating the presence of a strong vibrational interaction around $K = 5$ (Fig.4). Incorporating these empirical corrections in the model, it is now possible to evaluate the frequencies of all pump transitions and emission transitions for $n = 0$. A total of 26 emission lines pumped by 14 different pump transitions to states with $J \leq 26$ have been identified, with an r.m.s. deviation between calculated and measured frequencies of $0.005\ cm^{-1}$. For higher J the model remains useful, but the deviations rapidly become larger due to higher order terms in the energy expansion.

The Interacting Vibrational State

Having established indirectly the presence of a vibrational state interacting with the C-O stretch, an interesting question is whether the optical pumping technique can also provide more direct information about the perturbing levels. Somewhat by chance attention was drawn towards a curious resemblance between certain emission lines pumped by 9P34 and 9P38:

9P34 lines	55.43	53.91	39.43	37.79	cm^{-1}
9P38 lines	51.76	50.30	35.86	34.22	-
difference	3.67	3.61	3.57	3.57	

A possible explanation could be that the two pump lines lead to laser emission on transitions to the same lower laser levels, the upper laser levels being separated by about $3.6\ cm^{-1}$. Initial Stark effect investigations[9] suggested that 9P38 was pumping a level with $K = 6$ and $J = 10$, and subsequent more extensive measurements[10] confirmed the J assignment, but fixed the K-value at 5 . This

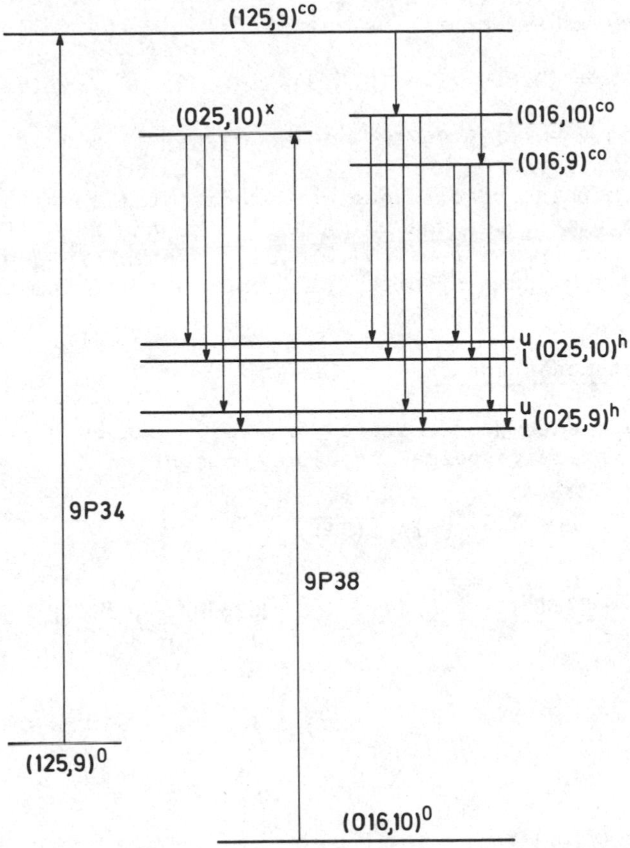

Fig. 5. Connection between emission lines pumped by 9P34 and 9P38. Levels labeled $(025,J)^{h}$ are bybridizations of $(025,J)^{co}$ and $(034,J)^{x}$, where x is presumed to be the CH_3 in-plane rock mode.

strengthened the suspicion that the states involved were also those responsible for the interaction, and since 9P34 had long been known to pump a K = 5 state, everything, including four additional 9P34 lines, could finally be brought together in the scheme shown in Fig.5. 9P34 pumps a torsionally excited C-O stretch state, and produces strong laser emission on $n = 1 \to 0$ transitions. The lines listed above arise from subsequent cascade emission to levels which are hybridizations of the C-O stretch states and the interacting states, labeled x . 9P38 pumps essentially pure x-states, and laser emission occurs down to the hybridized states. The validity of the scheme was verified a posteriori when frequency measurements of all the lines involved were carried out at NBS, Boulder.[2] A number of combination relations can be constructed which must be exactly fulfilled, and in all cases they hold to within the 1 MHz accuracy of the frequency measurements, as exemplified in the following:

$$(025,10)^x \rightarrow (025,9^u)^h \rightarrow (016,9)^{co} \rightarrow (025,9^\ell)^h \rightarrow (025,10)^x$$

$$1\ 509\ 040.2 - 1\ 136\ 942.0 + 1\ 180\ 092.5 - 1\ 552\ 190.1 = 0.6 \text{ MHz}$$

Based on these results one can predict the energies of the x-states for other J , and this has led to the identification of additional pump transitions to $(025,J)^x$ states at J = 8 , pumping with 9P10, J = 9, pumping with 9P12, and J = 25, pumping with 10R34.

Dipole Moment Measurements

The Stark energy of a state $(n\tau K,J)^v$ can be expressed in terms of an effective permanent dipole moment as

$$W_{Stark} = -\varepsilon\mu_{eff}\frac{MK}{J(J+1)} = -\varepsilon MK\left[\frac{A}{J(J+1)} + B\right]$$

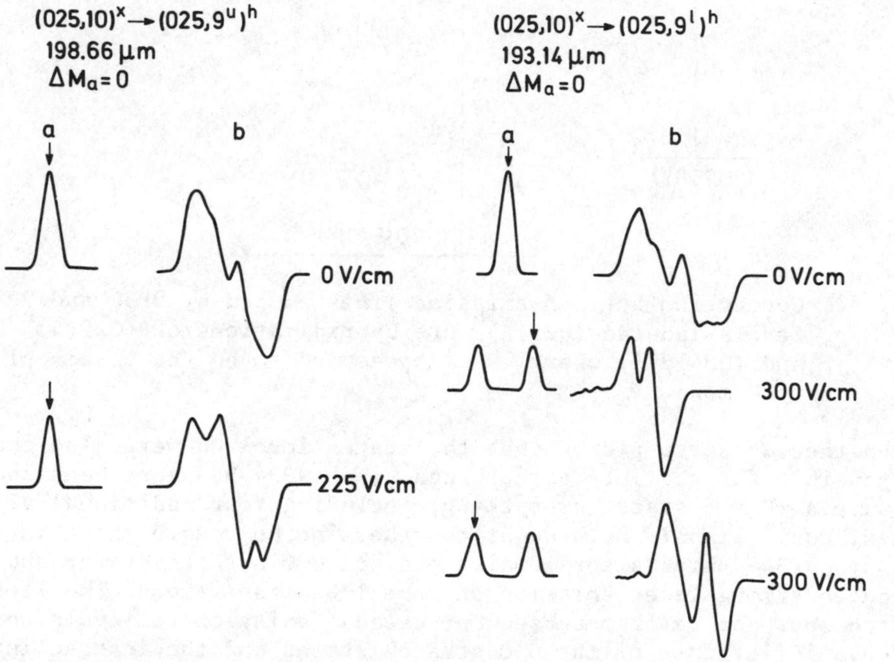

Fig. 6. Simultaneous Stark effect on pump and emission for two lines pumped by 9P38. Curves (a) show the far infrared emission for fixed pump frequency as the resonator length is scanned. Curves (b) show the derivative of the far infrared emission with respect to pump frequency as the pump is scanned over the CO_2 laser mode, with the far infrared resonator set at the arrows.

where ε is the electric field, and where A and B are state dependent.[11] The Stark shifts that are experimentally observed are quite susceptible to this state dependence, and therefore some care should be exercised in the interpretation, in particular for lines like those pumped by 9P38, where the effective dipole moment in both the upper and the lower laser level is in principle unknown. To eliminate this uncertainty, contact can be established to the ground state with its known dipole moment, by simultaneously monitoring the Stark effect on the pump transition and on the far infrared emission. Owing to its large Doppler width, the pump transition calls for a saturation technique, but fortunately enough, the necessary features are automatically built in. Pumping is effected in a resonator so that forward and backward running waves are simultaneously present, and at the line center a saturation dip in the pump absorption causes a dip in far infrared output, denoted a transferred Lamb dip (TLD).[12]

Some results obtained by this technique[10] for the twin lines $(025,10)^x \rightarrow (025,9^{u,\ell})^h$ at 198.66 µm and 193.14 µm are given in Fig.6. The pump polarization is parallel to the static electric field, implying $\Delta M_a = 0$, while the emission polarization, as always, is parallel to the plates, so that $\Delta M_e = \pm 1$. From the transition matrix elements it follows that in this configuration the strongest contribution to the far infrared signal arises from $M = J$ for $\Delta M_e = -1$, and $M = -J$ for $\Delta M_e = +1$. For the 198.66 µm the dipole moments happen to be such, that for these M, the Stark shift is almost the same in the upper and the lower laser level. Consequently, the far infrared line does not split, and both signals remain observable at the initial resonator setting as the electric field is increased from zero. For the pump transition, however, the Stark shift is proportional to M, and when the pump is scanned, two TLD are observed, moving apart linearly in the electric field at rates of ±31 MHz/(kV/cm). In contrast to the 198.66 µm line, the 193.14 µm line is split by the electric field, and signals from $M = J$ and from $M = -J$ can no longer be observed simultaneously. With the far infrared resonator tuned to the upshifted peak, a pump scan results in a single TLD corresponding to $M = -J$, moving at a rate of −31 MHz/(kV/cm), while tuning the far infrared resonator to the downshifted peak results in a TLD arising from $M = +J$, moving at a rate of +31 MHz/(kV/cm). The identical results for the Stark effect on the pump transition confirm that the two lines are pumped by the same transition, and from the known ground state dipole moment and the measured Stark coefficient, it can be inferred, that the level pumped has $K = 5$. The difference in the Stark effect of the two emission lines must be associated with the effective dipole moment of the lower laser levels. It can be understood if they are assumed to be hybridizations of the states $(025,9)^{co}$ and $(034,9)^x$, which must then be nearly degenerate.

From the known x-state energies for $K = 4$ and $K = 5$, it

can be estimated that the band origin is located at about 1075 cm^{-1}. This is very close to the frequency of the CH_3 in-plane rock mode, as observed by conventional infrared spectroscopy.[13]

Conclusion

While it is beyond dispute that the optical pumping method has proved productive for methanol, one might well question its general applicability. After all, no other single molecule has produced a similar number of far infrared lines. It should be kept in mind, however, that the entire analysis has been based on data collected by pumping with low pressure lasers using the normal $^{12}C^{16}O_2$ isotope. Considering that many other isotopes are available, and that the use of waveguide lasers or high pressure TEA lasers can significantly expand the tuning range, it seems safe to conclude that many other molecules can be successfully studied by the same technique.

References

1. T.Y. Chang and T.J. Bridges, Opt.Com. 1: 423 (1970).
2. F.R. Petersen, K.M. Evenson, D.A. Jennings, and A.Scalabrin, IEEE J.Quantum Electron. QE-16: 319 (1980).
3. R.M. Lees and J.G. Baker, J.Chem.Phys. 48: 5299 (1968).
4. Y.Y. Kwan and D.M. Dennison, J.Mol.Spec. 43: 291 (1972).
5. F. Strumia, this book.
6. J.O. Henningsen, "Spectroscopy of Molecules by Far Infrared Laser Emission" in:"Infrared and Millimeter Waves", Vol.5, Ed. K.J. Button, Academic Press, New York.
7. J.P. Sattler, T.L. Worchesky, and W.A. Riessler, Infrared Phys. 19: 217 (1979).
8. J.O. Henningsen, J.Mol.Spectr. 85: 282 (1981).
9. J.O. Henningsen, J.Mol.Spectr. 83: 70 (1980).
10. J.O. Henningsen, J.Mol.Spectr. in press.
11. E.V. Ivash and D.M. Dennison, J.Chem.Phys. 21: 1804 (1953).
12. M. Inguscio, this book.
13. A. Serrallach, R. Meyer, and Hs.H. Günthard, J.Mol.Spectr. 52: 94 (1974).

UNIMOLECULAR FRAGMENTATION KINETICS FOLLOWING LASER-INDUCED MULTIPHOTON IONIZATION OF POLYATOMIC MOLECULES

F. Rebentrost and K.-L. Kompa

Max-Planck-Institut für Quantenoptik

8046 Garching/Munich, Germany

Abstract

Multiphoton ionization (MPI) and fragmentation, induced by (preferably UV) lasers can provide both dynamical information about the process itself and shows prospects as an efficient ion source e.g. in mass spectrometry. Within a certain experimental parameter range a generalized picture of the mechanism has evolved which this paper intends to summarize and to substantiate by showing some of the experimental evidence. It may also be argued that there is hope for laser specific unimolecular chemistry under special conditions.

I. Introduction

Among the various pump-probe experiments which can be done to study the energy flow in polyatomic molecules upon laser excita tion the scheme shown in Fig. 1 has received a lot of interest within the last few years. Pumping by a UV laser or by a combina tion of IR and UV laser sources may populate a specific state having both electronic and vibrational energy. Now the decay modes of this state which may be radiative (fluorescence) or non-radia tive (photochemical, internal coupling) are of interest. The time sequence of these dynamic events may be interrupted by the second pulsed ionizing radiation which could be applied with or without a time delay. Under these condi-tions ionization may occur with a large cross section and a correspondingly high rate. The ion yield and its dependence on laser wavelength and intensity will then reflect the energy distribution in the molecule. The obvious experimental advantage connected with the detection of charged

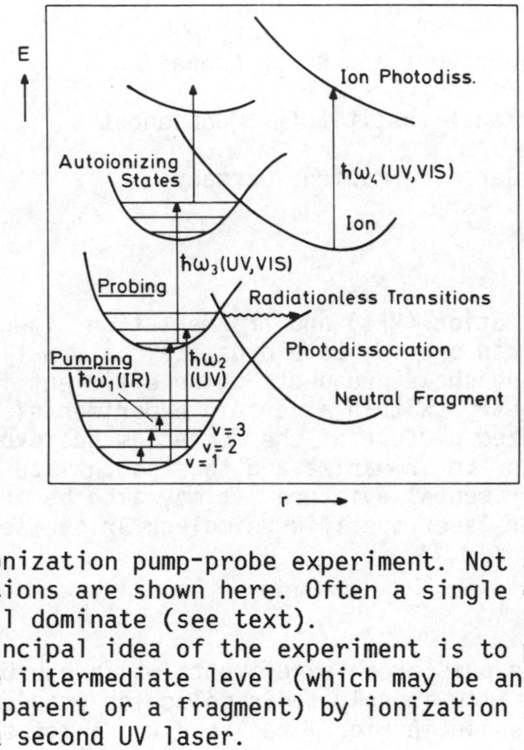

Fig. 1: Photoionization pump-probe experiment. Not all the possible transitions are shown here. Often a single excitation channel will dominate (see text).
The principal idea of the experiment is to probe the decay of some intermediate level (which may be an electronic level of the parent or a fragment) by ionization spectroscopy using a second UV laser.

particles in this scheme should be noted. Complications may arise if the ions formed undergo successive secondary absorption and photodissociation phenomena. On the other hand this may be desirable if the photochemical behaviour of such ions is of interest, too. Very important here is the spectral structure (and perhaps also the polarization) of the molecular absorptions which often may open the possibility to distinguish different absorbing species in a mixture, if tunable lasers are used for the excitation.

It is obvious that a mechanistic interpretation which can possibly be applied to the whole manifold of polyatomic molecules is desired here. To this end several schemes have been discussed in the literature [1] which can briefly be summarized with reference to Fig. 1.

There are basically three possibilities, namely

(1) Neutral photodissociation $h\omega_1$, $h\omega_2$ followed by ionization of fragments (not shown in the figure).
(2) Molecular photoionization followed by ion photodissociation, $h\omega_1$, $h\omega_2$, $h\omega_3$, $h\omega_4$.
(3) Direct multiphoton excitation of sequential autoionizing levels followed by fragmentation into ions and neutrals.

II. Some experimental results

a) There are documented cases for all three MPI fragmentation routes, mentioned before in the literature [1-11]. Scheme (1) applies for instance to the KrF laser induced photoionization of acetaldehyde versus butyraldehyde [4]. The comparison of these two compounds is interesting because for CH_3CHO the ionization potential is 10.22 ± 0.03 eV and thus slightly higher than the energy of two 2485 Å KrF photons while for $CH_3CH_2CH_2CHO$ the corresponding energy is 9.80 ± 0.07 eV and therefore in reach for 2-photon ionization. One can clearly see the effect of this small energy difference in the MPI mass spectra of the two aldehydes. Without any detailed discussion some results are reproduced in Fig. 2. The results imply that the resonant intermediate state reached by the first photon n,π transition has a short life time in comparison to the rate of excitation. Thus a 2-photon excitation can win or at least compete with the rate of photodissociation as is the case for butyraldehyde.The respective 3-photon excitation rate is too slow under these conditions as the absence of the parent molecular ion peak for acetal-dehyde indicates. There is no direct determination of the intermediate state lifetime possible from these observations and ref. 4 states only that the n,π decay must occur in a time shorter than the laser pulse duration (20 nsec), a result already known from other studies [7].

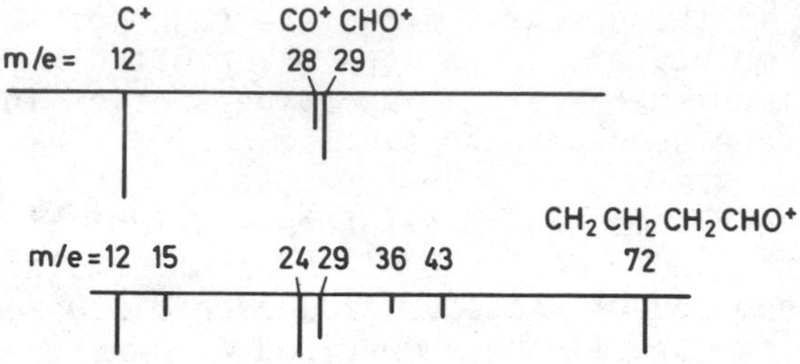

Fig 2: Time-of-flight KrF laser photoionization mass spectra of acetaldehyde (upper picture) and butyraldehyde (lower picture). The spectra as displayed here are strongly laser intensity dependent and relate only to a restricted experimental parameter range. The mass peaks at m/e = 12, 28 and 29 are clearly assignable to C^+, CO^+ and HCO^+ in the CH_3CHO spectrum. The parent molecule peak at m/e = 44 never appears as the ionization potential exceeds the energy of two KrF photons. This situation changes in going to $CH_3CH_2CH_2CHO$ (lower picture) with obvious interpretation.

b) As stated above another mechanism for MPI/MPF involves the rapid stepwise absorption up an autoionization ladder of states above the ionization potential. This has been claimed to apply for instance to the photoionization of hydrogen sulfide, H_2S [11]. A mechanism involving dissociation of parent H_2S^+ ions by absorption of three or more additional photons could not be completely ruled out, however. So far this example seems to stand out as an exception among all the known cases of UF laser MPI.

c) The third mechanism involves ionization followed by ion fragmentation. Benzene is the most thoroughly studied case here, although even for this molecule many important questions remain unanswered.

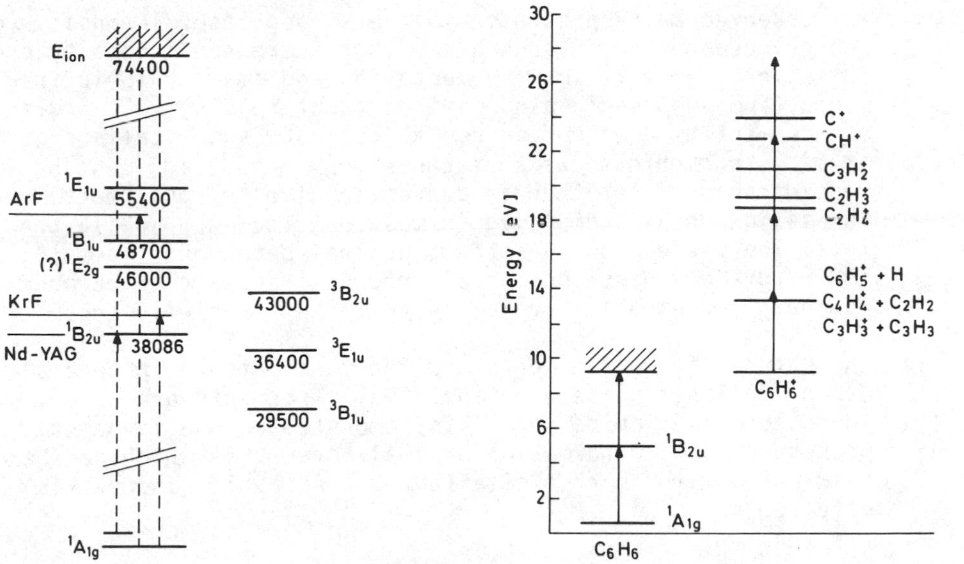

Fig. 3: Ladder switching mechanism for UV laser induced multiphoton ionization exemplified here for the benzene (C_6H_6) molecule.

Some of the main findings which are characteristic of many polyatomic molecules may be listed as follows:

(1) The ion yield can be very high (in the case of benzene it may approach 100%). It is of course dependent on resonance enhancement by intermediate states and on the transition moments relating the relevant states.

(2) There is a spectral dependence in the ionization yield which very closely resembles the groundstate one-photon absorption spectrum. This shows that the higher absorption steps leading into the ionization continuum occur with sufficiently large cross section. An ionization spectrum (for a limited spectral range) of benzene is displayed in Fig. 4 for illustration of this point. This spectral dependence makes this a selective detector for molecules in combination with a mass analyzer.

(3) The observed mass spectra exhibit a strong laser intensity (or energy) dependence in such a way that increased intensity produces more and more ion fragmentation and smaller ionic fragments (and supposedly also neutral counterparts). A calculation of the minimum energy required to generate such small (e.g. atomic) fragments shows that the energy deposition must be considerably higher than in conventional electron impact ionization usually employed in mass spectrometry. While the total ion yield shows a strong spectral dependence (see [3]) the nature and distribution of fragments appears to be predominantly controlled by intensity rather than wavelength.

(4) The nature of the fragments and the mass pattern in general depend on laser pulse duration. Small fragments are absent for psec pulse ionization. This demonstrates that dynamic processes lead to formation of small fragments which can be eliminated with short excitation pulses. Fig. 5 shows this effect.

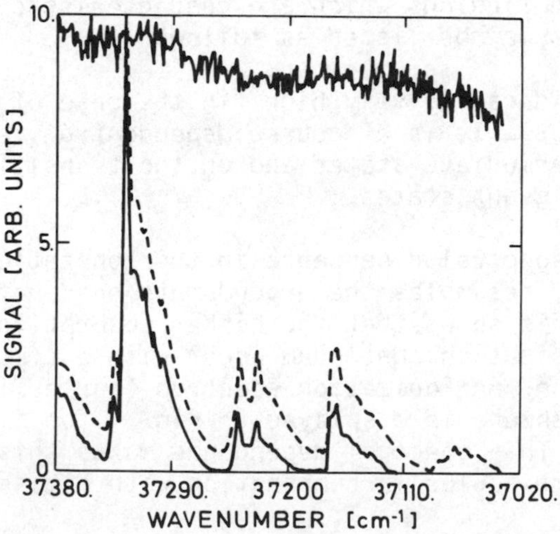

Fig. 4: Groundstate absorption and ionization spectrum of benzene. Dashed curve $S_0 \rightarrow S_1$ absorption, solid curve $S_0 \rightarrow S_1 \rightarrow$ ionization, top trace UV laser energy [12].

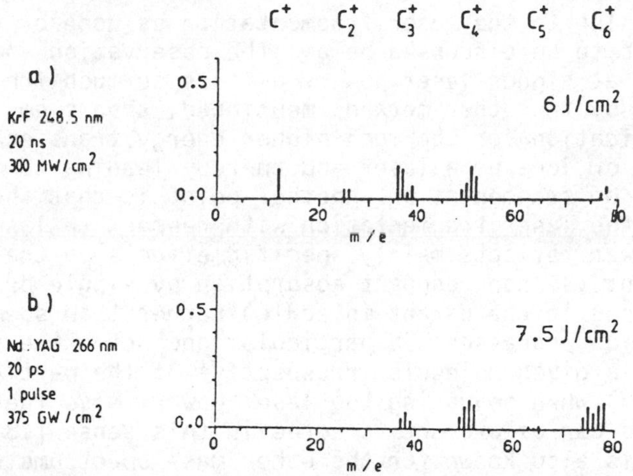

Fig. 5: Benzene MPI mass spectra with different laser pulse lengths.
a) TOF mass spectrum as obtained by ref. [5] with a 20 nsec (300 MW/cm²) KrF laser pulse (λ = 248.5 nm) showing fragmentation down to atomic ions.
b) Same mass spectral measurement, however, with 20 psec Nd-YAG laser pulse (375 GW/cm², λ = 266 nm). Small fragments are now absent indicating that the total timescale of fragmentation is longer than the laser pulse duration: Thus the sequence of absorption/fragmentation events may be stopped now at some intermediate level [6].

An interpretation and the formulation of a general model has then be attempted by various authors as discussed in the following section.

III. Statistical Models for MPI Fragmentation

Two statistical models have been developed to predict overall and detailed features of MPI fragmentation patterns (FP) [13-16]. These models involve only a minimal set of input data (e.g. heats of formation, partition functions, densities of states) and require a low computational effort. They are flexible enough to be applied to large classes of molecules and by this one can hope to put the rapidly increasing wealth of experimental information on a common basis.

One can argue that the experimental FP reflect mainly statistical features and only to a lesser extent specific dynamical effects. Thus similar trends in the kind of species and their intensities are observed in low-power laser fragmentation as in the case of other methods like electron impact mass spectrometry, charge transfer ionization or photoionization. As in these cases it should therefore be possible to apply the statistical theory of mass spectra [17] also to the laser fragmentation as done by the AMF and MAF models to be discussed below. The observation that laser fragmentation at higher laser powers differs so much from the fragmentation by the other methods mentioned, then seems to be simply an indication for the much higher energy transferable to the molecules or ions by a laser and thereby leading to a much higher degree of fragmentation. Another point is that the apparent specifity in the laser fragmentation with respect to laser wavelength and power reflects mainly specific effects in the early steps (resonant vs. non-resonant absorption by single or multiphoton processes in the parent molecule) rather than specifity in the dissociation processes. In particular one notes the similarity of the FP for a given molecule irrespective of the particular laser (visible or UV) when adjusting the laser power. Also the FP of different isomers can be brought to agree in this sense [18,19], a feature that is also known for the other mass spectrometric methods (where it is assigned to rapid energy and bond scrambling in polyatomic ions).

Statistical models can be used as a reference against which experimental FP can be compared. Deviations then signify specific dynamical effects needed in addition. This latter idea is pursued in the maximum entropy model [13] that has also been termed as the extreme statistical limit for the FP since it involves no assumptions on the mechanism leading to fragmentation.

a) The Maximum Entropy Model [13,14]

The most probable distribution for the particle numbers X_j of the species j and the probabilities x_{ij} of the particular level i is obtained by maximizing the entropy S in an ensemble of independent systems

$$S = - R \sum_j X_j \ln X_j - R \sum_j X_j \sum_i x_{ij} \ln x_{ij} + R\, X \ln X$$

Here $X = \Sigma X_j$ is the total number of particles and the maximum is searched subject to the constraints of conservation of the atoms constituting the species, total charge and probability $\Sigma_i x_{ij} = 1$. Besides it is assumed that the system is specified by a mean energy $\langle E \rangle$ absorbed from the laser. This latter constraint introduces a temperature $T = 1/R\beta$. The resulting particle numbers are then given by

$$X_j = X\, Q_j(\beta) \exp\left(\sum_k \gamma_k a_{kj} \right)$$

in terms of the partition functions Q_j. The γ_k are the Lagrange multipliers corresponding to atomic number and charge conservation and a_{kj} is the number of atomic sort k in the stoichiometric formula of j. With this one obtains also the FP as a function of $\langle E \rangle$ that will be monotonically related to the absorbed laser energy. Successful applications have been given for molecules like benzene, toluene, t-butylbenzene and tri-ethylenediamine [14].

b) The Absorption/Multiple Fragmentation (AMF) and Multiple Absorption/Fragmentation (MAF) Models [15,16].

Certainly it is also of interest to look in more detail into possible mechanisms for MPI fragmentation. In the AMF model one assumes that the primary action of the laser is the ionization of the parent mole cule by molecule and laser specific absorption processes. The important step for the following laser fragmentation is the second stage where the parent ions

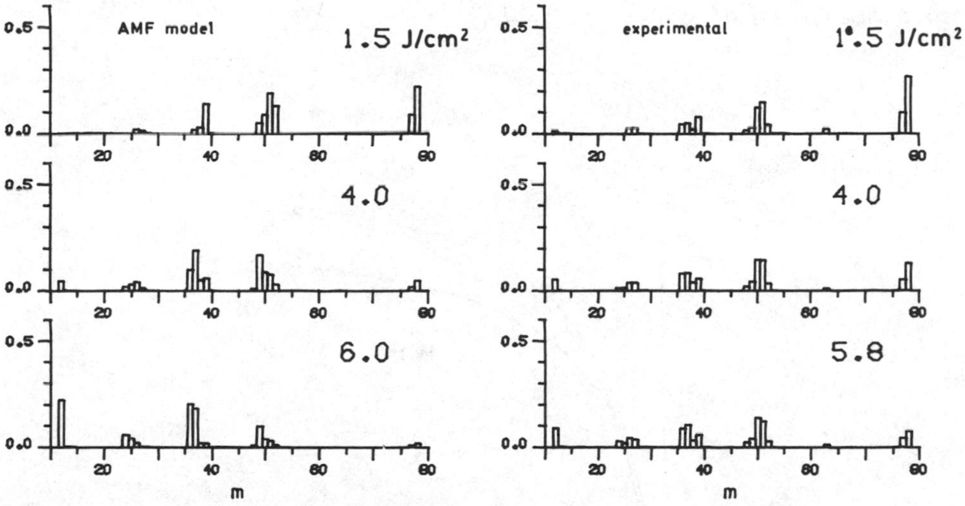

Fig. 6: MPI fragmentation patterns of benzene for excitation by a KrF-laser at various fluences (J/cm^2): a) AMF model calculations [15]; b) Experimental results [5].

receive further energy with rapid conversion of electronic to vibrational energy by radiationless transitions. Ions with sufficient excess energy then undergo unimolecular decay. The fragment mass distributions are then calculated using phase space considerations for a chain of multiple parallel and consecutive branchings. Since most ionic fragmentations seem to involve loose transition complexes these calculations can be performed on the basis of the products phase space model, by which all fragmentation rates, branching ratios and internal energy distributions are expressed in terms of the heats of formation and the densities of states of the fragments. Within this approach the quantitiy of principal interest is the probability $P_{ij}(E_i,E_j)$ to obtain a fragment ion j^+ with energy E_j in a single fragmentation step with a barrier E^o_{ij} from i^+ with energy E_i

$$P_{ij}(E_i,E_j) \propto \alpha_{ij}\, \rho_j(E_j)\, \rho_{ij}(E_i-E^o_{ij}-E_j)$$

Here α_{ij} is the reaction path degeneracy, and ρ_j, ρ_{ij} are the internal state densities of j^+ and the complementary neutral (including the translational state density), respectively. Fig. 6 gives a comparison between calculated and experimental FP. The model allows to make also other predictions, e.g. the prominent route for formation of C^+ should be the sequence

$$C_6H_6^+ \rightarrow C_4H_4^+ \rightarrow C_4H_3^+ \rightarrow C_4H_2^+ \rightarrow C_4H^+ \rightarrow C_3H^+ \rightarrow C_3^+ \rightarrow C^+$$

Also the kinetic and internal energies of ions formed by series of consecutive fragmentations will differ significantly in contrast to a mechanism of parallel fragmentations [20].

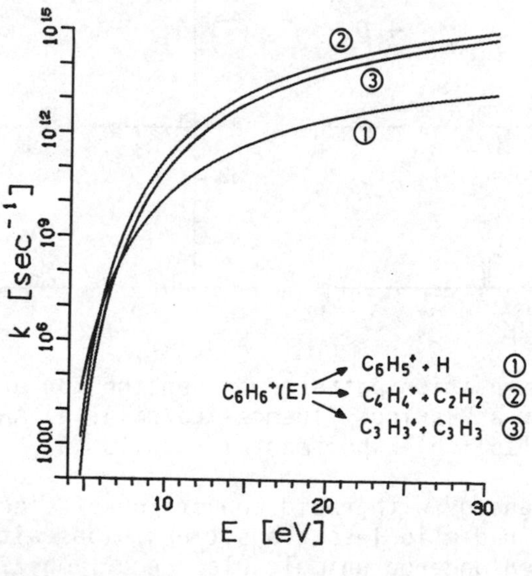

Fig. 7: RRKM fragmentation rates versus excess energy of $C_6H_6^+$ [16].

Calculations with the AMF model for benzene show that rather large excess energies must be deposited in the parent ion to obtain the smaller fragments (e.g. > 40 eV for C^+). This is mainly due to the large amount of energy obtained by the neutral fragments in the first fragmentation steps. On the other hand one concludes from RRKM rates (Fig. 7) that the decay rates of parent ions with excess energies much above 10 eV become so large that further absorption before fragmentation is unlikely for typical laser powers. Therefore absorption by secondary fragments will also occur and is necessary to overcome the energetical and kinetic barriers for the smaller fragments. The mechanism then involves multiple absorption/fragmentation steps. A quantitative study of such an MAF model requires additional data on ion absorption coefficients which are not available.

IV. Conclusion

The agreement between experiment and theory is surprisingly good considering the uncertainties in both the experimental (fragment energies, detailed dynamics) and theoretical (heats of formation, fragment spectra) input data. Obviously more work is needed to test the general validity of the model descriptions. Future practical applications of this ionization technique in mass spectrometry are conceivable. Principally it is possible with high flux laser sources to achieve very large excitation rates which may open the way to higher and more specific dissociation channels than are normally encountered in more traditional photochemical experiments.

References:

1) V.S. Antonov and V.S. Letokhov, Appl. Phys. 24, 89 106, 1981
2) L. Zandee and R.B. Bernstein, J. Chem. Phys. 71, 1359, 1979
3) U. Boesl, H.J. Neusser and E.W. Schlag, J. Chem. Phys. 72, 4327, 1980
4) J. P. Reilly and K.-L. Kompa, Advances in Mass Spectr. Vol. 8, 1979
5) J. P. Reilly and K.-L. Kompa, J. Chem. Phys. 73(11), 5468, 1980
6) P. Hering, A.G.M. Maaswinkel and K.-L. Kompa, Chem. Phys. Lett. 83, 222, 1981
7) B. A. Heath, M.B. Robin, N.A. Kuebler, G.J. Fisanick and T.S. Eichelberger IV, J. Chem. Phys. 72, 5565, 1980
8) G.J. Fisanick, T.S. Eichelberger IV, B.A. Heath and M.B. Robin, J. Chem. Phys. 72, 5571, 1980
9) D. Proch, D.M. Ridder and R.N. Zare, J. Chem. Phys. (in print)
10) M. Seaver, J.W. Hudgens and J.J. DeCorpo, J. Mass Spectr. Ion Phys. 34, 159, 1980
11) T.E. Carney and T. Baer, J. Chem. Phys. 75, 4422, 1981
12) W.E. Schmid, G. Müller and K.-L. Kompa, to be published
13) J. Silberstein and R.D. Levine, Chem. Phys. Lett. 74, 6, 1980

14) J. Silberstein and R.D. Levine, J. Chem. Phys. 75, 5735, 1981
15) F. Rebentrost, K.-L. Kompa and A. Ben-Shaul, Chem. Phys. Lett. 77, 394, 1981
16) F. Rebentrost and A. Ben-Shaul, J. Chem. Phys. 74, 3255, 1981
17) G. Forst, Theory of Unimolecular Reactions, (Academic, New York 1973)
18) D. Lubmann, J. Chem. Phys. (in press)
19) I.W. Hudgens, M. Seaver and J.J. De Corpo, J. Phys. Chem. 85, 761, 1981
20) T.E. Carney and T. Baer (submitted to J. Chem. Phys.)

LASER SNOW EFFECT IN CS_2 VAPOUR

Krzysztof Ernst

Institute of Experimental Physics
Warsaw University
Hoza 69
Warsaw, Poland

The formation of micron-size particles in a gas system induced by laser radiation was reported for the first time by Tam, Moe and Happer [1] in 1975. In their experiment the argon laser beam passing through a gas mixture of caesium vapour, hydrogen and helium induced the formation of small white particles identified as caesium-hydride crystals. The observed phenomenon was called "laser snow." Since then the laser snow effect has been observed in several gases illuminated by different laser lines [2-11]. Due to chemical reactions induced by laser light in a given gas system and subsequent condensation and coagulation processes small particles of size varying from 0.5 μ to 4 μ are formed. The particles cause intense scattering of laser light and they can be easily recognized by eye in the form of falling snow. A typical velocity of fall is of the order of mm/s.

In the case of CS_2 vapour the formation of laser snow particles was observed with either 337 nm line of nitrogen laser [6] or 351 nm and 357 nm lines of krypton laser [7] used as a source of excitation. The experimental set-up consisted of the cyllindrical cell containing CS_2 vapour excited by the nitrogen (or krypton) laser along the axis of the cell. The kinetics of the formation process was detected by measuring the intensity of the scattered He-Ne laser light. The presence of particles could be verified by eye looking at the scattered He-Ne laser light passing through the cell. Such a scattering is shown on Fig. 1. The only difference between two photographs is that on the lower picture the central part of the cell has a bigger diameter than its lateral parts. These pictures show clearly that the production rate of the particles depends on the geometry of the experiment. It is due to convection process. The effect of convection will also be discussed later.

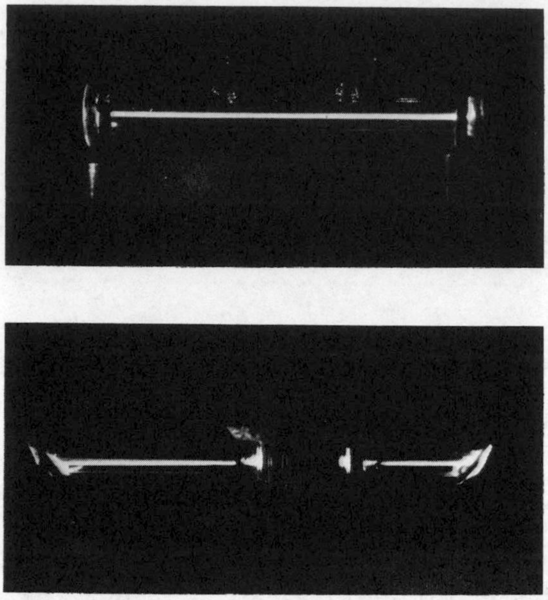

Fig.1 Scattering of the He-Ne laser beam by laser snow particles observed in two different cells.

One of the important features of the laser snow effect is the chemical reaction chain leading to the formation of the end product. The following sequence of reactions in the case of CS_2 vapour is proposed

$$CS_2 + h\nu \rightarrow CS_2^* \quad (1)$$

$$CS_2^* + CS_2 \rightarrow 2CS_2 \quad (2)$$

$$CS_2^* + CS_2 \rightarrow 2CS + S_2 \quad (3)$$

$$nCS \rightarrow (CS)_n \quad (4)$$

This process is possible due to the fact that laser lines used for excitation coincide with absorption lines belonging to one of two overlapping absorption bands (290-380 nm, 330-430 nm) of CS_2 vapour [12]. The process of particles formation is in fact associated with the fluorescence of CS_2 excited molecules. The violet lines (407, 413 and 415 nm) of krypton laser which could potentially coincide with one of lines belonging to the absorption band lying between 330 and 430 nm gave neither fluorescence nor the particles formation what confirmes that electronically excited CS_2 molecules are needed for the process of particles formation. In our experimental conditions the direct dissociation $CS_2 + h\nu \rightarrow CS + S$ is impossible.

The minimum photon energy needed for this process is equal to 4.46 eV (~230 nm) [13].

The aerosol particles falling down form a brown deposit on the bottom of the cell. In order to verify the proposed model the mass analysis of the deposited powder was performed. This analysis indicated the presence of CS, $(CS)_2$ as well as various forms of sulfur (S_2,..,S_8). Such a result agrees with the proposed model of chemical reaction chain leading to the particles formation.

As far as kinetics of the formation process is concerned it was established that laser snow particles can be created only above the treshold value of CS_2 vapour pressure depending on the intensity of laser beam and on the geometry of the experiment. The detailed analysis which takes in consideration diffusion and convection has been done [6,7].

The most interesting feature of the kinetics of particles formation was revealed by following temporal evolution of the intensity of scattered light [6]. One of recordings corresponding to the pressure of CS_2 vapour equal to 30 Tr is shown on Fig.2. The oscillatory character of the temporal dependence of the scattered signal is well pronounced. The positions of the maxima and their values depend on the CS_2 vapour pressure. With increasing the pressure periods of oscillations become longer. This oscillatory behaviour of the scattered light may be explained qualitatively as follows. The laser

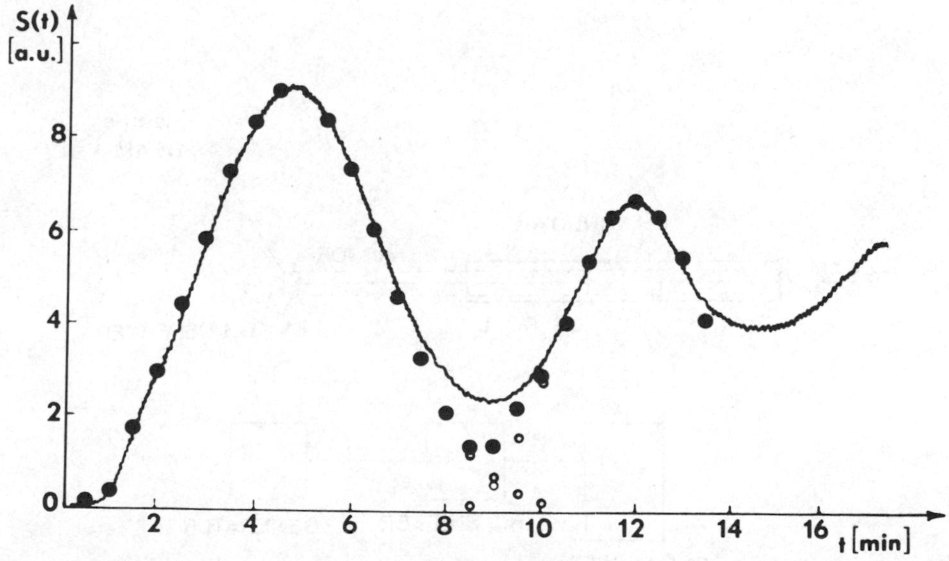

Fig.2 Temporal evolution of the scattered signal recorded for 30 Tr of CS_2 vapour pressure compared with theoretical prediction (solid points).

light induces a photochemical reaction with products giving rise to subsequent particles formation. As the particles grow in size due to condensation and coagulation the scattering signal increases. Since larger and heavier particles fall down faster they are removed preferentially from the production volume which corresponds to a decrease of the signal. When the largest particles are removed the small ones start to grow giving rise to a new cycle. It has to be emphasize that no nucleation of new particles takes place in the presence of large particles already formed which act as sinks for molecules created by photochemical reaction.

The quantitative model for one cycle of oscillations was given too [6]. It is based on three following assumptions: (i) decrease of particles concentration depends on the velocity of particles fall, (ii) the volume of the particle grows with a rate proportional to its surface area, (iii) intensity of the scattered light is proportional to a total surface of particles. In Fig.2 solid points represent the theoretical prediction of the equation describing the time dependence of the scattered light intensity for two cycles of oscillations. The agreement with experimental results is very good except close to the minimum. This can be explained by noting that in our theoretical model all particles are removed from the scattering volume at the end of the first cycle what is not satysfied experimentally. In order to verify our model explaining the oscillatory character of the scattered signal the temporal evolution of particle

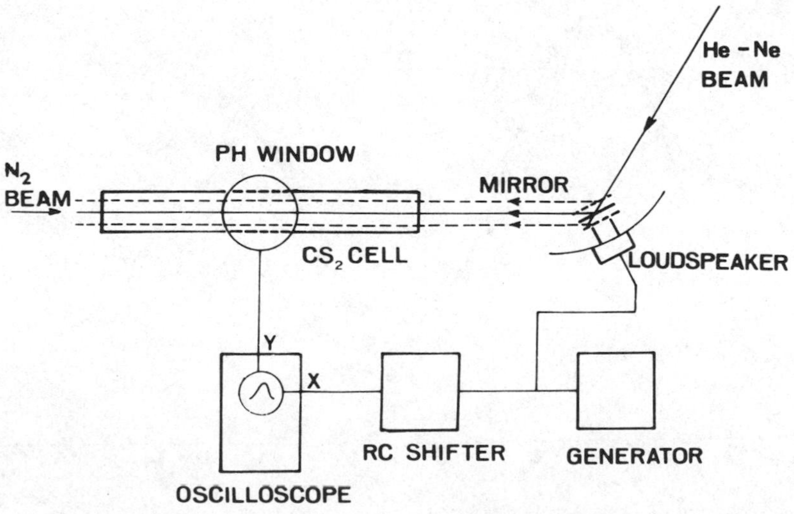

Fig.3 Experimental set-up for measuring the spatial distribution of particles concentration.

size was studied by measuring the angular distribution [6] and decay times [14] of the scattered light intensity. These measurements showed in fact that size of particles is also oscillating in time with an average size of the order of 1 μ.

The proposed model is principally based on the fact that oscillations are due to periodic falling down of particles. The most direct evidence for this can be obtained by measuring the spatial distribution of particles concentration and its temporal evolution. Such an experiment was performed [15] using a simple set-up shown on Fig.3. The He-Ne laser beam used as a monitoring beam was reflected by the mirror attached to the membrane of the loudspeaker. The aim of such a system was to transfer the membrane vibrations into vibrations of the beam along the vertical diameter of the cell. The intensity of the scattered light in the direction perpendicular to the plane of the beam oscillations was detected using a photomultiplier. The signal from the photomultiplier was applied to the Y plates of the oscilloscope while the X plates were connected to the acoustic generator. Such a system allowed to observe the spatial distribution of aerosol particles and to follow its temporal evolution directly on the oscilloscope screen. The results are shown on Fig.4. A sequence of 16 pictures is chosen with the time interval between two consecutive ones equal to 1 min. On each of the pictures the intensity of the signal is recorded as a function of the height in the cell. The central part of the picture corresponds to the signal received from the volume around the cell axis. The right hand part of it corresponds to the lower part of the cell. Following the sequence of pictures we can see that the signal increases first inside the excitation volume close to the axis of the cell. As the particles fall down the maximum of the signal moves toward the bottom of the cell. Finally, the maximum decreases because the particles are being removed from the detection volume. When a minimum is reached the signal starts to increase again close to the axis of the cell giving rise to the second cycle.

Oscillatory behaviour of the signal scattered by laser snow particles was also observed in UF_6 [16] and in the mixture of caesium vapour and deuterium [2] excited by a dye laser. In the latter case, however, period of oscillations was of the order of 1 s only.

Laser snow being itself an interesting object of studies can also offer various applications. To give an example let us consider a typical experimental set-up with a cylindrical cell containing CS_2 vapour illuminated by nitrogen laser beam along the cell axis. In such a situation the heating of gas by light absorption in the volume illuminated by laser light and concentrated along the axis of the cell leads to the temperature and density gradient pointing outwardly from that volume. Due to buoyancy force acting on the heated gas the convection pattern in the form of two rotating rings is observed. This pattern was visualized using a He-Ne laser beam. Its cross-section was transformed by means of a set of lenses to the rectangular

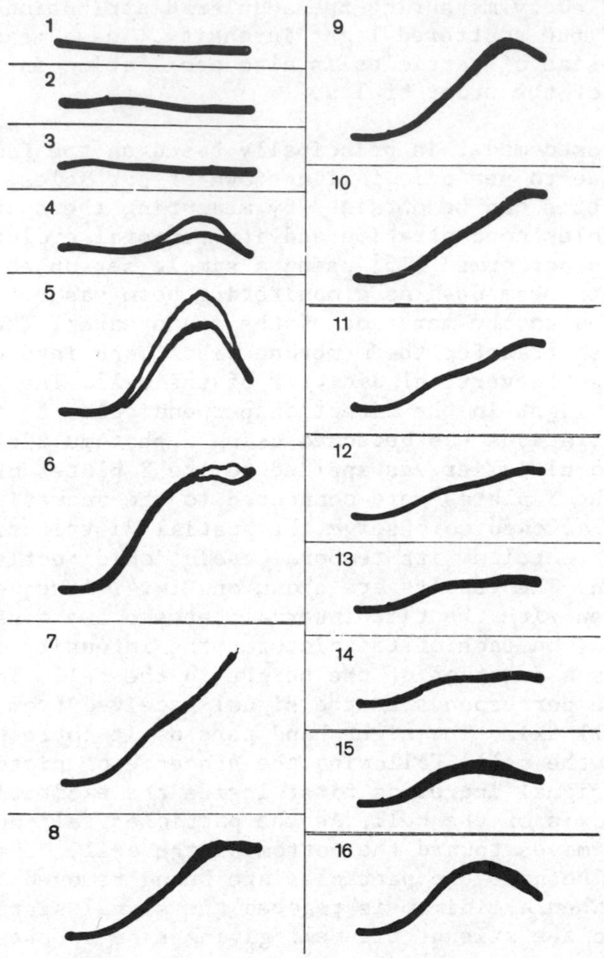

Fig.4 Sequence of recordings showing the temporal evolution of the intensity of the scattered signal as a function of the height in the cell. Double trace on some pictures is due to instabilities of generated frequency.

form with the thickness of about 0.2 mm. The beam intersected the cell perpendicularly to its axis allowing to visualize the convective pattern corresponding to the intersection plane (Fig.5). In such a configuration the problem of convection instability was investigated [17]. It was found that convection pattern in the form of rings was observed only above a treshold value of the laser pulse energy depending on the gas pressure.

The experiment described above shows that the laser snow effect can be conveniently used for visualizing and monitoring various phenomena occuring in gas systems.

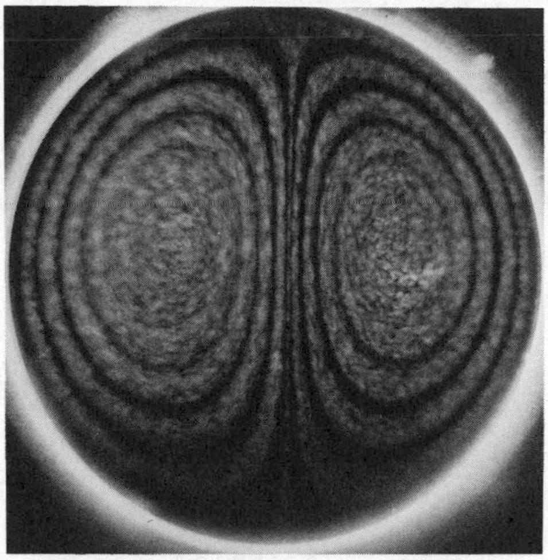

Fig.5 Convection pattern visualized by means of laser snow particles.

References

[1] A.Tam, G.Moe, W.Happer, Phys.Rev.Lett., 35,1630(1975).
[2] A.Tam, W.Happer, D.Siano, Chem.Phys.Lett., 49,320(1977).
[3] J.Picque, J.Verges, R.Vetter, J.Phys.Lett., 41,L305(1980).
[4] K.Iwamoto, N.Presser, J.Ross, J.Chem.Phys., 68,663(1978).
[5] S.Lin, A.Ronn, Chem.Phys.Lett., 56,414(1978).
[6] K.Ernst, J.Hoffman, Chem.Phys.Lett., 68,40(1979).
[7] N.Beverini, K.Ernst, M.Inguscio, F.Strumia, Appl.Phys., B26, 57(1981).
[8] A.Ronn, B.Earl, Chem.Phys.Lett., 45,556(1977).
[9] E.Borsella, F.Catoni, G.Freddi, J.Chem.Phys., 73,316(1980).
[10] T.Yabuzaki, T.Sato, T.Ogawa, J.Chem.Phys., 73,2780(1980).
[11] S.Chin, Can.J.Chem., 54,2341(1976).
[12] G.Gattow, W.Behrendt, in: Topics in Sulfur Chemistry, vol.2, ed. A.Senning (Thieme, Stuttgart, 1977).
[13] S.Silvers, M.NcKeever, G.Chawla, in: Advances in Laser Chemistry, ed. A.Zewail (Springer, Berlin, 1978) p.449.
[14] K.Ernst, J.Hoffman, Proc. IX Conf. on Quantum Electronics and Nonlinear Optics (Poznan, April 1980).
[15] K.Ernst, J.Hoffman, Chem.Phys.Lett., 75,388(1980).
[16] E.Borsella, F.Catoni, G.Freddi, raport Comitato Nazionale Energia Nucleare nr RT/FI(79)23.
[17] K.Ernst, J.Hoffman, Phys.Lett. (to be published).

AN INTRODUCTION TO PICOSECOND SPECTROSCOPY

Alfred Laubereau

Physikalisches Institut der Universität Bayreuth
Bayreuth, W. Germany

I. INTRODUCTION

Intense laser pulses with a time duration of several 10^{-12} s were first observed in 1966.[1] It was immediately recognized by investigators in the field that these picosecond light pulses would allow direct investigations of ultrafast molecular processes in condensed phases. In fact, the first application of these pulses to a study of radiationless transitions of dye molecules was reported soon after.[2] Numerous investigations were conducted by several laboratories on a wide variety of problems in recent years.[3] Considerable effort has been spent on the reproducible generation of well-defined pulses. As a result, quantitative investigations can be carried out on the time scale of picoseconds and subpicoseconds with carefully analysed pulses.

In the first section of this paper we discuss briefly the generation of picosecond pulses in mode-locked lasers. Special emphasis is laid on passive mode-locking in solid state systems. The subsequent sections are devoted to the applications of picosecond light pulses to relaxation phenomena of large molecules and to time resolved vibrational spectroscopy of small molecules and of solids.

II. GENERATION OF PICOSECOND PULSES

We first discuss a model of the pulse generation in a passively mode-locked laser. This model was first presented by Letokhov.[4] It reveals the inherent limitations

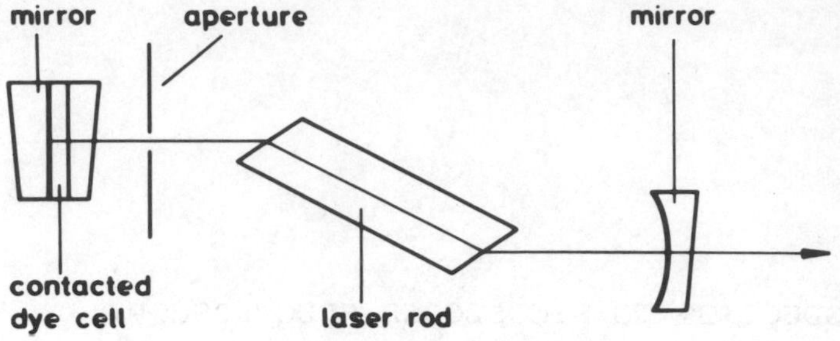

Fig. 1. Schematic of a passively mode-locked laser.

of the mode-locking technique and accounts for various difficulties, e.g. production of satellite pulses, background radiation and lack of reproducibility which may plague the experimenter.

In Fig. 1 the setup of a mode-locked laser is depicted schematically. The three essential components are: the optical cavity, the laser medium and the dye cell containing the nonlinear absorber. A solid state laser material is considered here, e.g. Nd:glass or Nd:YAG. Concave mirrors reduce the stringent requirements of stability of the laser cavity and facilitate optical alignment. A small aperture achieves pure TEM_{oo}-mode radiation. Longitudinal mode selection should be carefully avoided with the help of Brewster-angles at the laser rod and wedged plates for dye cell and laser mirrors.

Build-up of an ultrashort laser pulse proceeds in several steps. At the start, spontaneous fluorescence at the laser transition is amplified by the optically pumped laser rod. On account of the fluorescence bandwidth $\Delta\omega$ a large number of longitudinal modes is initially excited. For a typical cavity length of $\sim$ 1.5 m one calculates $\sim 6\times10^4$ cavity modes in Nd:glass with $\Delta\omega/2\pi c \approx 200\ cm^{-1}$. For Nd:YAG ($\Delta\omega/2\pi c \approx 6\ cm^{-1}$) one estimates 2×10^3 cavity modes. During this initial stage, population inversion of the laser transition and the transmission of the bleachable dye are not affected by the radiation. As a result, linear amplification of the fluorescence light takes place for numerous cavity round trips during this first stage.

Interference of the various laser modes with random phase relations leads to fluctuations of the light intensity. The spectral width $\Delta\omega$ of the fluorescence implies a duration of the fluorescence peaks of approximately $\Delta\omega^{-1}$. For Nd:glass (Nd:YAG) a value of 10^{-13} (3×10^{-12})s is estimated at the beginning of the linear amplification.

Although the total number of fluctuations is large, there is only a small number of intensity peaks exceeding the average intensity significantly. Calculations show that the most intense fluctuations exceed the average level by a factor of $\sim \ln m$, where m is the number of longitudinal modes excited.[5] This point facilitates the selection of one intensity peak (or at least of a small number) in the subsequent nonlinear stage.

The following numbers are typical for the linear amplification process of the Nd:glass laser: assuming a net gain of several per cent per cavity round trip at the centre of the laser line, the linear stage comprises $\sim 10^3$ cavity transits necessitating a build-up time of 10^{-5} s. The light intensity rises by many orders of ten to approximately 10^7 W cm^{-2}. The initial peak duration of 10^{-13} s increases to values of several picoseconds due to gain narrowing.

When the intensity peaks in the laser cavity approach values of the saturation intensity I_s of the nonlinear absorber, the further development of the laser radiation is determined by intensity-dependent cavity losses. In the vicinity of I_s the transmission increases, i.e. the dye becomes transparent. A second important parameter of the dye is the recovery time of the ground state population. For Nd-lasers, dyes with a recovery time of several ps and $I_s \sim 10^8$ W/cm^2 are available. The short relaxation time of the dye molecules is most desirable for the generation of picosecond pulses.

We note two significant processes acting together: (i) selection of one peak fluctuation (or of a small number). The most intense fluctuations which were built up during the linear amplification stage preferentially bleach the dye and grow quickly in intensity. The large number of smaller fluctuations, on the contrary, encounters larger absorption in the dye cell and is effectively suppressed; (ii) the fast dye relaxation tends to shorten the pulse duration and affects the shape of the light pulse. The wings of the pulse are more strongly absorbed than the peak. The pulse shape displays rapidly rising wings with a slight asymmetry on account of the finite relaxation time.

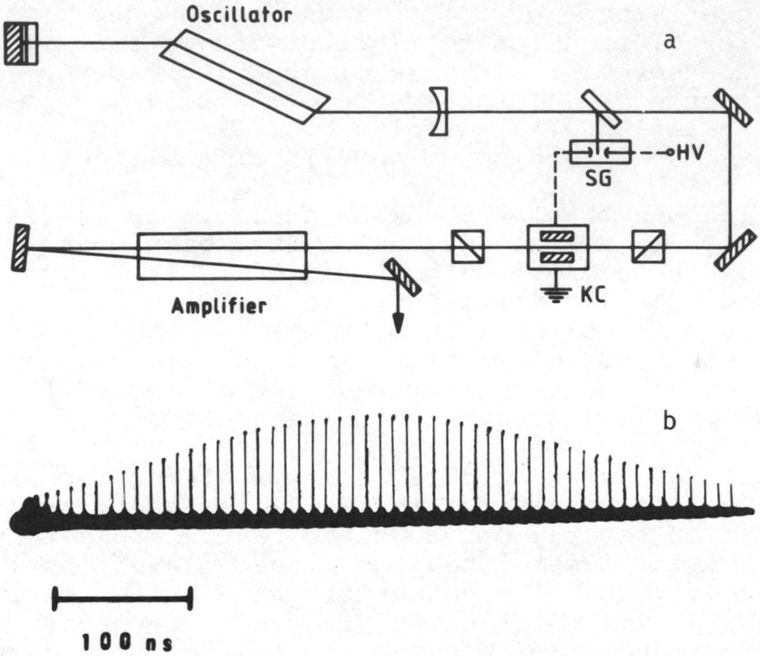

Fig. 2. a) Schematic of the mode-locked laser system with pulse selector and amplifier stage
b) Picosecond pulse train as observed with nanosecond time resolution. The missing pulse was selected and amplified.

The formation of light pulses in the laser cavity depends on the initial statistical intensity distribution. There is an 80 % probability for one intense fluctuation to exceed the following ones by at least 10 %. An initial difference of 10 % is sufficient for a discrimination of the corresponding light pulses by a factor of ten after multiple passes through the system. These numbers illustrate that theoretically we have a good chance for well mode-locked pulses.[5]

The generated pulse is rapidly amplified within a small number of cavity transits to an intensity level exceeding 10^9 W/cm^2. At this level various nonlinear effects occur which strongly effect the generated pulse. The nonlinear refractive index of the glass rod and of the dye solvent gives rise to self-phase modulation.[6,7] The dispersion of the laser components decomposes the frequency modulated pulse. Measurements of individual pulses as a function of position within the train clearly showed the change of pulse properties as a function of the number of cavity round trips.[8] Pulses positioned several hundred nanoseconds before the maximum of the pulse train display optimum properties.

The experimental system for the generation of single picosecond pulses by a Nd:glass laser is shown in Fig. 2a. The setup consists of the mode-locked laser oscillator, an electro-optic shutter to select a single pulse from the mode-locked pulse train and an amplifier system. A spherical cavity configuration and a contacted dye cell for the nonlinear absorber are used. The laser is pumped close to threshold for optimum performance. Fig. 2b shows an example of the pulse train detected with the help of a fast photodiode and a broad-band oscilloscope (total rise time ≈ 0.8 ns).

lasersystem	duration [ps]	tuning [cm^{-1}]	energy [J]	rep. rate [Hz]
Nd: glass	1 - 8	-	10^{-4} (10^{-2} - 1)	10 (0.5)
Nd: YAG	20 - 100	-	10^{-3} (10^{-1})	10 (5)
dye laser (passive, flash lamps)	2 - 5	1000	10^{-4} (10^{-3})	1 (1)
dye laser (passive, cw - Argon)	0.1 - 2	0 - 500	10^{-10} (10^{-3})	10^{8} (5)
dye laser (synchronous coupling)	1 - 20	2000	10^{-10} (10^{-4})	10^{8} (10^{2})
parametric generator	1 - 10	5000	10^{-4}	0.2 - 5

Fig. 3. Comparison of experimental systems for ultrashort pulse generation. The number in brackets refers to laser systems with amplifier stages.

For the selection of the single pulse we use a fast Kerr shutter which is operated by a laser-triggered spark gap.[9] A light pulse is cut from the leading part of the train envelope. Two passes through an Nd:glass rod (20 cm of length) provide an amplification by a factor of ∿ 200. Care has to be taken to avoid disturbing nonlinear effects in the Nd:glass, e.g. self-focusing and self-phase modulation.

A summary of the pulse properties obtained with several pulse generation devices is shown in Fig. 3. Two solid state laser systems and three dye laser configurations are considered: (i) passive mode-locking and flash lamp pumping, (ii) passive mode-locking and CW Ar laser pumping and (iii) synchronously coupled with a mode-locked Ar laser (see first column). Pulse generation by stimulated parametric amplification using picosecond pulses of a Nd:YAG or Nd:glass laser is included in Fig. 3. Typical numbers for the pulse duration, tuning range, single pulse energy and pulse repetition rate are listed in the following columns. The numbers in brackets in columns 4 and 5 refer to systems with amplifier stages. It is obvious from Fig. 3 that a unique system combining optimum numbers for all properties does not exist. For specific applications the laser system has to be carefully selected, according to the special needs of the experiment.

III. PICOSECOND SPECTROSCOPY OF LARGE MOLECULES

Substantial efforts have been made in recent years to study the dynamics of large organic molecules in liquid solutions. Fig. 4 presents schematically the energy levels and the various radiative and radiationless transitions of a dye molecule. Starting from the singlet ground state, absorption of visible light of suitable frequency promotes the molecule to the first excited singlet state with vibrational multiplicity. The large number of vibronic levels is represented in the Fig. by typical vibrational frequencies. There are several processes possible for the molecule on its way to the ground state: radiative decay by fluorescence with time constant τ_{fl}, or phosphorescence with time constant τ_{ph} (arrows) and radiationless transitions (wavy arrows), e.g. vibrational relaxation (lifetime τ_{vib}), intersystem crossing (τ_{ic}) and internal conversion τ_{int}. Additional processes involving solvent molecules are not included in the figure. The complex level diagram reveals the difficulties to separate individual processes.

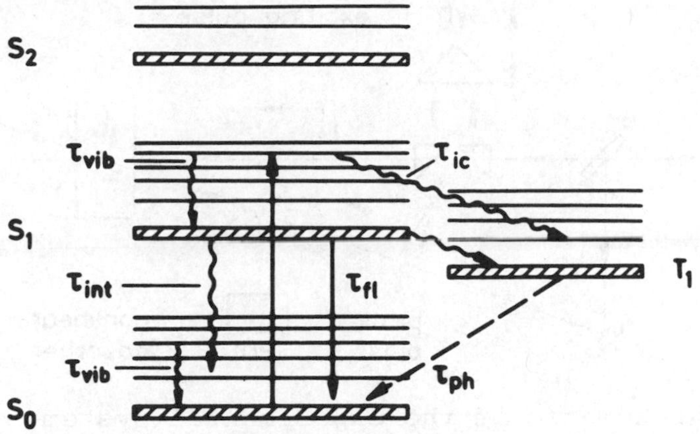

Fig. 4. Schematic of the energy diagram of dye molecules; horizontal bars represent the vibronic manifold of singlet and triplet states; radiative (straight lines) and radiationless transitions (wavy lines) are indicated.

In the first section of this article the importance of a short relaxation time of the nonlinear absorber for passive mode-locking was pointed out. This time constant is called ground state recovery time and is determined by internal conversion from S_1 to S_0 and by vibrational relaxation in the S_1 (and S_0).

The experimental setup measuring transient absorption changes is depicted schematically in Fig. 5.[2,10-12] The technique applied here gives a typical example for the two pulse method to excite and probe a sample as is commonly used in picosecond spectroscopy. An intense light pulse enters the dye cell and excites the absorbing molecules to a higher electronic state. For pulses of high peak intensity, a large number of molecules make an electronic transition and the ground state is notably depleted. A second weak pulse is generated by a beam splitter and serves as a probing pulse. It is properly delayed by a variable prism setup, shown in the Fig. Varying the geometrical path by a few mm, the delay time with respect to the pump pulse is varied by several ps. The probe pulse is directed into the sample and measures the instantaneous transmission of the dye cell. The probe intensity is set at a low level in order not to bleach the

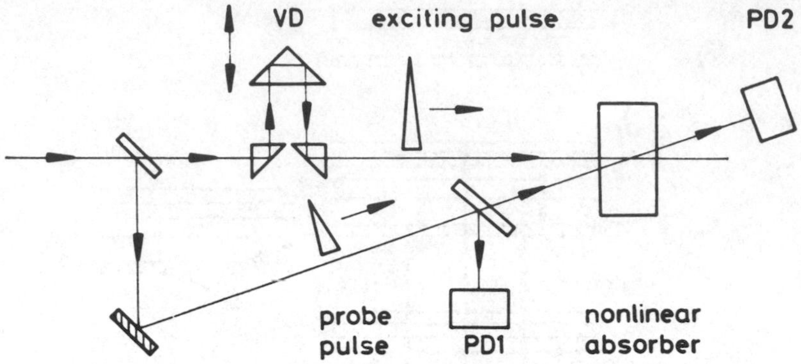

Fig. 5. Schematic of the experimental system to measure transient absorption changes for the determination of the ground state recovery time.

dye additionally. When the probe pulse follows immediately after the bleaching pulse, a reduced absorption is measured. Measuring the transmission as a function of the probe delay, the recovery of the ground state population is directly obtained. Several nonlinear absorbers currently used in Q-switch laser systems have been investigated.[10-13]

An example is presented in Figs. 6 and 7. The investigated molecule is the switching dye No. 15, the structure of which is depicted in Fig. 6a. The absorption spectrum of the molecule displays a strong band in the vicinity of the emission wave length of Nd-lasers at $\sim$ 1.06 μm (broken curve of Fig. 6b). The transmission of the probe pulse is plotted on a logarithmic scale in Fig. 7 versus delay time t_D between pump and probe pulses. $t_D = 0$ marks maximum overlap between the two pulses. The result for the ground state recovery time is $\tau = 4 \pm 1$ ps.[13]

Transient absorption changes may be also used to study intersystem crossing. The experimental setup is depicted in the upper part of Fig. 8. Part of the incident laser pulse (1060 nm) is converted to the 3rd harmonic at 353 nm by a two step process (second harmonic production and subsequent sum frequency generation) and used as excitation pulse. The second part of the laser emission is used for probing purposes. It produces a broad band emission in the visible ("picosecond continuum") via parametric four photon amplification and other nonlinear

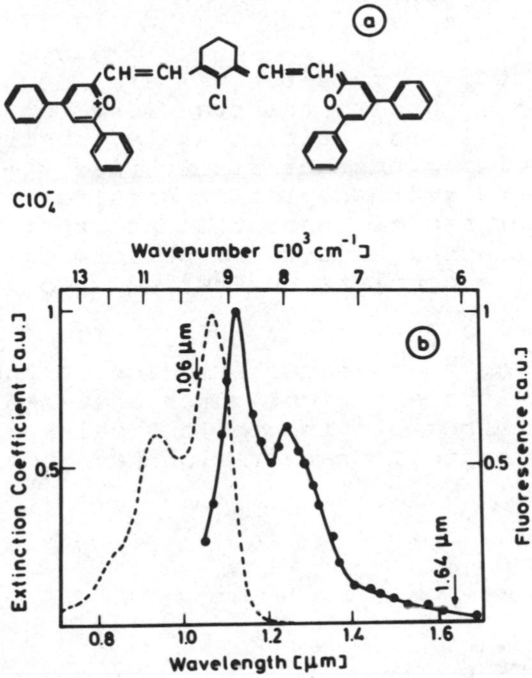

Fig. 6.
a) Investigated molecule no. 15
b) Fluorescence and absorption spectrum (after ref. 13).

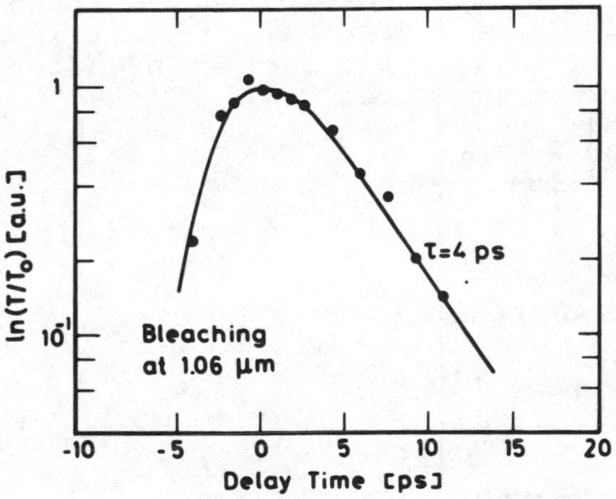

Fig. 7.
Transient bleaching of dye no. 15 versus delay time (after ref. 13).

processes.[14-17] The probe continuum allows transmission measurements in a vast spectral range with ∿ 10 ps time resolution.

Experimental results for trans-retinal are shown in the lower part of Fig. 8.[18] Due to the transient population of the bottom of the lowest triplet state, absorption increases are built up with a characteristic rise time of $\tau_{ic} \simeq 34.$ ps. Since ground state depletion simultaneously produces a transient decrease of absorption at other frequency positions, measurements have to be performed in a large spectral range for a complete understanding of the signal transients.

For investigations of spontaneous emission on the picosecond time scale, ultrafast light gates are highly desirable. Optical shutters employing the optical Kerr effect or the nonlinear optical transmission have been

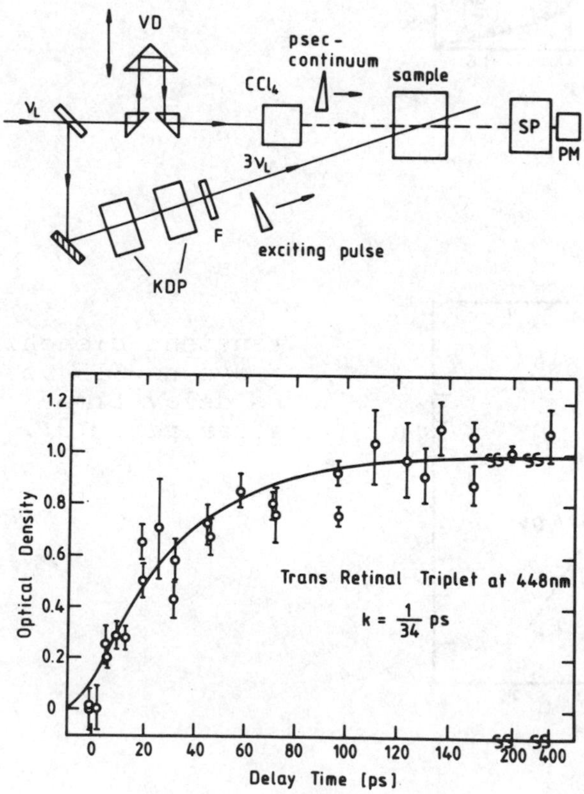

Fig. 8.
Top: Schematic of experimental setup to study intersystem crossing.
Lower part: Transient absorption of trans-retinal due to triplet formation (after ref. 18).

demonstrated to be practicable in this time domain. A versatile device is the ultrafast Kerr shutter which benefits from the rapid reorientation of anisotropic molecules in liquids.[19] A more recent device for ps light gating applies parametric frequency mixing in a nonlinear crystal. The experimental system is shown in Fig. 9.[20-22] When the operating pulse is present in the crystal, parametric sum (or difference) frequency generation of the investigated spontaneous emission of the sample produces new light which is detected by a photomultiplier. The specific advantage of the mechanism is its prompt response, since the parametric interaction is of electronic origin with a response time < 10^{-13} s. The fluorescence emission of the sample is induced by the frequency doubled laser pulse.

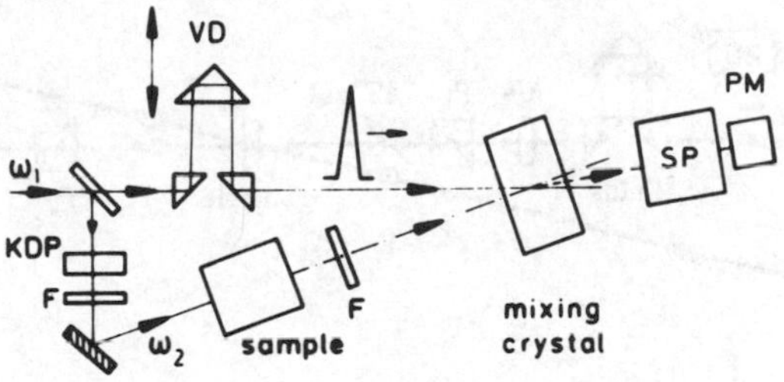

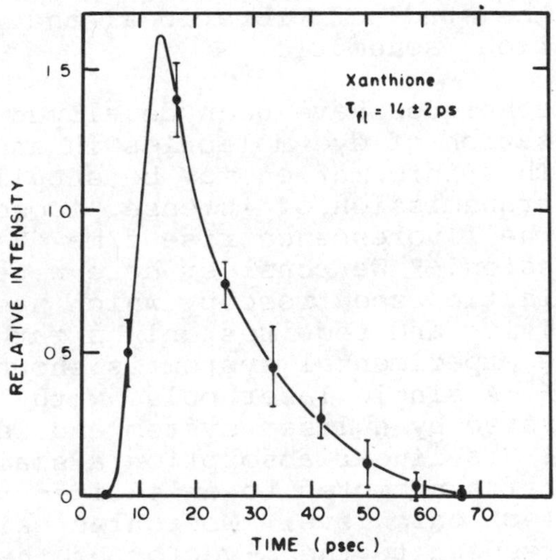

Fig. 9. Top: Schematic of experimental setup to study spontaneous emission with a ps light gate. Lower part: Fluorescence emission of xanthione versus delay time between excitation pulse and reference pulse (after ref. 21).

Data on the dye xanthione are shown in the lower part of Fig. 9. After the termination of the excitation process one observes a rapid decrease of the fluorescence emission; i.e. the excited state population decays with the fluorescence lifetime of $\tau_{fl} = 14 \pm 2$ ps.[21]

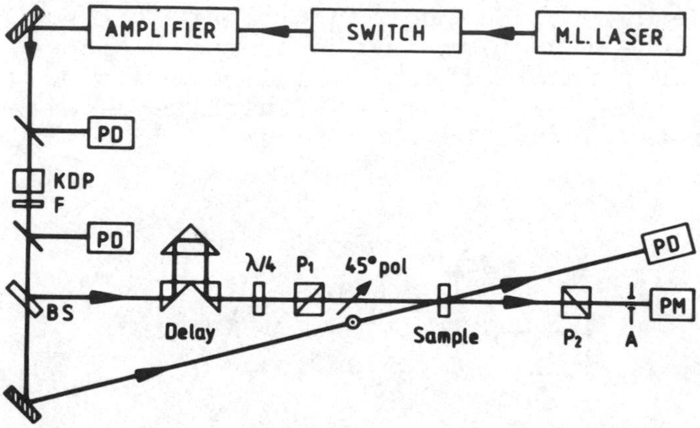

Fig. 10. Experimental setup to measure induced dichroism for the study of vibrational and rotational relaxation (schematic).

Several techniques have been developed to study vibrational relaxation of dye molecules in an excited electronic state. This information may be supplied by measuring the nonlinear transmission of intense picosecond pulses[23], by a study of the fluorescence rise time [24,25], or via stimulated emission.[26] We consider here a new technique based on polarization spectroscopy which has the advantage of high sensitivity and requires only a rather weak excitation.[27] The experimental system is shown schematically in Fig. 10.[27,28] A single laser pulse with linear polarization is generated by a laser system and serves as excitation pulse. Via linear absorption a small fraction of dye molecules is promoted to an excited vibronic state in the upper electronic level. Molecules being favourably oriented with respect to the E-vector of the light pulse, interact more intensively so that the rotational distributions of the molecules remaining in the ground state and of the molecules in the upper electronic state are not in equilibrium.

The subsequent vibrational and rotational relaxation is monitored by a linearly polarized probing pulse, the E-vector of which is rotated at 45° with respect to the excitation field. The amplitude of the probe pulse is measured behind a crossed polarizer as a function of delay time. If the excitation pulse is blocked, practically no signal is observed. With excitation process, the induced anisotropy of the molecular system leads to a probe transmission signal. The phenomenon is called induced dichroism. An example is shown in Fig. 11. The laser dye phenoxazone 9 in the solid solvent polystyrene is investigated.[27]

The probe signal rapidly rises to a maximum which is due to the transient population of excited vibronic levels and decreases rapidly by a factor of 2.5. A certain contribution to the peak around $t_D = 0$ also stems from a four wave mixing effect between pump and probe pulse. Comparison with calculations taking also into account excited state absorption yields $\tau_{vib} = 0.8 \pm 0.2$ ps.

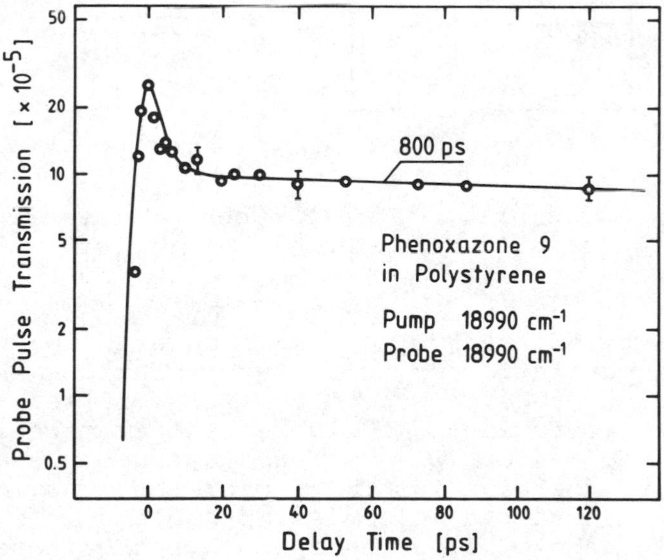

Fig. 11. Induced dichroism of phenoxazone 9 in solid polystyrene (after ref. 27).

The asymptotic time behaviour is very slow since rotational motion is strongly hindered in the solid solution and since the lifetime of the molecules at the bottom of the S_1 level is $\sim 10^{-9}$ sec. In liquid solutions, the asymptotic slope yields quantitative information on rotational relaxation times.[27-30]

For large molecules there is urgent need for new spectroscopic tools to supply additional information on these complex systems. A novel technique of this sort is the two-pulse-spectroscopy[31-33] with tunable ultrashort light pulses. The molecular transitions are depicted in Fig. 12.

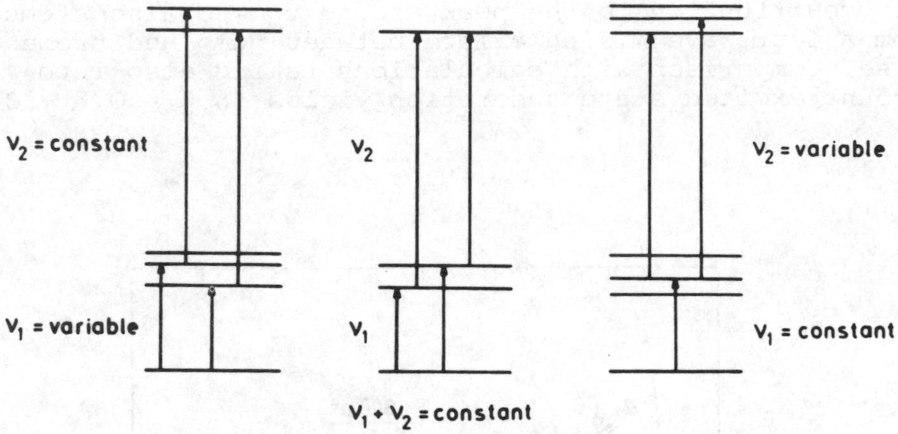

Fig. 12. Schematic of three types of investigations using the two-pulse spectroscopy.

A first infrared pulse (ν_1) generates an instantaneous excess population of molecular vibrations in the electronic ground state. A second (visible) pulse (ν_2) promotes the excited molecules to a fluorescent electronic state. Time delays of a few 10^{-12} seconds between the exciting and probe pulses give information on the dynamics of the vibrationally excited molecules. Tunable pulses allow additional kinds of spectroscopic investigations.

Tuning the infrared frequency ν_1 and holding the value of ν_2 constant, the excited vibrational states in the electronic ground state are changed and transitions to different Franck-Condon states in S_1 occur. The situation is simpler when the total energy $h\nu_1 + h\nu_2$ is held constant and when both frequencies are varied to investigate different vibrational states in S_o. In a third set of experiments the infrared frequency is held constant and the frequency ν_2 is tuned through the bottom of S_1.

In Fig. 13b, a two-pulse spectrum of nileblue-A-oxazone is shown.[31] The frequency of the first pulse was held at $\nu_1 = 2935\ cm^{-1}$ while the frequency of the second pulse was tuned from 15600 to 16100 cm^{-1}. The detailed two-pulse spectrum should be compared with the smooth conventional absorption spectrum of Fig. 13a, taken over the same total energy range. The structure of Fig. 13b is tentatively assigned to transitions involving mixed intermediate states consisting of skeletal modes with large Franck-Condon factors and of low frequency bending modes contributing large dipole moments. It is desirable to have a three-dimensional picture, signal versus ν_1 and ν_2 and versus time delay between the two pulses. In this way new information for the interpretation of peak positions and on energy redistribution processes is available.

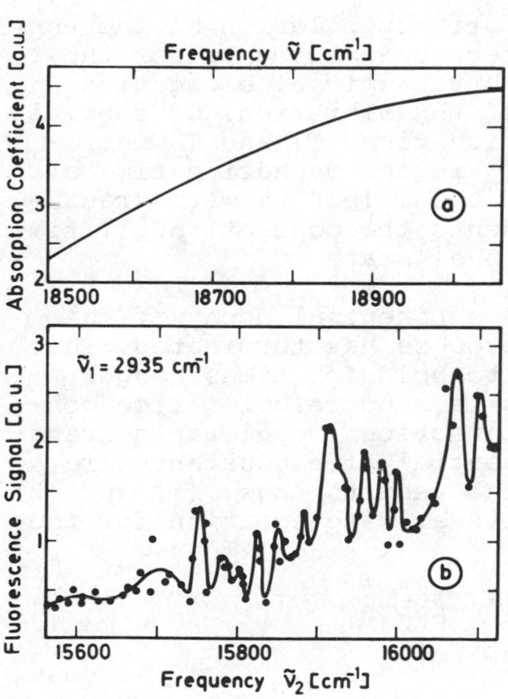

Fig. 13.
a) Part of the absorption spectrum of nileblue-A-oxazone in the visible.
b) Ultrafast two-pulse spectrum of nileblue-A-oxazone (10^{-4} M in CCl_4). The observed fluorescence is plotted versus the frequency of the visible pulse; the frequency of the infrared pulse is held constant at 2935 cm^{-1}. The abscissa in Figs. a and b allows a ready comparison of equal values of the total excitation (after ref. 31).

IV. VIBRATIONAL RELAXATION OF SMALL LIQUID MOLECULES AND OF SOLIDS

Theoretical estimates and experimental data of the infrared and Raman linewidths suggest that the relaxation times of normal modes of simple polyatomic molecules in liquids are of the order of picoseconds. With the introduction of ultrashort pulses and pulse probing techniques, direct investigations of elementary vibrational excitations could be started. The experiments give new insight in the dynamics of vibrational processes in liquids and solids. The measurements of liquids are discussed first.

The vibrational excitation of the molecules is described by the expectation value of the normal mode operators, $\langle q \rangle$, and the excited state population, n. For the two level approximation the equations of motions have the form[34,35]

$$\left(\frac{\partial^2}{\partial t^2} + \frac{2}{T_2}\frac{\partial}{\partial t} + \omega_o^2\right)\langle q \rangle = \frac{1}{2m}\frac{\partial \alpha}{\partial q} E^2 \left[1-2\,n\right] \tag{1}$$

$$\left(\frac{\partial}{\partial t} + \frac{1}{T_1}\right)(n-\bar{n}) = \frac{1}{2\hbar\,\omega_o}\frac{\partial \alpha}{\partial q} E^2 \frac{\partial}{\partial t}\langle q \rangle \tag{2}$$

$\bar{n}$ is the thermal equilibrium population of the upper levels. The coupling parameter $\partial\alpha/\partial q$ accounts for the change of polarizability with vibrational elongation. m denotes the reduced mass of the vibration. Of special interest are the two relaxation times T_1 and T_2 introduced in Eqs. (1) and (2). T_2 is the dephasing time of the vibrational amplitude of the molecules with transition frequency ω_o. T_1 represents the population lifetime of the first excited vibrational state.

Eqs. (2) and (3) assume exponential decay of $\langle q \rangle$ and n after the excitation process has terminated. This time dependence is expected to hold for times longer than τ_c where $\tau_c \sim 10^{-13}$ s is a correlation time connected to intermolecular interactions, e.g. rapid translational motion.[36] The exponential time constants are well supported by experimental results. Eqs. (1) and (2) are supplemented by the nonlinear wave equation for the light field:

$$\Delta E - \frac{1}{c^2}\frac{\partial^2}{\partial t^2}(\mu^2 E) = \frac{4\pi}{c^2}\frac{\partial^2}{\partial t^2} P^{NL} \tag{3}$$

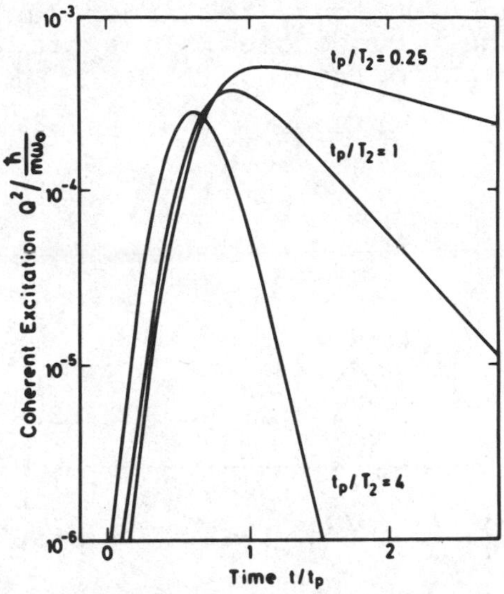

Fig. 14. Calculated coherent vibrational excitation versus time for several values of tp/T_2.

μ denotes the refractive index of the medium; the nonlinear polarization P^{NL} represents the stimulated scattering off the vibrational mode of interest. The coherent amplitude $\langle q \rangle$ leads to the induced dipole moment $\partial\alpha/\partial q \langle q \rangle E$ of the indiviudal molecule. Summing over the molecules with number density N yields

$$P^{NL} = N \frac{\partial\alpha}{\partial q} E \langle q \rangle \tag{4}$$

Eqs. (1) to (4) represent a complete set describing the generation of an intense Stokes pulse and of a coherent material excitation by stimulated scattering.

The time evolution of the vibrational system is depicted in Fig. 14. For the homogeneously broadened system under consideration, simple exponential decay with time constant $T_2/2$ is predicted for the coherent excitation $|\langle q \rangle|^2$.

The coherent excitation is monitored by a delayed probe pulse. The amplitude $\langle q \rangle$ gives rise to a macroscopic polarization $P = N\partial\alpha/\partial q \; E_p \langle q \rangle$, where E_p denotes the electromagnetic field of the probe pulse.[35] The polarization emits a scattering emission, which is shifted by the vibrational frequency to larger (anti-Stokes) or smaller (Stokes) frequencies. This process is called coherent Raman scattering.

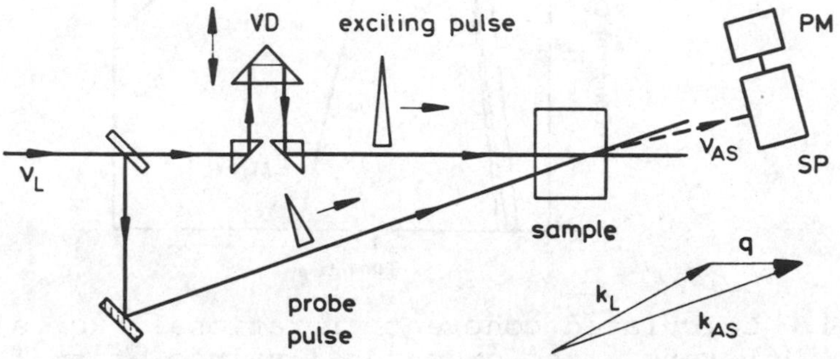

Fig. 15. Experimental set-up for measuring the dephasing time T_2 of a molecular vibration in liquids; the inset illustrates the k-matching condition of the coherent probe scattering.

Fig. 15 shows schematically part of the experimental system for the measurement of the vibrational dephasing time T_2.[37] A single short light pulse traverses the liquid sample and generates the vibrational excitation of interest via stimulated Raman scattering. A second probe pulse of variable delay time t_D passes through the excited volume, and the coherent anti-Stokes scattering signal is observed with the help of a spectrometer and a photomultiplier. Of special importance in this experiment is the k-matching condition between the material excitation of wave vector q (which is generated by the pumping pulse), the probe pulse of wave vector k_L and the scattered light k_{AS}. The inset of Fig. 15 illustrates the required k-matching geometry.

An experimental result for the simple case of homogeneous line broadening is shown in Fig. 16. The fundamental vibrational mode of liquid N_2 at 2326 cm^{-1} is investigated.[38] From the decay of the signal curve we deduce a dephasing time of T_2 = 150 ps.

It is interesting to relate the exponential relaxation with time T_2 to the line shape observed in conventional Raman spectroscopy. For the homogeneous system the exponential decay corresponds to a Lorentzian line with full width (FWHM in units of cm^{-1})

$$\delta\tilde{\nu}_{hom} = (\pi c T_2)^{-1} \tag{5}$$

For liquid N_2 we deduce from Eq. (5) and the measured T_2 a dynamic line broadening of 0.070 cm^{-1}. This number is in excellent agreement with the linewidth of the isotropic scattering component observed in spontaneous Raman spectroscopy;[39] i.e. the dephasing time T_2 fully accounts for the spectroscopic line broadening. It is interesting to note that an expectionally long population lifetime T_1 of the fundamental mode of approximately one minute was reported by several authors.[40] As a result, population decay does not effect the dephasing process of liquid N_2.

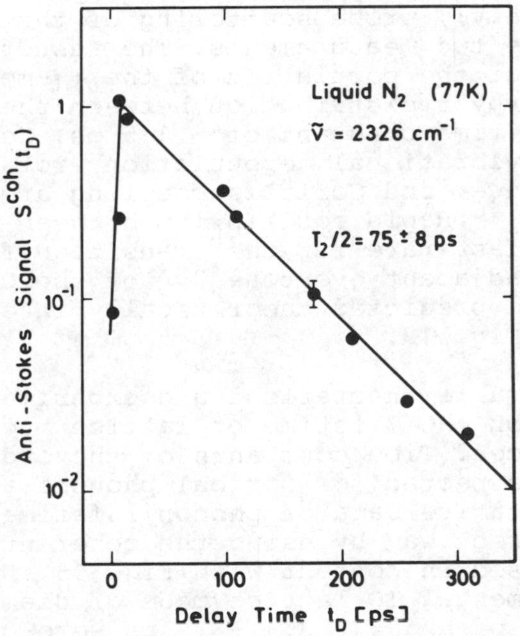

Fig. 16. Time resolved coherent Raman scattering of the fundamental mode of N_2 versus delay time (after ref. 38).

T_1-measurements will be discussed next. Spontaneous anti-Stokes probe scattering allows the study of the instantaneous population of a vibrational level. With this technique it was possible, for the first time, to observe population lifetimes, energy transfer and energy redistribution.[41] Time constants between 1 ps and 100 ps were measured for different dynamical processes in a number of polyatomic molecules.[41-45] The experimental method consists of two steps. First, a powerful short light pulse traverses the sample and excites the vibrational mode of interest. Stimulated Raman scattering of a visible pulse of infrared absorption of a resonant IR pulse can produce an excess population of the excited vibrational state. A second weak pulse of different frequency probes the vibrational excitation via spontaneous anti-Stokes Raman scattering. The scattering intensity is observed in a large solid angle of acceptance at approximately 90° scattering angle.

We have investigated the population lifetime and energy relaxation processes in a number of pure liquids and in liquid mixtures. Fig. 17 shows data of a solution of CH_3I.[44] The symmetric ν_1 or asymmetric ν_4 CH_3-stretching mode is excited by resonant absorption (Figs. a and b, respectively); probe scattering of the ν_1-mode is observed in the two measurements. The results indicate rapid decay of the population of the ν_1-mode with $T_1 =$ 1 ps and energy redistribution between the two neighboring modes with a time constant of $\sim$ 1.5 ps. For an explanation of the fast vibrational depopulation, rotational coupling, Fermi resonance, and Coriolis coupling are expected to be important.[46] Rapid equilibrium between the ν_1 and ν_4 modes and a fast rate for the transition from the ν_1 mode to the adjacent overtone $2\nu_5$ of the CH_3 bending vibration are predicted theoretically in accordance with the data of Fig. 17.

Considerable theoretical and experimental information exists on the lifetime of lattice modes in crystalline substances. Time constants of the order of 10^{-13} to 10^{-11} s are expected for optical phonons at room temperature. Quantitative data of phonon lifetimes are obtained in a very direct way by using the coherent probing technique discussed in context with Fig. 15. For the lifetime of the fundamental TO lattice mode of diamond values of $\tau = 2.9 \pm .3$ ps and $\tau = 3.4 \pm .3$ ps were observed at 295 and 77 K, respectively.[47] Comparison of the measured τ-values with results inferred from the linewidth of spontaneous Raman scattering gives good agreement bet-

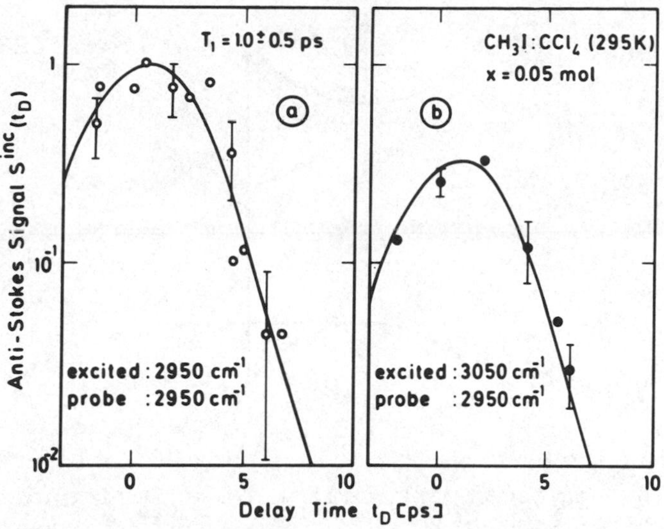

Fig. 17. Incoherent anti-Stokes probe scattering S^{inc} (t_D) vs. delay time of the symmetric CH_3-stretching mode (ν = 2950 cm^{-1}) of CH_3I dissolved in CCl_4 (mole fraction 0.05). (a) The mode at 2950 cm^{-1} is directly populated by the resonant infrared excitation pulse. (b) The asymmetric CH_3-stretching mode at 3050 cm^{-1} is excited by the tunable pump pulse; the excess population of the vibration at 2950 cm^{-1} is observed indicating rapid energy redistribution between the neighbouring modes (after ref. 44).

ween the two techniques. Similar results were also reported for the A_{1g} CO_3-mode of $CaCO_3$.[48]

In a different kind of experiment Raman scattering is used to detect the nonequilibrium LO-phonons released during the relaxation of hot carriers in a semiconductor.[49] The excitation process is illustrated by Fig. 18. Via single photon absorption using short pulses of a synchronously pumped Rhodamine 6G laser (λ = 575 nm) electrons are promoted from the valence band to the conduction band in GaAs. Since the photon energy exceeds the band gap, higher states (k > 0) are populated with subsequent electronic relaxation. In this process LO phonons are

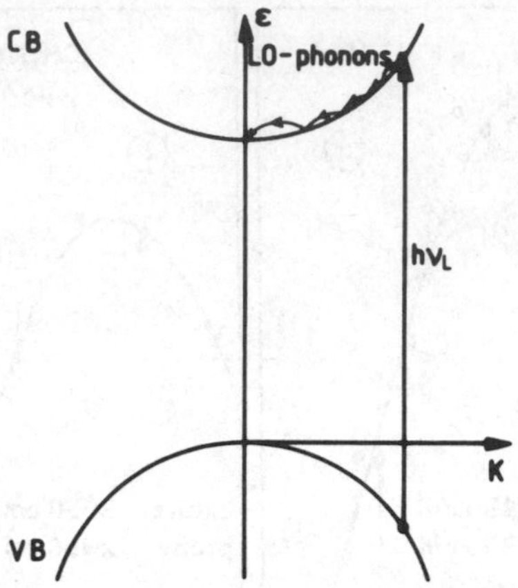

Fig. 18. Optical excitation of a semiconductor (schematic).

released, which are observed via enhanced spontaneous Raman scattering.

For this purpose, the laser output is divided into two equally intense portions, a variable time delay is introduced, and the two beams are recombined at the surface of the GaAs crystal. Raman scattering of the dye laser pulses is detected in backward direction with a photon counting spectrometer.

Adjusting perpendicular polarization between the two pulses, Raman scattering produced by the first exciting pulse may be blocked by a polarizer in front of the spectrometer and the observed anti-Stokes signal is given by the convolution of the probe pulse with the phonon population N(t). Experimental results are shown in Fig. 19.[49] The rise and decay of the phonon burst is measured.

Estimates suggest that the hot carriers lose more than 90 % of the excess energy in about 5 ps, in agreement with the observed rise time in Fig. 19. The decay of the anti-Stokes signal represents the decay of the excess population, and τ = 7 ps is the relaxation time of the non-equilibrium LO-phonons.

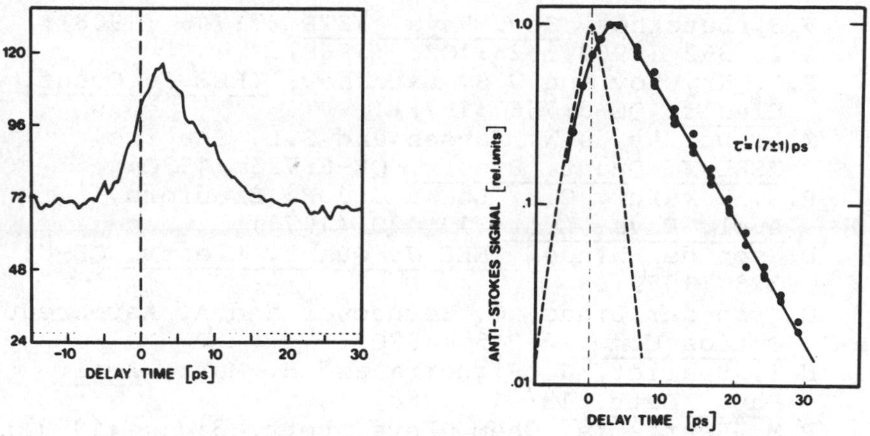

Fig. 19. Anti-Stokes intensity of Raman scattering from LO-phonons in GaAs versus delay time (after ref. 49).

V. CONCLUSIONS

Picosecond pulses have found applications in numerous physical problems. Available pulses allow measurements with a time resolution of a fraction of a picosecond. Time resolution will be shifted to even smaller values in the near future. This chapter discusses several techniques which study ultrafast processes in liquids and solids. It is shown that new experimental methods have been devised which benefit from the specific material excitation provided with high quality picosecond light pulses. New and detailed information on different aspects of molecular and lattice dynamics is obtained.

REFERENCES

1. A.J. DeMaria, D.A. Stetser, and H. Heyman, Appl. Phys. Lett., 8:174 (1966).
2. I.W. Shelton and J.A. Armstrong, IEEE J. Quantum Electron., QE-3:696 (1967).
3. See, for example, Topics in Applied Physics, vol. 18, "Ultrashort Light Pulses", S.L. Shapiro, ed., Springer, Berlin (1977), and the literature cited here.

4. V.S. Letokhov, Sov. Phys. JETP 27:746 (1968); 28:562 (1969); 28:1026 (1969).
5. P.G. Kryukov and V.S. Letokhov, IEEE J. Quant. Electr. QE-8:766 (1972).
6. M.A. Duguay, J.W. Hansen and S.L. Shapiro, IEEE J. Quant. Electr. QE-6:725 (1970).
7. R.C. Eckardt, C.H. Lee and J.N. Bradford, Appl. Phys. Lett. 19:420 (1971).
8. D. von der Linde, IEEE J. Quant. Electr. QE-8: 328 (1972).
9. D. von der Linde, O. Bernecker and A. Laubereau, Optics Comm. 2:215 (1970).
10. E.I. Scarlet, J. Figueira and H. Mahr, Appl. Phys. Lett. 13:71 (1968).
11. P.M. Rentzepis, Chem. Phys. Lett. 3:717 (1969); M.M. Malley and P.M. Rentzepis, ibid 3:534 (1969).
12. D. von der Linde and K.F. Rodgers, IEEE QE-9: 960 (1973).
13. B. Kopainsky, W. Kaiser and K.H. Drexhage, Opt. Comm. 32:451 (1980); B. Kopainsky, A. Seilmeier and W. Kaiser, in "Picosecond Phenomena II", Springer Series in Chemical Physics, vol. 14, R.M. Hochstrasser, W. Kaiser and C.V. Shank, eds., Springer, Berlin (1980).
14. S.A. Akhmanov, R.V. Khokhlov and A.P. Sukhorukov, in "Laser Handbook", F.T. Arecchi and E.O. Schulz-DuBois, eds., North-Holland, Amsterdam (1972).
15. A. Penzkofer, A. Laubereau and W. Kaiser, Phys. Rev. Lett. 31:863 (1973).
16. A. Penzkofer, A. Seilmeier and W. Kaiser, Opt. Comm. 14:363 (1975).
17. W.C. Smith, P. Liu and N. Bloembergen, Phys. Rev. A 15:2396 (1977).
18. R.M. Hochstrasser, D.L. Narva and A.C. Nelson, Chem. Phys. Lett. 43:15 (1976).
19. M.A. Duguay and J.W. Hansen, Appl. Phys. Lett. 15:192-194 (1969).
20. H. Mahr and M.D. Hirsch, Opt. Comm. 13:96 (1975).
21. L.A. Halliday and M.R. Topp, Chem. Phys. Lett. 46:8 (1977).
22. B. Kopainsky and W. Kaiser, Chem. Phys. Lett. 66:39 (1979).
23. A. Penzkofer, W. Falkenstein and W. Kaiser, Chem. Phys. Lett. 44:82 (1976).
24. G. Mourou and M.M. Malley, Chem. Phys. Lett. 32:476 (1975).
25. B. Kopainsky and W. Kaiser, Opt. Comm. 26:219 (1978).
26. D. Ricard, J. Chem. Phys. 63:3841 (1975).
27. D. Reiser and A. Laubereau, Appl. Phys. B (1982) in press.

28. C.V. Shank and E.P. Ippen, Appl. Phys. Lett. 26:62 (1975).
29. H.E. Lessing, A. von Jena and M. Reichardt, Chem. Phys. Lett. 36:517 (1975).
30. D. Waldeck, A.J. Cross, Jr., D.B. McDonald and G.R. Fleming, J. Chem. Phys. 74:3381 (1981).
31. A. Seilmeier, W. Kaiser, A. Laubereau, S.F. Fischer, Chem. Phys. Lett. 58:225 (1978).
32. A. Seilmeier, W. Kaiser, A. Laubereau, Optics Comm. 26:441 (1978).
33. J.P. Maier, A. Seilmeier and W. Kaiser, Chem. Phys. Lett. 70:591 (1980).
34. J.A. Giordmaine, W. Kaiser, Phys. Rev. 144:676 (1966).
35. A. Laubereau and W. Kaiser, Rev. Mod. Phys. 50: 607 (1978).
36. C.P. Slichter, in "Principles of Magnetic Resonance", Harper and Row, eds., New York (1963); S.F. Fischer, A. Laubereau, Chem. Phys. Lett. 35:6 (1975); W.G. Rothschild, J. Chem. Phys. 65:2958 (1976).
37. D. von der Linde, A. Laubereau and W. Kaiser, Phys. Rev. Lett. 26:954 (1971).
38. A. Laubereau, Chem. Phys. Lett. 27:600 (1974).
39. W.R.L. Clements, B.P. Stoicheff, Appl. Phys. Lett. 12:246 (1968); M. Scotto, J. Chem. Phys. 49: 5362 (1968).
40. W.F. Calaway and G.E. Ewing, Chem. Phys. Lett. 30:485 (1975); J. Chem. Phys. 63:2842 (1975); S.R. Brueck, R.M. Osgood, Jr., Chem. Phys. Lett. 39:568 (1976).
41. A. Laubereau, D. von der Linde and W. Kaiser, Phys. Rev. Lett. 28:1162 (1972).
42. R.R. Alfano and S.L. Shapiro, Phys. Rev. Lett. 29:1655 (1972); A. Laubereau, L. Kirschner and W. Kaiser, Opt. Comm. 19:182 (1973).
43. A. Laubereau, G. Kehl and W. Kaiser, Optics Comm. 11:74 (1974); P.R. Monson, L. Patumtevapibal, K.J. Kaufmann and P.W. Robinson, Chem. Phys. Lett. 28:312 (1974).
44. K. Spanner, A. Laubereau and W. Kaiser, Chem. Phys. Lett. 44:88 (1976).
45. A. Laubereau, S.F. Fischer, K. Spanner and W. Kaiser, Chem. Phys. 31:335 (1978); A. Fendt, S.F. Fischer and W. Kaiser, Chem. Phys. 57:55 (1981); Chem. Phys. Lett. 82:350 (1981).
46. A. Miklavc and S.F. Fischer, Chem. Phys. Lett. 44: 209 (1976); 69:281 (1978); R. Zygan-Maus and S.F. Fischer, Chem. Phys. 41:319 (1979).

47. A. Laubereau, D. von der Linde and W. Kaiser, Phys. Rev. Lett. 27:802 (1971).
48. A. Laubereau, G. Wochner and W. Kaiser, Opt. Comm. 14:75 (1975).
49. D. von der Linde, J. Kuhl and H. Klingenberg, Phys. Rev. Lett. 44:1505 (1980).

NEW PHENOMENA IN COHERENT OPTICAL TRANSIENTS

Richard G. Brewer

IBM Research Laboratory
San Jose, California 95193

These lectures embrace two topics: (1) the laser spectroscopy of solids and (2) the effect of elastic collisions on atoms that are in a coherent superposition state due to preparation by a laser field.

LASER SPECTROSCOPY OF SOLIDS

Remarkably narrow optical homogeneous linewidths[1,2] of the order of 1 kHz have now been observed in low temperature zero-phonon transitions of dilute impurity ion crystals, such as Pr^{3+} in LaF_3. Novel nonlinear optical resonance techniques have been devised for this purpose using ultrastable frequency-locked cw dye lasers where the measurements are performed either in the frequency domain (hole burning) or in the time domain (coherent optical transients). These studies effectively bring the Mossbauer effect into the optical region. The high sensitivity and spectral resolution allow precise measurements of hyperfine structure, detailed investigations of the dynamic line broadening mechanisms and perhaps in the future time dilation studies. Hence, the observed linewidths are no longer limited by inhomogeneous strain broadening (~5 GHz) or even by static local fields due to neighboring spins (~100 kHz). However, weak magnetic field fluctuations from local spins are readily detected. As an example, spin decoupling and line narrowing,[2] which are well known in NMR, are observed for the first time in an optical transition of Pr^{3+}:LaF_3 at 2°K where the ^{19}F-^{19}F dipolar interaction is quenched and the optical linewidth drops from 10 to 2 kHz, clearly demonstrating the spin broadening mechanism. Results are discussed in terms of a Monte Carlo line broadening theory.[3]

ELASTIC COLLISIONS

One of the initial applications of coherent optical transients has been the study of gas phase elastic collisions and their effect on atoms in a coherent superposition state. The first work[4,5] carried out in the infrared region for a vibration-rotation transition of CH_3F showed that photon echoes display a characteristic time dependence due to elastic collisions of the velocity-changing type.[6] This mechanism can be viewed as a molecular diffusion in velocity space where the internal state of the molecule is unaffected. A solution of the quantum mechanical transport equation,[4,5] which utilizes an appropriate collision kernel, fully confirms this mechanism.

Traditional line broadening theory[7] encompasses a different regime where elastic collisions are state-dependent and introduce random phase interruptions. Their effect is analyzed here for the case where atoms are initially in a superposition state. Recent optical free induction decay measurements for the visible I_2 transitions[8,9] support the concept of a phase interrupting elastic collision and provide no evidence that velocity changing elastic collisions occur.

Still more recent photon echo measurements[10] show that light atoms such as Li allow one to identify both velocity-changing and phase interrupting elastic collisions where the former corresponds to large impact parameters and the latter to close encounters. A scattering theory[11] predicts three different time regimes, but only two have been observed so far for any one species.

REFERENCES

1. R. G. DeVoe, A. Szabo, S. C. Rand and R. G. Brewer, Phys. Rev. Lett. 23, 1560 (1979).
2. S. C. Rand, A. Wokaun, R. G. DeVoe and R. G. Brewer, Phys. Rev. Lett. 43, 1868 (1979).
3. R. G. DeVoe, A. Wokaun, S. C. Rand and R. G. Brewer, Phys. Rev. B 23, 3125 (1981).
4. J. Schmidt, P. R. Berman and R. G. Brewer, Phys. Rev. Lett. 31, 1103 (1973).
5. P. R. Berman, J. M. Levy and R. G. Brewer, Phys. Rev. A 11, 1668 (1975).
6. See also B. Comasky, R. E. Scotti and R. L. Shoemaker, Opt. Lett. 6, 45 (1981).
7. R. G. Breene, Jr., The Shift and Shape of Spectral Lines (Pergamon Press, Inc., New York, 1961) for example.
8. A. Z. Genack and R. G. Brewer, Phys. Rev. A 17, 1463 (1978).

9. R. G. Brewer and S. S. Kano in Nonlinear Behaviour of Molecules, Atoms and Ions in Electric, Magnetic or Electromagnetic Fields (Elsevier, Amsterdam, 1979), edited by L. Neel, p. 45.
10. T. W. Mossberg, R. Kachru and S. R. Hartmann, Phys. Rev. Lett. 44, 73 (1980).
11. P. R. Berman, T. W. Mossberg and S. R. Hartmann, Phys. Rev. A (submitted).

OPTICAL BISTABILITY

S. Desmond Smith and Eitan Abraham

Department of Physics
Heriot-Watt University
Edinburgh, EH14 4AS, U.K.

INTRODUCTION

The replacement of electrical currents by optical beams in information processing has become feasible due to the discovery of optical bistability and related devices+. An optically bistable device is one which can exhibit two steady transmission states for the same input intensity. A fabry-Perot resonator containing a medium with a nonlinear refractive index constitutes the simplest example of such a system. As a laser light is irradiated on the cavity and increased from zero to a maximum and back to zero, the output-input characteristic can show a hysteresis cycle: the lower and upper branches (Fig. 1) are locally stable and hence the system is said to be bistable. Under suitable choice of parameters a device of this kind can function e.g. as a memory, as an optical transistor and can be used in optical logic [2,3]. From a theoretical viewpoint, optical bistability constitutes an example of a first-order phase transition in a far-from-equilibrium system [4]. In this paper we shall concentrate on the device aspects.

Optical bistability and the principles behind 'optical circuit' elements were first put forward by Szoke et al [5]. In their paper it was predicted that bistability would occur at exact resonance if a Fabry-Perot resonator was filled with a saturable absorber in which the absorption coefficient is a decreasing function of local intensity. Szoke et al presented

\+ For a comprehensive review on optical bistability and related devices see ref. [1].

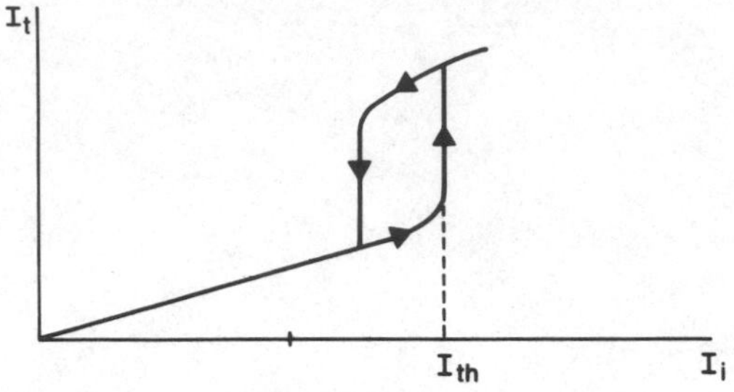

Figure 1. Characteristic curve of a nonlinear optical cavity. As the input intensity I_i varies from zero to $I_i > I_{th}$ and back to zero, the transmitted intensity I_t can exhibit a hysteresis cycle.

results for a resonator containing SF_6 gas irradiated by CO_2 laser a 10.6 μm: nonlinear behaviour was shown but this absorptive bistability was not demonstrated. The first numerical study of absorptive bistability by McCall [6] became the initial step towards the first observation of optical bistability by Gibbs et al [7] using N_a vapour; the role of nonlinear refraction was discovered. The first dynamical study of a Fabry-Perot cavity containing Kerr media (nitrobenzene, MBBA and CS_2) was done by Bischofberger and Shen [8]. Observation of semiconductor optical bistability was first reported simultaneously (but independently) by Miller, S.D. Smith and Johnston [9] in InSb and by Gibbs et al [10] in GaAs. In all these experiments the predominant mechanism is nonlinear refraction; it was not until the experiment by Weyer et al [11] that absorptive OB (in Na) was observed. The above devices are termed intrinsic as the feedback (provided by the cavity) is optical; hybrid devices are those in which feedback is achieved by means of some external electronic circuit. The latter were first constructed by P.W. Smith and Turner [12] using an electro-optic crystal inside a resonator. The concept was advanced by Garmire et al [13] in which the resonator was replaced by using the feedback to a crystal placed between crossed polarizers. Certainly the most promising device uses come from semiconducting materials, which, for reasons we explain below, allow the construction of micron-dimension optical resonators.

THEORETICAL BACKGROUND

Almost simultaneously with the experiment of Gibbs et al [7], Felber and Marburger [14] reported the following, rather simple, plane-wave theory of a Fabry-Perot resonator filled with a cubic non-absorptive nonlinearity which explained optical bistability. Suppose we have a medium in a cavity (Fig 2) with its forward (E_F) and backward (E_B) fields inside it. The transmission τ of the

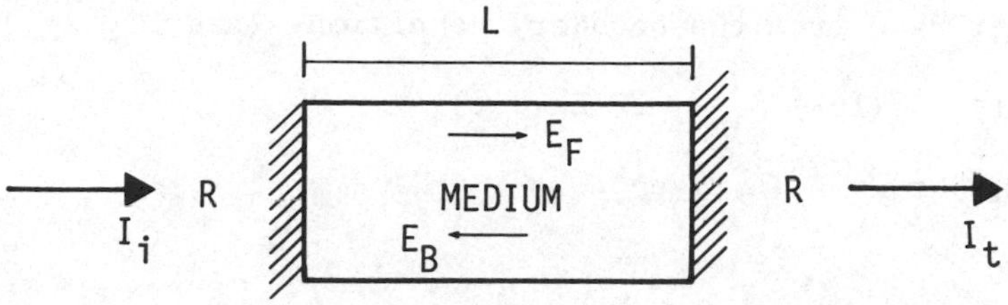

Figure 2. Fabry-Perot interferometer of length L and mirror reflectivity R; I_i and I_t as in Fig 1; E_F and E_B are the forward and backward fields respectively.

cavity, i.e. the ratio between the transmitted intensity I_t and the input intensity I_i, is given by

$$\tau = \frac{It}{Ii} = \frac{1}{1 + F \sin^2\Phi/_2} \tag{1}$$

where $F = 4R/(1-R)^2$ is the finesse, R is the mirror reflectivity, $\Phi = 2\pi n_0 L/(\lambda/2)$ is the round-trip phase shift, λ is the wavelength of the radiation and n_0 is the refractive index of the medium (assumed linear). The Airy function (1) is periodic in Φ and becomes unity (full transmission) whenever the optical path n_0L is an integral number of $\lambda/2$. If the medium is nonlinear n_0 is replaced by $n(I_c) = n_0 + n_2 I_c$, where I_c is the intracavity intensity; the transmission $\tau(I_c)$ is also an Airy function (Fig. 3),

$$\Phi(I_c) = 4\pi n_0 \frac{L}{\lambda} + 4\pi n_2 \frac{L}{\lambda} I_c = \theta\,(\mathrm{mod}\ 2\pi) + 4\pi n_2 \frac{L}{\lambda} I_c \tag{2}$$

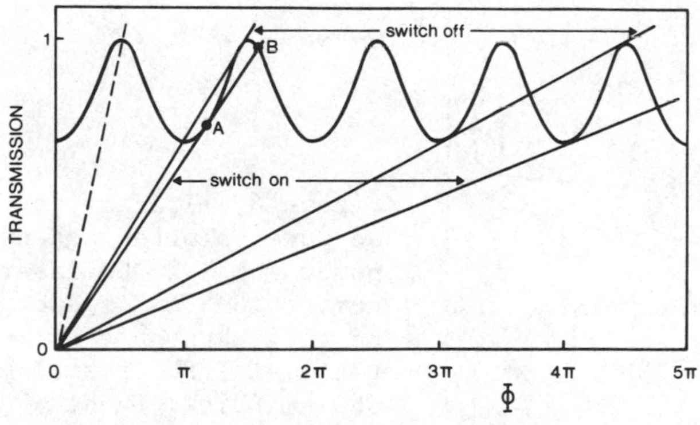

Figure 3. Intersection between relations (1) and (4) for $\theta = \pi$ and $F = 0.5$ demonstrating no bistability in first order [3]. Only from the intersection point A progressively wider bistable regions begin to appear. Straight lines of shallower gradient correspond to higher input intensities I_i.

On the other hand from the boundary conditions (see Fig 2)

$$E_F(o,t) = \sqrt{(1-R)}\, E_i + \sqrt{R}\, E_B(o,t)$$
$$E_B(L,t) = \sqrt{R}\, E_F(L,t)\, e^{i\theta} \qquad (3)$$

and with

$$I_t = (1-R)\, |E_F|^2 \text{ and } I_c = |E_B|^2 + |E_F|^2 = (1+R)|E_F|^2$$

it is possible to obtain a second expression for τ which relates external and internal intensities;

$$\tau = \frac{I_c}{I_i}\,\frac{(1-R)}{(1+R)} \qquad (4)$$

Equations (1) and (4) must satisfy self consistency and therefore the values of I_c will emerge from a graphical solution as shown in Fig. 3 using data from ref. [9]. We follow the intersection point as I_i is varied from zero to some maximum. At low I_i the point moves smoothly along the curve until it reaches A; here a small change in I_i causes a 'jump' to B, i.e. a sudden increase in the transmission that occurs at $I_i = I_{th}$ in Fig. 1. Upon decreasing I_i the point moves smoothly from B to C where it can only jump to D. Clearly, when the two discontinuous changes occur at different values of I_i optical hysteresis is obtained. If from B we increase I_i even further, successive bistable regions (multistability) can appear as it was first observed by Miller, S.D. Smith et al [10,15] in InSb; one of such curves we show in Fig. 4 where different modes of operation can be readily identified: bistable, transistor (slope > 1), and limiter (plateau).

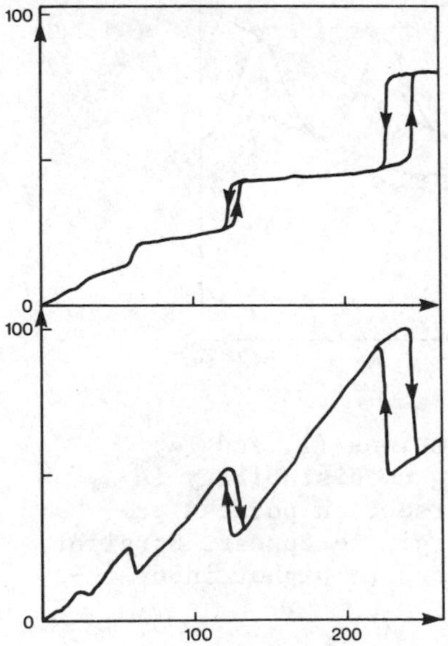

Figure 4. Transmitted and reflected power plotted against incident power for a cw CO laser beam (wavenumber 1895 cm^{-1}, spot size ∿ 180μm) passing through a polished plane-parallel InSb crystal [5 x 5mm x 560 μm thick, N_D-N_A ∿ 3 x 10^{14} cm^{-3} (n-type)] at ∿ 5K (from ref [15].)

A more complete picture for practical considerations entails the inclusion of linear absorption (α) in the criteria for making devices: the question of cavity optimization arises as follows. We wish to have a high finesse cavity to make the Airy function (1) steep as for bistability multiple intersections with the straight line (4) are necessary. From (2) we conclude that the longer the cavity the smaller should be I_c in n_2LI_c, but the linear absorption will effectively suppress the finesse. Consequently, the minimum (critical) input intensity I_{crit} for a first bistable region should result from a compromise between nonlinearity (n_2), linear absorption (α) and mirror reflection (R). By demanding the condition for multiple intersection

$$\frac{d\tau}{dI_c} \gtrsim \frac{\tau}{I_c} \tag{5}$$

(i.e. the derivative of the Airy function at the inflection point greater than the slope of the straight line) a plane-wave calculation [16] shows that

$$I_{crit} = \frac{1}{\beta} \cdot \frac{1}{\mu_o} \tag{6}$$

where $\beta = 3n_2/\lambda\alpha$ is the figure of merit of the material and $\mu_o \simeq 0.65\,\pi\,(1-R+A)$ is the figure of merit of the cavity respectively; $A = 1-e^{-\alpha L}$ is absorption per pass. The critical initial detuning must be $\theta_{crit} = 2.72/\pi(1-R+A)$ and the optimization condition is $A = 1 - R$. This simple discussion covers most of the basic experimental effects and concepts of practical importance to date.

Most of the theoretical development of optical bistability has been done for atomic systems by using the envelope Maxwell-Bloch equations⁺. The model consists of either a Fabry Perot or a ring cavity filled with N(N >>1) two-level atoms and the field is considered classical

$$E(z,t) = E_F(z,t)\, e^{i(kz-\omega t)} + E_B\, e^{-i(kz+\omega t)} + c.c \tag{7}$$

where $k = 2\pi/\lambda$, and $\omega=kc$, and the envelopes E_F and E_B are assumed to change very little in an interval k^{-1} or in a time ω^{-1}. (slowly varying envelope approximation). The atomic dynamics are dictated by the Schroedinger equation, which combined with the Maxwell wave equation, gives a system of first-order nonlinear partial differential equations known as Maxwell-Bloch equations. This model has its limitations, e.g. a) it does not consider the Gaussians profile of a laser beam, b) in the time dependent

+ For derivation and details see [1].

standing wave effects are neglected and this gives an incorrect distribution of fields [17] in the cavity; in the steady-state these effects can readily be included [6,18-20]. However, in the case of the ring cavity, standing-wave difficulties do not arise (no backward wave) and the Maxwell-Bloch equations are more tractable. Using a ring geometry it is possible to show that positive-slope branches can become unstable giving rise to self-pulsing [21] in absorptive bistability and period-doubling leading to chaos in refractive bistability [22]; the latter has been observed [23] in a hybrid device. Recently, Firth [24] and Abraham et al [25] found that a Fabry-Perot containing a cubic nonlinearity with linear absorption whose dynamics is governed by

$$\tau \frac{\partial \chi_{NL}}{\partial t} + \chi_{NL} = n_2(|E_F|^2 + |E_B|^2)$$

$$\frac{\partial E_F}{\partial z} + \frac{n_o}{c}\frac{\partial E_F}{\partial t} = (i\,\chi_{NL} - \frac{\alpha}{2})\,E_F \qquad (8)$$

$$-\frac{\partial E_B}{\partial z} + \frac{n_o}{c}\frac{\partial E_B}{\partial t} = (i\chi_{NL} - \frac{\alpha}{2})E_B$$

(in (8) χ_{NL} (z,t) is the non linear 'susceptibility' of the medium and τ its relaxation time) has unstable positive slope regions in whcih periodic motion can degenerate into an aperiodic one. (Fig. 5.)

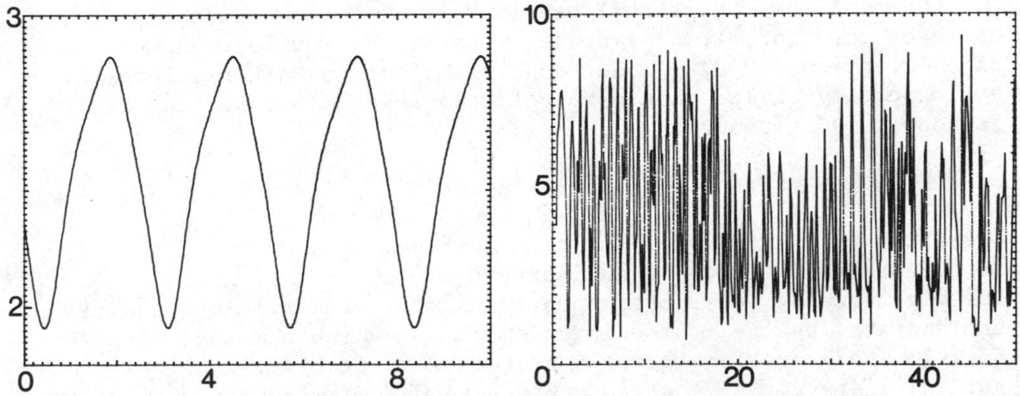

Fig. 5. Numerical integration of equations (8). Positive slope branches of the corresponding steady state equation show instabilities (a) periodic and (b) aperiodic motion of the output intensity.

The Maxwell-Bloch equations can be greatly simplified when the fields are replaced by their spatial averages. This is the so-called mean field approximation [4] which is valid when the atomic density, the mirror transmissivity and the cavity detuning approach zero.

EXPERIMENTS ON SEMICONDUCTORS

The pioneering experiments on bistability in semiconductors were achieved on InSb by Miller, S D Smith et al [9,15] and on GaAs by Gibbs, McCall et al [10]. These semiconducting materials have been known to exhibit large passive non-resonant $\chi^{(3)}$ around 10^{-8} - 10^{-11} esu, but the discovery of strong non-linear refraction in the region just below the optical bandgap in both materials, gives susceptibilities many orders of magnitude higher with $\chi^{(3)}$ in the region 10^{-2} - 1 esu. In both cases the giant effect can be explained by saturation mechanisms: nonlinearity in GaAs relying on the existence of excitons whilst that in InSb is ascribed to inter-band transitions (for a detailed discussion of non-linear effects in semiconductors see ref. [26]).

The advantages of using semiconductors in optically bistable systems are that: (i) the fabrication technology is highly developed; (ii) semiconductor devices are small - in the range of 4 - 100 μm - and have short calculated cavity build-up times (1.2 ps for GaAs and 20 ps for InSb already demonstrated); (iii) the observed non-linearities are large ($n_2 \sim 4 \times 10^{-4}$ cm^2/kW in GaAs and 1.0 cm^2/kW in InSb at 77 K with $\chi^{(3)} \sim 1$ esu) so that power densities as low as 10 W/cm^2 or 8 mW incident power can be used; (iv) while switch-off times may be limited by material relaxation times, switch-on times, in principle, are not; this has been demonstrated to some extent in GaAs [10]. Also, material relaxation times can be engineered to a considerable extent by e.g. doping, particularly for the non-linearity in InSb which is present also in impure material. These factors combined suggest that small, low-power, low-energy, fast switching, integrable devices may ultimately be possible with semiconductors.

Miller et al [27] derived a semi-empirical theory - which fits measurements well both in absolute magnitude and wavelength dependence - for bandgap resonant non-linear refraction in InSb and possibly in other III-V semiconductors. The theory is one extreme limit of the density matrix approach with intraband relaxation being sufficiently strong to thermalize the carrier distribution in times much less than the recombination time, and it rests on the following mechanism:-

a) A fraction of optical absorption results in the creation of free electrons and holes in the semiconductor bands.

b) The electrons and holes thus created relax rapidly to thermal distributions, which then decay with a single interband recombination time τ_R.

c) The thermal populations of the band states partially block (i.e. saturate) the absorption which would otherwise exist in the spectral region above the band-gap energy; this alteration of the absorption spectrum in turn affects the refractive index seen by the incident photons.

Two further simplifications are made: every absorbed photon creates a free electron and a hole, and the Fermi-Dirac distributions are approximated by Boltzmann distributions.

The general populations of electrons and holes in the steady state is

$$\Delta N = \alpha(\hbar\omega) I\, \tau_R / \hbar\omega \qquad (9)$$

where $\alpha(\hbar\omega)$ is the linear absorption coefficient, I is the pump intensity, and $\hbar\omega$ the photon energy. The direct interband absorption coefficient for III-V compound semiconductors for $\hbar\omega' > E_G$ (E_G: bandgap energy) in the parabolic approximation is

$$\alpha_d(h\omega') = \frac{(8\sqrt{2})m^{\frac{1}{2}}e^2}{3\hbar^2 c}\left[\frac{\mu}{m}\right]^{3/2}\frac{1}{n_0}\frac{mP^2}{\hbar^2}\frac{(\hbar\omega'-E_G)^{\frac{1}{2}}}{\hbar\omega'}[1-f_e(E_c)-f_h(E_v)] \quad \ldots (10)$$

where $\mu = m_c m_v/(m_c + m_v)$ (both m_c and m_v are the conduction- and valence band effective masses respectively) and P the momentum matrix element, $f_e(E_c)$ and $f_h(E_v)$ are the occupation probabilities of electrons and holes in the conduction and valence bands, respectively, at the energies appropriate for the direct transitions at photon energy $\hbar\omega'$. The presence of the field causes a change in the distributions given by

$$\Delta f_e(E_c) = 4\pi^{3/2}\left[\frac{\hbar^2}{2m_c}\right]^{3/2}\frac{\Delta N}{(kT)^{3/2}}e^{-E_c/kT} \; ;$$
$$\Delta f_h(E_v) = 4\pi^{3/2}\left[\frac{\hbar^2}{2m_v}\right]^{3/2}\frac{\Delta N}{(kT)^{3/2}}e^{-E_v/kT} \qquad (11)$$

which in turn causes a change in α_d

$$\Delta\alpha_d(\hbar\omega') \simeq -\frac{16\pi^{3/2}}{3}\frac{e^2\hbar}{m_c}\left[\frac{\mu}{m_c}\right]^{3/2}\frac{mP^2}{\hbar^2}\frac{1}{n_0}\frac{\Delta N}{(kT)^{3/2}}\frac{(\hbar\omega' - E_G)^{\frac{1}{2}}}{\hbar\omega'} \quad \ldots$$

$$\ldots \quad \exp\left[\frac{-\mu(\hbar\omega' - E_G)}{m_c kT}\right] \tag{12}$$

From the Kramers-Kronig relations the change in refractive index is

$$\Delta n(h\omega) = \frac{\hbar c}{\pi}\int_0^{\infty}\frac{\Delta\alpha_d(\hbar\omega')}{(\hbar\omega')^2 - (\hbar\omega)^2}\, d(\hbar\omega') \tag{13}$$

hence using (9), (12) and (13) one can write n_2 ($\equiv \Delta n/I$ in this case) as

$$n_2 \simeq \frac{-8\sqrt{\pi}}{3}\,\frac{e^2\hbar^2}{m}\,\frac{mP^2}{\hbar^2}\,\frac{1}{n_o}\,\frac{1}{kT}\,\frac{\alpha(\hbar\omega)\tau_R}{(\hbar\omega)^3}\,J\left(\frac{\mu(\hbar\omega - E_G)}{m_c kT}\right) \tag{14}$$

where

$$J(a) = \int_0^{\infty}\frac{x^{\frac{1}{2}}e^{-x}}{x - a}\,dx$$

and it is assumed that $|h\omega - E_G| << E_G$ and $kT << E_G$; note that here $n_2 < 0$ which corresponds to self-defocussing. In Fig. 6 we show a comparison between theory and experiment using several samples of InSb (n-type, 4×10^{14} cm^{-3}) at 77°K; n_2 was measured with the use of the method of beam profile distortion [28] due to self-defocussing.

In the previous section we have shown multistability in InSb at 5 K in which $E_G \sim 1900$ cm^{-1}. Miller et al also observed bistability at 77°K (Fig. 7) for which $E_G \sim 1840$ cm^{-1}; in this case 70% reflecting coating was used whereas in previous cases the natural reflectivity of the crystal was made use of. The sample was 130 μm thick and made of impure polycrystalline InSb. Miller and Smith [29] observed optical transistor action namely that a weak intensity laser controlled the output of a strong input beam; the device was termed 'transphasor' as it operates by transfer of phase (cf. previous section).

Staupendhal and Schindler [30] observed bistability in Tellurium. Their experimental scheme was intended for optical-optical modulation, that is, modulation of a laser beam by optical tuning of a Fabry-Perot interferometer. Here, the variation of refractive index depends on the change of density of electron-hole pairs, which in turn are created by multiphoton absorption - this is the origin of the non-linearity. The plane parallel slab of Tellurium is irradiated with a CO_2 laser ($\lambda = 10.6\mu m$, $\tau_p \sim 300ns$)

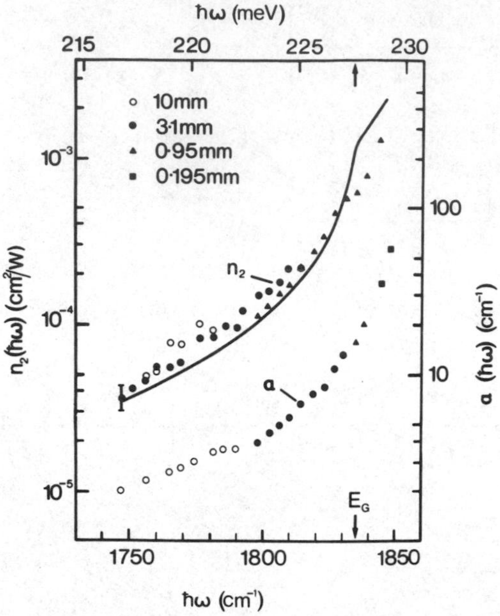

Fig. 6 Measured n_2 and absorption coefficient α as a function of photon energy ħω compared with theory for n_2. The error in α measurements is of the order of the point size. The solid line is calculated with use of the measured α (smoothed for clarity) from Eq. (11). All the parameters in this theoretical curve are based on measured values.

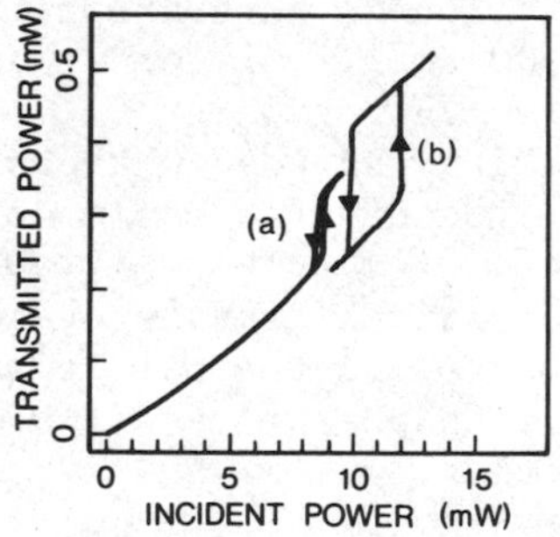

Fig. 7 Transmitted power plotted against incident power for cw CO laser beam (wavenumber 1827 cm^{-1}, spot size ∿ 150 μm) passing through a polished polycrystalline InSb slice (5 x 5 mm x 130 μm thick) coated to ∿ 70 percent reflectivity on both faces, held at ∿ 77 K. Onset of bistability is seen at ∿ 8 mW [trace (a)] with clear bistability at slightly higher powers with different cavity detuning [trace (b)]. (From ref. [15]).

of peak intensity 6.6 MW/cm^2. The experimental set-up we are interested in is what these authors call self-controlled: the laser beam that changes the carrier concentration (the control beam) and the beam to be modulated (the probe beam) are exactly the same (this is essentially Fig. 2). The slab was temperature-stabilized in the region of 300°K.

CONCLUDING REMARKS

If devices derived from the family of bistable devices, but particularly non-linear Fabry-Perot interferometers, are to be practically useful in optical processing and computing, switching and response times need to approach picoseconds. A scaling example is the following: the experiments on InSb gave a change in refractive index $\Delta n = 10^{-17}$ for 1 $photon/cm^3$ and this is the most favourable case to consider at present. The area d^2 of the sample is limited by diffraction so d must satisfy $d^2 > (\lambda/n_o)^2$; then for $\lambda = 5$ μm, $n_o = 4$, we have d ∿ 2.0 μm. For a thickness of 6 μm, R = 95%, the cavity build-up time $t_c \sim 2$ ps. With the former choice of parameters the resulting switching pulse energy is 1 pJ. Sufficient carriers can be excited with a ∿ mW input power for less than 1 ns 'switch-up' time; the latter can be reduced provided the power is increased in order to keep the same 1 pJ pulse energy. Recovery is determined by the recombination time τ_R and for the mechanisms discussed in [1] it is shown that n_2 is proportional to τ_R. For example if $n_2 = 1.0\ cm^2/kW$ ($\chi^{(3)} \sim 1$ esu) when $\tau_R = 100$ ns, then for $n_2 \simeq 10^{-5}\ cm^{-2}/kW$($\chi^{(3)} \sim 10^{-5}$ esu) for $\tau_R \sim 1$ ps; this last value of n_2 still represents a very large non-linearity. The size of τ_R itself can be reduced for faster recovery by making use of the 'micro engineering' methods discussed in [1]. Consequently there is hope that optically bistable devices may be realisable on picosecond time scales. The question as to whether they can compete with present devices to process information can then be considered.

REFERENCES

1. E. Abraham and S.D. Smith, Rep. on Prog. in Phys. (1982), to appear.
2. E. Abraham and S.D. Smith, J. Phys. E. 15: 33 (1982).
3. E. Abraham, C.T. Seaton and S.D. Smith, Scientific American (1982), to appear.
4. R. Bonifacio and L.A. Lugiato, Opt. Comm. 19: 172 (1976); see also Optical Bistability, eds. C.M. Bowden, M. Ciftan and H.R. Robl (Plenum, New York, 1981).
5. A Szöke, V. Daneu, J. Goldhar and N. A. Kurnit, Appl. Phys. Lett. 15: 376 (1969).
6. S.L. McCall, Phys. Rev. A9: 1515 (1974).
7. H.M. Gibbs, S.L. McCall and T.N.C. Venkatesan, Phys. Rev. Lett. 36: 1135 (1976).
8. T. Bischofberger and Y.R. Shen, Appl. Phys. Lett. 32: 156

(1978); Opt. Lett. 4: 40 (1979); Phys. Rev. A19: 1169 (1979).
9. D.A.B. Miller, S.D. Smith and A.M. Johnston, Appl. Phys. Lett. 35: 658 (1979).
10. H.M. Gibbs, S.L. McCall, T.N.C. Venkatesan, A.C. Gossard, A. Passner and W. Wiegman, Appl. Phys. Lett. 35: 451 (1979).
11. K.G. Weyer, H. Wiedenman, M. Rateike, W.R. MacGillivray, P. Meystre and H. Walther, Opt. Comm. 37: 426 (1981).
12. P.W. Smith and E.H. Turner, Appl. Phys. Lett. 30: 280 (1977).
13. E. Garmire, J.H. Marburger and S.D. Allen, Appl. Phys. Lett. 32: 320 (1978).
14. F.S. Felber and J.H. Marburger, Appl. Phys. Lett. 28: 731 (1976).
15. D.A.B. Miller, S.D. Smith and C.T. Seaton, IEEE J. Quant. Elec. QE-17: 312 (1981).
16. D.A.B. Miller, IEEE J. Quant. Elec. QE-17: 306 (1981).
17. E. Abraham, R.K. Bullough and S.S. Hassan, Opt. Comm. 29: 109 (1979).
18. H.J. Carmichael, Optica Acta 27: 147 (1980).
19. J. Goll and H. Haken, Phys. Stat. Solidi B 101: 489 (1980).
20. S.S. Hassan, S.D. Tewari and E. Abraham, Opt. Comm. (1982), to appear.
21. R. Bonifacio, M. Gronchi and L.A. Lugiato, Opt. Comm. 30: 129 (1979).
22. K. Ikeda, Opt. Comm. 30: 257 (1979); K. Ikeda, H. Daido and O. Akimoto, Phys. Rev. Lett. 45: 709 (1980).
23. H.M. Gibbs, F.A. Hopf, D.L. Kaplan and R.L. Shoemaker, Phys. Rev. Lett. 46: 474 (1981).
24. W.J. Firth, Opt. Comm. 39: 343 (1981).
25. E. Abraham, W.J. Firth and E.M. Wright, Proc. V National Quant, Elect. Conf., Hull, U.K. (1981), ed. P.L. Knight (Wiley and Sons), to appear.
26. A. Miller, D.A.B. Miller and S.D. Smith, Adv. in Physics, (1982), to appear.
27. D.A.B. Miller, C.T. Seaton, M.E. Prise and S.D. Smith, Phys. Rev. Lett. 47: 197 (1981).
28. D. Weaire, B.S. Wherrett, D.A.B. Miller and S.D. Smith, Opt. Lett. 4: 331 (1979).
29. D.A.B. Miller and S.D. Smith, Opt. Comm. 31: 101 (1979).
30. G. Staupendhal and K.A. Schindler, Proc. II Int. Symp., Reinhardsbrunn, GDR, 1980, Ultrafast Phenomena in Spectroscopy Vol. 2, eds. J. Schwarz, W. Triebel and G. Wiederhold, p. 437.

THE SEARCH FOR PARITY NON-CONSERVATION IN ATOMS

P. G. H. Sandars

Clarendon Laboratory
Oxford University

LECTURE I BASIC FEATURES

I.I Introduction

In the traditional picture of an atom the only interaction of concern is the electromagnetic, acting between the nucleons and the electrons and among the electrons themselves. As every undergraduate knows this interaction is invariant under space inversion; the atomic Hamiltonian therefore commutes with the parity operator and in the absence of degeneracy the atomic eigenstates have definite parity. The well-known parity selection rules then follow directly.

Perhaps surprisingly, this clearcut picture was not altered significantly by the discovery in 1957 of the non-conservation of the parity (PNC) in the weak interactions. The effects observed could be fully interpreted in terms of so-called 'charged currents' involving the presence of massive charged intermediate bosons $W^{\pm}$. While exchange of such a charged boson can lead to an additional interaction between a neutron and a proton, there is no corresponding interaction within an atom because the exchange is forbidden by baryon and lepton conservation assuming, as all theories do, that the boson has zero baryon and lepton number. Thus the post 1957 charged current theory of the weak interactions did not predict any atomic interaction first order in the weak interactions. Second order interactions were possible but were not likely to be experimentally significant.

This clearcut situation in which an atom is expected to have well-defined parity was rudely shattered in the early 70's following the development of the gauge theories unifying the weak with the electromagnetic interaction.[1] A central feature of these theories is the existence of a heavy neutral boson Z^o - a partner to the photon. Because it is neutral such a boson will lead to additional interactions in atoms. Since the weak interactions violate parity conservation in a fundamental way these new 'neutral current' interactions are expected to do so too.

The first evidence for these neutral current phenomena was obtained from high energy neutrino experiments and over the years a great deal of data has been accumulated in excellent agreement with the standard Weinberg-Salam model.[2a] More directly relevant to atoms is the beautiful experiment on polarized electrons carried out at the Stanford Linear Accelerator (SLAC).[2b] This demonstrated in an unambiguous way that PNC interactions exist between electrons and nucleons at the level expected on the basis of the Weinberg-Salam model. There is therefore little doubt that the atomic Hamiltonian contains PNC terms at the level of G_F the weak interaction constant;[3] the search for experimental evidence of these PNC interactions is the subject of this course of lectures.

The plan of these lectures is as follows: In the rest of this lecture we discuss at a qualitative level the form and magnitude of the PNC term in the atomic Hamiltonian and outline in general terms the strategy one must adopt to amplify the resultant PNC effects to an observable level. We note the importance both of the fractional size of the PNC effect and of its absolute magnitude. In the second lecture we discuss in some detail the various types of experiment which are sensitive to atomic PNC effects with particular emphasis on optical rotation and the electric field induced phenomena used in current experiments. The third lecture deals with the current experimental situations in bismuth; we describe in some detail the main features of and the differences between the various experiments. Finally in the fourth lecture we describe the experiments under way on Tℓ and Cs and the work on hydrogen is reported. We conclude with a brief discussion of future prospects. Greater detail on many points may be found in the review references (a) to (e).

I.2 Form of PNC Atomic Hamiltonian

Because the weak interaction is due to the exchange of a heavy particle it's range will be extremely short on an atomic scale. One can show in very general terms that the most general P violating short range interaction between two Fermions which is also T invariant has the form of a product between an axial-vector current and a polar vector current. There are therefore just five independent interactions possible in an atom, as set out below:

Table 1

Current	Particles				
Axial	e	e	p	n	e
Vector	p	n	e	e	e
Coupling Constant	C_{1p}	C_{1n}	C_{2p}	C_{2n}	C_{3e}

Each of these interactions can be specified by a real coupling constant as indicated.

The full form of the interaction can be readily written down in terms of the usual Dirac operators. However it is more convenient to take the non-relativistic limit as far as the nucleons are concerned and then to retain only the terms of interest to us which are those which are odd parity in the electronic part of the operator. On doing this we obtain:

$$H_{PNC} = H^{(1)}_{PNC} + H^{(2)}_{PNC} + H^{(3)}_{PNC} \tag{I.1}$$

$$= \frac{G_F}{\sqrt{2}} \sum_{eN} \left[C_{1N}\, \gamma^5_e\, \delta(\underline{r}_{eN}) + C_{2N}\, \underline{\sigma}_N \cdot \underline{\alpha}_e\, \delta(\underline{r}_{eN})\right]$$

$$+ \frac{G_F}{\sqrt{2}} \sum_{e>e'} C_{3e} \left[\gamma^5_e - \underline{\sigma}_e \cdot \underline{\alpha}_{e'} - \underline{\alpha}_e \cdot \underline{\sigma}_{e'}\right] \delta(\underline{r}_{ee'})$$

where N stands for n or p and the sum extends over all electrons and nucleons in the atom.

While the expression (I.1) above is quite general and does not presume any specific weak interaction theory, it does of course contain the Weinberg-Salam model as a special case with the coefficients

$$C_{1p} = \frac{1}{2}(1-4\sin^2\theta_w) \qquad C_{2p} = \frac{1}{2}(1-4\sin^2\theta_w)$$

$$C_{1n} = -\frac{1}{2} \qquad C_{2n} = -\frac{1}{2}(1-4\sin^2\theta_w) \tag{I.2}$$

$$C_{3e} = -\frac{1}{2}(1-4\sin^2\theta_w)$$

Here $\sin^2\theta_w$ is obtained from high energy experiments; the best current value[2a] being $\sin^2\theta_w = 0.23 \pm 0.01$.

Several points follow directly from (I.1 and (I.2).

(i) Because the electronic wave-function is relatively uniform over the nuclear volume the first term in (I.1) will be simply additive for all nucleons and to a good approximation can be re-written:

$$H^{(1)}_{PNC} = \frac{G_F}{2\sqrt{2}} Q_w \, \rho_N \, (\underline{r}_e) \, \gamma^5_e \tag{I.3}$$

where $\rho_N(r_e)$ is a normalized nuclear density $\int \rho_N(r_e)\, d\tau_e = 1$ and the so-called weak charge Q_w is defined by[3]

$$Q_w = Z(1-4 \sin^2\theta_w) - N. \tag{I.4}$$

(ii) As will be evident from (I.4) Q_w increases approximately in proportion to the size of the nucleus, in contrast to $H^{(2)}_{PNC}$ which through the spin term σ_N remains roughly constant throughout the periodic table. Thus we have the very important feature that for the heaviest nuclei $H^{(1)}_{PNC} \simeq 100\, H^{(1)}_{PNC}$. For this reason most experimental effort to date has been concerned with $H^{(1)}_{PNC}$.

(iii) Not only is $H^{(2)}_{PNC}$ smaller because it lacks the additive sum over nucleons, it is also proportional to $(1-4 \sin^2\theta_w) \simeq 0.08$ with the presently accepted value of the Weinberg angle. This near cancellation does however make the value of this term extremely sensitive to the exact value of the angle and ultimately this may be a point of significance. Similarly, higher order corrections will presumably modify all the coefficients and these changes may be much more visible in C_{2N} than in C_{1N}. Thus the pursuit of experiments sensitive to these coefficients is of great importance.

(iv) The electron-electron interaction is also both proportional to $(1-4 \sin^2\theta_w)$ and lacks the full additive feature of $H^{(1)}_{PNC}$. There are very great difficulties in devising an experiment to isolate its effect from $H^{(1)}_{PNC}$. But success would be of great importance since the PNC $e^- - e^-$ interaction is unlikely to be observed in high energy physics in the near future.

As a final point in this section we note that although it is essential to treat the electron relativistically in heavy atoms it is useful to have the non-relativistic limit of I.1 available explicitly: (we neglect the electron-electron part)

$$H^{NR}_{PNC} = \frac{G_F \alpha}{2\sqrt{2}} \left[\sum_{eN} \delta(r_{eN}) \left[C_{1N}\, \underline{\sigma}_e \cdot \underline{P}_e - C_{2N} \left(\underline{\sigma}_N \cdot \underline{P}_e + i \underline{\sigma}_N \cdot \underline{\sigma}_e \times \underline{P}_e \right) \right] + \text{h.c.} \right]$$

I.3 Electronic Matrix Elements

Accurate determination of the PNC constants C_{1N}, C_{2N} and C_{3e} from experiments will require detailed and reliable theories of the effect of the PNC operators in many electron systems. This is a major subject which is outside the scope of these lectures, and the interested student is referred to the review articles by the author[d] and by Fortson and Wilets[e]. Nonetheless it is useful to have some very approximate ideas about the various matrix elements and we therefore briefly summarize some points made in more detail by Bouchiat and Bouchiat.[3]

The first point to note is that the presence of the $\delta(r_{eN})$ restricts the matrix elements of H^{NR}_{PNC} to those between s and p functions which have the required non-zero density at the nucleus. The first term is a scalar in electronic space and so its non-zero matrix elements are between $s_{1/2}$ and $p_{1/2}$; the second term is a vector and can therefore have a non-zero element to the $p_{3/2}$ state as well.

We can get an order of magnitude estimate of the magnitude of the PNC matrix element by use of the effective quantum number results

$$\psi_s(0) \simeq \left(\frac{Z}{n^{*3}}\right)^{\frac{1}{2}} \qquad p\psi_p \simeq \left(\frac{Z}{n^{*3}}\right)^{\frac{1}{2}}$$

so that

$$\langle s_{1/2} \mid H^{(1)}_{PNC} | p_{1/2}\rangle \simeq G_F \frac{Q_w Z^2 \alpha}{(n_s^* n_p{}^2)^{3/2}} \tag{I.5}$$

$$\langle s_{1/2} \mid H^{(2)}_{PNC} | p_{1/2}, {}_{3/2}\rangle \simeq G_F \frac{Z \alpha}{(n_s^* n_p^2)^{3/2}}$$

Here n*, the effective quantum number, is of order 2 in heavy atoms.

On substituting the value $G_F = 2.22 \times 10^{-14}$a.u. we obtain as crude estimates

$$\langle s_{1/2} | H^{(1)}_{PNC} | p_{1/2}\rangle \simeq 10^{-16}\ Z^3 \text{ a.u.}$$

$$\langle s_{1/2} | H^{(2)}_{PNC} | p_{1/2}\rangle \simeq 10^{-16}\ Z^2 \text{ a.u.} \tag{I.6}$$

(I.6) illustrates the by now well-known Z^3 law for $H^{(1)}_{PNC}$.

We have not included the e - e interaction in (I.6) because it is not of immediate experimental interest. Preliminary estimates suggest that its matrix elements will be of order 10^{-16} Z^2 a.u. for $s_{1/2}$ to $p_{1/2}$ and 10^{-16} Z for $p_{3/2}$ to $d_{3/2}$.

I.4 Enhancement

In view of the very small values for PNC matrix elements in (I.6) it is not surprising that considerable ingenuity has been expended in finding circumstances in which the PNC effects are amplified sufficiently that they can be observed experimentally. The basic principle is as follows. Due to the presence of the PNC term in the Hamiltonian, both the initial and final states in a transition of interest are parity mixed

$$|\tilde{i}\rangle = |i\rangle + |i^{PNC}\rangle$$

$$|\tilde{f}\rangle = |f\rangle + |f^{PNC}\rangle$$

As a result the electromagnetic transition amplitude contains two terms (neglecting terms of order $(G_F)^2$)

$$A = \langle\tilde{f}|T|\tilde{i}\rangle = \langle f|T|i\rangle + \{\langle f_{PNC}|T|i\rangle + \langle f|T|i_{PNC}\rangle\} = A_{reg} + A_{PNC}$$

The corresponding physical processes - absorption, emission or refractive index - will also contain two terms

$$I = |A_{reg}|^2 + \{A_{reg}\, A^*_{PNC} + A_{PNC}\, A^*_{reg}\}$$

The fractional size of the PNC part is just

$$F \simeq \frac{|A_{PNC}|}{|A_{reg}|}$$

Clearly the magnitude of this PNC fraction can be enhanced by making A_{PNC} as large as possible and by utilizing a transition for which the regular amplitude is weak. The order of magnitude situation for current experiments is set out in table I.2 below.

In H, the near degeneracy of the $2s_{1/2}$ and $2p_{1/2}$ levels which are split by $\Delta W \simeq 10^{-7}$ a.u. means that a PNC E1 transition between $2s_{1/2}$ states, nominally of the same parity, is enhanced by a factor of order 10^7 relative to (I.6). The regular transition here is an allowed M1 transition of strength α atomic units. In Bi and Pb one also makes use of a normal M1 transition but the PNC E1 is enhanced by the Z^3 factor rather than by near degeneracy. In the highly forbidden transitions in Cs and Tℓ, the M1 amplitude is of order α^3 a.u.,

Table I.2

Element/ Transition	PNC Amplitude	Reg Amplitude	A_{reg} A_{PNC} (Atomic units)	$F = \frac{\lvert A_{PNC}\rvert}{\lvert A_{reg}\rvert}$
H $2s \to 2s$	Degeneracy enhanced E1	M1	10^{-11}	10^{-7}
Bi, Pb $6p^n \to 6p^n$	Z^3 enhanced E1	M1	10^{-12}	10^{-8}
Cs $6s \to 7s$; Tℓ $6p_{\frac{1}{2}} \to 7p_{\frac{1}{2}}$	Z^3 enhanced E1	Forbidden M1	10^{-17}	10^{-3}

with again the Z^3 enhancement, the result being a PNC fraction of order 10^{-3}.

At first sight the higher PNC fraction for the Cs and Tℓ experiments in highly forbidden M1 transitions might seem to imply that these experiments are the best so far devised. This is not necessarily so because as the final column of the table shows the high PNC fraction is at the expense of a very small magnitude for the PNC effect in absolute terms. Clearly in the limit one can obtain as high PNC fraction as one likes by having a sufficiently forbidden regular transition. But in practise this is at the expense of the absolute magnitude to such an effect that intensity and background problems become so severe that the high fractional effect is not in fact usable. This is indeed the situation in the Cs and Tℓ experiments where background problems have made the low signal levels implied by the use of such highly forbidden transitions impossible to work with and both experiments now use an E1 induced by an external electric field in place of the M1 as the regular transition amplitude, with corresponding increase in overall signal but decrease in PNC fraction.

LECTURE II PNC PHENOMENA

In the previous lecture we showed that one can enhance the small effects of the PNC interaction to measurable size. In this lecture we discuss in more detail the specific PNC phenomena which are of experimental interest. We follow the approach of Bouchiat, Poitier and Bouchiat[4] and list the possible phenomena in terms of the PNC invariants which are involved. We then show with a couple of examples how this approach is related to a calculation involving atomic matrix elements. Finally we discuss in general terms the problem of spurious PC effects which 'mimic' the PNC effects of interest.

II.1 PNC Invariants

Our aim here is to construct from the observable quantities which define the experiment the possible rotational invariants which are odd under parity and even under time reversal. A list of such experimental quantities together with their P and T transformations is given below.

		P	T
Photon momentum	$\underline{k}$	−	−
Linear polarization	$\underline{\varepsilon}$	−	+
Circular polarization	η	−	+
Magnetic field	$\underline{B}$	+	−
Electric field	$\underline{E}$	−	+
Angular momentum	$\underline{J}$	+	−

We restrict our attention first to a transition induced in an atom in the absence of external fields ($\underline{B} = 0$, $\underline{E} = 0$). The only PNC invariant on which the result of an experiment can depend is then the circular polarization η, giving rise to the well known phenomena of optical rotation and circular dichroism.

When we extend our attention to $\underline{E}$ field induced transitions the possibilities widen somewhat. Taking the case of circularly polarized radiation first, the only PNC combination (linear in E) which does not involve atomic polarization is seen to be

$$\eta(\underline{k}.\underline{E}\times\underline{B}).$$

This process is a close analogy to the Faraday effect $\eta\,\underline{k}.\underline{B}$ except that $\underline{B}$ is replaced by $\underline{E}\times\underline{B}$. In a similar way we see that the only term available for plane polarized light is

$$(\varepsilon.\underline{B})(\underline{\varepsilon}.\underline{E}\times\underline{B})$$

We note that this process is analogous to magnetic birefringence which is described by $(\varepsilon.B)(\varepsilon.B)$ except that again we have a replacement $\underline{B} \rightarrow \underline{E}\times\underline{B}$.

If we now allow experiments sensitive to atomic polarization then terms in which $\underline{J}$ replaces $\underline{B}$ are allowed since they have the same transformation properties. An important example is

$$\eta(\underline{k}\times\underline{E}.\underline{J})$$

where J can be atomic polarization either in the initial or final state. The experiments on Cs and Tℓ are of this type. If as in these experiments one produces or analyses the polarization $\underline{J}$ by circular polarization in a second PC transition, one can describe the overall process by a PNC invariant

$$\eta_1(\underline{k}_1\times\underline{E}.\underline{k}_2)\eta_2.$$

In more complex experiments many other PNC possibilities arise, a detailed analysis of the radio-frequency hydrogen experiments in these terms has been given by Dunford, Lewis and Williams.[5]

II.2 Atomic Matrix Elements

While the PNC invariants are very useful in thinking about possible types of experiment, detailed and specific consideration is also important both to estimate the magnitude of the effect and to be sure of the precise circumstances in which it occurs. We illustrate this with two examples.

We consider first circular polarization dependence in an M1 transition with a small admixed PNC E1 amplitude. The strength for a circularly polarized photon travelling in the $\hat{z}$ direction will be proportional to

$$I_\eta = \left|\sum_{M'M} \langle J'M'|-\underline{\mu}.\underline{B}_\eta - \underline{D}.\underline{E}_\eta|JM\rangle\right|^2$$

where $\underline{\mu}$ and $\underline{D}$ are the magnetic and electric dipole operators and the fields B_η and E_η are given by

$$E_\eta = \frac{(\hat{x} - i\,\eta\,\hat{y})}{\sqrt{2}}, \quad B_\eta = i\eta\frac{(\hat{x} - i\,\eta\,\hat{y})}{\sqrt{2}}$$

$\eta = \pm 1$ represent right and left circularly polarized light ($-\eta$ = helicity). The PC part of the transition will have the form

$$I_\eta^{PC} = \sum_{MM'}\left|\langle J'M'|\frac{(\mu_x - i\;\mu_y)}{\sqrt{2}}|JM\rangle\right|^2$$

while the PNC term will be

$$I_\eta^{PNC}=2\text{ Real Part}\sum_{MM'}\langle J'M'|D_x - i\eta\,D_y|JM\rangle_{PNC}\langle J'M'|i\eta(\mu_x - i\eta\,\mu_y)|JM\rangle^*.$$

Bearing in mind that the PNC matrix elements are pure imaginary in the usual phase convention we see that terms proportional to η survive in I^{PNC}, for example

$$\sum_{MM'}\langle J'M'|D_x|JM\rangle_{PNC}\langle J'M'|i\eta\,\mu_x|JM\rangle^*$$

It is convenient to carry out the sum over magnetic quantum numbers and to write the PNC effect in the more transparent form

$$\frac{I_+ - I_-}{I_+ + I_-} = 2\ \text{Imag.}\ \frac{E1^{PNC}}{M1}$$

where $E1^{PNC}$ and M1 are the respective reduced matrix elements

$$M1 = <J'||\mu||J>, \quad E1^{PNC} = <J'||D||J>.$$

Of course the real theoretical problem remains the calculation of $E1^{PNC}$ but this is outside the scope of these lectures.[e]

As a second example we consider the electric field induced transition leading to $\eta\ \underline{k}.\underline{E}\times\underline{B}$.. We choose $\underline{B}$ in the z direction, $\underline{E}$ in the x direction and $\underline{k}$ in the y direction. The PNC interference term linear in $\underline{E}$ will then be proportional to terms of the form

$$\text{Real Part} <J'M'|D_z - i\eta D_x|JM>_{PNC} <J'M'|D_z - i\eta D_x|J''M''><J''M''|D_x|JM>$$

where $|J''M''>$ is a state admixed into $|JM>$ by the static electric field. Bearing in mind that the PNC E1 elements are pure imaginary we see that real terms survive of the form (in an obvious notation)

$$D_z^{PNC}\ i\eta\ D_x\ D_x^E \quad \text{and} \quad i\eta\ D_x^{PNC}\ D_z\ D_x^E$$

respectively for M' - M=0 and ±1.

An important feature of this interference term is that it vanishes if we sum over M and M'. This can most easily be seen by writing D in spherical tensors and writing out the magnetic dependence of for instance the first term above:

$$(-1)^{J'-M'} \begin{pmatrix} J' & 1 & J \\ -M & 0 & M \end{pmatrix} (-1)^{J'-M'} \begin{pmatrix} J' & 1 & J'' \\ -M' & q & M \end{pmatrix} (-1)^{J''\ M''} \begin{pmatrix} J'' & 1 & J \\ -M'' & -q & M \end{pmatrix}$$

It follows immediately from the symmetry properties of the 3j symbols that the contribution from opposite sign of M and q values cancel identically. Thus in order for this term to be non-zero the magnetic field must be sufficiently great that the different magnetic transitions are resolved. This important role of $\underline{B}$ is hardly surprising since the PNC invarient $\eta\ \underline{k}.\underline{E}\times\underline{B}$ clearly requires $\underline{B}$ to be physically significant.

II.3 Spurious Effects

A major problem in all PNC experiments is the presence of spurious effects which to some extent 'mimic' the small PNC signal of interest. In particular, the atomic levels of opposite parity which are mixed by the PNC Hamiltonian can also in general be mixed by the interaction with any unwanted apparatus electric fields. The types of phenomena induced by such spurious mixtures, which are clearly of central importance, can conveniently be discussed in the same way as the PNC effects above.

In the table below we list the PNC phenomena of interest together with some (not all) PC effects which can cause experimental difficulty.

PNC Effect	'PC mimics'
$\eta = \underline{\sigma}.\underline{k}$	$\eta\ \underline{k}.\underline{B};\ (\varepsilon.B)^2;\ (\varepsilon.E)^2$
$\eta\ \underline{k}.\underline{E}\times\underline{B}$	$\eta(\underline{k}\times\underline{E}.\underline{E}\times\underline{B});\ \eta(\underline{k}.\underline{B})E^2$
$\underline{\varepsilon}.\underline{B}\ \underline{\varepsilon}.\underline{E}\times\underline{B}$	$(\underline{\varepsilon}.\underline{E}\times\underline{B})^2;\ (\varepsilon.E)^2B^2$
$\eta_1\ k_1.E\times k_2\eta_2$	$\eta_1\ (k_1\times E).(E\times k_2)\eta_2$

An important advantage of the optical rotation/circular dichroism experiments is apparent from the table. In the absence of 'intentional' electric and magnetic fields the only first order spurious effect is the Faraday effect in a residual magnetic field - there can be no term linear in an unwanted electric field. This is not the case for the case of E field induced phenomena where the additional terms due to an unwanted D field can constitute a significant problem. It is particularly important to note that such effects limit the usefulness of 'chance' coincidences between states of opposite parity since both the PNC and the spurious effect are proportional to the energy difference ΔW.

While the table above indicates the possible form of spurious effects detailed calculation in each case is required to establish the absolute magnitude. In particular it is sometimes possible to choose the detailed operating conditions such that the spurious effect is either reduced in relative magnitude or in a favourable case removed entirely.[4]

II.4 Wavelength Dependence

Thus far we have discussed the form of the PNC phenomena but not their dependence on wavelength. This dependence can be very important for distinguishing the effect from background and also on occasions from mimicking effects which may have a different wavelength dependence. We discuss one example, others are described in the paper by Bouchiat, Poitier and Bouchiat.[4]

The dependence of refractive index on circular polarization η leads to an optical rotation angle

$$\phi = \frac{\pi L}{\lambda} \frac{(n^+ - n^-)}{} = - \frac{4\pi(n-1)L}{\lambda} \text{ Imag. } \frac{E1^{PNC}}{M1}$$

Thus the wavelength dependence has the dispersive form of the refractive index and is antisymmetric about the resonant frequency. It is easy to show that the corresponding Faraday effect has a symmetric form and hence can be readily distinguished experimentally. One can also readily show that at an atomic density which gives one absorption length at line centre this angle becomes $\pm R/_2$ at the peaks, where

$$R = \text{Imag. } \frac{E1^{PNC}}{M1}$$

LECTURE III BISMUTH OPTICAL ROTATION

III.1 Introduction

The most sensitive parity experiments have proved to be those looking for optical rotation. The reason is a happy coincidence between technique and atomic properites. The straight-through geometry of the measurement system and the availability of excellent polarisers allows full use of the high intensity and wavelength control of modern CW lasers. At the same time the optimum optical depth of an atom under study is one absorption length and for an allowed M1 transition this can be readily achieved at workable vapour pressures in an oven some tens of centimetres long.

Although an experiment on Pb is under preparation all experiments to date have worked with Bi. The reason is that the element is heavy, as required by the Z^3 law, and has two allowed M1 transitions in convenient spectral regime for modern CW lasers:

$$\text{Bi:} \quad 6p^3 \rightarrow 6p^3 \quad J = 3/2 \rightarrow J = 3/2 \quad 876 \text{ nm}$$

$$J = 3/2 \quad J = 5/2 \quad 648 \text{ nm}$$

A major difference between these transitions is that the one at 648 nm is overlaid with molecular absorption about equal in strength to the atomic M1 line whereas the 876 nm is essentially free. Indeed this absorption appears to rule out experiments on the other Bi transi tions at higher frequencies.

Table III.1 Calculations of $R = \text{Im E1}^{PNC}/\text{M1}$ for Bi

	648 nm	876 nm
	$R \times 10^8$	
Semi-empirical (Novikov et al)	-17	-13
Dirac Hartree-Fock (Carter and Kelly)	-22	-16
Parametric potential plus shielding (Harris, Loving and Sandars)	-13	-11
Dirac Hartree-Fock plus shielding (Martensson et al)	-10	- 8

As has been pointed out in lecture II, the optical rotation to be expected is proportional to and of order of the dimensionless ratio

$$R = \frac{\text{Imag.E1}^{PNC}}{\text{M1}}$$

A great deal of work has been put into an attempt to make reliable calculations of R for the two transitions. This work has been reviewed by the author[e] and by Fortson and Wilets[d] and we simply report the present position in the table above.

The reasonable agreement among the theoretical estimates may well be fortuitous and is certainly not shared by the experiments. There are at present five Bi experiments on the two lines and they are listed below with their latest published results:

Table III.2 Current experimental results for Bi experiments

	$R \times 10^8$	Reference	Figure
Oxford 648 nm	-8.9 ± 2	(5)	(3)
Novosibirsk 648 nm	-20.2 ± 2.7	(6)	(4)
Moscow 648 nm	-2.4 ± 1.3	(7)	(5)
Seattle 876 nm	-10.4 ± 1.7	(8)	(6)
Oxford 876 nm	(not yet available)	(5)	(7)

Description of the individual experiments are available in the original papers (5) to (8) or in the reviews (c) and (d). The disagreement over results means that a detailed study of the similarities and differences between the experiments is essential and in this lecture we attempt to throw some light on this by making a point by point comparison of all five experiments. We do this in terms of five main headings: angle measurement, wavelength dependence, Bi dependence, removal of Faraday effect and special checks. A brief summary is found in table III.3, and schematic diagrams in figs. (1), (2), (3), (4), (5).

III.2 Angle Measurement

All the experiments use the same basic method of angle measurement: the intensity through crossed polarizers is given by

$$I = I_o\{(\phi_M+\phi)^2 + b\}$$

where ϕ_M is an impressed modulation angle, ϕ is to be measured, b ($\simeq 10^{-6}$) is the background transmission when the polarizers are fully crossed. A term linear in ϕ is obtained by extracting the cross-term

$$I_M = 2\phi_M \, \phi \, I_o.$$

With

$$\phi_M \simeq (b)^{\frac{1}{2}} \simeq 10^{-3}$$

we see that

$$\frac{I_M}{I} \simeq \frac{\phi}{10^{-3}},$$

thus an angle of the expected magnitude, $\phi \simeq 10^{-7}$, gives rise to a fractional intensity change of order 10^{-4}. This 'amplification' by the use of good crossed polarizers is an essential feature of the method. Indeed the angle measurement sensitivity of all the experiments is so good ($\sim 10^{-8}$ in 5 minutes) that other experimental problems dominate.

The specific means used to extract the cross-term I_M and make the angle measurement varies from experiment to experiment. Seattle, Oxford 876 and Moscow use a Faraday modulation cell to produce a sinusoidal variation, $\phi_M \cos \omega t$, and the cross-term I_M which also varies at ω is extracted by phase sensitive detection. In the Seattle and Oxford 876 nm experiments the dependence of I_M on intensity variations is removed by electronically dividing the light transmitted through the polarizer against the full beam incident on

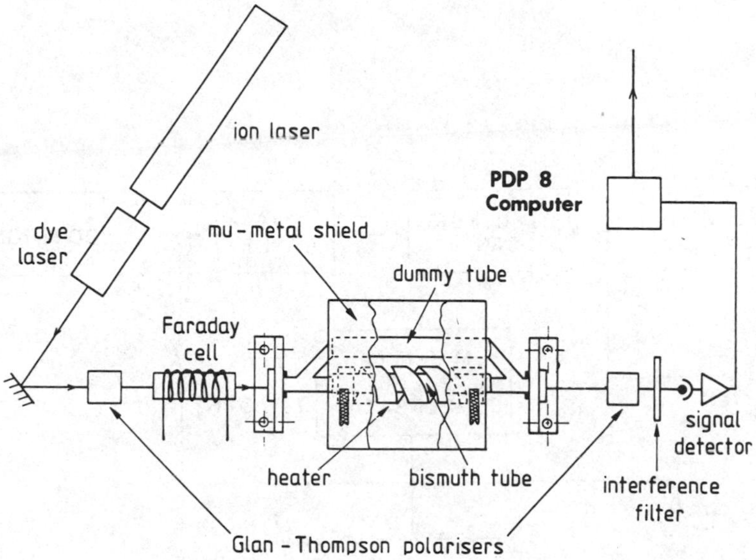

Fig. 1. The Oxford 648 nm experiment.

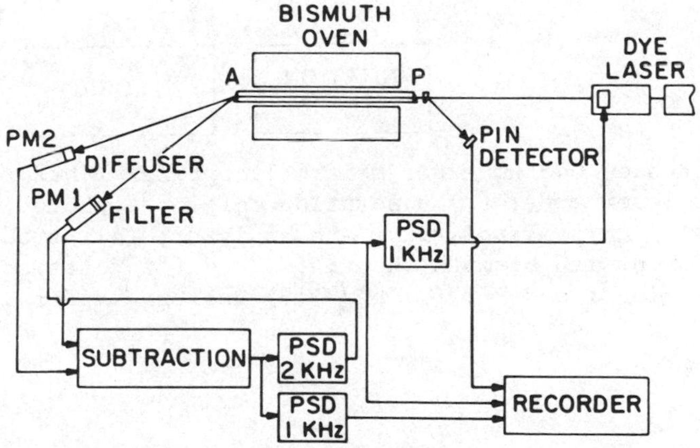

Fig. 2. The Novosibirsk 648 nm experiment.

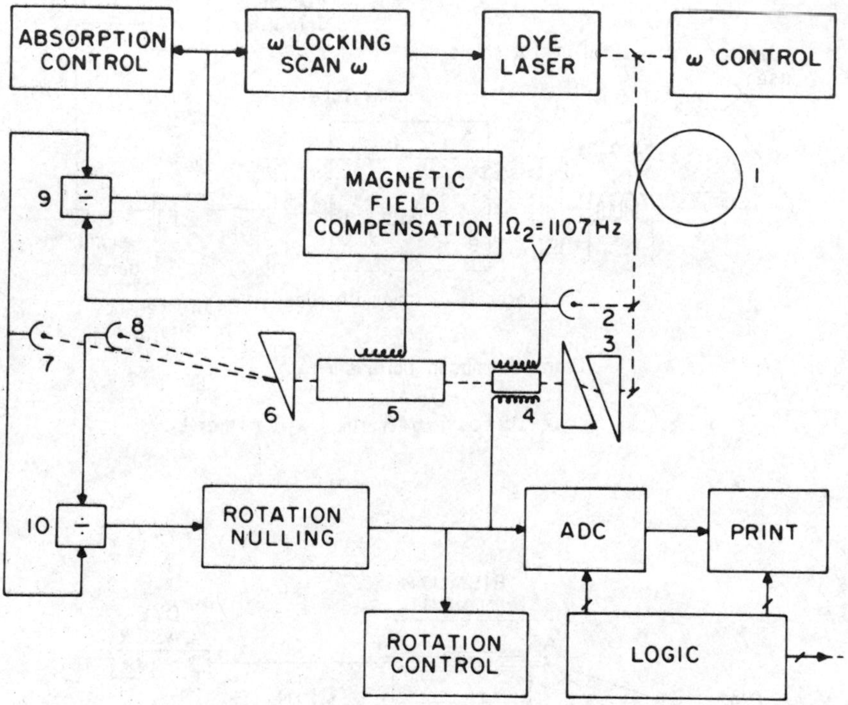

Fig. 3. The Moscow 648 nm experiment: (1) optical fiber to transmit pure laser mode; (2) absorption reference photodiode; (3) and (6) crystal polarizer and analyzer; (4) Faraday cell; (5) oven with bismuth vapor; (7) and (8) rotation reference photodiode; and PM; (9) and (10) analog dividers.

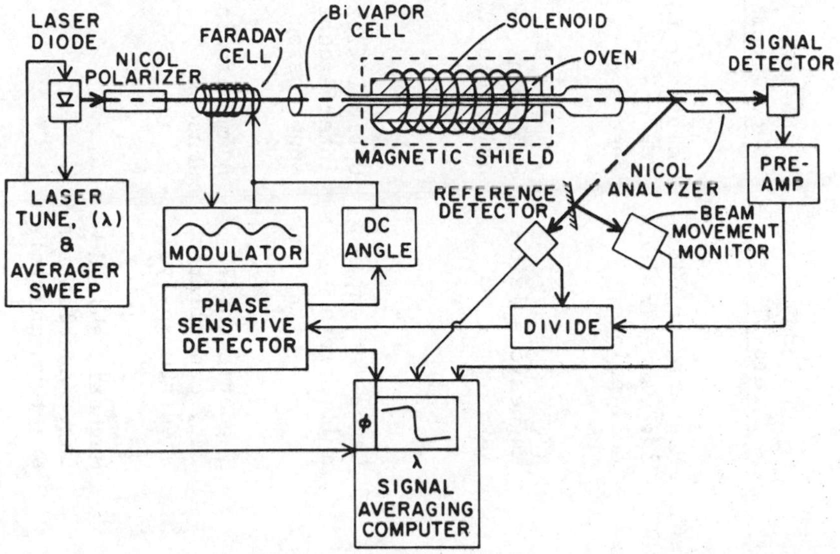

Fig. 4. The Seattle 876 nm experiment.

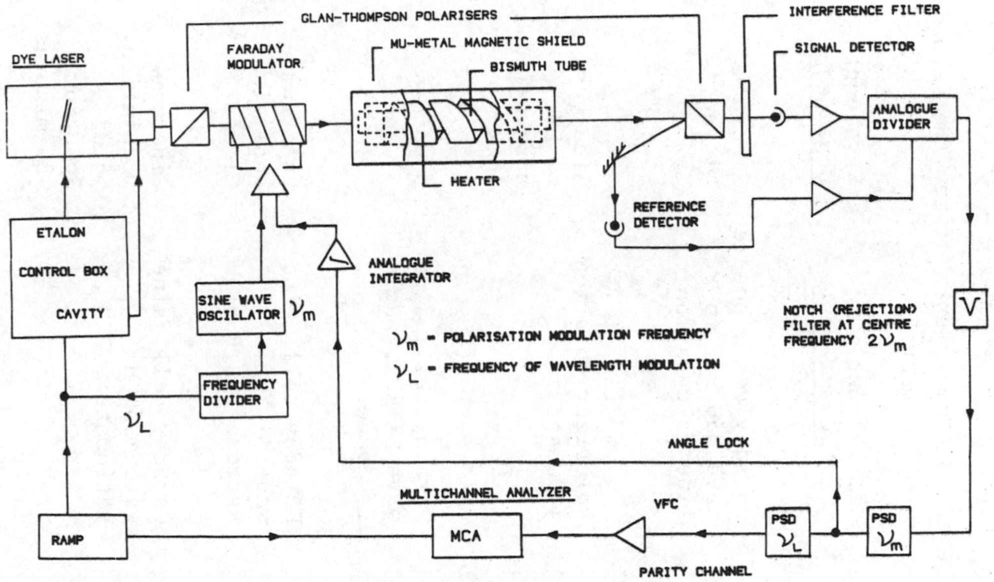

Fig. 5. The Oxford 876 nm experiment.

Table III.3 Comparison between the methods used in the different bismuth experiments

Experiment	Angle measurement	Wavelength dependence	Bismuth dependence	Faraday removal	Checks/ special points
Oxford					
648 nm Dye laser	Faraday cell, square wave 350 Hz, on-line calculation	2 points ∿1 sec	Double oven, automatic interchange ∿1 minute	Magnetic shielding cancellation null points	Polarizer rotation on-line checks
Novosibirsk					
648 nm Dye laser	Polarizer rotation ∿100 secs	Sine wave ∿1 kHz absorption lock		Magnetic shielding	Molecular, E2 null points
Moscow					
648 nm Dye laser	Faraday cell, sine wave ∿1 kHz, angle feedback	3 points ∿1 sec	Oven temperature ∿½ hour	Cancellation, null points	Optical fibre, interference suppression
Seattle					
876 nm Diode laser	Faraday cell, sine wave ∿1 kHz Ratiometer, PSD	Spectrum analysis ∿1 sec	Bismuth density	Magnetic shielding, cancellation spectrum analysis	Beam position sensors
Oxford					
876 nm Dye laser	Faraday cell, sine wave ∿350 Hz Ratiometer, PSD	Sine wave modulation ∿6 Hz Spectrum ∿20 minutes	Bismuth density	Magnetic shielding, cancellation spectrum analysis	

the second polarizer. Moscow, on the other hand, uses direct feedback to keep the nett angle between the polarizers zero and uses the feedback angle as the measured quantity. Oxford 648 also uses a Faraday modulator but ϕ_M has a square wave form consisting of three periods of order a millisecond duration with angles +A, 0, -A respectively. The corresponding intensities are digitised and then the angle is calculated in real time from

$$\frac{2\phi}{A} = \frac{I_1 - I_3}{I_1 + I_3 - 2I_2} .$$

The main advantage of this method is that only a single detector is required and the normalization against intensity changes is carried out digitally.

The Novosibirsk experiment does not use a Faraday modulator and ϕ_M is produced by a mechanical off-set of the second polarizer. However the method of angle measurement is so closely tied up with the procedure for changing wave-length that we defer discussion to the next section.

While the crossed polarizer method of angle measurement is very sensitive, a number of potential systematic effects involving non-ideal polarizers combined with beam movement and with interference have been identified. Steps taken to eliminate these problems will be outlined in section (III).

III.3 Wavelength Dependence

The fundamental difficulty inherent in the Bi optical rotation experiments is the lack of any means of changing the size or sign of the effect, as is possible in the experiments using electric field induced transitions. Besides the density of the Bi itself, the only discriminant available to remove background effects is the dependence of the optical rotation on the laser wavelength.

For a single transition this dependence has a simple dispersive form but in Bi there is a much more complicated spectrum due to the presence of hyperfine structure on the transition line. However the form of this spectrum is well-understood since the parameters on which it depends can be determined from a detailed study of the corresponding spectrum of the Faraday effect under the same experimental conditions. Such a study has been made for all experiments and the agreement with theory is good, giving confidence in the predicted form of the optical rotation spectrum (figs. 6 ,7).

Two basic methods have been adopted to utilize the wave-length dependence. In the Seattle 876 and the Oxford 876 experiments a large portion of the hyperfine structure is scanned repetitively

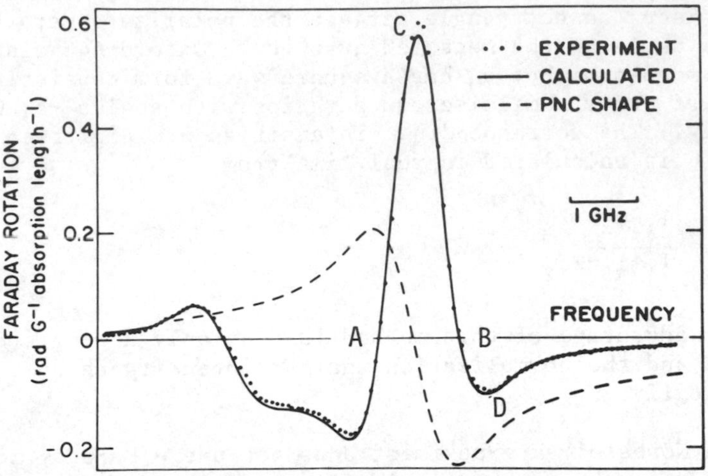

Fig. 6. Theoretical PNC and Faraday curves for the (6, 7) hyperfine component at 876 nm in bismuth. The experimental Faraday points taken at Oxford show excellent agreement with the theory.

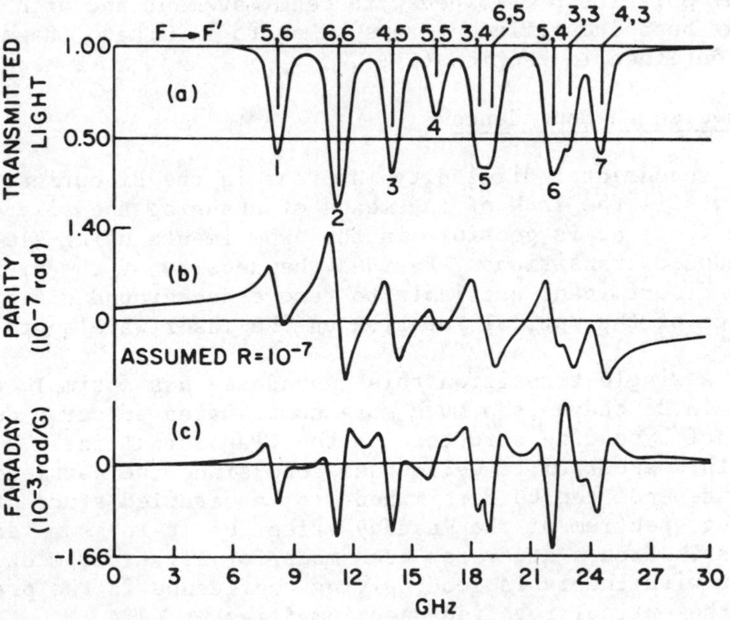

Fig. 7. Theoretical absorption, parity and Faraday curves for the 876 nm transition.

($\sim$ secs) and the angle $\phi(\lambda)$ is stored. The resulting time averaged ($\sim$ 1 hour) wavelength spectrum is fitted to a linear combination of (a) a Faraday curve to eliminate the effect of residual magnetic fields, (b) a predicted ϕ^{PNC} curve and (c) various smooth linear and quadratic terms to allow for variation in apparatus parameters. Each such fit gives an amplitude for the PNC angle which is regarded as a single measurement, to be repeated many times under varying conditions.

The Oxford 648 and Moscow experiments have concentrated on the isolated $F = 6 \rightarrow 7$ component of the 648 nm hyperfine spectrum and compare angles at two points A and B (fig. 6) in the case of Oxford and A, B and C in Moscow. This wavelength jumping has a repetition rate of order 1 sec and is achieved through the available electronic control mechanisms on the commercial dye laser systems. The resultant wavelength dependent angles are accumulated digitally in both cases.

The Novosibirsk system is rather different and requires separate discussion in that the wavelength change is more rapid than the angle measurement, strictly defined. We outline now the principle of the current Novosibirsk experiment; the earlier experiment differs only in the practical detail that the laser wavelength could change only in discrete steps rather than continuously.

The basic feature of the Novosibirsk experiment (fig. 2) is that the second polarizer splits the light beam into two parts, one with the same polarization as the first polarizer and the other much smaller beam with perpendicular polarization. We can write these two intensities in the form

$$I_{11} = I_o A(\lambda)$$

$$I = I_o A(\lambda)\ \{\phi_M^2 + 2\phi_M\ \phi_{PNC}(\lambda)\}$$

where A (λ) is the wavelength dependent absorption in the bismuth. Use is made of the fact that the absorption is an approximately symmetric function of wavelength about the absorption line while the PNC angle ϕ_{PNC} has an antisymmetric dispersive shape. The wavelength is modulated sinusoidally $\lambda = \lambda_o + \delta\lambda \cos\omega t$, with $2\pi\omega \sim 1$ kHz. If λ_o is chosen so that I_{11} contains no first harmonic at ω, I can only contain such a term through the presence of the ϕ_{PNC} term. In practice λ_o is fixed by a feedback system so that $I_{11}(\omega) = 0$ and I (ω) is measured by phase-sensitive detection for angles ϕ_M and $-\phi_M$, the angle difference being produced by mechanical rotation of the second polarizer. ϕ_{PNC} can now be determined from $\Delta = I^+ - I^-$ provided ϕ_M and the intensity are independently determined.

The main advantage of the Novosibirsk method is that the rapid change of wavelength makes the system insensitive to angle noise slower than the modulation frequency $\sim$1 kHz, in contrast to all the other experiments where the wavelength changes are on a period of order seconds. A major disadvantage is that the central laser frequency is forced to be at those specific points on the absorption curve where the first harmonic is zero.

III.4 Bismuth Dependence

An important feature of all the optical rotation experiments is the existence of spurious wave-length dependent angles of the same order of magnitude as the expected Bi effect. While the origin of these angles in beam movement and interference is gradually becoming understood is is clearly of the utmost importance to separate out experimentally that part of the wavelength dependent angle which depends on the presence of the bismuth.

In the case of the Seattle and Oxford 876 experiments which have continuous wavelength scan, the problem is perhaps not so severe because wavelength dependence of the background angle won't 'mimic' the extremely complex PNC hyperfine spectrum and therefore won't contribute appreciably to the PNC term in the fitting procedure. Thus in these experiments no special steps are taken to establish the Bi dependence directly, though experiments are carried out on different bismuth densities to check that the measured quantity has the correct dependence on optical path length.

The problem is more severe when only a few wavelength points are used and for this reason the Oxford group has developed a sophisticated automatic double oven system in which the bismuth is rapidly ($\sim$secs) inserted into or removed from the polarimeter without movement of any optical component. A similar insertion or removal of the bismuth can be achieved, though less rapidly ($\simeq$ 30 minutes), in the Moscow 648 experiment by varying the vapour pressure of bismuth by heating or cooling the oven.

The problem of extracting the bismuth dependent, wavelength dependent, angle is complicated in the Novosibirsk 648 experiment by the fact that their method intrinsically requires the presence of the bismuth in the feedback which sets the first harmonic on the parallel signal to zero. Thus it is not possible for them directly to compare the signal with and without bismuth. The alternative procedure which they use is to take a number of 'null' points which are similar to the $F = 6 \rightarrow 7$ component in their absorption but have no PNC angle predicted. The apparatus set up is considered satisfactory when these null points give zero and a measurement on the $6 \rightarrow 7$ peak is then carried out. In effect, by using several points on the curve the Novosibirsk group argue (as is done for the scanning experiments) that there is no plausible reason for the wavelength dependent angle to be fortuitously large at and only at the position of the $6 \rightarrow 7$ component.

We note at this point that detection of a bismuth dependent angle with the correct wavelength dependence does not in itself constitute evidence for a PNC effect. This is because the PNC wavelength dependence is identical to that of the refractive index and it is possible to envisage a mechanism whereby light is deflected, perhaps by density variations, by an amount which depends on the refractive index and this deflection in turn might lead to a change in the measured angle of polarization. Calculation suggests that this effect could possibly be comparable to the PNC angle. In the Washington experiment the beam movement is monitored with a position sensor and the component with the dispersive form of ϕ_{PNC} is extracted; this is found to be too small to be a danger. In the Oxford experiment the second polarizer is rotated at regular intervals ($\sim$ 1 hour) since this should show up and average out deflection effects. In addition the experiment has been done with the polarization rotated through 90°. Apart from one early experiment, no significant effects have been found. The Moscow experiment relies on careful design of the oven to make it as symmetric as possible together with the use of very high quality polarizers which should minimize the effects of beam movements.

III.5 The Faraday Effect

The Faraday effect plays a centrol role in the bismuth experiments. On the plus side it forms a useful indicator for wavelength position and an invaluable method for determining the bismuth optical depth. But on the negative side, it is a potentially serious source of systematic error.

Taking the normalization question first, one can readily see that the ratio of PNC angle to Faraday angle at a known magnetic field is directly proportional to $R = \text{Imag.}E1^{PNC}/M1$, the experimental quantity to be determined. This is used in a direct way in the scanning experiments (Washington, Oxford 876) by superimposing on the bismuth a known longitudinal $\underline{B}$ field, the resulting pattern being accumulated and analysed in the same way as for the PNC data runs. The Oxford and Moscow experiments also use the same measurement procedure for their PNC and Faraday measurements; the difference being that the PNC effect is measured between A and B the Faraday effect is measured between C and D (fig.6). Faraday normalization is rather more difficult in the Novosibirsk experiment because the centre frequency is fixed at a point where the first harmonic is small; the measurement has to be made on another absorption line, which is chosen to be more sensitive to the Faraday effect.

Because the Faraday effect is potentially large compared to the expected PNC angles (10^{-4} at 1 gauss compared to 10^{-7}) considerable effort has been expended in ensuring that it does not lead to systematic error. The Washington, both Oxford and the Novosibirsk experiments have μ-metal shielded ovens in which the residual field is a few milligauss and this is further cancelled by an applied field. The Moscow experiment has no shielding, though the longitudinal field can be very accurately cancelled. In the scanning experiments, any residual Faraday effect is recognized and removed by the curve fitting procedure. Since the contribution is only about the same magnitude as the expected PNC effect, this should be entirely adequate. The Oxford and Moscow 648 experiments exploit the fact that the maximum difference $\Delta\phi^{PNC}$ coincides with zero $\Delta\phi^{F}$ (fig. 6), thus greatly reducing the sensitivity to external fields. In the Oxford experiment this is carried further by using a small reversing applied field to lock the central laser wavelength to the points where the Faraday angles exactly cancel. Tests with applied fields show that here too the Faraday effect is not a danger. This option is not available in the Novosibirsk experiment since the wavelength is fixed by the absorption curve. Thus one has to rely primarily on very good shielding and cancellation.

We note here for completeness that in addition to the Faraday effect linear in a longitudinal magnetic field there is also magnetic birefingence proportional to B^2 and that this can in certain circumstances mimic the optical rotation signal. Calculation suggests that this effect is negligible in the experiments with shielded ovens but could conceivably be significant in the present version of the Moscow experiment.

A rather more subtle possible systematic error which the Faraday effect might produce can occur if there is an AC magnetic field on the sample (presumably at line frequency) and the laser frequency is modulated at the same frequency in which case one has a term which is antisymmetric about line centre and mimics the PNC angle. All the groups are aware of this potential difficulty and at Oxford and Seattle measurements have been made which show that it is too small to be significant.

III.6 Results and Discussion

The most obvious point to note about the results listed in table III.2 is that the three experiments on the 648 nm line differ by appreciably more than their quoted errors. At least two, and possibly all three, must suffer from significant systematic errors. The potential for such effects is also indicated by the difference between the present Oxford 648 nm result and that previously published ($R = +2.7 \pm 4.7 \times 10^{-8}$). Nor is the situation necessarily better on the 876 nm line where the present Washington result differs from one previously published ($R = -1.1 \pm 1.9 \times 10^{-8}$). In view of this, a result from the Oxford 876 nm experiment is awaited with considerable interest.

In such circumstances the study and elimination of systematic errors is of crucial importance. In this respect, the wide diversity of detailed method between the experimenrs as set out in table III.3 is of considerable advantage in the task of identifying the source of the errors. These can be of two types. False effects which mimic a PNC signal and add or subtract from the real one. Or a normalization fault in which a real effect is measured but is given the wrong absolute value through, for example, an incorrect measurement of the Bi optical depth. Many such effects are possible. Perhaps the most obvious example of the first class is the bending of the light beam by refractive index gradients in the Bi, combined with non-ideal polarizers which was mentioned earlier. Such an effect should be detectable either directly through beam movement or by repeated experiments with different orientations of the second polarizer with respect to the Bi oven. Normalization errors on the other hand, should not be present because of the ease with which a known Faraday effect can be measured with the same detection system as the PNC signal.

What then can be said at the present time about PNC optical rotation in bismuth? Perhaps a significant point here is that the Oxford 648 nm and the Washington 876 nm experiments are second generation and include a large number of improvements, particularly in the form of checks against known or suspected systematic errors. The two groups concerned believe that their experiments are now reliable. Whether this is so will perhaps be demonstrated when the results from the next generation of experiments from Novosibirsk and Moscow become available in the near future.

Even when agreement is reached work on bismuth will continue. The excellent statistical sensitivity means that one should be able to determine the nuclear spin dependent term through a variation at the 1% level of the observed PNC angle for different hyperfine components.

LECTURE IV CAESIUM, THALLIUM, HYDROGEN AND OTHER POSSIBILITIES

IV.1 Highly Forbidden M1 Transitions

(i) Circular Dichroism in Cs. In their pioneering 1974 paper the Bouchiat proposed a very elegant experiment to search for circular polarization in the highly forbidden 6s → 7s M1 transition of Cs. The fractional circular polarization is given by $\delta = 2\ \mathrm{Imag.}\ E1^{PNC}/M1$. $E1_{PNC}$ has been calculated both by the Bouchiats and by the author with satisfactory agreement on a value $-i \times 1.7 \times 10^{-11} |e| a_0$. The M1 matrix element was initially unkown but has been measured by the Bouchiat group who find $M1 = -4.24 \times 10^{-5}\ \mu_B$. These numbers give a theoretical prediction $\delta = 2.2 \times 10^{-4}$ for the circular polarization.

The PNC fraction is satisfactorily large but this is achieved at the expense of a transition probability $\sim 10^{-12}$ of that for a normal atomic transition. Nonetheless calculations indicate that present laser intensities are in principle sufficient to give adequate signal to noise. Unfortunately, the Paris group has found in practice that at usable Cs pressures the background from collision induced transitions seriously exceeds the M1 transition rate and precludes straightforwars application of the originally proposed method. It may be possible to avoid the difficulty while retaining sufficient signal to noise by using reduced Cs vapour pressure or an atomic beam, but the Bouchiat team have adopted an alternative approach.

(ii) E1 Field Induced Polarization in Cs $6s_{1/2} \rightarrow 7s_{1/2}$. They look at the same $6s \rightarrow 7s$ transition, but induced by the presence of static electric field E_0. As discussed in Lecture II, interference between $E1^{PNC}$ and $E1^{ind}$ then leads to non-zero polarization. This is given by

$$\underline{P}_e = \frac{8\,F\,(F+1)}{3(2I+1)^2}\,\text{Imag.}\,\frac{E1^{PNC}}{E1^{ind}}\,\eta_1\,(\underline{k}_1 \times \underline{E}_0) \qquad \text{(IV.1)}$$

which is proportional to the circular polarization, η_1 of the incoming photon and is directed perpendicular to both $\underline{E}_0$ and $\underline{k}$. The presence of $\underline{P}_e$ leads to circular polarization in fluorescence observed along $k \times E_0$. Thus the experiment looks for the pseudo-scalar quantity

$$\eta_1\,(\underline{k}_1 \times \underline{E}_0 \cdot \underline{k}_2)\,\eta_2$$

This new scheme illustrated in fig. 8 has four advantages
(i) that the magnitude of the transition can be increased with E_0, though this is at the expense of fractional polarization and there is a practical limit set by electric field breakdown.
(ii) The effect reverses with both $\underline{E}_0$ field direction and with the circular polarization of both incoming and outgoing photons.
(iii) The effect can be enhanced using a multi-pass system.
(iv) Calibration is also straighforward since a magnetic field parallel to E_0 produces a polarization in a well understood manner.

The apparatus has been constructed and performs well, the only limitation being signal to noise. The present published result is

$$\left|\frac{E1^{PNC}}{M1}\right| = (0.56 \pm 1.8) \times 10^{-3} \qquad 90\% \text{ confidence level}$$

c.f. theory (1.1×10^{-4})

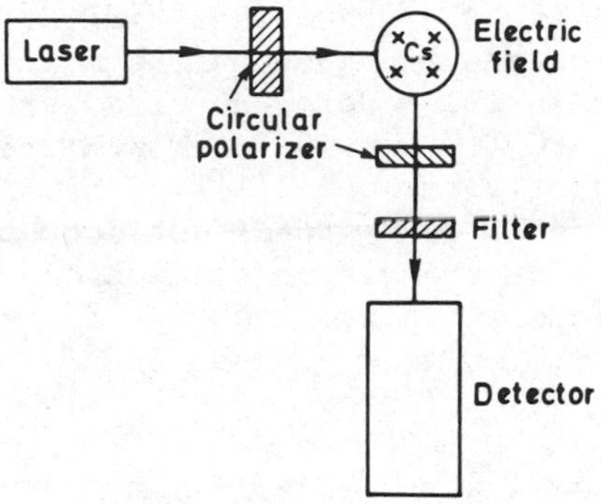

Fig. 8. The Paris cesium experiment.

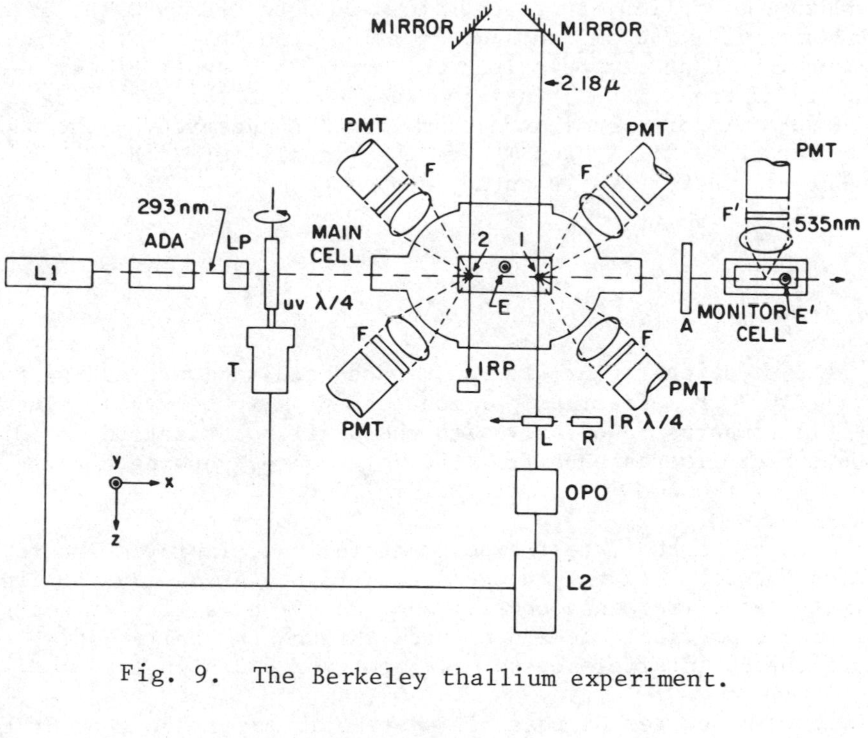

Fig. 9. The Berkeley thallium experiment.

but the sensitivity has reached the required level[9] and a significant result is expected in the near future.

(iii) $6p_{\frac{1}{2}} \rightarrow 7p_{\frac{1}{2}}$ in Tℓ. A somewhat similar experiment is being carried out by Commins and his group at Berkeley on the 293 nm $6p_{\frac{1}{2}} \rightarrow 7p_{\frac{1}{2}}$ transition in Tℓ.[10] Here the circular polarization is expected to be large PC $\approx 2.5 \times 10^{-3}$. The M1 element has been determined experimentally: $M1 = (-2.11 + 0.030) \times 10^{-5}\ \mu_B$. As in the Bouchiat experiment this M1 turns out too small to use directly as the parity conserving transition element and an auxiliary electric field is applied. This again produces a polarization of the form

$$\eta_1(\underline{k}_1 \times \underline{E}_o)$$

proportional to the helicity of the 293 nm photon and directed perpendicular to both the photon direction and the electric field.

A schematic of the Berkely thallium experiment is given in fig. 9. The main difference from the Paris equipment is the way that the PNC polarization $\underline{P}$ is measured. This polarization which is in a direction perpendicular to the 293 nm light from the laser L1 and to the applied electric field $\underline{E}$ is detected by means of a strong circularly polarized beam at 2.18μ, the $7p_{1/2} \rightarrow 8s_{1/2}$ transition frequency. The atomic polarization in the $7p_{1/2}$ state determines the intensity of the 2.18 μ transition and hence of the resulting $8s_{1/2} \rightarrow 6p_{3/2}$ fluorescence at 323 nm which is measured by the photomultipliers PMT. If I_+ and I_- are the signals for 2.18 μ photons with $\eta_2 = \pm 1$ then there is an asymetry

$$\Delta = \frac{I_+ - I_-}{I_+ + I_-} = 0.7\ P$$

where the dilution factor 0.7 can be accurately determined from the larger ($\sim 10^{-2}$) PC effect proportional to $\eta_1(\underline{k} \times \underline{E}_o \times \underline{k}_2)$. The value 0.7 compares favourably with the small polarization (≈ 0.08 P) of the direct fluorescence from the $7p_{1/2}$ state, showing clearly the advantage of the two laser method.

Features of the latest experiment include: improved lasers, use of a Pockels cell to produce circularly polarized 293 nm light, automatic laser frequency control and the use of a mirror (not shown in figure) to reflect the 293 nm back through the cell. This reduces the PC interference term but not the PNC signal.

The main sources of possible systematic error are thought to be (A) Effects due to imperfect UV circular polarization which can give a false parity signal through incomplete cancellation of the PC effect proportional to $(\underline{k}_1 \times \underline{E}_o \cdot \underline{k}_2)\eta_2$.

(B) A false signal due to the presence of a spurious electric field combined with incomplete cancellation of the quadratic effect proportional to $\eta_1(\underline{K}_1 \times \underline{E} \cdot \underline{k}_2)\eta_2$. All the quantities required to calculate these effects can be, and are, measured experimentally so that the appropriate corrections can be made with confidence.

The data from 11 runs (∿400 hours total) are as follows.

	Parity Asymmetries Δ_p	
	with mirror	without mirror
Prior to correction for stray electric field and polarization effects	0.93×10^{-5}	1.18×10^{-5}
After correction	1.43×10^{-5}	1.55×10^{-5}

The final normalized experimental result is

$$\delta_{exp} = (2.8 \pm^{1.0}_{0.9}) \times 10^{-3}$$

which is to be compared with the theoretical prediction

$$\delta_{theor} = (2.1 \pm 0.7) \times 10^{-3}.$$

IV 2 Hydrogen $2s_{1/2} \rightarrow 2s_{1/2}$

PNC experiments on atomic hydrogen are potentially of great importance because the atomic theory is free from the uncertainty due to many-body effects inherent in the heavy atom case. But at the same time the experiments are very difficult. While the PNC is certainly enhanced by the near degeneracy of the $2s_{1/2}$ and $2p_{1/2}$ levels, spurious effects due to unwanted electric fields are also enhanced. As a result the experiments are still some way from the sensitivity necessary to see the predicted PNC phenomena and even when this level of sensitivity has become routine the problem of systematic effects may still be a serious constraint, as the bismuth case has shown. For this reason, and because the experiments use radio-frequencies rather than the optical frequencies which are the subject of this school, we will be much briefer in our description of this work than the amount of effort expanded and the progress made would justify.

While a number of other possibilities have been examined the best experimental scheme in hydrogen appears to be a radio-frequency transition among the Zeeman and hyperfine substates of $2s_{1/2}$. Three groups, at Michigan, Washington and Yale, are setting up radio-frequency experiments on hydrogen of the general type in which one observes a PNC E1 matrix element in interference with a PC matrix element, either M1, or E1 induced by an external electric field. The interference is observed through the change in intensity on 180° reversal of the relative phase of the two rf fields which interact with the PNC and the PC matrix elements. In principle one can make the PNC fraction large by reducing the magnitude of the PNC transition field, though this is at the expense of transition probability and a limit is set by background effects.

A basic feature of two of the experiments is to work at a magnetic field (569 Gauss or 1180 Gauss) for which the $2p_{1/2}$ and $2s_{1/2}$ levels cross. The PNC admixture of $2p_{1/2}$ into $2s_{1/2}$ then becomes:

$$\delta = \frac{\langle 2s_{1/2}\ m_j\ m_I | H_1^{PNC} + H_2^{PNC} | 2p_{1/2}\ m_j'\ m_I' \rangle}{W(2s_{1/2}\ m_j\ m_I) - W(2p_{1/2}\ m_j'\ m_I') + i\,\Gamma_{2p_{1/2}}} \qquad \text{(IV.2)}$$

where we have included the nuclear spin quantum number m_I which is necessary because H_2^{PNC} is nuclear spin dependent. It is useful to note that because H_1^{PNC} is spin independent ($\Delta m_I = 0$) its contribution to δ vanishes at the low field crossing point for which $\Delta m_I = 1$.

The main advantage of the crossing point is not, as might be thought at first sight, the enhanced size of δ because of the small energy denominator; this is counter-balanced by increased electric field quenching and a smaller electric field must be used with the same ultimate sensitivity. It is: first that there is a (relatively) rapid variation in PNC transition probability with magnetic field which helps in background discrimination and second that because of the presence of the $i\Gamma_{2p_{1/2}}$ term due to the natural decay of the admixed $2p_{1/2}$ state, δ can be real or imaginary depending on the value of the magnetic field, allowing flexibility in the types of matrix element which can be used in interference with the PNC E1 element.

We now outline briefly the most significant features of the various experiments.

<u>Michigan</u>.[4a, 11] A schematic of the Michigan apparatus is shown in fig. 10. A beam or order 10^{14} hydrogen metastables per sec. is produced in a duoplasmatron source followed by a cesium charge exchange cell. The atoms are prepared in the $m_J = +\frac{1}{2}$ $m_I = -\frac{1}{2}$ state by familiar quenching plus transition methods. The 1600 MHz transition $J_j = \frac{1}{2}$ $m_I = -\frac{1}{2}$ $m_j = -\frac{1}{2}$ $m_I = \frac{1}{2}$ is induced at the high field crossing point by a combination of transverse DC electric field E_x and transverse rf electric field R_x giving a matrix element $\alpha\ E_x R_x$ for the $\Delta M = 0$ transition. Atoms which have undergone the transition are detected by quenching in a Lyman α detector.

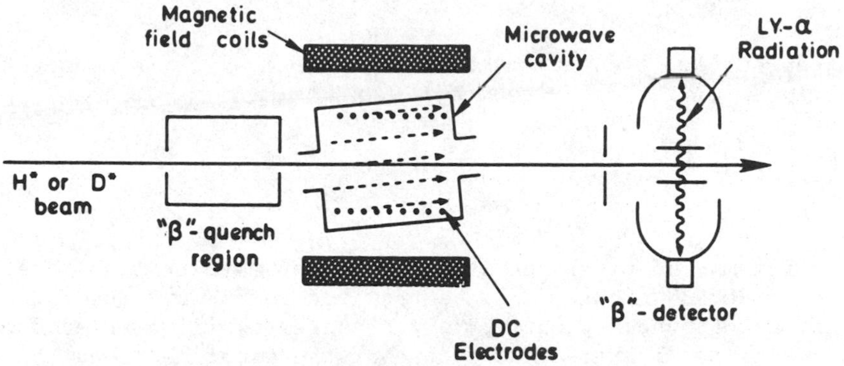

Fig. 10. The Michigan hydrogen experiment.

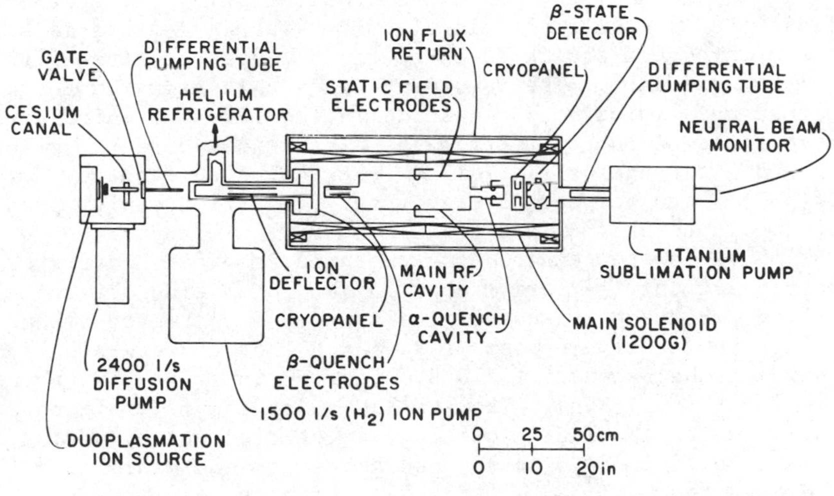

Fig. 11. The Washington hydrogen experiment.

The PNC E1 matrix element interacts with the Z component of the rf field R_Z. The PNC fraction is then given by

$$F = \frac{E1^{PNC} \; R_Z}{\alpha E_x \; R_x} .$$

This can be 'improved' by making $\frac{R_z}{R_x}$ large which is achieved by slightly tilting the cavity so that

$$\frac{R_z}{R_x} \sim 10.$$

Williams and his group at Michigan estimate that an integration time of order 30 minutes will be required to observe the PNC signal against background. Clearly, very great care will be needed to remove spurious background effects but the experiment has the great advantage that the signal of interest must change sign with E_x, B, R_x/R_z as well as changing in a known manner as a function of magnetic field. The PNC invariant can be seen to be i(R.B)($\underline{R}\times\underline{B}.\underline{E}\times\underline{B}$) where the i which ensures time reversal invariance comes from iΓ in IV.2 above.

Washington.[12] In this experiment too the $\frac{1}{2}$-$\frac{1}{2}$ ↔ (-$\frac{1}{2}$,$\frac{1}{2}$) transition is observed at the 569 Gauss crossing point. The apparatus is illustrated schematically in fig. 11. The 2s metastable beam is passed through a static electric field where the atoms with $m_j = -\frac{1}{2}$ are quenched. The remaining matastable atoms in the $m_j = +\frac{1}{2}$ state then pass through two cavities in the first of which there is a static field E_Z^I and an rf field R_Z^I along the Z direction, and in the second a stronger r.f. field R_Z^{II} at the same frequency but 90° out of phase with respect to the first. The first cavity produces the PC amplitude $\alpha \, E_Z^I \, R_Z^I$ and the second PNC amplitude $\beta \, R_Z^{II}$ where the 90° phase has compensated for the pure imaginary nature of the PNC matrix element. A term linear in PNC is obtained by detecting interference between these amplitudes which is achieved either by reversing the relative phases of R_Z^I and R_Z^{II} or by reversing E_Z^I or by reversing the magnetic field B. The PNC invariant for the experiment is i($\underline{R}.\underline{B}$)($\underline{R}.\underline{B}$)($\underline{E}.\underline{B}$). The setting up of the relative phase of R_Z^I and R_Z^{II} which is clearly critical is carried out by adding an additional Dc field E_{Zc2}^{II} to the second cavity and minimising the interference term $\alpha_I E_Z^I \, R_Z^I \, \alpha_{II} \, E_Z^{II} \, R_Z^{II}$ which is zero when the two terms are 90° out of phase. Atoms which have undergone a transition ($\frac{1}{2}$-$\frac{1}{2}$) → (-$\frac{1}{2}$,$\frac{1}{2}$) in these two cavities are detected as follows. The beam is first passed through cavity III where the rf drives all those atoms remaining in the $m_j = \frac{1}{2}$ state to the $2p_{1/2}$ state where they decay rapidly to the ground state. The reamining maetastable atoms in the (-$\frac{1}{2}$+$\frac{1}{2}$) state are detected by quenching the beam in a perpendicular electric field inside a Lyman α detector.

The major difference between the two experiments is that the Washington group use a double cavity system. This has some advantages in flexibility in that the PNC cavity can be designed to minimise PC transition amplitudes, these are induced in the first cavity. On the other hand new complications involving fringing fields and the realtive phase between the cavities are introduced. Only itme will decide w-ich is the more advantageous. Clearly there are advantages at this stage in investigating a variety of possibilities.

Yale.[13] Hinds has carried out a detailed analysis of the hydrogen experiments and has argued that far from the $2s_{1/2}$, $2p_{1/2}$ crossing points having advantages, there is clear advantage in carrying out the experiments at low or zero magnetic field in order to minimise spurious systematic effect. No details of the experiment have yet been published.

IV.3 Other Serious Possibilities

(i) Optical rotation in Pb and Tℓ.[14] An optical rotation experiment on the 1.28 nm J = 0 → 1 transition in Pb is in progress at the University of Washington using essentially the same techniques as in their Bi experiment. A similar experiment will be carried out on the 1.28 nm line J = $1/2 \rightarrow 3/2$ in Tℓ in the near future.

(ii) Circular dichroism in Tℓ.[15] The $6p_{1/2} \rightarrow 7p_{3/2}$ transition is not highly forbidden (M1 = $-0.04\ \mu_B$) as the $6p_{1/2} \rightarrow 7p_{1/2}$ and a circular dichroism experiment is feasible. A problem here is an E2 transition amplitude which is much larger than the M1. However difficulties from this source can be avoided by use of hyperfine transitions for which F = 0 ↔ 1 where the E2 component vanishes.

(iii) Linear polarization in crossed E and B fields in Tℓ.[4] The Berkeley group are setting up an experiment to look for an $(\underline{\varepsilon}.\underline{B})(\underline{\varepsilon}.\underline{E}\times\underline{B})$ PNC effect in the $6p_{1/2} \rightarrow 7p_{1/2}$ transition in Tℓ.

(iv) J = 0 → J = 0 in Pb.[16] An important experiment in the next generation will be a search for circular dichroism in a J = 0 → J = 0 transition which is sensitive only to the nuclear spin-dependent terms. The best example is probably the 401 nm transition in Pb.

(v) Two photon $1s_{1/2} \rightarrow 2p_{1/2}$ in hydrogen.[17] Calculation suggests that interference between M1 and $E1^{PNC}$ amplitudes in a two photon $1s_{1/2} \rightarrow 2p_{1/2}$ transition in atomic hydrogen will lead to a circular polarization dependence of order 10^{-4} which may be measurable.

(iv) Optical rotation in diatomic molecules.[18] There is considable interest in the possibility of looking for optical rotation due to diatomic molecules. Not only can the effect be large in certain circumstances due to the closeness of states of opposite parity (rotational or λ doublets) but also a wide range of molecules have potentially usable transitions in the visible regions of the spectrum convenient for CW lasers.

(vii) Near degeneracy in the rare earths. In the rare earths one can find a number of pairs of levels of the same angualr momentum but opposite parity separated by only a few cm^{-1} or less. The majority of these correspond to the substitution p → d which at first sight might be very advantageious for the detection of H_{3e}^{PNC} since H_{1N}^{PNC} and H_{2N}^{PNC} have zero p → d matrix elements because the electron density vanishes at the nucleus. Unfortunately initial calculations suggest that the matrix element of H_{3e}^{PNC} is so small that the non-zero effects from H_{1N}^{PNC} due to cinfiguration interaction will dominate. This difficulty in separating H_{3e}^{PNC} is a general one for which there may be no solution.

IV.4 Summary and Conclusion

Experimental evidence is rapidly accumulating that PNC effects exist in heavy atoms at about the level expected on current unified theories of the electromagnetic and weak interactions. But as we have seen in Bi there is still some disagreement between experiments which must be resolved before one can be completely certain. The Bi situation suggests that there is considerable advantage in having additional experiments of the Cs and Tℓ type to check the Paris and Berkeley work.

Once this stage has been concluded and reliable results are being obtained in the current round of experiments, three immediate priorities suggest themselves. First, experiments of the present type but of high absolute accuracy, together with more reliable calculations which hopefully will be available, will allow a direct measurement of the strength of the electro-weak interaction at the very low energies obtaining in atomic systems. Second, new or more refined experiments which will allow detection of (a) the nuclear spin dependent term and (b) the dependence of the spin independent term on N and Z separately. Third, completion of the hydrogen programme to the point where the individual coupling constants can be obtained free from the theoretical uncertainties inherent in the heavy atom work.

Finally, and probably most difficult of all, perhaps one will be able to devise an experiment able to measure the electron-electron PNC interaction.

Books

(i) Gauge theories of weak interactions. J. C. Taylor, Cambridge (1976)
(ii) Weak interactions. D. Bailin, Sussex (1977)

Review References

a. Proceeding of Internation Workshop on Neutral Current Interactions in Atoms, ed. W. L. Williams, Cargèse 1979.
b. Sandars, P. G. H. in 'Exotic Atoms' ed. Crowe, K. et al, Plenum Press, 1980
c. Commins, E. D. and Bucksbaum, P. H., Ann. Rev. Nucl. Sci. 30; 1 (1980)
d. Fortson, E. N. and Wilets, L., Advances in Atomic and Molecular Physics 16; 319 (1980)
e. Sandars, P. G. H., Physica Scripta 21; 284 (1980)

Other References

1. Dydak, F., Phil. Trans. Roy. Soc. in press (1981).
2. Prescott, C. Y. et al, Phys. Letters 77B; 347 (1978).
3. Bouchiat, M. A. and Bouchiat, C. C., Phys. Letters B 48; 11 (1974).
4. Bouchiat, M. A., Poirier, M. and Bouchiat, C. C., Journal de Physique 40; 1127 (1979)
4a. Dunford, R. W., Lewis, R. R. and Williams, W. L., Phys. Rev. A 18, 2421 (1978).
5. Baird, P. E. G. et al, Paper presented to VII Vavilov Conference on Non-Linear Optics, Novosibirsk, (1981).
6. Barkov, L. M. and Zolotorev, M. S., J.E.T.P. Lett. 27; 357 (1978). Phys. Letters 85B; 308 (1979).
7. Bogdanov, Y. V., Sobel'man, I. I., Sorokin, V. N. and Struck I. I., J.E.T.P. Letters 32; 214 (1979).
8. Hollister, J. H. et al, Phys. Rev. Letters 46; 643 (1981).
9. Bouchiat, M. A. and Pottier, L., in reference (a) p.122 (1980).
10. Bucksbaum, P., Commins, E. and Hunter L., Phys. Rev. Letters 46; 640 (1981).
11. Wieman, C. A. in ref. (a) p.213 (1979).
12. Trainor, T. A. in ref. (a) p.231 (1979).
13. Hinds, E. A., Phys. Rev. Letters 44; 374 (1980).
14. Fortson, E. N., Private communication (1981).
15. Bucksbaum, P., in reference (a).
16. Bouchiat, M. A. and Bouchiat, C. C., J. Physique 35; 899 (1974).
17. Drukarev, E. G. and Moskalev, A. N., J.E.T.P. 46; 1078 (1977).
18. Sushkov, O. P. and Flambaum, V. V., J.E.T.P. 48; 608 (1978).

LASER PHOTOIONIZATION SELECTIVE DETECTION OF ATOMS, MOLECULES AND MOLECULAR BONDS

Vladilen S. Letokhov

Institute of Spectroscopy, USSR Academy of Sciences, 142092, Troitsk, Moscow Region, USSR

INTRODUCTION

The methods of nonlinear laser excitation of atoms and molecules[1] are very effective in various problems of laser spectroscopy. These lectures concern the application of the methods of multistep and multiphoton ionization of atoms, molecules and molecular bonds for their detection. This is of great importance in many applications.

The first proposals of the kind were made in work[2,3] for detection of atoms, in [3,4] for detection of molecules in a mass-spectrometer. Then they were discussed in my lectures in 74-75 [5,6]. In 1975 a conception was put forward [7] to detect molecular bonds in a photoionic projector on the basis of step-wise selective photoionization of molecules on the surface. Since then a great number of experiments have been performed which prove such an approach to be promising in spectroscopy. In essence, it was the photoionization approach that allowed ultimate sensitivity or detecting single atoms and molecules. What is more, in this case there is principal possibility for direct observation of the molecular structure.

1. DETECTION OF SINGLE ATOMS

The main advantage of the photoionization method of atomic detection is very high sensitivity which can reach its ultimate value, that is to detect each atom being in resonance with laser radiation. To attain such ultimate sensitivity it is necessary that, first, the maximum yield of multistep photoionization should be realized. According to multistep photoionization method (Chapter 3) in [1] the atoms are excited by laser radiation into the intermediate state by one or several steps and then ionization of excited atoms is accomplished. Conventionally two approaches can be outlined depending on the way of the atom ionization from the intermediate state. They are: 1) non-resonant ionization, 2) resonant ionization. In the first case the excited atom is ionized by the additional or by the same laser radiation. Low cross-sections ($\sigma_{ph.i} = 10^{-17} \div 10^{-19}\ cm^{-2}$) are characteristic for such non-resonant photoionization and corresponding saturation energy densities are $E_{sat} = (0.01\text{-}1)\ J/cm^{-2}$. In the second case the atom is excited either into the autoionization state having little lifetime or it is excited into the highlying Rydberg state with subsequent ionization by the electric field pulse. These cross-sections exceeds the non-resonant ionization cross-section into continuum by several orders.

Morever, the conditions of detection of each atom during its crossing of the volume of interaction τ_{cr} with the radiation dictate a required period of laser pulse repetition $T \lesssim \tau_{cr}$. This results in a requirement on necessary average radiation power at k-n transition of excitation: $P^{kn}_{av} = E^{kn}_{sat}\ A/T$, where A is the laser beam cross-section.

The first experiments on single atom detection by multi-step photoionization were successfully carried out in [8,9]. In work [8] cesium atoms in a buffer gas were excited and ionized by the same laser pulse. Electron-ion pairs formed were detected by a proportional counter. In such experiment the flight time of detected atoms through the irradiated volume is long enough since it is determined by the atom diffusion in a buffer gas. However, it is impossible to achieve high spectral resolution in such experiment since the buffer gas leads to considerable collisional broadening of absorption lines. In the cases when maximum spectral resolution is to be realized, detection should be carried out in a vacuum in the atomic beam as in experiment[9]. In this case it is possible to attain a high degree of selectivity in detecting rare isotopes by using isotope shifts at several successive excitation steps. The atomic beam method enables us also to combine it with a mass spectrometer which allows the selectivity of detection of single atoms to be increased.

Detection of single atoms in a beam has been carried out and studied in detail in [9,10] with Na and in [10,11] with Yb atoms. With the use of samples containing about 10^{10} atoms the isotope and hyperfine structure of the 6^1S_0- 6^3P_1 transition of the Yb has been studied in work[12]. Now we are going to consider the results of this work in more details to illustrate the potentialities of this method.

In the experiment tantalum foil was used with about 10^{10} ions of stable isotopes implanted in it. The ions of stable Yb isotopes were produced at the output of a mass-separator which separates the products of the nuclear reactions taking place in the target under the irradiation by a beam of accelerated protons. The separated ions with their energy of 30 KeV penetrate into the foil to a depth of 100 Å and become neutralized. As a result, the atoms get insulated from the outside medium and under normal conditions can be preserved in the foil for a long time. Such a technique of ion implantation into foil is a convenient way of accumulating, storing and transporting of small amounts of atoms. When the foil is heated to a temperature of 1200°C there is a distinct diffusion of ytterbium atoms from tantalum. To form a directed atomic beam the foil was placed into a tantalum cricible with a narrow cylindric channel.

The Yb atoms in the beam were excited to Rydberg states in steps by the radiation of pulsed dye lasers with at a tunable frequency (Fig. 1). The duration of each pulse was 7 ns, the spectrum width $\Delta\upsilon_L = 1\ cm^{-1}$. All the three dye lasers were pumped by a N_2- laser with its pulse rate of 12 Hz. The laser beams crossed the atomic beam

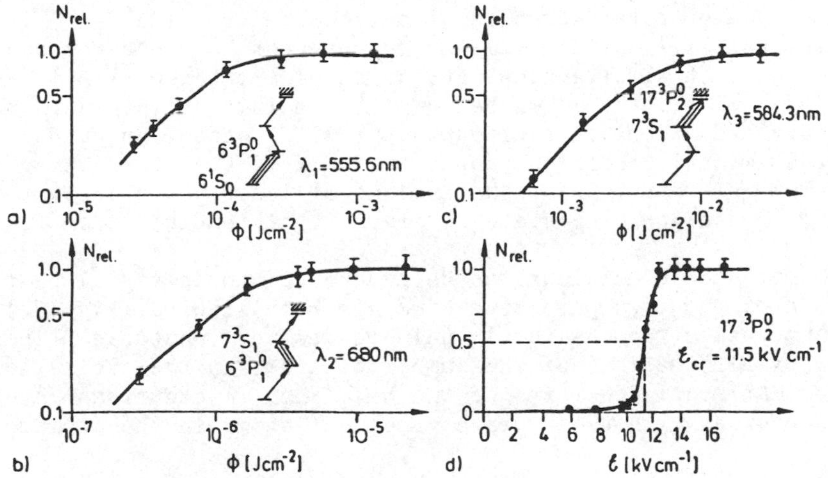

Fig. 1. Dependence of Yb ion yield on laser pulse energy fluences for first, second and third excitation steps (a, b, c respectively) and strength of pulsed electric field (d) (from [11]).

between two electrodes. An electric field pulse was fed to them with a delay of 20 to 50 ns about laser pulses. The resulting ions were removed through the hole in one of the electrodes and recorded by a secondary electron multiplier (SEM). The geometry of the laser beams, the atomic beam and the hole made it possible to extract almost all ions from the spacing between electrodes. In the experimental conditions the recording efficiency of ions by a secondary electron multiplier was close to unity.

For effective excitation of atoms to high-lying states it is necessary to saturate all quantum transitions in use. The strength of pulsed electric field must be sufficient to ionize Rydberg atoms with a near-unity probability. Only with these two conditions satisfied, the regime of maximum ion yield can be realized. Therefore the coice of the principal quantum number in such experiments is very important. The cross-section of atomic excitation at the last step decreases considerably as the principal quantum number increases ($\sigma_{exc} \sim n^{-3}$). On the other hand, the critical field, i.e. a field where the ion yield is close to unity, builds up sharply with a decrease in n. The case is considered optimal when the cross-section of high-lying state excitation is as maximum as possible and the electric field strength essential for effective ionization can be easily obtained under laboratory conditions. For ytterbium this is satisfied by the state $17\,^3P_2^0$ (E_{cr} = 11.5 kV/cm).

Fig. 1 shows the dependences of ytterbium ion yield on the pulse energy flux of the first, second and third excitation steps (a, b, c,) and on the strength of pulsed electric field (d) in exciting the $17\,^3P_2^0$ state (similar dependences have been obtained for sodium). At simultaneous saturation of all three transitions about 5/12 of all the atoms being in the volume of interaction with the laser radiation are excited to the $17\,^3P_2^0$ state. So, under optimal experimental conditions about one half of the atoms being in the volume of detecting at the instant of laser pulse arrival become excited to a high-lying state and each highly excited atom is ionized.

Under such conditions the dependence of ionic signal on the temperature of the oven pure ytterbium has been studied (Fig. 2a). As the temperature reduces the signal becomes very unstable. This is caused by fluctuations of the number of atoms in the excitation volume. At small densities of the beam such fluctuations may be of the same order with the average number of atoms in the excited volume.

The atomic beam density may be made so low that in most cases there is no more than one atom in the volume of detecting. Under such conditions of ion count the statistics of occurrence of atoms in the volume of detecting has been studied (Fig. 2 b,c). The distribution obtained is close to the Poisson. Much similar experiments have been performed on tantalum foil with implanted atoms of Yb. The experiment has shown that the technique of detection in an atomic beam formed by atoms diffusible from foil can be realized successfully with dye lasers having a low pulse rate. In this case

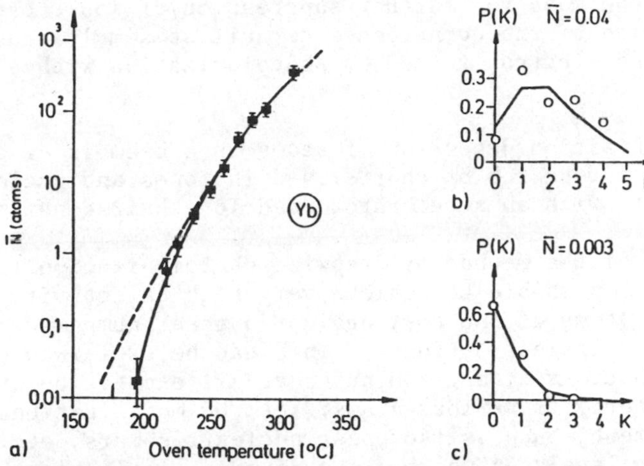

Fig. 2. Temperature dependence of ytterbium ion signal and fluctuations of the number of atoms in the area of detection: a) dependence of ytterbium ion yield on oven temperature under maximum ion yield (the dashed curve-calculation, the solid curve-experiment); b) distribution of the number of ytterbium ions recorded during T = 10s, with the average number of atoms in the excitation volume $\bar{N}$ = 0.04; c) the same with T = 20s, $\bar{N}$ = 0,003 (from [11]).

it is possible to produce spectra of elements accessible in amounts of 10^{10} to 10^9 atoms. The sensitivity of the method can be radically increased by using a copper-vapor laser with a pulse rate repetition from 10 to 20 kHz to pump the dye lasers. This will enable every second atom crossing the excitation are to be ionized. With the geometry of the scheme improved, 10^5 to 10^6 atoms in the foil will suffice to measure the HFS spectrum.

There is the problem of ionic background always arising in photoionization detection of atoms with a high sensitivity. It is caused by ions not connected with the selective action of laser radiation on atoms to be detected. Such ions may be formed in the volume under external or experimental conditions.

As atoms are detected in a beam in vacuum a significant background is set up due to thermal ions and electrons. Their source is a high-temperature oven. A set of several diaphragms being under potentials of different signs and amplitudes makes it possible to reduce the background level to single ions and electrons. Its further discrimination is carried out using the technique of signal gating. Such a method makes it possible to get rid of the ions formed in the volume under the action of cosmic radiation during detecting the useful signal.

One of the ways for further suppression of ionic background is the application of the coincidence circuit at simultaneous detecting of the ion and electron formed by photoionization with electron multipliers[12].

Photoionization detection of atoms in a beam is of great interest for systematic research of short-lived isotopes and nuclear isotopes in experiments with an accelerator used for their generation.

The use of the method of stepwise photoionization for detecting atoms makes it possible to achieve very high selectivity (up to 10^{19}) in detecting atoms of one sort against a great number of atoms of another sort (Cs against Ar in [13]). This can be explained by a great difference in transition frequencies at all excitation steps and in ionization energy of various atoms. Yet the modifications of one and the same element, such as isotopes, nuclear isomers, etc., the atomic energy levels of which are shifted a little, can be excited and ionized with much smaller selectivity. This is associated with inevitable absorption in the line wings. For many atoms however, we can choose such a sequence of quantum transitions that there will be two or three intermediate levels subjected to shifts [14]. Since the processes of excitation at each step are independent the total selectivity of excitation and ionization will be equal to the product of selectivities at each step.

The method of multistep resonant photoionization in atomic beam makes it possible to increase materially the detection selectivity of single atoms of a rare isotope (isomer) among $10^{13} \div 10^{18}$ atoms of abundant isotopes. Potential high selectivity of detection by multistep photoionization has a great interest for detection of very rare cosmogeneous isotope (^{10}Be, ^{14}C, ^{26}Al, etc.) It is necessary, of course, in this case to use some modifications of the photoionization method which provide very high detection selectivity. The measurement accuracy being 10%, it is necessary that the S selectivity should range from 10^{+9} for ^{10}Be to 10^{+15} for ^{26}Al. These required values are substantially higher than the limiting value of selectivity restricted by overlapping of spectral line wings. It is possible, for example, to use multistep excitation with an isotope shift at two or even at three steps of excitation by narrowband laser radiation. It is also possible to use isotopically selective ionization in combination with mass-spectroscopy of produced ions.

Now the most difficult thing for laser selective detection of the above-said cosmogenous isotopes is to create lasers which would provide resonant excitation of the first transition of these atoms. The wavelengths of the transitions from the ground state to the first excited state lie in the UV and UUV. Progress in tunable lasers of UV and UUV regions will lead to development and application of new lasers methods for detection of rare cosmogeneous isotopes. This holds true to the same extent for anthropogeneous very rare isotopes resulted by development and spread of nuclear technology all over the world.

2. DETECTION OF MOLECULES BY PHOTOIONIZATION MASS-SPECTROMETRY

It is much more difficult to detect single molecules than single atoms. Even a simpler task, such as selective detection of traces of polyatomic molecules at a level of 10^6 molecules, for example, has not been solved yet by physical methods. The basic method to identify trace quantities of molecules now is mass-spectrometric analysis. The characteristic efficiency of transformation of molecules into ions with their subsequent detection on present-day mass spectrometers amounts to 10^{-5} of the number of molecules[15], and hence is far from being ultimate. The case is somewhat worse with selectivity. As a rule, molecular mass spectra consist of a lot of fragment ions where there are often no molecular ions at all. Despite the fact that the analysis of fragments gives valuable information on the molecular structure, it is rather difficult to analyze multicomponent mixtures. Due to the overlap of mass spectra it is especially difficult to analyze molecular components at a low percentage. Preliminary separation of mixture components in a chromatograph can partially overcome this problem [15] but not completely yet. Therefore it seems interesting to try to synthesize the existing methods for detection and identification of complex molecules, particularly mass-spectroscopic analysis, with the methods of selective photoionization of molecules.

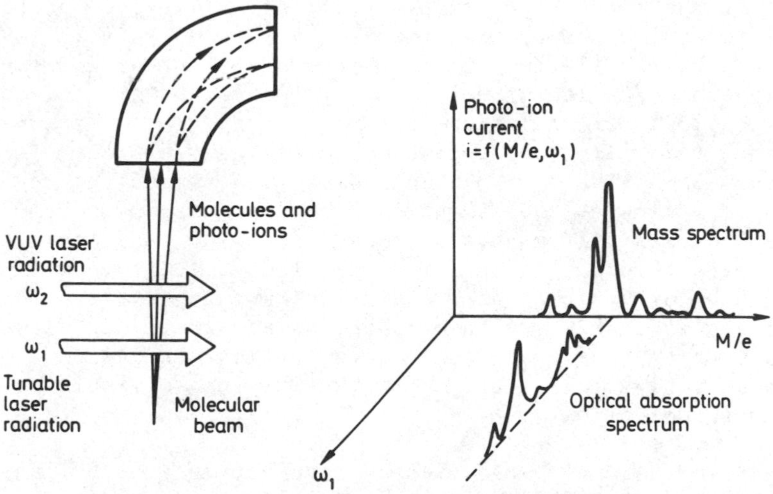

Fig. 3. Simplified scheme of a two-dimensional mass-optical spectrometer (from [5,6]).

Idea of the use of selective two-step IR-VUV molecular photoionization in mass spectrometers has been proposed in works[3,4]. On the basis of this idea the possibility of creation a universal detector for trace amounts of complex molecules as a so-called two-dimensional mass-optical spectrometer has been discussed many times[5,6]. This is illustrated in Fig. 3. A laser with the tunable frequency ω_1 excites selectively the vibrational (electronic for some molecules) state of molecules. Due to such excitation the photoionization band-edge usually lying in the VUV region shifts by a small amount. The second laser operating in the VUV range performs photoionization of molecules, with its frequency ω_2 in the region of maximum slope of the photoionization band-edge. In this case the selective preexcitation of molecules by the tunable laser (a rather small value of E_{exc} = 0.1 ÷ 0.5 eV) brings about a detectable photoionization cross-section (10^{-1} to 10^{-2}% with molecular distribution over rotational states allowed for), i.e. a change in photocurrent. The photoions enter a standard mass-spectrometer where the mass spectrum, i.e. the photocurrent i=f (M/e), is measured. Besides, this modification of a mass spectrometer is able to measure the photocurrent for a given value of M/e as a function of the tunable laser frequency ω_1. In this case the IR spectrum of traces of complex molecules will be measured since, the tunable laser frequency and the frequency of molecular absorption are coincident, the molecules will be excited and hence the ionic photocurrent amplitude will vary. The laser mass-spectrometer with selective molecular photoionization, instead of usual nonselective ionization by an electron beam (or by wide-band VUV radiation), will produce simultaneously an optical (IR and visible) absorption spectrum and mass spectrum.

After the first successful works[16,17] on two-step photoionization of molecules through intermediate electronic states in a two-frequency (UV + VUV) laser field there were a lot of works on multiphoton photoionization of polyatomic molecules with one-frequency UV or visible laser radiation (see review [18]). These experiments have shown that it is possible to attain high yield of molecular photoionization (10 to 100%). This has given impetus to realization of more simple "two-dimensional optical mass spectrometer" to detect ultrasmall amounts of molecules using photoionization of molecules through the intermediate electronic state with one-frequency UV or visible laser radiation. Even though the potential selectivity of such a method of photoionization is much lower than that in case of two-frequency photoionization, this method may provide very high sensitivity of detection (up to single molecules) at moderate selectivity.

At irradiation of molecules by resonant intense UV laser radiation there is a possibility for molecule (or its fragments) to absorb a greater number of photons than it is necessary for a simple ionization. It results in a highly fragmented mass spectrum, and the information on initial molecules may be lost. Ultimate ionization yield with moderate molecular fragmentation can be attained in two-step photoionization only when the intermediate electron-excited state is stable during a laser pulse duration and the laser quantum energy at the second step exceeds the vertical ionization potential of molecule from excited state. The latter corresponds to a lower yield of molecular dissociation after it absorbs a second photon as compared to ionization.

It is under these optimal conditions that the detection of single polyatomic molecules is demonstrated in work[19]. The experiment was performed on naphthalene molecules with the use of an excimer KrF laser with λ = 249 nm to excite and ionize the molecules from an intermediate electron state. The photoions were recorded with a time-of-flight mass spectrometer specially designed for laser-photoionization experiments.

Fig. 4 illustrates the dependence of total ion signal of naphthalene (curve 1) and separately for a molecular ions (curve 2) on energy fluence of laser radiation pulses as well as mass spectra of resulting photoions. In the region of low energy fluences the dependence of ionization yield of naphthalene is quadratic with respect to energy, and only molecular ions can be observed in the mass spectrum (oscillogram a, Fig. 4). With laser radiation energies over 0.05 J/cm^2, both the molecular ions and total ion signals of naphthalene undergo saturation. In the region from 0.15 to 0.20 J/cm^2 the dependence of total ion signal tends to a plateau which is followed with somewhat increase of the fraction of fragments in the mass spectrum. According to [17], a slight drop in molecular ions yield may take place due to their dissociation in the laser field.

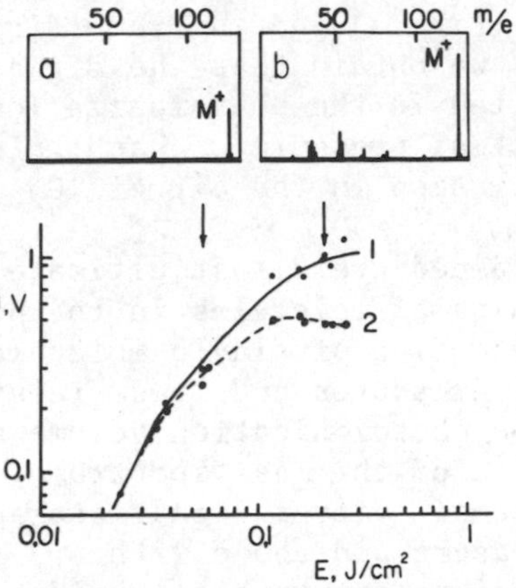

Fig. 4. Dependences of total photoions (curve 1) and molecular ions signal (curve 2) of naphthalene on energy fluence of laser radiation pulses at λ = 249 nm (from [19]).

The quadratic behavior of the photoion yield in the low energy region is explained by a two-step molecular photoionization since the potential of naphthalene I = 8.12 eV is smaller than the energy of two laser quanta 2 $\hbar\omega$ = 10 eV. The saturation of naphthalene ionization yield in the region from 0.05 to 0.10 J/cm^2 is qualitatively consistent with the saturation of photon absorption at the first step, calculated from the known molecular absorption cross-section in gas phase. The fact that at pulse energies from 0.15 to 0.20 J/cm^2 the total ion signal tends to a plateau shows that the photon absorption at the second step reaches saturation, too. As the energy of two laser photons exceeds essentially the molecular ionization potential and the photoionization is connected with the elimination on nonlocalized π-electron, then the competition of molecular dissociation against ionization after absorption of the second photon may be neglected. Thus, the photoionization efficiency at such laser radiation energies is almost 100%.

At these conditions mass spectrum of naphthalene molecular mass at a laser pulse energy fluence of 0.15 J/cm^2 when the photoionization efficiency of this molecule is close to 100% was detected with high realibility. To attain the lowest partial pressure of naphthalene the mass spectrometer was evacuated within a week. The ion peaks in the oscillogram (Fig. 4) correspond to single-ion pulses at the output of the electron multiplier. Along with naphthalene molecular ions there are several peaks observed in the oscillogram which are related to photoionization of organic impurities in the air.

The average value of the naphthalene molecular ions signal in this experiment was about 1 ion/pulse. Taking into account the fact that the yield of naphthalene molecular ions at 100% ionization efficiency is about 50% (Fig. 4) and the recording efficiency of photoions is equal to 50% we obtain that the signal observed corresponds to 4 molecules detected in the photoionization volume. This value corresponds to a partial pressure of naphthalene of 10^{-14} Torr and its relative concentration in the air of 10^{-9}.

The results obtained are almost ultimate with respect to the efficiency of recording of molecules in the photoionization volume and correspond to detection of single molecules per a pulse. To record lower partial pressures and lower relative concentrations it is necessary that the photoionization volume and the total gas pressure in the ion source of the mass spectrometer should be increased. To obtain a higher sensitivity a signal storage technique can be used. The use of tunable lasers and those with two frequency independently tunable laser pulses must substantially widen the class of molecules to be detected.

Thus, the studies carried out support the new approach to the problem of analyzing trace amounts of complex molecules put forward ten years ago[3]. It is based on selective multistep photoionization of molecules by laser radiation with subsequent mass-spectroscopic analysis of ionization products.

The method of multistep photoionization of polyatomic molecules in the mass spectrometer was developed from the very beginning to detect trace amounts of complex molecules. Using one or several lasers tuned in resonance to appropriate molecular transitions, including vibrational ones, it is possible to excite successively the molecule to any state with very high selectivity from which it is ionized with high efficiency (see review [18]).

3. LASER VISUALIZATION OF MOLECULES AND SPATIAL LOCALIZATION OF MOLECULAR BONDS

One of the most difficult problems in molecular physics is to obtain direct information on spatial-chemical structure of complex organic molecules. X-ray crystallographic analysis is the most effective method which under favourable conditions makes it possible to obtain information on spatial and chemical structures of macromolecules. Yet the possibilities of X-ray optics are limited and have not enabled us so far to develop an X-ray microscope for direct visualization of molecules. Electronic microscopy of biological macromolecules constructed mainly of light atoms does not provide atomic resolution. Only after preliminary labelling by heavy atoms of them it is possible to attain the resolution 10 to 20 Å. After field ionization electron, and then ion microscopy was discovered (see [20]) some attempts were made to produce an image of some organic molecules through their absorption on the emitter of electronic or ionic projectors. The resolution of field ionization electronic microscopy, however, is no better than 20 Å which is not sufficient to resolve the atomic details of molecules. Even an easier goal, to obtain the configuration of molecular contours at atomic level, has proved almost inaccessible. Field ionization ion microscopy calls for such strong electric fields that macromolecules evaporate in them and their microscopy also entails great difficulties which have not been overcome yet.

Let us consider some principal modifications of the methods of field ionization electron and ion microscopy are under discussion which give every reason to hope for effective application of these methods to observe the structure of macromolecules.

The principle of a photoelectron laser microscope for molecular bonds can be understood from Fig. 5 which illustrates its basic scheme proposed in works[7]. Under action of ultra-short pulses with different freqencies ω_1, ω_2 ... successive multistep excitation of the quantum levels responsible for the chosen molecular bond AB takes

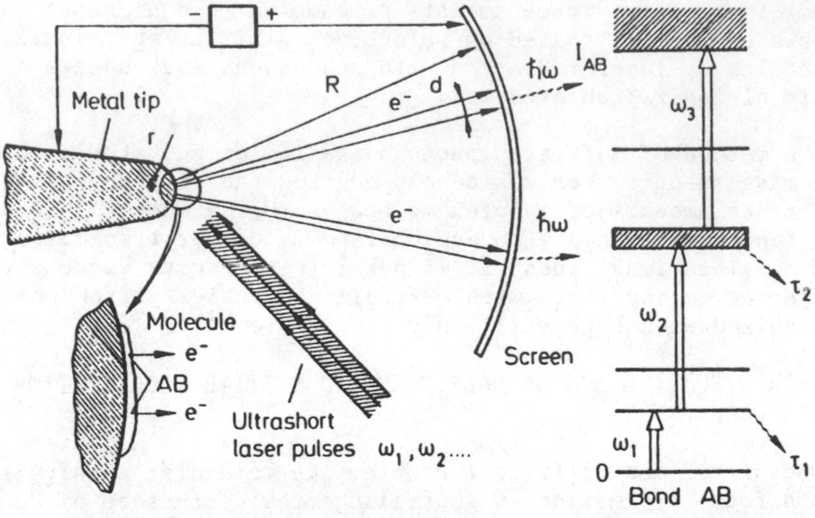

Fig. 5. Basic scheme of photoelectronic (photoionic) laser microscope: 1) semispherical needle tip; 2) macromolecule absorbed on the tip surphase; 3) screen; 4) high-voltage source (in the photoionic microscope the polarity has the other sign); 5) laser beams with ω_1, ω_2, frequencies. In the photoionic microscope the laser radiation must photodetach positive molecular ions instead of electrons. Right: scheme of multistep selective photoionization of the selected molecular bond (from work [7]).

place and subsequent ionization, i.e. to extract of an electron from the section of the molecule with the bond AB. Acted upon by a strong electric field near the cathode, the photoelectron moves along the radical lines towards the screen. Like in the field ionization electronic Muller's microscope, the screen displays an magnified image of the section of the cathode emitting electron. The difference consists, first of all, in the fact that the electric field serves only to transfer photoelectrons to the anode rather than to extract them out of the cathode. Such a "soft" action allows ionizing a certain sort of molecules or particular bonds (parts) of a molecule on the cathode surface without involving the cathode electrons.

The idea of combination of laser selective extraction of photoelectrons and formation of the emitting surface image in an electric field is rather general by character. It may be applied to the case of photoion generation. As known, the field ionization ionic microscope is based on display of the point surface by the ions formed in field ionization of the atoms of residual (sample) gas in the projector chamber at the instant they approach the point surface. The tip surface image is formed on the screen by positive ions flying normally to its surface. The magnification of a field ionization ion microscope M reaches 10^6 and the resolution is 2 to 3 Å. Enhancement of resolution of ion microscope, compared to that of electron one, is achieved on account of a heavy mass of ion and a corresponding decrease in ion localization indeterminacy. The attempts to use a field ionization ion microscope to produce an image of biological molecules have not lived up to expectations. The main obstacle in producing ionic images of large organic molecules is that it is necessary to set up strong electric fields. In such fields the molecules decompose and evaporate, and different distortions take place[20].

The ion microscope may have two potential laser modifications[7]. First, the strength of the electric field required for atomic ionization near the tip can be substantially reduced if the atoms are excited to a state near the ionization limit. The critical strength of electric field can be, probably, reduced to values below 10^7 V/cm. At such strengths a strong electric field can completely eliminate molecular evaporation and decomposition. Second, attempts can be made to detach selectively chosen functional groups of macromolecules absorbed on the tip surface as ionic fragments using laser radiation. The detached positive molecular photoions can be directed to the screen using again an electric field of moderate strength. Such a version of photoionic microscopy has no analogue in field ionization ionic microscopy[20] since photoions must be emitted right from the tip surface where molecules are absorbed in the absence of sample (residual) gas. The idea of such a laser photoionic emission microscope is quite similar to the idea of a laser photoelectron microscope[7]. It is interesting to estimate and compare the ultimate spatial resolution of emission photoelectronic and photoionic microscopes.

These simple estimations[7,21] show that using a photoionic microscope it is possible to reach resolution of several Å which may be sufficient for visualization of macromolecules. For this purpose, however, photoions should be detached from the surface.

As for direct photodetachment of molecular ions from the surface, the most interesting for the photoion microscope, until recently there was no experimental data on this process. Recently in work[22] the effect of photodetachment of molecular ions from the surface of molecular crystals has been discovered. The experiment consisted in mass-spectroscopic analysis of photoionization products as the surface of molecular crystals of anthracene or five different nucleic acid bases is irradiated by excimer laser UV radiation (λ = 249 nm, 308 nm, 337 nm). The specific feature of the experiments was that they employed the UV radiation in the bands of electron absorption of molecules in solid phase.

Fig. 6 shows the mass spectra of the two bases of nucleic acids and anthracene with the use of the shortest-wave radiation in the

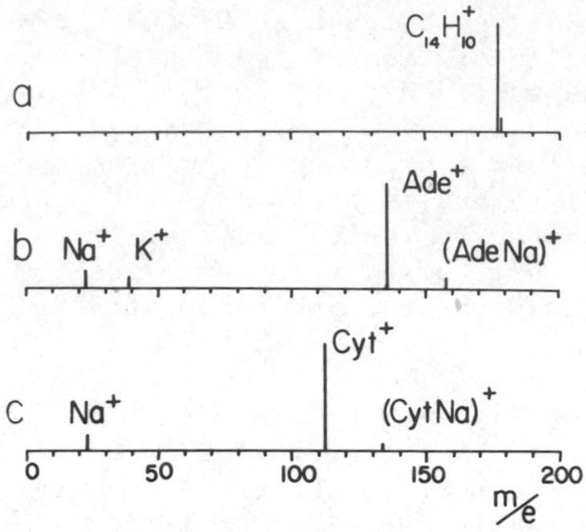

Fig. 6. Mass-spectra of photoions produced from irradiating the surface of molecular crystals at λ = 249 nm: a) anthracene (150 kW/cm^2), b) adenine (25 kW/cm^2), c) cytosine (1 MW/cm^2). In the brackets are the irradiation intensities used (from work[22]).

experiments, λ = 249 nm. In this case the photoion yield was maximum. In all the cases the molecular ion is one of the most intensive components in the mass spectrum produced with the use of a time-of-flight mass spectrum with its resolution of 200. As the wavelength increases the radiation power must be increased greatly since the radiation becomes out of resonance with the electron absorption band of the molecules being irradiated. This leads to strong fragmentation of molecular ions and, finally, gives rise to a mass spectrum typical of the case of plasma formation by laser radiation.

The main peculiarity of mass spectra is a sharp increase of molecular ions with decreasing radiation wavelength and the absence of fragmentary ions as the crystals are irradiated by short-wave radiation with its power much lower than the threshold of thermal action on the surface and especially that of plasma formation. In more detail this effect was studied[23] on photodetachment of the molecular ions of adenine and anthracene from the surface of molecular crystals and anthracene adsorbed on the metal surface. It has been shown that this effect is photophysical by nature and nonlinearly dependent of laser energy fluence which points to the fact that two or even three UV photons are required to photodetach molecular ions.

For successful realization of the photoion microscope project the photodetachment of molecular ions by laser radiation must be selective with respect to molecular bonds or molecular groups. It is just in this case that we may expect direct visualization of large molecules to be carried out. For this purpose, a suggestion was made to realize multistep selective excitation of selected types of bonds or groups by successive picosecond laser pulses of high intensity[7] as shown in Fig. 5, the right side. The problem of extraction of a molecular ion fragment with Å-spatial resolution is, however, very difficult for the following fundamental reasons. Picosecond laser radiation is able to deposite significant energy to some definite parts of a molecule in a short time $\tau_{exc} = 10^{-12} \div 10^{-13}$ s determined by laser light intensity. But the molecular fragment with its thermal velocity $v_0 = 10^4$ cm/s will push from the molecule and the surface by the distance $\delta r = 1$ Å in a time no shorter than 10^{-12} s. However, during this time fast transfer of electron excitation from the excited parts of a large molecule to its other parts may occur. Recent direct picosecond experiments[24], for example, have shown that the time of electron excitation energy transfer between different parts of a large molecule is shorter than 10^{-12} s. This is a serious obstacle to selective photodetachment of photoions even in using picosecond laser pulses with a tunable frequency.

To overcome this problem and provide photoion (photoelectron) detachment from certain parts of a molecule, in work[7] an obvious solution was suggested which consisted in preliminary chemical labelling of certain bonds or groups of a macromolecule by chromophores. It is possible to select a chromophore group with its electron excitation energy lower than in the adjacent parts of the molecule. This can, first, afford highly selective laser excitation of a labelled part, second, prevent transfer of electron excitation to the adjacent parts of the molecule and, third, make easier photodetachment of the excited chromophore groups at a low kinetic energy of photoions. These arguments point to the fact that chemical labelling of macromolecules holds much promise for the development of a laser photoionic microscope with Å-spatial resolution. It should be emphasized that some methods of labelling of large biomolecules by heavy metals with a dimension of the order of Å have been developed to produce their image in an electron microscope. For example, some parts of polynucleotides can be labelled with Os[25] and other heavy atoms. In case of a photoionic microscope the requirements to a chemical label are simpler than in case of an electron microscope since there is no need for a high electron density provided only by heavy metals. It is sufficient that a compact label should have a longer-wave electron absorption band compared to the adjacent parts of the molecule.

The effect of photodetachment of molecular ions from the surface without their significant fragmentation, i.e. with their molecular individuality preserved, discovered in work[22] provides us with one more possibility of creating a photoionic microscope. Indeed, it is possible to get rid of selective formation of molecular ions and identify their type by measuring their mass in a device like a two-dimensional time-of-flight mass-spectrometer. Time resolving of observation of an image in the photoionic projector makes it possible to fix the spatial structure of a macromolecule with simultaneous time-of-flight mass analysis of the fragments producing the image. This alternative way of creating a photoionic microscope seems promising.

So, the wave-corpuscular photoion microscopy can attain simultaneously high spatial and spectral resolution. The possibilities available here, however, also apply to the problem of picosecond time resolution.

Let there be a device in which the picosecond laser pulses are able to photoionize certain parts of a molecule and visualize the points of their photodetachment on the photoelectron (photoionic)

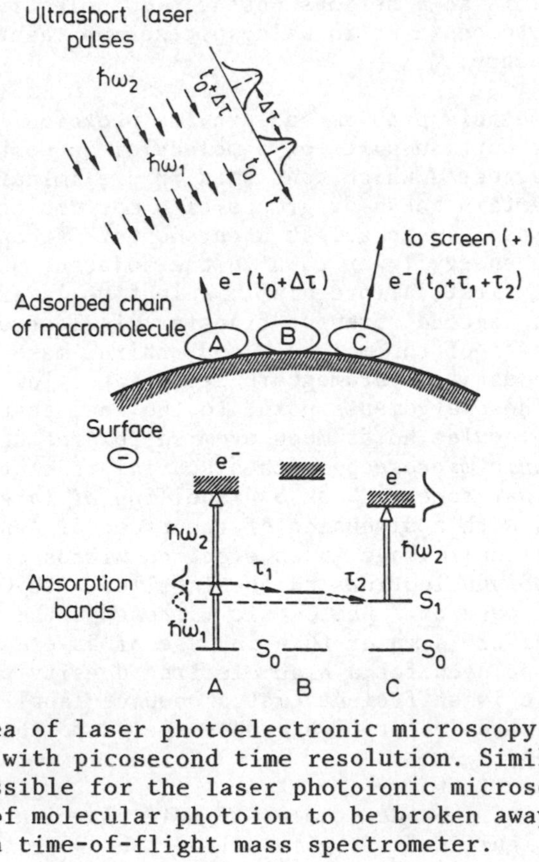

Fig. 7. An idea of laser photoelectronic microscopy of molecular bonds with picosecond time resolution. Similar modification is possible for the laser photoionic microscope with the type of molecular photoion to be broken away identified with a time-of-flight mass spectrometer.

projector screen by the method of multistep excitation. It is not difficult to imagine its modification which enables us, for example, to observe directly the energy transfer of electron excitation along the chain ABC of the molecule absorbed on the surface of the projector needle tip (Fig. 7). For this purpose, it is necessary to excite the electron state S_1 of the selected part of the molecule (A, for example, Fig. 7) by a picosecond laser pulse at the ω_1 frequency with their duration shorter than the time of electron excitation transfer τ_1. Now it is possible to excite the molecule by another laser pulse at the frequency ω_2 corresponding to stepwise photoionization of another part of the molecule (C, for example, Fig. 7) and observe the appearance of photoelectrons (photoions) as the delay between pulses $\Delta\tau$ changes. In this way we can have information on the time of electron excitation transfer along a molecular chain. Such an approach, of course, calls for diagrams of successive quantum transitions for selective stepwise photoionization of photodetachment of ions. If it is impossible to realize it is advisable to attain information on the place of photoion detachment not from the known irradiation frequencies ω_1, ω_2 ... but from mass analysis of molecular ions to be broken away.

Thus, the schemes of photoelectronic and photoionic microscopes can be modified to a device which will allow studies of molecular structures and molecular dynamics with high spatial, spectral (energy) and time resolution at the same time.

Finishing this discussion of the problems of laser visualization of molecules and localization of molecular bonds with the use of laser radiation we should note that actually the problem is more complicated. It is sufficient to say that such complex questions as manipulation (adsorption, etc.) with macromolecules on the surface, their complex three-dimensional structure, multiple measurements at irreversible photodestruction of molecules under the action of the very first laser pulse, etc. have remained beyond discussion. On the other hand, new interesting experimental results in this new field of laser photophysics and photochemistry of molecules on the surface may be expected. As an example we may refer to the recent observations[26] of the breakaway of such a complex molecule as rhodamin GG (A = 479) from the surface in the form of a molecular ion at a very low degree of fragmentation (the masses of the ions vary from 410 to 440) under powerful picosecond laser pulses with $\lambda = 530$ nm.

Thus, there is the possibility of combination of wave (optical) and corpuscular (electronic) microscopy. It is based on photodetachment of photoelectrons or photoions from certain parts of macromolecules using laser radiation. This laser wavecorpuscular types of microscopy makes it possible, in principle, to realize high spatial (to several Å), spectral (to $0.001 \div 0.01$ eV) and time (to picoseconds) resolution at the same time.

REFERENCES

1. V. S. Letokhov, Nonlinear Laser Chemistry, Springer Series in Chemical Physics, Springer Verlag, Berlin-Heidelberg-New York (1982).
2. V. S. Letokhov, Soviet Patent No. 784679, Appl. in 30.03. 1970, Publ. in 1.08. 1980.
3. R. V. Ambartsumyan, V. S. Letokhov, Appl. Optics 11, 354 (1972).
4. C. B. Moore, Accounts of Chem. Res. 6, 323 (1973).
5. V. S. Letokhov in Frontiers in Laser Spectroscopy, Proceedings of Les Nouches Summer School on Theoretical Physics, July 1975, France (North Holland Publ. Co., 1977), Vol. 2, pp. 771-907.
6. a) V. S. Letokhov in Tunable Lasers and Applications, Proceedings of the Loen Conference (Norway, 6-11 June, 1976), ed. by A. Mooradian, T. Jaeger and P. Stokseth, Springer Series in Optical Sciences, Vol. 3 (Springer Verlag, Berlin-Heidelberg-New York), 1976, pp. 122-139.
   b) W. S. Letochow, Laserspektroskopie, Akademie-Verlag Berlin, Vieweg-Braunschweig (1977).
7. V. S. Letokhov, Kvant. Elektron. (Russian) 2, 930 (1975), Phys. Lett. 51A, 231 (1975).
8. G. S. Hurst, M. H. Nayafeh, J. P. Young, Appl. Phys. Lett. 30, 299 (1977).
9. G. I. Bekov, V. S. Letokhov, B. I. Mishin, Pis'ma Zh. Eks. Teor. Fiz. (Russian) 27, 52 (1978).
10. G. I. Bekov, V. S. Letokhov, O. I. Matveyev, V. I. Mishin, Zh. Eksp. Teor. Fiz. (Russian) 75, 2092 (1978).
11. G. I. Bekov, V. S. Letokhov, O. I. Matveyev, V. I. Mishin, Optics Lett. 3, 159 (1978).
12. G. I. Bekov, E.P. Vidolova-Angelova, V. S. Letokhov, V. I. Mishin, in Laser Spectroscopy IV, Proceedings of the Fourth Intern. Conf. (Rottach-Egern, FRG, June 11-15, 1979) ed. by H. Walther and R.W. Rothe, Springer Series in Optical Sciences, Vol. 21 (Springer-Verlag, Berlin-Heidelberg-New York), 1979, pp. 283-295.
13. G. S. Hurst, M. H. Naufeh, J. P. Young, M.G. Payne, L. W. Grossman, in Laser Spectroscopy III, Proceedings of the Third Intern. Conf. (USA, Jackson Lake Lodge, Wyoming, July 4-8, 1977) ed. by. J. L. Hall and J. L. Carlsten, in Springer Series in Optical Sciences, Vol. 7 (Springer Verlag, Berlin-Heidelberg-New York), 1977, pp. 44-55.
14. V. S. Letokov, V. I. Mishin, Optics Comm. 29, 168 (1979).
15. W. V. Ligon, Science 205, 151 (1979).
16. V. S. Antonov, I. N. Knyazev, V. S. Letokhov, V. M. Matiuk, V. H. Movshev, V. K. Potapov. Optics Lett. 3, 37, (1978).
17. V. S. Antonov, V. S. Letokhov, A. N. Shibanov, Zh. Eksp. Teor. Fiz. (Russian) 78, 2222 (1980).

18. V. S. Antonov, V. S. Letokhov, Appl. Phys. 24, 89 (1981).
19. V. S. Antonov, V. S. Letokhov, A. N. Shibanov. Optics Comm. 38, 182 (1981).
20. a) E. W. Muller and T.T. Tsong, Field Ion Microscopy. Principles and Applications (N.Y., L. Amsterdam: Elsevier Publ. Co. 1969).
b) E. W. Muller, T. T. Tsong, Field Ion Microscopy. Field Ionization and Field Evaporation, in Progress is Surface Science, Vol. 4, part (Pergamon Press, 1973).
21. V. S. Letokhov, Comm. on Atomic and Molecular Physics..., (1982).
22. V. S. Antonov, V. S. Letokhov, A. N. Shibanov, Pis'ma Zh. Eksp. Teor. Fiz. (Russian), 31, 471 (1980).
23. V. S. Antonov, V. S. Letokhov, A. N. Shibanov, Appl. Phys. 25 71 (1981).
24. B. Kopainsky, W. Kaiser, and F. P. Schäfer, Chem. Phys. Lett. 56, 458 (1978).
25. M. D. Cole, J. W. Wiggins, and M. Beer, J. Mol. Biol. 117, 387 (1977).
26. V. S. Letokhov, V. G. Movshev, S. V. Chekalin, Zh. Eksp. Teor. Fiz. (Russian) 81, 480 (1981).

OBSERVATION, STATE REDUCTION AND THE QUANTUM ERASER IN QUANTUM OPTICS

Pierre Meystre and Marlan O. Scully

Max-Planck Institut für Quantenoptik
D-8046 Garching bei München, West Germany

and

Institute for Modern Optics
Department of Physics and Astronomy
University of New Mexico
Albuquerque, New Mexico 87131, USA

INTRODUCTION

The role of observation[1] and the attendant information lie at the heart of the problem of measurement and state reduction in quantum mechanics.[2] For the past few years we have been interested in specific calculations[3] associated with this type of problem and the search for potentially realizable experiments probing the influence of an observer. In the following discussion we propose and analyze an experiment such that the presence of information accessible to an observer would qualitatively change the outcome of an experiment.

In an attempt to prepare the reader for what follows, we summarize our results in the next few lines. Specifically, we consider the interference between light scattered from atoms located at sites 1 and 2 as in Fig. 1a. These atoms have three levels (see Fig. 1b), are pumped from c to a by pulse ℓ_1, and interference fringes between the γ photons[4] emitted by atoms 1 and 2 are sought. An absence of interference between photons γ_1 and γ_2 of Fig. 1a is predicted when the states b and c are distinguishable. This is as would be expected since an atom in the b state has left information as to "which path" the photon took i.e. which atom it was scattered from.

A paper dealing with these and other issues will appear shortly in the Physical Review.[4] The present notes are meant to treat several points which people in the quantum optics community have raised concerning the present problem. For example, the preparation of the radiation field which excites our atoms (e.g. a single photon pulse) might seem artificial. Further questions dealing with the necessity of quantum field theory in the present problem have been raised (see discussion at the end of Section 3).

In Section 2, we give a detailed description of an excitation mechanism which precisely achieves the desired goal, i.e. the single photon excitation of our atoms. We consider a simple "source" two-level atom initially excited in its upper state. This provides a physically realizable model for a very weak light source, and could in fact be built in the laboratory by using ion traps. We compute the scattering of the field emitted by this source by atoms 1 and 2 and show that the result is qualitatively the same as that predicted for the case of an instantaneous excitation mechanism of atoms 1 and 2. There are, however, quantitative differences in trivial geometrical factors and in the temporal behavior of the emitted photons.

In Section 3, we address the question of the necessity of using a fully quantum mechanical description for the present problem. We show explicitly that a semi-classical approach is not appropriate in the present problem. The argument according to which the disappearance of fringes is due to vacuum fluctuations which wash out the fringes in the three-level atom case is shown to be misleading. In fact, the semi-classical approximation predicts an absence of fringes even in the two-level atom case.

SCATTERING OF LIGHT EMITTED BY A SOURCE ATOM BY TWO-LEVEL ATOMS

In past work, we have considered a "quantum eraser" model where the atoms are excited instantaneously. In practice, this could be achieved with a pulse of duration much smaller than the relaxation times of the atoms. However, it is instructive to analyze what happens if this condition is not fulfilled.

A simple model which allows us to answer this question is the following. We consider a two-level atom source excited in its upper level $|A\rangle$ at time $t = 0$. The two "scattering" atoms are located some distance away, and are initially in their ground state $|b\rangle$, so that the initial state vector is

$$|\psi(0)\rangle = |Ab_1b_2\rangle|0\rangle \qquad (1)$$

By $|0\rangle$ we mean that the radiation field is initially in the vacuum state $|0\rangle \equiv |0_k\rangle$. As time goes on, the source atom decays to its ground state by spontaneous emission. In this process, it emits a wave packet (photon) which propagates away and, after a long enough time (propagation time) reaches the scattering atoms. It may then excite these two-level atoms. Since we consider only one source atom, there is only one quantum $\hbar\omega_0$ of energy available in this wave packet, where ω_0 is the energy of the A → B transition. If the energy of the transition in the "scattering" atoms is also $\hbar\omega_0$ at most one atom can be excited by conservation of energy.

This model of excitation will closely simulate the conditions discussed in Ref. 4 provided that the spontaneous decay rate of the source atom is very fast.

We can compute the dynamics of this system in a way similar to that of Ref. 4. Since all three atoms are two-level systems, the state-vector at time t is of the form

$$|\psi(t)\rangle = A(t)|Ab_1b_2\rangle|0\rangle + \sum_{\vec{k}} B_{\vec{k}}(t)|Bb_1b_2\rangle|1_k\rangle$$

$$+ C_1(t)|Ba_1b_2\rangle|0\rangle + C_2(t)|Bb_1a_2\rangle|0\rangle \quad (2)$$

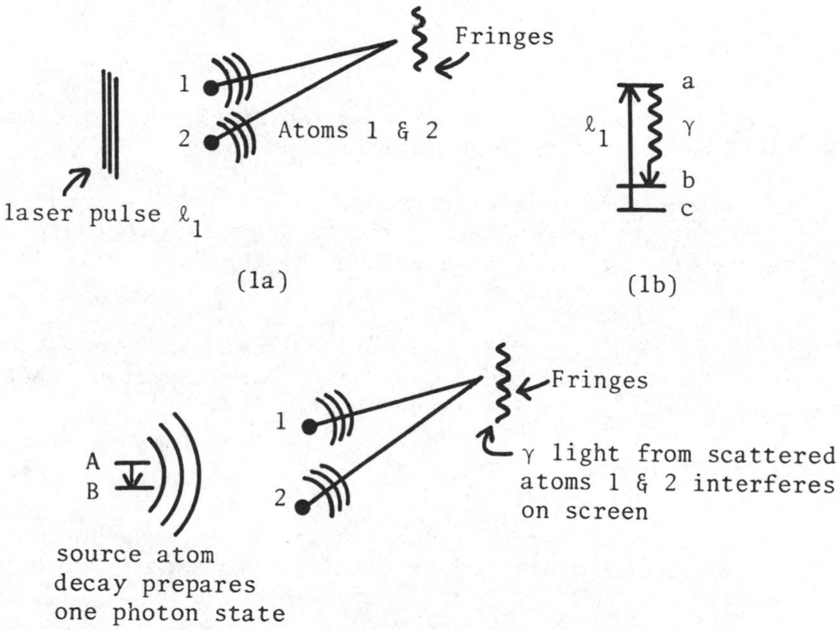

(1a) (1b)

(1c)

FIGURE 1

In the interaction picture, the Schrödinger equations of motion for the various amplitudes are:

$$\dot{A} = - i \sum_{\vec{k}} \kappa^*_{\vec{k}} e^{i(\Delta_k t + \vec{k}\cdot\vec{r}_0)} B_{\vec{k}} , \tag{3a}$$

$$\dot{B}_{\vec{k}} = - i\kappa_{\vec{k}} A e^{-i(\Delta_k t + \vec{k}\cdot\vec{r}_0)} + C_1 e^{-i(\Delta_k t + \vec{k}\cdot\vec{r}_1)}$$

$$+ C_2 e^{-i(\Delta_k t + \vec{k}\,\vec{r}_2)} \tag{3b}$$

$$\dot{C}_j = - i \sum_{\vec{k}} e^{i(\Delta_k t + \vec{k}\cdot\vec{r}_j)} B_{\vec{k}} , \qquad j = 1,2 \tag{3c}$$

where

$$\Delta_k = \omega_0 - ck . \tag{4}$$

The coupling constant $\mu_{\vec{k}}$ depends on the dipole matrix element corresponding to the mode $\vec{k}$ of polarization $\varepsilon_{\vec{k}}$.

It is tedious, but straightforward, to solve these equations provided that one neglects multiple scattering (e.g. reabsorption by the "source atom" of radiation emitted by the "scattering" atoms).

Under these conditions one can perform the Wigner-Weisskopf approximation on the "source" atom and finds, as expected, that it decays exponentially to its ground state, i.e.

$$A(t) = - \gamma_s A(t) , \tag{5}$$

where γ_s is then its spontaneous decay rate. In this paper, we shall always neglect the irrelevant level shifts associated with this process.

The calculation then proceeds then as follows: One inserts Eq. (5) in Eq. (3b), integrates it formally and inserts the result in Eq. (3c). It is then quite easy to identify the terms due to multiple scattering and these are neglected. One can at this point, perform the Wigner-Weisskopf approximation for the "scattering" atoms. The only operation left is a sum over $\vec{k}$ which can be transformed to an integral and evaluated following well-known techniques. When all this is done, one finds that

$$C_j(t) = + \frac{i(2\pi)^2|\kappa|^2}{(\gamma - \gamma_s)} \left(\frac{k_0}{r_{j0}}\right)\left[e^{-\gamma_s(t-r_{j0}/c)} - e^{-\gamma(t-r_{j0}/c)}\right] \times e^{ik_0 r_{j0}}\theta(t-r_{j0}/c) \tag{6}$$

where

$$r_{j0} = |\vec{r}_j - \vec{r}_0| \tag{7}$$

is the distance between the source and the "scatterer" atom j, and γ is the spontaneous decay rate of the "scattering" atoms.

Eq. (6) gives precisely the result that one would have expected intuitively. For times shorter than $t = r_{j0}/c$, the wave-packet emitted by the source has not reached the atom at $\vec{r}_j$ so that it remains in its ground state. For $t > r_{j0}/c$ the probability of being in the upper state is given by the combined effects of the incident wave packet ($\exp(-\gamma_s t)$) and spontaneous decay ($-\exp(-\gamma t)$). In Fig. 2, we show the combined effect of these two processes on $C_j(t)$.

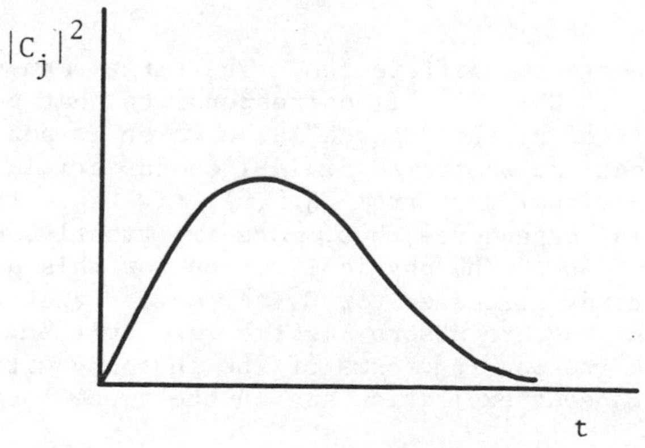

FIGURE 2

Note that this is also analogous to the dynamics that one would expect if the atoms were excited by a wave-packet emitted by a pulsed laser.

We are now in a position to study the interference pattern produced by atoms excited in this fashion. It is straightforward to determine what the photons emitted by our system look like, by inserting Eq. (5) and (6) in Eq. (3b) and performing a simple integration. The general form of $B_{\vec{k}}(t)$ is however rather complicated, so that we won't reproduce it here. If one notes that terms of the form exp $i\omega(t + r/c)$ correspond to incoming waves and don't play a role, one can immediately take the limit $t \gg 1/\gamma$, $1/\gamma_s$, that is we consider only the asymptotic limit of $B_{\vec{k}}$. Nonbelievers can keep the explicit time-dependence in $B_{\vec{k}}$, do all the integrals, and <u>then</u> take the limit $t \to \infty$, and will find that indeed, the contributions of these terms cancel exactly. For $t \to \infty$ one finds

$$B_{\vec{k}}(t \to \infty) = \frac{-i\kappa_{\vec{k}}}{i\Delta_{\vec{k}} + \gamma_s}$$

$$- (2\pi)^2 \kappa_k |\kappa|^2 \left\{ \sum_{j=1,2} \left(\frac{k_0}{r_{j0}}\right) \frac{e^{-i(\vec{k}\cdot\vec{r}_j - \vec{k}\cdot\vec{r}_{j0})}}{(\Delta_k - i\gamma)(\Delta_k - i\gamma_s)} \theta(t - r_{j0}) \right\} \tag{8}$$

It is instructive to compare B_k to the corresponding form for instantaneous excitation as in Ref. 1. In this case, the asymptotic form of the wave-packet takes the form

$$B_{k,inst}(t \to \infty) = \sum_{\vec{k}} \frac{i\kappa_k e^{-i\vec{k}\cdot\vec{r}_i}}{i\Delta_{\vec{k}} + \gamma} |1_{\vec{k}}\rangle \quad . \tag{9}$$

We immediately note two differences. The first term in Eq. (8) does not appear in Eq. (9). It corresponds to that part of the wave packet emitted by the "source" atom which is not scattered. We also note that the scattered field (second term in Eq. (8)) has a different functional form from Eq. (9). It has a trivially different spatial dependence, but, more importantly, exhibits two poles instead of one. The physical reason for this difference is the same as already discussed for $C_j(t)$, namely that the "scattering" atoms first have to absorb light before emitted . In Fig. 3, we sketch the corresponding forms of the photons emitting in the case of instantaneous excitation and in the present case.

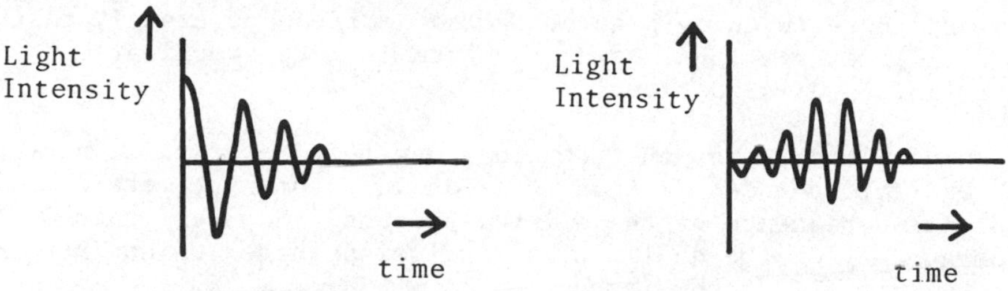

Instantaneous Excitation Excitation by Source Atom

FIGURE 3

Once one knows $B_{\vec{k}}$, it is straightforward to obtain the form of the interference patterns at a detector located at r. A lengthy calculation gives

$$\langle\psi|\hat{E}^-(r)\hat{E}^+(r)|\psi\rangle = \Bigg| -(2\pi)^2\kappa\left(\frac{k_0}{r}\right)e^{-(i\omega_0+\gamma_s)(t-r/c)}\,\theta(t-r/c)$$

$$+\Bigg\{-(2\pi)^4 i\,\frac{\kappa|\kappa|^2}{\gamma-\gamma_s}\Bigg[\left(\frac{k_0^2}{R_1 r_{10}}\right)e^{ik_0 r_{10}}\,e^{(i\omega_0-\gamma_s)[t-\frac{R_1+r_{10}}{c}]}$$

$$\times\,\theta\left(t-\frac{R+r}{c}\right) - \text{same}(\gamma\to\gamma_s)\Bigg] + \text{same}(1\leftrightarrow 2)\Bigg\}\Bigg|^2, \qquad (10)$$

where

$$R_1 = |\vec{r}-\vec{r}_1|.$$

Again, it is instructive to compare this with the result of instantaneous excitation, in which case one finds

$$\langle\psi|\hat{E}^-\hat{E}^+|\psi\rangle \propto \Bigg\{\left[\theta\left(t-\frac{r_1}{c}\right)e^{-\gamma(t-r_1/c)}/r_1\right]^2$$

$$+\,\theta(t-r_1/c)\theta(t-r_2/c)e^{-\gamma(t-r_1/c)}e^{-\gamma(t-r_2/c)}e^{ik_0(r_1-r_2)}/r_1 r_2\Bigg\}$$

$$+\ \text{same}\ (1\leftrightarrow 2) \qquad (11)$$

We first note that we now have a contribution due to the fact that the incident wave might not be scattered by the atoms (1st term in Eq. (10)). In practice, this contribution could be eliminated rather simply by appropriate masks. Thus the relevant terms are there in curly brackets. They correspond exactly to the result (11), except that again, the form is more complicated due to the finite build-up time.

We see, then, that the principal new feature of this "source atom" excitation model is that it leads to a somewhat more complicated time dependence of the emitted photons. However, this does not change any of the qualitative conclusions based on an instantaneous excitation scheme.

SEMICLASSICAL VS QUANTUM FIELD TREATMENTS OF PRESENT PROBLEM

Now that we are content that we understand the effects of the excitation by a wave-packet, let us return for a moment to questions of principle. In particular, it has been argued that the difference (fringes or no fringes) between the two- and three-level system scaterers is completely understandable semi-classically. The proponents of this point of view argue that it is the presence of "vacuum fluctuations" that washes out the fringe pattern between the photons γ_1 and γ_2 emitted by our three-level atoms. That is γ_1 and γ_2 might be expected to have an essentially random phase since they have been "stimulated" by vacuum fluctuations. This need not be the case for scattering from two-level atoms since the incident pulse can leave the atoms 1 and 2 in a coherent superposition of states and the emitted radiation thus has a well defined phase.

This is not correct. In fact the semiclassical theory predicts an absence of fringes even in the case of scattering by two-level atoms. This is discussed in the following paragraphs.

Over the past decade quantum optics has seen the successful application of semiclassical radiation theory over and over again. In this formulation of the radiation-matter interaction one couples the atomic (Schrödinger) and radiation (Maxwell) equations of motion to give a self-consistent description of the radiative phenomenon. This is summarized in Fig. 4. According to this very successful approach to the problem the ith atom acquires a microscopic dipole mement

$$\vec{p}^{\,i}(t) = \langle \psi^i(t) | e\vec{r}^{\,i} | \psi^i(t) \rangle \qquad (12)$$

which in turn acts as a tiny radiating dipole. Summing over the dipoles in a small volume provides the source term in Maxwell's equations.

This approach has enjoyed a wide ranging success and is in fact adequate for the overwhelming bulk of problems in modern optics. Of course, there are cases where this semiclassical theory

Dipoles Device

Field

Field Indices

Dipoles

$$i\hbar \frac{\partial \psi}{\partial t} = (H_0 + e\vec{r}\cdot\vec{E})\psi$$

$$\Box^2\vec{E} = -\mu_0 \sum_i \frac{\partial^2}{\partial t^2} \langle \psi^i | e\vec{r}^{\,i} | \psi^i \rangle$$

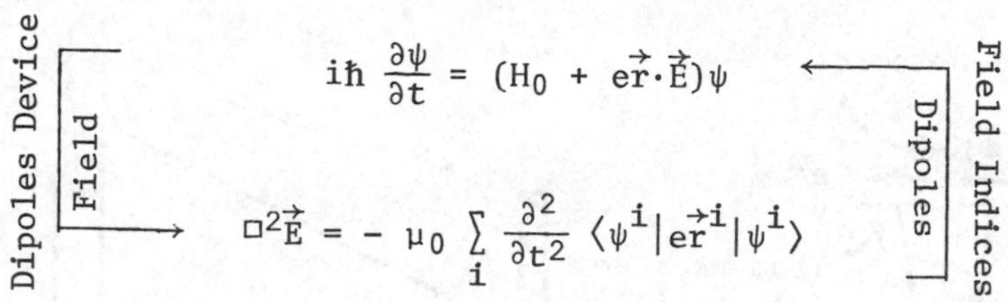

Figure depicting coupled
Schrödinger–Maxwell equations

FIGURE 4

fails to explain the observed physics. Examples of such failure
include: spontaneous emission, the Lamb shift, the laser linewidth,
etc. However, even in these cases one can still understand the
basic physics of such q.e.d. effects by including a random noise
source (representing the effects of vacuum fluctuations) in the
Maxwell–Schrödinger equations.

Thus it is that one may have a comfortable feeling that this
simple and physically appealing semiclassical theory (plus a little
noise as needed) is all that is really needed in quantum optics.
However, as we argue below, this is not completely true. There are
interesting situations relating to the foundations of quantum
mechanics (e.g., involving the subtleties of observation and state
reduction) which can only be understood via the complete machinery
of a fully quantized radiation theory. That is, in these cases,
the semiclassical (+ noise) approach leads to incorrect conclusions
even in very simple problems. One such case is precisely the
"quantum eraser" experiment.

In order to see the sense in which semiclassical theory fails,
consider the gedanken experiment sketched in Fig. 5. In this
figure we consider the passage of a beam of "light" excited atoms
(L.H.S. of Fig. 5) past two two-level atoms 1 and 2. These excitation atoms recoil when a photon is emitted as they decay from the
excited state $|A\rangle$ to the ground state $|B\rangle$. If an atom is detected
in the atomic detector, it signals that a photon has been emitted
in the direction of atoms 1 and 2.

Before interacting with the photon emitted in the $|A\rangle \rightarrow |B\rangle$
transition, the state of atoms 1 and 2 and photon χ is

$$|\psi\rangle = |b_1 b_2\rangle|\chi\rangle. \tag{13}$$

after interaction, we have

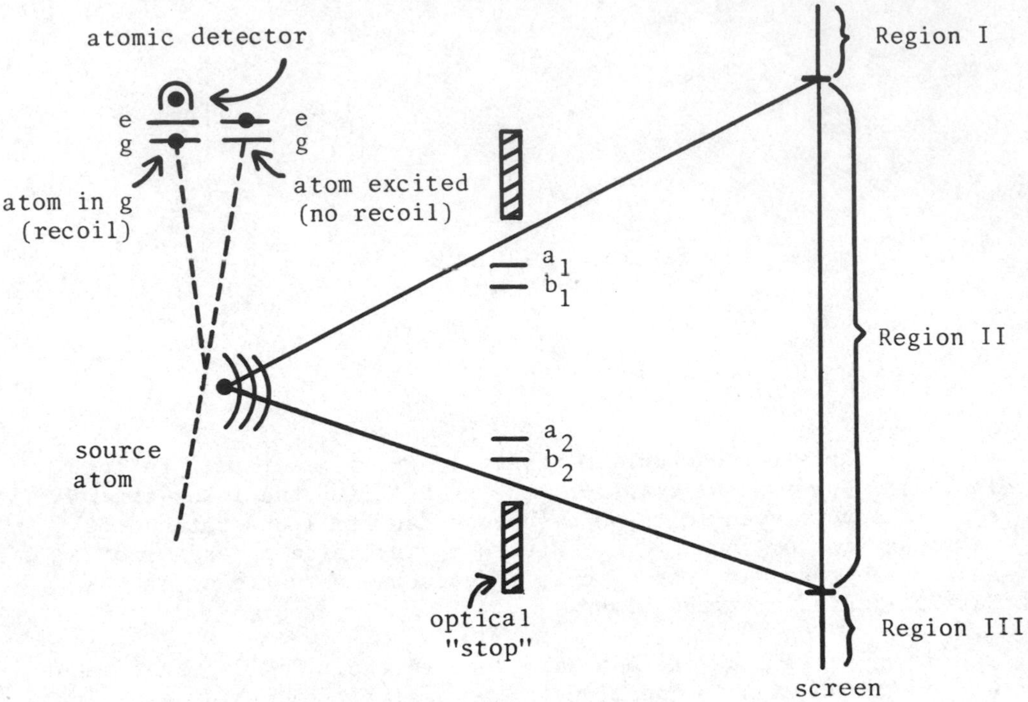

FIGURE 5

Gedanken setup designed to prepare the state $\psi = \frac{1}{\sqrt{2}} [|a_1 b_2\rangle + |b_1 a_2\rangle]$

$$|\psi\rangle = \alpha[|a_1 b_2\rangle + |b_1 a_2\rangle]|0\rangle + \beta|b_1 b_2\rangle|\chi\rangle, \tag{14}$$

which should be compared with Eq. (2), where we have however explicitely included the state of the source. However, in the present case, this is not necessary, since we know (because we saw a click in the atomic detector) that the photon χ is incident on atoms 1 and 2.

Now, if we see a count in a detector located in region II, we can infer that atoms 1 and 2 are still in b state (provided that we neglect the scattering by atoms 1 and 2 in this direction, which is clearly a higher order effect). However, if we see no count in the region II detector (but recall that we did see a count in the "excitation atom" detector so that we know that a photon χ was emitted into region II) then the χ photon was certainly observed by either atom 1 or 2; in such a case we have prepared an atomic state

$$|\psi\rangle = \frac{1}{\sqrt{2}} [|a_1 b_2\rangle + |b_1 a_2\rangle] \tag{15}$$

Here, we have assumed for simplicity that both scatterers are equidistant from the source. Let us next look for interference fringes in the regions I and III. Consider first a semiclassical "calculation" of the field in these regions. The total dipole

$$\vec{p}(t) = \vec{p}_1(t) + \vec{p}_2(t) \tag{16}$$

implies the radiated field

$$\vec{E}(t,\vec{r}) = \vec{E}_1(t,\vec{r}) + \vec{E}_2(t,\vec{r}) \tag{17}$$

where the field phases ϕ_1 and ϕ_2 are determined by the dipoles 1 and 2 respectively. The total dipole is given by

$$\vec{p}(t) = \langle\psi|e\vec{r}_1 + e\vec{r}_2|\psi\rangle \tag{18}$$

where $|\psi\rangle$ is the state defined in Fig. 5.

Substituting the explicit form for $|\psi\rangle$ we find

$$\vec{p}(t) = \frac{1}{2} [\langle a_1|e\vec{r}_1|b_1\rangle\langle b_2|a_2\rangle + \langle a_2|er_2|b_2\rangle\langle b_1|a_2\rangle + \text{c.c.}]$$

$$= 0 \quad , \tag{19}$$

since $|b\rangle$ and $|a\rangle$ are orthogonal.

This is not unexpected, since we are dealing with spontaneous emission. That is, we must include vacuum fluctuations or "noise" into the problem in order to induce a small dipole and get the atom to emit (spontaneously). The effect of this noise is to produce a state vector of the form

$$|\psi\rangle = \alpha_1|a_1 b_2\rangle + \alpha_2|b_1 a_2\rangle + \beta|b_1 b_2\rangle \ , \tag{20}$$

which leads to an atomic polarization

$$\vec{p} = \vec{p}_0[\alpha_1^*\beta + \alpha_2^*\beta + \text{c.c.}]$$

$$= \vec{p}_0[e^{-i(\omega_0 t + \phi_1)} + e^{-i(\omega_0 t + \phi_2)} + \text{c.c.}] \ , \tag{21}$$

where ϕ_1 and ϕ_2 are the (random) phases resulting from the action of "vacuum noise" on atoms 1 and 2. This implies that the field distribution in Regions I and III will be of the form

$$\vec{E}_1 + \vec{E}_2 = \varepsilon_0 \frac{e^{i[k|r-r_1|-i\omega t-i\phi_1]}}{|r-r_1|} + \frac{e^{i[k|r-r_2|-i\omega t-i\phi]}}{|r-r_2|} + \text{c.c} \qquad (22)$$

and the interference cross term is given by

$$E_1{}^*E_2 = \frac{\varepsilon_0{}^2\, e^{ik(\Delta r_1-\Delta r_2)}\left\langle e^{i(\phi_1-\phi_2)}\right\rangle}{(\Delta r)^2} + \text{c.c.} \qquad (23)$$

Now the random phases ϕ_1 and ϕ_2 are only correlated over dimensions of order λ, but we are always taking our atoms to be spaced far apart compared to λ. Thus the expectation value

$$\left\langle e^{i(\phi_1-\phi_2)}\right\rangle = 0 \qquad (24)$$

for this geometry (well separated atoms).

To summarize, a semiclassical (plus phenomenological vacuum noise) approach to this problem clearly leads to the conclusion that there will be no interference fringes observed in the experimental arrangement of Fig. 2. This is in obvious disagreement with the results of Section 2 and Ref. 4.

In conclusion, we find that a semiclassical vacuum noise philosophy leads to a qualitatively different conclusion from that of quantum field theory, even for the case of two-level atoms. This result is interesting in providing an unusual example in which such a "random electrodynamics" disagrees dramatically with the quantum theory of radiation, and also in providing a stimulus for further deeper questions which will be published elsewhere.

REFERENCES

1. E. P. Wigner, Am. J. Phys. 31:6 (1963; see also E. Wigner, "The Scientist speculates", ed. I. Good, London, Heinemann, (1962), p. 284.
2. Our operational approach to the problem of measurement in quantum mechanics (i.e. invision an experiment and carry through the theory of this particular measurement in detail) is the result of many helpful conversations with Prof. Willis Lamb. In this context see especially W. E. Lamb, Jr., Phys. Today, 22:3 (1969).

3. M. O. Scully, R. Shea and J. D. McCullen, Phys. Rep. 43:486 (1978). This paper is also to be found in "W. E. Lamb, Jr. a Festschrift", D. Ter Harr and M. O. Scully, eds., North Holland, Amsterdam (1978).
4. M. Scully and K. Drühl, Phys. Rev. (to be published); M. Scully, V Laser Spectroscopy Conference (Jasper) proceedings to be published.

A TWO-PHOTON EXPERIMENT CONCERNING DELAYED CHOICE AND HIDDEN VARIABLES IN QUANTUM MECHANICS

Kai Drühl

Institute for Modern Optics
University of New Mexico
Albuquerque, New Mexico 87131 USA

INTRODUCTION

In this lecture we discuss in more detail two aspects of the photon scattering experiment which was presented in Prof. Scully's lectures[1]. This experiment is a quantum optical version of the two slit experiment[2] which is often used to clarify important questions concerning the process of measurement in quantum mechanics. In our version the two slits are replaced by two atoms at fixed positions, which are allowed to absorb and re-emitt photons. In this simple scattering system all aspects of the scattering process can be treated by explicit calculation. This makes it possible to formulate precisely and hence put into focus some basic quantum mechanical assertions about the measurement process which continue to be of interest today. Furthermore, our system is close enough to reality to allow for actual experiments to be performed, which in itself is an attractive feature.

In the first section we shall describe the basic experimental setup and the way in which features of the scattered radiation depend upon the level structure of the scattering atoms. This summarizes some important points from the lecture's of Prof. Scully and mainly serves to make this lecture self-contained. In the second section we extend the original arrangement to include a second photon to be emitted in each scattering event considered, and discuss aspects of "delayed choice" for correlation experiments on the two photon system.

In the third section we finally exploit the fact that two photon correlation experiments provide a testing ground for theories which aim at a more detailed description of physical phenomena than is provided by quantum mechanics ("realistic" or "hidden variables" theories). This is done by means of Bell's inequalities. We give a brief introduction to this topic which will establish contact between experiments performed or suggested for this purpose and the experiment we propose here. Of course such introduction cannot give an account of all the subtle aspects of the problem at hand, and we refer the interested reader to some of the articles published on this topic, which may serve as a starting point for further study.

THE SCATTERING EXPERIMENT

In our version of the two slit experiment we propose to observe radiation emitted from two atoms. These atoms are supposed to be supported (e.g. by a macroscopic crystal) at fixed positions such that radiative recoil may be neglected. Radiation is emitted from these atoms as they return to the groundstate from an excited upper level, and we assume that both atoms were coherently excited by one single photon for each scattering event. This assumption is important for our discussion of the basic quantum mechanics underlying this experiment.

In this respect the situation is quite similar to the two slit experiment, where just one single electron per event is passing through the two slit arrangement. Let us finally remark that there is also a non-vanishing probability for the incoming photon not to excite any of the atoms. However, this event can be detected by suitable means, and Prof. Scully has described a corresponding Gedanken experiment in his lecture. We shall therefore discard the corresponding components of the quantum mechanical state, and start from a situation where the incoming photon has left the two atoms in a coherent superposition of states wherein one atom is excited while the other atom is in the groundstate and vice versa.

Having thus prepared our system we now look for the radiation emitted as the excited atoms spontaneously return to the groundstate. In particular, we may put a photo-detector at a specified position on an observation screen and count the number of photons detected there for a large ensemble of scattering events. The distribution of photon counts over the screen will show specific features, which are depending on the level structure of the atoms used, and it is this dependence which we want to analyze here.

Let us first consider the case of scattering by two level atoms. The atomic state of interest are the groundstate $|b\rangle$ and an excited state $|a\rangle$ which is radiatively connected to the groundstate. After excitation by an incoming photon resonant to the transition frequency ω_{ab} the aystem of excited atoms is described by a state vector:

$$|\psi\rangle = \frac{1}{\sqrt{2}} \{|0;a_1b_2 \pm |0;b_1a_2\rangle\} \qquad (1)$$

Here the subscripts 1, 2 refer to the atoms considered; the state label 0 indicates that no photon is present. In general the two components of $|\psi\rangle$ could have coefficients different from those in (1), depending on the symmetry properties of the incoming photon wavefunction with respect to reflections at the plane of symmetry (which is orthogonal to the axis joining the two atoms, and at equal distance from both). The choices made in (1) are meant to represent two typical cases.

Let us consider the first component of $|\psi\rangle$ in (1). After a time long compared to the lifetime of the excited level a_1 the atom 1 will have returned to the groundstate b_1. The photon emitted will be found in a pure quantum state $|\gamma_1\rangle$. The corresponding wavefunction may be calculated in the Wigner Weisskopf approximation[3]. Neglecting the small probability that the photon may be reabsorbed from atom 2 we obtain:

$$|\gamma_1\rangle = \sum_{\underset{\sim}{k},\varepsilon} \frac{w(\underset{\sim}{k},\varepsilon)}{\omega_{ab} - \nu_k - i\gamma} e^{-i\underset{\sim}{k}\cdot\underset{\sim}{R}_1} |\underset{\sim}{k},\varepsilon\rangle \qquad (2)$$

Here $|\underset{\sim}{k},\varepsilon\rangle$ is a photon plane wave of wavevector $\underset{\sim}{k}$ and polarization vector ε, $\underset{\sim}{R}_1$ is the position of atom 1, ω_{ab} and ν_k are the frequencies of the atomic and the emitted photon and γ is the decay rate of the transition $a \to b$. $w(\underset{\sim}{k},\varepsilon)$ is proportional to the matrix element of the atom field interaction, and the well known relation between γ and w can be derived from the condition that the state $|\gamma_1\rangle$ in (2) be normalized to unity . Note that we are working in the Heizenberg picture throughout where the states are independent of time. For the second component of $|\psi\rangle$ in (1) the same analysis applies with atom 1 replaced by atom 2. Due to the linearity of the quantum mechanical equations of motion the final state $|\psi;\text{out}\rangle$ evolving from $|\psi\rangle$ after times long compared to the decay time is obtained by adding both contributions

$$|\psi;\text{out}\rangle = \frac{1}{\sqrt{2}} \{|\gamma_1;b_1b_2\rangle \pm |\gamma_2;b_1b_2\rangle\}$$

$$= \frac{1}{\sqrt{2}} \{|\gamma_1\rangle \pm |\gamma_2\rangle\}|b_1b_2\rangle \qquad (3)$$

$$= |\gamma_{\pm}>|b_1 b_2> \quad . \tag{3}$$

Note that in this case the scattering process leaves both atoms in the groundstate, while the photon emitted is in a coherent superposition of states $|\gamma_1>$ and $|\gamma_2>$ which correspond to emission from each of the atoms 1 and 2.

As in the case of the two slit experiment interference will occur between the corresponding probability amplitudes, and can be observed e.g. by counting with a photo detector the number of photons emitted into a given solid angle.

Let $\hat{I}$ be an operator corresponding to such an observation. The number of counts is then given by the expectation value of $\hat{I}$ in the state $|\psi;out>$. The matrix elements needed are given by:

$$<\gamma_j|\hat{I}|\gamma_k> = e^{ik_0(r_j - r_k)} I; \qquad j,k = 1,2 \ . \tag{4}$$

Here I is a constant, ck_0 is the atomic transition frequency ω_{ab} and r_j is the distance from the j-th atom to the photodetector. Explicit expressions for I have been given in Ref. 1 for the case where $\hat{I}$ is the electric field intensity operator at a given space time point. From (3) and (4) we obtain the expectation value

$$<\psi;out|\hat{I}|\psi;out> = <b_1 b_2|b_1 b_2> I[1 \pm \cos k_0(r_1 - r_2)], \tag{5}$$

showing that an interference pattern is observed in the number of photon counts as the detector is moved across the observation screen. From the position of minima in this pattern information about the relative phases of the two components of the photon wave function can be obtained.

So far out discussion has been quite parallel to the well known discussion of the two slit experiment. The question then arises whether our experiment admits for a situation in which information about the "photon path" could be obtained, and whether or not interference patterns could be observed in this case.

Indeed such a situation is found if we replace the two level atoms by atoms having three levels $|a>$, $|b>$ and $|c>$. Here the highest state $|a>$ is radiatively by connected to both the groundstate $|c>$ and the lower excited state $|b>$. In order to simplify the following discussion we may furthermore assume that $|b>$ is metastable with respect to transitions to $|c>$. Such a situation would occur for example if both $|c>$ and $|b>$ are s-states while $|a>$ is a p-state.

As in the discussion above our atoms are first coherently excited to level $|a>$ by a single photon. Subsequently they will decay to levels $|b>$ or $|c>$ via photon emission. Let us now arrange our detection system so that it is sensitive only to radiation emitted in the $a \rightarrow b$ transition, i.e. we ignore radiation from the $a \rightarrow c$ transition.

In this case the situation is quite different from the experiment above. Upon detection of a photon at frequency ω_{ab} we know from energy conservation that exactly one atom will be in the excited state $|b>$. By looking at the atoms we can find out which atom is excited and hence did the scattering. Having thus access to "photon path" information, analogy to the two slit experiment would suggest that no interference pattern is observed in the distribution of photon counts. This is confirmed by explicit calculation. Starting from the initial state

$$|\psi> = \frac{1}{\sqrt{2}}\{|0\, a_1 c_2> \pm |0\, c_1 a_2>\}$$

after emission of a photon at frequency ω_{ab} the system is in the state

$$|\psi;out> = \frac{1}{\sqrt{2}}\{|\gamma_1 b_1 c_2> \pm |\gamma_2 c_1 b_2>\}$$

$$= \frac{1}{\sqrt{2}}\{|\psi_1;out> \pm |\psi_2;out>\} \tag{6}$$

For the operator $\hat{I}$ corresponding to the photodetector at specified position we now find the matrix elements:

$$<\psi_j;out|\hat{I}|\psi_j;out> = <b_j c_k|b_j c_k><\gamma_j|\hat{I}|\gamma_j>, \quad j = 1,2; k \neq j$$

$$= I. \tag{7.1}$$

$$<\psi_1;out|\hat{I}|\psi_2;out> = <b_1 c_2|c_1 b_2><\gamma_1|\hat{I}|\gamma_2>$$

$$= 0 \tag{7.2}$$

While the diagonal matrix elements are the same as in the two level case (4) the off-diagonal elements now vanish, due to the orthogonality of the atomic states.

As a consequence the interference terms disappear from the expression for the expected number of photons counted:

$$<\psi;out|\hat{I}|\psi;out> = I \tag{8}$$

hence no interference pattern is observed in the distribution of photon counts.

It is at this point that our experiment differs essentially from the two slit experiment. In the latter information about the "electron path" is obtained by measuring the recoil each slit will experience as the electron passes through. For this to be feasible the uncertainty Δp of the slit momentum should be much less than the electron momentum p_e:

$$p_e >> \Delta p \geqslant \hbar/\Delta x$$

It follows that the corresponding uncertainty Δx in the position of the slit is much larger than the electron wave length λ. As a consequence no interference pattern (or rather a very weak pattern) will be found:

$$\Delta x >> \hbar/p_e = \lambda$$

From this point of view the absence of interference is due to the fact that the positions of the slits are ill defined from the very beginning. This in turn results by Heisenberg's uncertainty principle from the fact that we have chosen to specify their momentum rather sharply. Following Heisenberg's[4] early terminology we might argue that the fluctuations in position of the slits have disturbed phase relations in the wave function of the scattered electron in a way which in principle cannot be controlled.

However intuitive this picture may be for the discussion of the two slit experiment it offers little help in understanding our photon scattering experiment. In both the two level and the three level situation the scattering atoms are found in sharply defined quantum states before and after the scattering. Also their positions in both cases have been assumed to be as sharply defined as required for the observation of interference patterns.

The disappearence of interference patterns then is due to the fact that our arrangement allows for the possibility to observe which atom did the scattering. Let us emphasize here that it is not necessary to actually make such an observation. The mere possibility for this, as reflected by the orthogonality of the corresponding atomic states, is sufficient. [A similar situation is found for temporal quantum interference (quantum beats) of two transitions in a single atom. Here quantum beats are not observed if the transitions start from the same level and lead to two different groundstate levels[4,5]. However, in our case (spatial) interference for a single transition in two atoms) several new aspects arise, which we shall discuss in the following].

TWO PHOTON CORRELATION AND DELAYED CHOICE

There is another important aspect in which the two level and three level cases are different. In the two level case the scattered photon is left in a well defined quantum state, $|\gamma^{\pm}>$, while both atoms are left in the groundstate (3). On the other hand in the three level case it does not make sense to talk about the state of the scattered photon alone. Rather the atomic part and the photon part of both components of the wavefunction remain correlated after the scattering event (6) and the state does not factor into an atomic and a photon part.

It will be convenient to discuss this correlation in terms of a two photon correlation. This can be achieved for example in the following way. Let us assume that there is an intermediate level $|b'>$ radiatively connected to both the metastable level $|b>$ and the ground state $|c>$. We can then pump the population in level $|b>$ to level $|b'>$ by a laser π pulse, and subsequently observe the photons emitted in the transition $b' \rightarrow c$. Denoting such photons by the letter ϕ we can write the final quantum state as

$$|\psi;\text{out}> = \frac{1}{\sqrt{2}}\{|\gamma_1\phi_1;c_1c_2> \pm |\gamma_2\phi_2;c_1c_2>\}$$

$$= \frac{1}{\sqrt{2}}\{|\gamma_1\phi_1> \pm |\gamma_2\phi_2>\}|c_1c_2>$$

$$= |\chi_{\pm}^{\text{out}}>|c_1c_2> \tag{8}$$

Here ϕ_1 and ϕ_2 are the photons emitted from atoms 1 and 2. Their states are orthogonal. It appears to be obvious that our manipulations should not affect the results of any observation $\hat{I}_\gamma$ on the γ-photons. Hence we obtain the same result for the state (8) as we did for the state (6):

$$<\psi;\text{out}|\hat{I}_\gamma|\psi;\text{out}> = I_\gamma$$

Let us now introduce a special arrangement to observe the ϕ-photons emitted. By placing a selective photodetector which is sensitive to photons at the ϕ-frequency only at equal distance from both atoms we have an observation whose corresponding operator $\hat{I}_\phi$ has matrix-elements:

$$<\phi_j|\hat{I}_\phi|\phi_k> = I_\phi; \quad j,k = 1, 2 \tag{9}$$

For joint observation of both the ϕ and the γ-photons we now obtain from (8), (4) and (9)

$$<\chi_\pm;out|\hat{I}_\phi\hat{I}_\gamma|\chi_\pm;out> = I_\phi I_\gamma[1\pm \cos k_0(r_1-r_2)] \tag{10}$$

This is a very surprising result, since it appears that upon setting up a special arrangement to register the ϕ-photons an interference pattern in the distribution of γ-photons is again observed.

At first sight this may even seem paradoxical, since we are free to trigger the emission of ϕ-photons long after the γ-photons have been emitted. In such a situation no direct physical influence of the ϕ-photon count on the γ-photon count is possible.

In order to better understand the physics behind this let us recall that we are dealing with a two photon correlation experiment. Thus for any count of a γ-photon we have to check whether or not the ϕ-counter registered a count. Only those events are to be kept where both the γ- and the ϕ-counter registered a count. From this point of view we might expect more readily that interference patterns could reappear in a subensemble of scattering events.

In order to see this it is convenient to pass to a new basis of ϕ photon states:

$$|\phi_\pm> = \frac{1}{\sqrt{2}}\{|\phi_1> \pm |\phi_2>\} \tag{11}$$

The matrix elements of $\hat{I}_\phi$ between these states are found from (9)

$$<\phi_+|\hat{I}_\phi|\phi_+> = 2I_\phi$$

$$<\phi_-|\hat{I}_\phi|\phi_-> = <\phi_-|\hat{I}_\phi|\phi_+> = 0 \tag{12}$$

Thus the photo counter described by $\hat{I}_\phi$ will project out the $|\phi_+>$ component only from any ϕ photon state and reject the $|\phi_->$ component.

The final two photon state $|\chi;out>$ of our scattering system can now be decomposed as:

$$|\chi;\text{out}\rangle = \frac{1}{\sqrt{2}}\{|\gamma_1\phi_1\rangle + |\gamma_2\phi_2\rangle\} \tag{13a}$$

$$= \frac{1}{\sqrt{2}}\{|\gamma_+\phi_+\rangle + |\gamma_-\phi_-\rangle\} \tag{13b}$$

In the decomposition (13b) the corresponding coherent superpositions of photons emitted from both atoms are correlated with each other, and it is precisely the first component which contributes to the two photon counts as described in Eq. (10). Thus the total ensemble of scattering events is decomposed into two subensembles of equal size, depending on whether the ϕ counter was triggered or not. In the former ensemble an interference pattern is found, while in the latter ensembel the corresponding "negative pattern" occurs.

Alternatively we may choose an arrangement where only ϕ-photons emitted from atom 1 are counted. In this case (13a) gives the appropriate decomposition of the final quantum state and the corresponding decomposition of the ensemble of scattering events.

This leads us to an interesting type of "delayed choice" experiment[6]. In fact we may choose to select one or the other type of decomposition long after the emission of γ photons taken place, and even <u>after</u> the γ-photons <u>have been registered already</u>. In this way even an ensemble of observed events can "a posteriori" be made to bear out either wavelike (interference) or particle like (no interference) aspects. However strange this may seem to be it does not violate any principle of quantum mechanics. In particular, it is not possible to assign a given γ-photon count to a subensemble in both decompositions. This would require a simultaneous observation of both the $\phi_1 - \phi_2$ and the $\phi_+ - \phi_-$ aspects of the ϕ photon state, which is not possible according to quantum mechanics.

In fact the situation here is exactly the same as in the experiment discussed by Einstein, Podolsky and Rosen (EPR)[7], or the variant discussed by Bohm[8]. For all these cases we have a perfect correlation between the states of two particles which does exist even if these particles are spatially separated. It is this correlation which lead EPR to argue that for any single experiment the observables measured for both particles actually had a precise predetermined value. Since the same argument can be made for observations which are complementary in the sense of quantum mechanics (like position and momentum, or different spin directions), they concluded that the quantum mechanical description was incomplete.

In our case the complementary observations are related to the particle and wavelike aspects of the two photons, as expressed in the two decompositions (13a) and (13b) of the photon wave function.

In the next and final section of this lecture we shall discuss these aspects in more detail.

A NEW CASE FOR BELL'S INEQUALITIES

As we have seen in the last section, quantum mechanics predicts perfect correlations between the results of certain observations made on the two photons emitted in our scattering experiments. It is at this point that theories aiming at a more detailed description of physical systems than the one provided by quantum mechanics give substantially different predictions and can hence be tested experimentally. Such theories are commonly called "realistic". Let us briefly explain what is meant by this; for a more detailed discussion and a guide to the literature on the subject we refer the interested reader to the review article by Clauser and Shimony[9].

Quantum mechanical predictions are of statistical nature. From the point of view of "realistic" theories however there is a description of physical systems which allows to predict the outcome of every experiment with certainty. The statistical nature of quantum mechanical predictions then would be a consequence of the incomplete description of the system giving the quantum mechanical state as the statistical average over a certain ensemble of systems, each of which has a precise value assigned to all its observables.

This situation is familiar in classical statistical mechanics. Here a complete description is given by the "microstate" of the system which is just a point in phase space. Since information about the microstate is not available for a macroscopic system, the relevant properties are calculated by taking averages over the corresponding thermodynamical ensembles.

Exactly the same assumptions are made in a "deterministic hidden variable theory"[9]. Hence it is assumed that there exists states λ which assign a well defined value $A(\lambda)$ to any quantum mechanical observable A. These states are assumed to form a space Λ, in which probability measures can be defined. If $d\rho(\lambda)$ is the measure corresponding to the quantum state $|\psi\rangle$ under consideration, the expectation value of observable A is given by

$$\langle A\rangle = \int_{\Lambda} A(\lambda)d\rho(\lambda) = \langle\psi|A|\psi\rangle \tag{14}$$

Furthermore such a theory is called "local" if for two spatially separated particles the results of measurements on one particle are independent on whatever measurement is performed on the other particle. This means that for the product of two observable A and B associated to each of the particles we have:

$$(A\cdot B)(\lambda) = A(\lambda)\cdot B(\lambda)$$

$$\langle A\cdot B\rangle = \int_{\Lambda} A(\lambda)\cdot B(\lambda)d\rho(\lambda) \tag{15}$$

Let us now assume that we are dealing with a set of observables A_α labeled by a parameter α for which:

$$A_\alpha(\lambda) = \pm 1$$

$$\langle A_\alpha\cdot B_\alpha\rangle = \langle\psi|A_\alpha\cdot B_\alpha|\psi\rangle = -1 \tag{16}$$

A well known example are the observables associated with a spin 1/2 system, where α denotes the direction in space along which the component of spin is measured and ψ is a singlet state. From (15) and (16) we then conclude that:

$$A_\alpha(\lambda) = -B_\alpha(\lambda) \text{ for all } \lambda$$

For three different parameters α, β and γ we obtain:

$$\langle A_\alpha\cdot B_\beta\rangle - \langle A_\alpha B_\gamma\rangle = \int_{\Lambda} [A_\alpha(\lambda)A_\beta(\lambda) - A_\alpha(\lambda)A_\gamma(\lambda)]d\rho(\lambda)$$

$$= -\int A_\alpha(\lambda)A_\beta(\lambda)[1 - A_\beta(\lambda)A_\gamma(\lambda)]d\rho(\lambda)$$

And hence:

$$|\langle A_\alpha\cdot B_\beta\rangle - \langle A_\alpha\cdot B_\beta\rangle| \leqslant 1 + \langle A_\beta\cdot B_\gamma\rangle \tag{17}$$

Quantum mechanical predictions disagree with this inequality. If A_α denotes the spin direction in the x-y plane at an angle of α with the x-axis, the quantum mechanical prediction for a singlet state $|\psi\rangle$ is:

$$\langle\psi|A_\alpha B_\beta|\psi\rangle = -\cos(\alpha - \beta) \tag{18}$$

For $\beta = 0$, $\alpha = \pi/3$, $\gamma = -\pi/3$ we have

$$|\langle\psi|A_\alpha B_\beta|\psi\rangle - \langle\psi|A_\alpha B_\gamma|\psi\rangle| = 1 \quad \text{but} \quad 1 + \langle\psi|A_\beta B_\gamma|\psi\rangle = \tfrac{1}{2}$$

which is in contradiction to (17).

It is possible to prove similar inequalities under assumptions weaker than those put down in (14), (15), and (16). Let us assume the parameters α, β and γ correspond to orientation angles in a plane, and that in addition the correlated expectation value for two such orientations depends only on the relative angle:

$$\langle A_\alpha B_\beta\rangle = E(|\alpha - \beta|) \tag{19}$$

The function E can then be shown to satisfy the following inequality under very general conditions[9]:

$$|3E(\phi) - E(3\phi)| \leqslant 2. \tag{20}$$

In proving inequality (20) an inequality concerning correlations in a set of four different observables is derived by techniques similar to the one sketched above in the derivation of (18). The symmetry property (19) is then used to reduce this to (20) for a special arrangement of relative orientations.

In our case observables denoted by A would correspond to observations on the γ photons, while observables denoted by B correspond to ϕ photons. Let us furthermore identify the parameters α and β with the phase angles appearing in (4) for the matrix elements, with a modified expression for ϕ photons.

$$\langle\gamma_1|A_\alpha|\gamma_2\rangle = \langle\gamma_1|\hat{I}_\gamma|\gamma_2\rangle = I_\gamma e^{ik_0(r_1-r_2)} = I_\gamma e^{i\alpha}$$

$$\langle\gamma_j|A_\alpha|\gamma_j\rangle = I_\gamma \tag{21a}$$

$$\langle\phi_1|B_\beta|\phi_2\rangle = \langle\phi_1|\hat{I}_\phi|\phi_2\rangle = I_\phi e^{-i\beta}$$

$$\langle\phi_j|B_\beta|\phi_j\rangle = I_\phi \tag{21b}$$

For the correlated expectation values of A_α and B_β in the state $|\chi;\text{out}\rangle$ given by (13) we obtain:

$$\langle\chi;\text{out}|A_\alpha B_\beta|\chi;\text{out}\rangle = I_\gamma I_\phi[1 + \cos(\alpha-\beta)] \tag{22}$$

Hence the symmetry condition (19) is fulfilled, which was in fact our motivation for changing the sign of phase angle in (21b). In fact it is possible to show that our two photon system is equivalent to other well known two particle systems which have been studied in this context. Examples are the system of two spins in a singlet state, or a system of two photons in states of opposite circular polarization. For the normalization of operators chosen to obtain (20) the quantum mechanical result may then be summarized as ($I_\phi = I_\gamma = 1$):

$$\langle \chi;\text{out}|A_\alpha A_\beta|\chi;\text{out}\rangle = 1 + \cos(\alpha - \beta) = E_{QM}(|\alpha - \beta|) \qquad (23)$$

Inequality (20) is clearly violated by E_{QM} in a range of values for $\phi = \alpha - \beta$.

Our system is not only formally equivalent to other systems considered already. It also offers some new experimental and theoretical features. One of them is connected with the fact that no analysers are needed to sort out photons of specified polarization or spins in specified directions. This is an experimental simplification. In addition it eliminates some theoretical problems concerning assumptions about the efficiency of such analysers and its dependence on state of the particle analyzed.

Secondly, the observables discussed there are related to phase angles which are determined by the photon travel path, and not to directions of spin or planes of polarization. Hence, it will be possible to test Bell's inequalities for a different type of quantum system than has been considered until now. A more detailed discussion of theoretical and experimental problems in this experiment is beyond the scope of this lecture and will be given elsewhere.

REFERENCES

1. M. O. Scully and K. Drühl, The quantum eraser. A proposed photon correlation experiment concerning observation and "delayed choice" in quantum mechanics, to be published in Phys. Rev. Lett.
2. D. Bohm, "Quantum Theory", Prentice Hall, Englewood Cliffs, 1951, Chapter 6.2.
3. V. Weisskopf and E. Wigner, Z. Phys. 63:54 (1930).
4. S. Haroche, "High Resolution Laser Spectroscopy", K. Shimoda, ed., Springer, Berlin (1976).
5. J. A. Wheeler, "Problems in the Formulation of Physics, G. T. deFranzia, ed., North Holland, Amsterdam.
6. A. Einstein, B. Podolsky and N. Rosen, Phys. Rev. 47:777 (1935).
7. Ref. 2, Chapter 22.16.
8. J. F. Clauser and A. Shimony, Rep. Prog. Phys. 41:1881 (1981).

SUBNATURAL SPECTROSCOPY

A. Guzman*, P. Meystre*, and M. O. Scully*†

*Max-Planck-Institut für Quantenoptik
D-8046 Garching, West Germany

and

†Institute of Modern Optics
Department of Physics and Astronomy
University of New Mexico
Albuquerque, NM 85721, USA

I. INTRODUCTION

In optical experiments, the precision of the measurement is often limited by the broadening of the linewidth caused by the interaction of the system under investigation with its environment, such as Doppler effect, collisions and spontaneous decay. It might seem that, after the Doppler or collision-broadened width has been eliminated using one of the many schemes introduced in the past for that purpose,[1] the natural linewidth remains the ultimate limit to high-resolution spectroscopy.

Recently we have proposed and analyzed[2,3] experimental situations which provide line-narrowing beyond the natural linewidth. They involve time delayed measurements in a system of two-level atoms with energy separation ω driven by a monochromatic, tunable field. The system is prepared at time $t = 0$ in, say, the lower excited state $|a\rangle$, and coherently driven by a cw laser of frequency υ. The emitted radiation is then allowed to reach the detector only after a time delay. The narrow linewidth obtained in this delayed detection scheme is due to the fact that,[4] in the transient regime, the probability for induced transitions in a two-level system interacting with a monochromatic electromagnetic field is not weighted by a lorentzian of width $\gamma_{ab} = (\gamma_a + \gamma_b)/2$, but rather $\gamma_{ab} = (\gamma_a - \gamma_b)/2$. Our general analysis[4] involves a fully quantum-mechanical treatment and we have considered both the small signal and the strong signal regime. For the sake of simplicity we will describe here only the semiclassical approach, whose results coincide with the quantum-mechanical ones, as far as spontaneous decay directly between the two unstable states can be neglected.

Knight and Coleman[5] have shown that a narrowed absorption line (and consequently a narrowed fluorescence line) may be achieved with a closely related method, where the system of two-level atoms is weakly driven by an exponentially decaying laser pulse. The major advantage of this method is that it may be applied to systems with a stable lower level. The exponential decay of the pulse amplitude has the same effect on the fluorescence lineshape as the exponential decay of the lower unstable level in a system with 2 decaying levels.

These lecture notes are organized as follows. In Section 2, we present the semi-classical theory of transient line-narrowing, using a semi-classical approach and limiting ourselves to the weak-field limit. We then give in Section 3 a physical interpretation of this effect based on the well-known results of the Rabi problem. In Section 4, we briefly discuss the method of Knight and Coleman, based on the use of pulsed excitation. In Section 5, we consider the behavior of a system of two-level atoms under strong-signal excitation. We discuss the appearance of a dip at line center, which allows to perform "dip-spectroscopy" in homogeneously broadened systems. Finally, Section 6 discusses the possible advantages of time-delayed spectroscopy over conventional methods.

II. SEMICLASSICAL, SMALL-SIGNAL THEORY OF TRANSIENT LINE-NARROWING

We consider the following experimental situation.[2] An ensemble of two-level atoms with upper level $|a\rangle$ and lower level $|b\rangle$ decaying at rates γ_a and γ_b, respectively, is prepared at time $t = 0$ into the lower state $|b\rangle$ and coherently driven by a weak CW field from $|b\rangle$ to the upper level $|a\rangle$ (see Fig. 1).

The equations of motion of the two-level system density matrix, as given for example in Ref. 6, are

$$\dot{\rho}_{aa} = -\gamma_a\rho_{aa} + i(\mu E/\hbar)(\rho_{ba}-\rho_{ab}) , \tag{1a}$$

$$\dot{\rho}_{bb} = -\gamma_b\rho_{bb} + i(\mu E/\hbar)(\rho_{ba}-\rho_{ab}), \tag{1b}$$

$$\dot{\rho}_{ab} = -(i\omega+\gamma_{ab})\rho_{ab} - i(\mu E/\hbar)(\rho_{bb}-\rho_{aa}), \tag{1c}$$

where $E = E_0 \cos\nu t$ is the driving laser field, $\hbar\omega$ the energy difference between the two atomic levels, and μ the dipole moment of the transition. In the absence of collisional broadening, $\gamma_{ab} = (\gamma_a+\gamma_b)/2$.

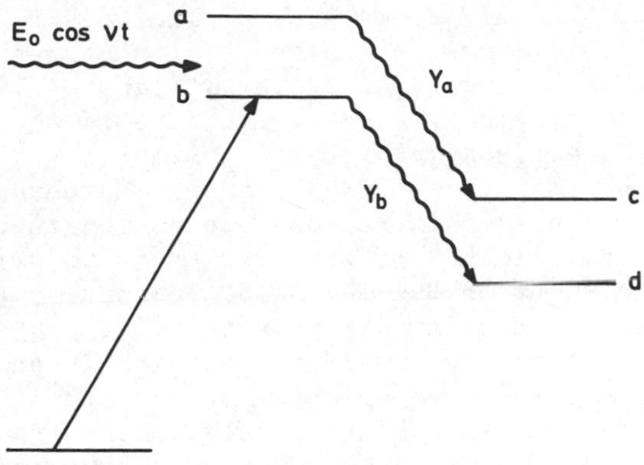

Fig. 1. Level diagram indicating the excitation from ground state $|g\rangle$ to state $|b\rangle$, the monochromatic incident field driving the $|a\rangle$ to $|b\rangle$ transition, and the decays of levels $|a\rangle$ and $|b\rangle$ to some distant levels $|c\rangle$ and $|d\rangle$.

For weak fields, it is possible to solve the system of Eqs. (1a) to (1c) iteratively, considering the field as a perturbation. To zeroth order, we obtain trivially

$$\rho_{bb}^{(0)}(t) = \exp(-\gamma_b t). \tag{2}$$

Introducing Eq. (2) in Eq. (1c) and making as usual the rotating-wave approximation, we obtain

$$\rho_{ab}^{(1)}(\Delta,t) = -i(\mu E_0/2\hbar)\frac{\exp(-i\nu t)}{i\Delta+\delta_{ab}}\left[e^{-\gamma_b t} - e^{-(i\Delta+\gamma_{ab})t}\right] \tag{3}$$

where $\Delta \equiv \omega - \nu$, and $\delta_{ab} = (\gamma_a - \gamma_b)/2$.

The probability $\rho_{aa}(\Delta,t)$ for the atoms to be in the upper level at time t as a function of the detuning Δ is then:

$$\rho_{aa}^{(1)}(\Delta,t) = \left(\frac{\mu E_0}{\hbar}\right)^2 \frac{1}{\Delta^2+\delta_{ab}^{\ 2}}\left[e^{-\gamma_a t} + e^{-\gamma_b t} - 2(\cos\Delta t)e^{-\gamma_{ab} t}\right] \tag{4}$$

This expression contains in the denominator δ_{ab} instead of γ_{ab}, as one might have expected.[7] In general, however, experiments are prepared in such a way that the fluorescence linewidth is not δ_{ab}, but rather γ_{ab}. In Ref. 2, we show explicitely that this is due to an integration of ρ_{aa} over <u>all times</u>, which corresponds to having atoms entering the interaction region at random times, detectors open for all times, etc. ... However, there is nothing fundamental about γ_{ab} and by a clever choice of excitation and detection, one may hope to take advantage of the narrower Lorentzian factor appearing in Eq. (4). Indeed, it is possible to obtain spectra narrowed beyond the natural width γ_{ab}, provided that the system is prepared by exciting the atoms in the $|b\rangle$ state instantaneously at t = 0, and that one does not count all photons emitted by the system. To illustrate this point, we consider specifically an experiment where we measure the time-delayed spectrum of the transition, i.e., the number of photons emitted spontaneously in the transition $|a\rangle \rightarrow |c\rangle$ from time $t = \theta$ on, with $\theta > 0$. Obviously, this <u>time-delayed-lineshape</u> is given by

$$N(\Delta,\theta) = \gamma_a \int_\theta^\infty \rho_{aa}(\Delta,t)dt. \tag{5}$$

Introducing $\rho_{aa}(\Delta,t)$ from Eq. (4) into Eq. (5) we obtain

$$N(\Delta,\theta) = \gamma_a (\mu E_0/\hbar)^2 \frac{1}{\Delta^2+\delta_{ab}^2} \left[\frac{e^{-\gamma_a\theta}}{\gamma_a} \quad \frac{e^{-\gamma_b\theta}}{\gamma_b} \right. \\ \left. + \frac{2e^{-\gamma_{ab}\theta}}{\Delta^2+\gamma_{ab}^2} (\Delta \sin\Delta\theta - \gamma_{ab} \cos\Delta\theta) \right] \tag{6}$$

In the limit $\theta \rightarrow 0$ we recover the familiar Lorentzian lineshape of width γ_{ab}:

$$N(\Delta,\theta=0) = \gamma_a (\mu E_0/\hbar)^2 \frac{1}{\Delta^2+\gamma_{ab}^2} \left(\frac{2\gamma_{ab}}{\gamma_a\gamma_b} \right). \tag{7}$$

For $\theta > 0$, however, the shape of the line is determined by the prefactor in Eq. (6), whenever the exponential in the last term in brackets decays fast enough to damp the oscillations introduced by the sinusoidal terms. Since it is δ_{ab} that appears in the prefactor, rather than γ_{ab}, the line can have a width less than γ_{ab}, but depending on θ. In Fig. 2 we have plotted the time-delayed spectrum as function of the detuning Δ, for several values of the delay (θ = 0, 2, 5, and 20, in units of γ_a^{-1}). In this case, $\gamma_a = 1$, and $\gamma_b = 1.001$. For the sake of clarity all curves have been normalized to 1. The broadest one corresponds to $\theta = 0$ and has the usual Lorentzian shape with full-width at half-maximum γ_{ab}. As θ is increased, the width of the measured line decreases, and for very

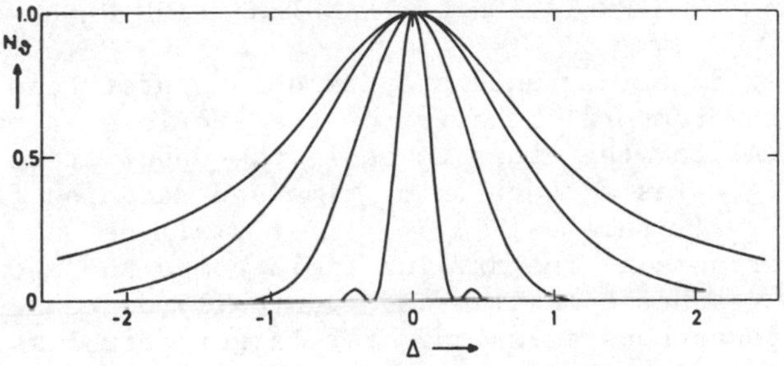

Fig. 2. Normalized time-delayed lineshape $N(\Delta,\theta)$ as a function of the detuning Δ and for various values of the delay θ (see text). Time in units of $1/\gamma_b$.

large delays ($\theta \cong 20\gamma_a^{-1}$) oscillations become observable. The minimum linewidth achievable in theory is δ_{ab}, as expected from Eq. (4).

In Fig. 3, we show the full-width at half-maximum of $N(\Delta,\theta)$ as a function of δ, and normalized to the linewidth for no delay (i.e., γ_{ab}, for γ_a/γ_b = 3, 2, and 1.01.). We observe that for $\gamma_a \cong \gamma_b$, there is almost no theoretical limit to the narrowing. However, since the signal intensity is exponentially decreasing with θ, there is a competition between narrowing of the linewidth and loss of signal intensity. This common feature of all time delayed spectroscopic techniques imposes a practical limit on this method.

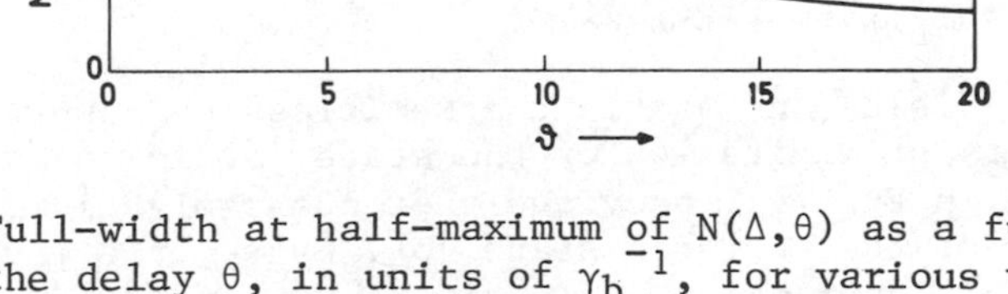

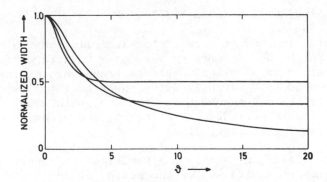

Fig. 3. Full-width at half-maximum of $N(\Delta,\theta)$ as a function of the delay θ, in units of γ_b^{-1}, for various values of $\delta_{ab}=(\gamma_a-\gamma_b)/2$. (See text).

III. PHYSICAL INTERPRETATION OF TRANSIENT LINE-NARROWING

In order to better understand and interpret this effect, we now discuss the temporal behavior of the two-level atoms. The probability $\rho_{aa}(\Delta)$ for the atoms to be in the upper state $|a\rangle$ is plotted in Fig. 4 as a function of time and detuning for the case $\gamma_a = 1.01$, $\gamma_b = 1$ (same as Fig. 2). For small detunings, $\rho_{aa}(\Delta,t)$ initially increases as the driving field pumps the atoms to level $|a\rangle$. It then reaches a maximum and eventually decreases to zero, because of spontaneous emission. For larger detunings, secondary maxima appear, while the principal one occurs at earlier times and has a smaller value. Let us now discuss these three points in more detail. First, we note that the extra maxima are due to the third term in brackets in Eq. (4), which is oscillating at frequency Δ. For small detunings, they are not observable, since the oscillations are so slow that spontaneous emission damps them away. The smaller value of the principal maximum as Δ increases reflects the well known fact that the driving field has a weaker effect on the atomic transition for larger detunings, and is not very important here. The third feature, i.e., the fact that the first maximum of $\rho_{aa}(t)$ occurs earlier for large Δ, is much more interesting. This is again a well-known characteristic of two-level system dynamics. What happens here is just the onset of Rabi oscillations, which are of course faster for larger detunings. The shift of the maximum to lower times for increasing Δ is the key point in understanding the line-narrowing effect. According to Eq. (5), the time-delayed spectrum $N(\Delta,\theta)$ is the area under the probability curve between $t = \theta$ and $t = \infty$. From Fig. 4 we can see that, for a given θ, this integral is a smaller fraction of the full integral (non-delayed spectrum) as Δ increases. This means that for a given time delay, the decrease of the number of photocounts with respect to the non-delayed spectrum is higher for larger Δ. The line-narrowing is a direct consequence of this behavior.

IV. LINE-NARROWING VIA PULSED EXCITATION

An alternative way to obtain transient line narrowing has been proposed by Knight and Coleman.[5] The major advantage of their method is that it is also applicable to ground state transitions. The basic idea is to coherently drive the transition between a (stable) state $|b\rangle$ and an unstable state $|a\rangle$ with a pulsed radiation field switched on suddenly at $t = 0$, and decaying exponentially at rate γ_L (see Fig. 5).

For weak fields and within the semiclassical approximation, the probability of excitation of the state $|a\rangle$ is formally identical to that given in Eq. (4), provided that one replaces γ_b, the spontaneous decay of the unstable state $|b\rangle$, by γ_L, the inverse time constant of the laser pulse. (In Ref. 5, $|b\rangle$ is taken to be stable.) The detuning Δ is, as before, the difference between the frequency

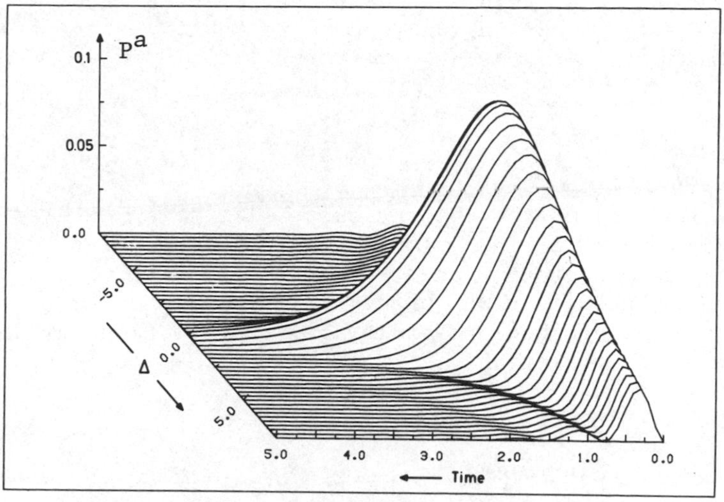

Fig. 4. Probability $P^a \equiv \rho_{aa}(\Delta, t)$ for the atom to be in the excited state $|a\rangle$ as a function of the detuning Δ and t, in units of γ_b^{-1}, for $\gamma_a = 1.01$ and $\gamma_b = 1$.

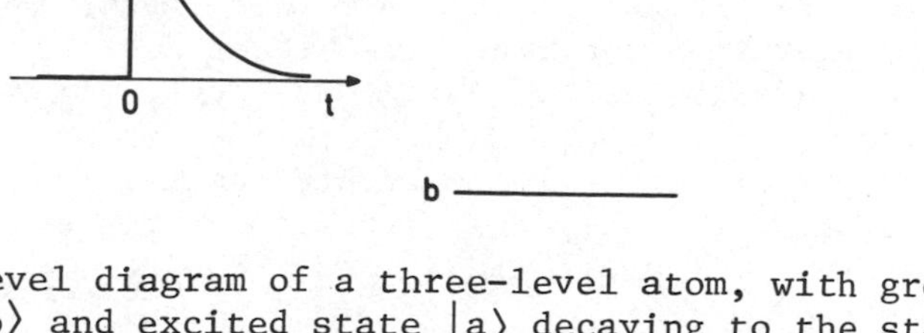

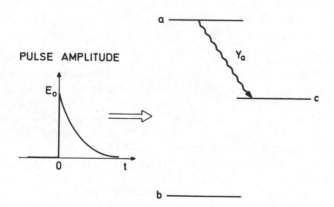

Fig. 5. Level diagram of a three-level atom, with ground state $|b\rangle$ and excited state $|a\rangle$ decaying to the state $|c\rangle$. The transition $|a\rangle - |b\rangle$ is driven by an exponentially decaying laser pulse.

of the atomic transition of the system and the laser frequency. A further advantage of this method is that one can match γ_L with the decay constant γ_a. By making $\gamma_L \cong \gamma_a$ in Eq. (4), the probability of exciting the state $|a\rangle$ at time t becomes

$$\rho_{aa}(\Delta,t) = (\mu E_0/\hbar)^2 \, e^{-\gamma_a t} \, \frac{\sin^2(\Delta t/2)}{(\Delta/2)^2} \, . \tag{8}$$

In this case most of the absorption at time t is centered within a band whose full width at half-maximum is on the order of $2\pi/t$. Therefore, the absorption spectra at times longer than a lifetime have a width narrower than the natural width, and decreasing with time. Thus, we can produce arbitrarily narrow lines by an adequate choice of γ_L and of delay θ.

V. TRANSIENT-DIP SPECTROSCOPY

Up to now, we have limited our discussion to the weak field limit. In this Section, we consider the effects related to the behavior of the system in the strong signal regime. This case can be analyzed by solving exactly (i.e., non-perturbatively) the equations of motion of the density matrix.[3] Again, the full quantum-mechanical results agree with the semiclassical ones,[4] provided that one neglects spontaneous emission directly from $|a\rangle$ to $|b\rangle$. The major result of this analysis is that under appropriate conditions, the combination of a strong field plus delayed detection can lead to the appearance of a dip at line-center of the time-delayed spectrum. The general expression for $N(\Delta,\theta)$ is rather complicated and will not be reproduced here. For the purposes of these notes, it is enough to describe the general behavior and discuss its physical meaning qualitively.

In Fig. 6 we have plotted the function $N(\Delta,\theta)$ for a fixed value of the laser intensity, but omitting the attenuating exponential factor $\exp(-\gamma_a\theta)$(see Ref. 3). The decay constants and Rabi frequency are taken to be respectively $\gamma_a = 3.0$, $\gamma_b = 1.0$, and $(\mu E_0/\hbar) = 2.0$. As expected, the spectrum has the well known power-broadened Lorentzian lineshape in the limit $\theta = 0$ (i.e., no time delay). As θ is increased, however, the line width first decreases (transient line-narrowing), and thereafter, broadens back while a narrow dip appears at line center. This dip becomes deeper and wider as θ is further increased. It is important to realize that the dip appears with a smaller delay, the stronger the field, and that it can have an arbitrarily small width (less than δ_{ab}).

In order to illustrate the origin of the transient-dip, we have plotted in Fig. 7 the probability $\rho_{aa}(\Delta,t)$ for the atoms to be in the upper state as a function of time and detuning Δ, for the same values as those of Fig. 6. (This figure is the strong field

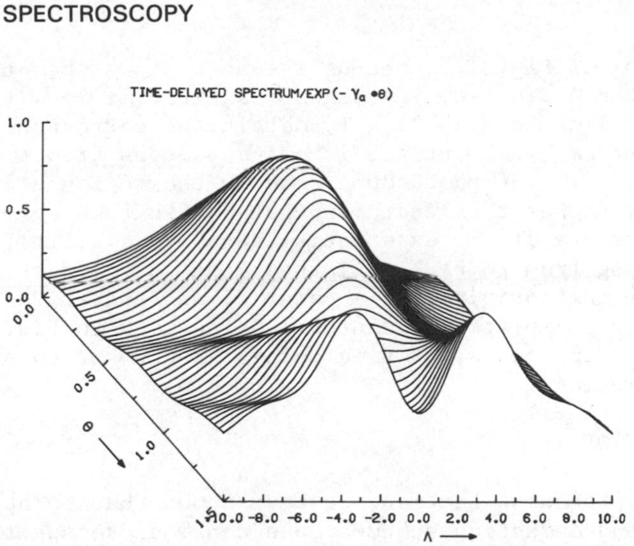

Fig. 6. Photocount distribution $N(\Delta,\theta)$, without the attenuating factor $\exp(-\gamma_a\theta)$, for a fixed laser intensity, as a function of Δ and θ (see text). Time in units of $\gamma_b{}^{-1}$.

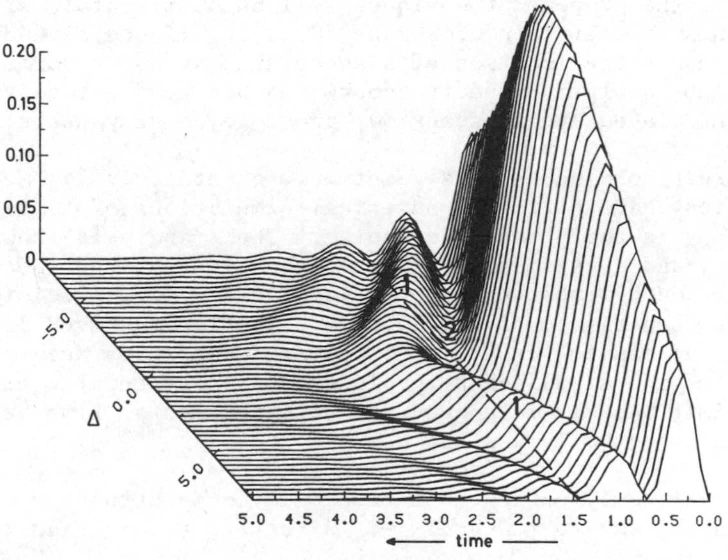

Fig. 7. Probability $P^a(\Delta,t) \equiv \rho_{aa}(\Delta,t)$ of the atom to be in the excited state, in the strong signal regime. Time in units of $\gamma_b{}^{-1}$ (see text).

correspondent of Fig. 4). Let us assume that we open the detector at $t = \theta_0$ when $\rho_{aa}(\Delta = 0,t)$ reaches its first minimum (region around the straight line in Fig. 7). To obtain the corresponding time-delayed spectrum, we count the incoming photons from $t = \theta_0$ on as a function of Δ. The photocount rate in the region around point 1 is obviously higher than that associated with the region around point 2. Because of the exponential decay of $\rho_{aa}(\Delta,t)$, the integration beginning from point 1 will then lead to a larger value of $N(\Delta,\theta_0)$ than that originating a point 2, and therefore to the appearance of a dip at line center. Going back to Fig. 6, we can see that the dip occurs for time delays around θ_0, consistent with this argument.

VI. DISCUSSION

In the preceding Section, we have shown that by appropriate excitation and detection schemes, one can measure spectral lines with a resolution higher than that given by the normal linewidth. However, the signals are now much weaker, and the question remains whether or not this is a useful technique.

There is no unique answer to this question. Whether transient line-narrowing and/or transient dip spectroscopy will be useful depends on the particular situation at hand. Obviously, if there is no noise source at all in the experiment (a highly unrealistic situation), the proposed techniques will be detrimental, since one has thrown away useful information. However, if one already has a good idea about the location of a spectral line (as is normally the case in high-resolution spectroscopy) and has to fight noise problems, then delayed spectroscopy may prove to be very useful.

In order to discuss this point more quantitatively, Metcalf and Phillips[8] have performed numerical simulations of time-delayed spectroscopy in the presence of noise. Their analysis shows that in general, and despite the fact that the time-delayed spectra are necessarily much weaker than non-delayed ones, "... precision measurement benefit considerably from this time-resolved line-narrowing, unless the experimenter has extraordinary accurate knowledge of the signal shape. Since unknown systematic effects usually limit the ultimate accuracy of experiments, time-resolution techniques are usually desirable".

We conclude by noting that time-delayed spectroscopy also provides a useful way to measure $\gamma_a-\gamma_b$ directly, rather than γ_a and γ_b separately.

ACKNOWLEDGMENTS:

We acknowledge many useful discussions with H. Metcalf, M. Sargent III, L. Wilcox, and H. Walther.

REFERENCES

1. For a recent review, see for instance "Laser Spectroscopy" IV, by H. Walther and K. W. Rothe, (Springer-Verlag, Berlin, Heidelberg, New York 1979).
2. P. Meystre, M. O. Scully, and H. Walther, Optics Com. 33, 153 (1980).
3. H. W. Lee, P. Meystre, and M. O. Scully, submitted to Phys. Rev. Lett.
4. H. W. Lee, P. Meystre, and M. O. Scully, to be published in Phys. Rev. A.
5. P. L. Knight and P. E. Coleman, J. Phys. B13, 4345 (1980).
6. M. Sargent III, M. O. Scully, and W. E. Lamb, Jr., "Laser Physics" (Addison-Wesley, 1974).
7. This is actually a very old result, which was first obtained by G. Breit, Rev. Mod. Phys. 5, 91 (1933), in the context of his theory of optical dispersion. However, its physical implications have been largely ignored, on the ground that "it is a non-observable transient effect", or some other arguments along the same lines.
8. H. Metcalf and W. Phillips, Optc. Lett. 5, 540 (1980).

SPECTROSCOPIC APPLICATIONS OF PHASE CONJUGATION

Murray Sargent III[§]

Max-Planck-Institut für Quantenoptik

D-8046 Garching, Federal Republic of Germany

Phase conjugation is an important process[1] that inverts a phase front in space so that it retraces the path through which it came. This can be accomplished either with "rubber mirrors" or with nonlinear optics. It has useful applications in propagation through turbulent media, through bad optics, and through optical fibers. As such, the method interests people in astronomy, military weapons, laser induced fusion, and optical communications. Alternatively, we can turn the technique around to use it to study properties of the conjugating medium. In this paper, we outline this last application, using nonlinear optical techniques. We consider the propagation of two, three, and four-wave electromagnetic fields through "single-photon" two-level media and through two-photon multilevel media. We consider cw fields at first, allowing later treatment of pulsed fields by careful application of Fourier analysis. The approach provides various ways of measuring dipole (T_2) and level (T_1) lifetimes, Stark shifts, and other parameters characterizing the responses of media.

The nonlinear techniques we consider all involve "pump" waves of arbitrary intensity (see Fig. 1) interfering with weak signal (or probe) waves in a nonlinear medium. The fringe pattern created by the interference between the signal and the downward-moving pump wave in Fig. 1 induces a grating, which scatters that pump's energy into the oncoming path of the signal. This same grating scatters the upward-moving pump wave backward with respect to the signal. The signal is, thus, retroreflected. The backward-moving wave is called a conjugate wave, since the signal's spatial factor $\exp(i\mathbf{K}_1 \cdot \mathbf{r})$ is conjugated, giving $\exp(-i\mathbf{K}_1 \cdot \mathbf{r})$. If the signal is described by a superposition of **K** vectors, i.e., a complicated phase front, the whole phase front is inverted or conjugated.

[§]Permanent address: Optical Sciences Center, University of Arizona, Tucson, AZ 85721. Work supported in part by the Alexander-von-Humboldt Stiftung, and in part by the United States Office of Naval Research under contract N00014-81-K-0754.

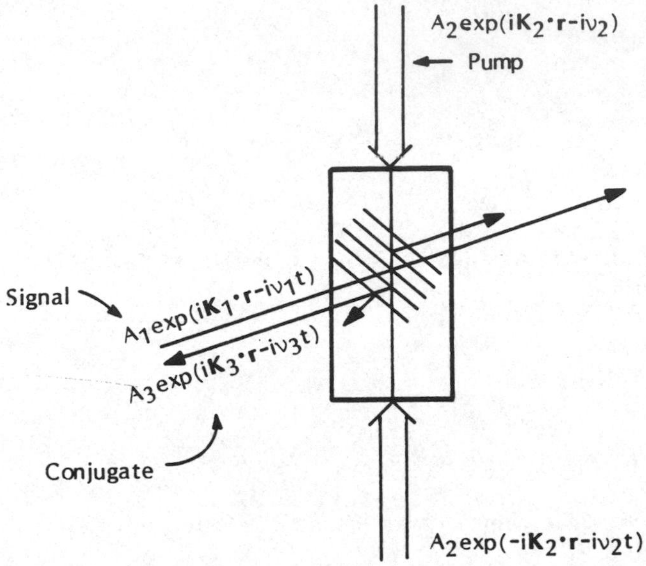

Fig. 1. Two and four-wave interactions in a nonlinear medium. One or two pump waves with amplitude A_2, frequency ν_2, and wave vectors $\mathbf{K}_2$ a $-\mathbf{K}_2$ interfere with weak signal (A_1) and conjugate (A_3) waves to induce gratings in the nonlinear medium. These gratings scatter the pump waves into the signal and conjugate wave paths.

For a simple two-wave interaction, the transmitted signal-wave amplitude A_1 is given by Beer's law, $A_1(z) = A_1(0)\exp(-\alpha_1 z)$, where the absorption coefficient α_1 is a nonlinear function of the pump amplitude A_2. For four-wave interactions, the signal and conjugate waves propagate according to a generalization of Beer's law known as the coupled-mode equations

$$dA_1^*/dz = -\alpha_1^* A_1^* + i\kappa_1 A_3 \tag{1}$$

$$dA_3/dz = \alpha_3 A_3 + i\kappa_3 A_1^*. \tag{2}$$

Provided the coefficients do not depend on A_1 and A_3, these equations can be solved in closed form as appropriate superpositions of terms like $\exp(\mu z)$. Transmission and reflection coefficients are easily derived from these general solutions.

Spectroscopy enters when we allow the signal frequency ν_1 to differ from the pump frequency ν_2. The corresponding fringe patterns then begin to "walk", with velocity $(\nu_1-\nu_2)/(K_1+K_2)$. If they walk too fast for the medium to respond, the gratings are washed out and no pump-wave scattering occurs. This produces marked changes in the Beer's-law absorption coefficient for two-wave interactions (see grating-dip spectroscopy[2,3]), and kills the conjugate-wave generation.[4] In a two-level medium, the limiting time constant is T_1, the decay time of the population difference. If $|\nu_2-\nu_1|>1/T_1$, the gratings are reduced substantially in size ($\kappa_n\to 0$). Hence one can measure T_1 using this fact. The reflection and transmission spectra look very different depending on the relative sizes of T_1 and the dipole lifetime T_2. If $T_2 << T_1$, the spectra are dominated by T_1. In this case, T_2 can be determined by transmission spectra with beat frequencies on the order of $1/T_2$. Associated with the moving gratings are population pulsations for the individual atoms. Essentially, the atomic populations act like detectors with finite bandwidths, and try to follow the pump-signal beat frequency. When this frequency exceeds their bandwidths ($\simeq 1/T_1$), the populations see only an average field, and no longer pulsate. The pulsations act, in turn, like modulators, putting sidebands on the pump. One of these sidebands oscillates at the signal frequency, one at the conjugate frequency. Still another way to look at these effects is in terms of the dynamic Stark effect. For that, one diagonalizes the Hamiltonian including the pump-wave interaction energy. In the present paper, we use the old energy eigenvalues and therefore encounter the pulsations.

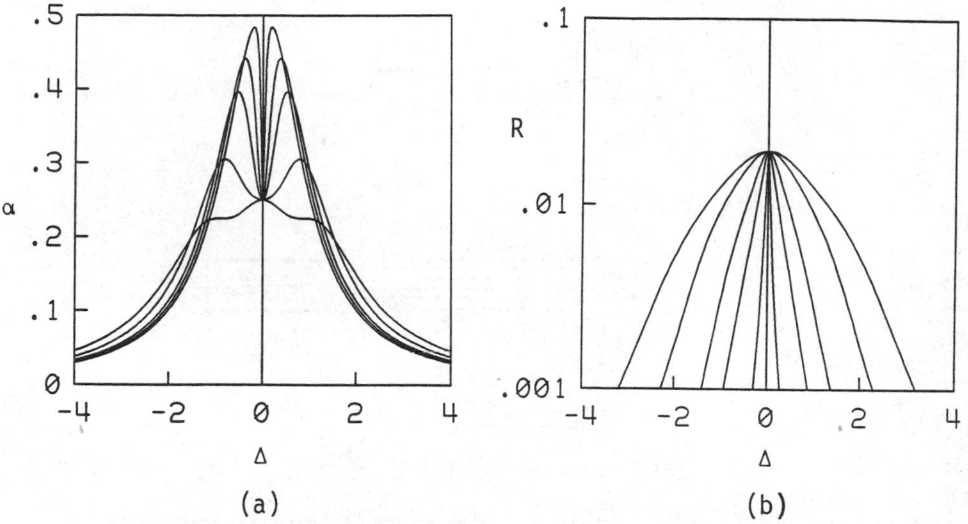

Fig. 2. (a) Two-wave absorption coefficient vs pump-signal detuning $\Delta = \nu_2 - \nu_1$ for pump intensity $I_2 = 1$ (equal to the saturation intensity) and $T_1/T_2 = 1, 2, 4, 8, 16, 32$ (broadest "dip" to narrowest). (b) Intensity reflection coefficient vs Δ for the same set of parameters (from Ref. 4).

Figure 2a shows the absorption coefficient for two-wave interactions for various values of T_1/T_2, and Fig. 2b shows the corresponding reflection spectra for four-wave mixing. The two-wave case shows a dip with width roughly proportional to $1/T_1$, while the four-wave case shows a peak with this dependence. While the absorption coefficient has grating-independent contributions, the reflection coefficient vanishes in the absence of gratings. This is conceptually simpler, but the formulas are more complicated. In particular, the projection of the induced polarization onto the signal and conjugate directions implies an average over the spatial holes burned by the standing-wave pump, yielding square roots of complex expressions.

Another factor that enters with the four-wave case is phase matching. The induced conjugate wave polarization has spatial dependence $\exp(-i\mathbf{K}_1 \cdot \mathbf{r})$, but its frequency differs from ν_1. To see this intuitively, note that the walking grating in Fig. 1 Doppler shifts the scattered downward-moving pump wave back into resonance with the oncoming signal, i.e, $\nu_2-(\nu_2-\nu_1) = \nu_1$. However, it Doppler shifts the upward-moving pump wave away from ν_1 to the frequency $\nu_3=\nu_2+(\nu_2-\nu_1)$. This phase mismatch can also lead to a narrow reflection spectrum,[5] particularly for off-resonance (n_2) phase conjugation.

Phase fronts can also be conjugated by transmission as diagrammed in Fig. 3a. The coupled-mode equations are the same as (1) and (2) with α_3 replaced by $-\alpha_3$, and no averaging over pump spatial holes (pump is a running wave). A phase mismatch occurs for noncollinear operation. This mismatch can be avoided by using one

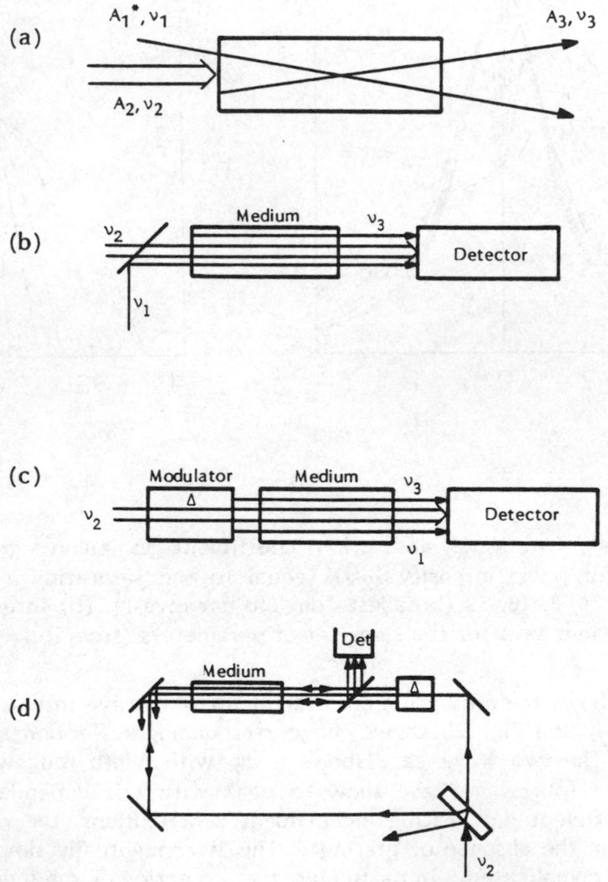

Fig. 3. (a) Three-wave transmission phase conjugation. (b) Heterodyne phase conjugation techniques used to study the response of nonlinear media. (b) and (c) are reviewed in Ref. 3; (d) was introduced in Ref. 6.

of the collinear heterodyne configurations shown in Fig. 3b, 3c, or 3d. Because one studies the intensity of the signal at the beat frequency Δ, the detector must have a bandwidth greater than Δ. If the pump wave is tuned to the atomic line center, the amplitudes A_1 and A_3 become equal, causing the coupled-mode equations to reduce

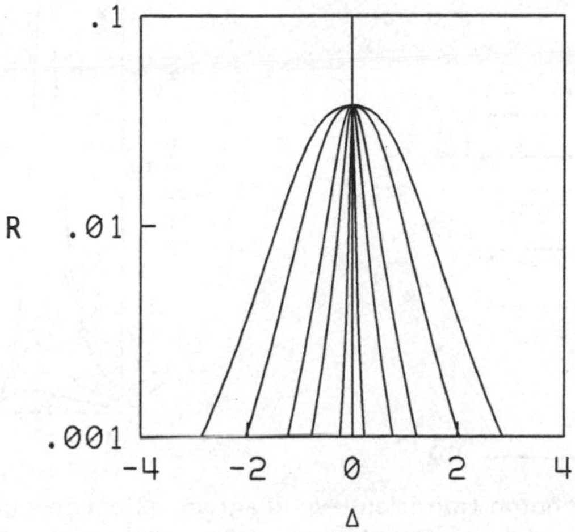

Fig. 4. Three-wave reflection coefficient vs pump-signal detuning for centrally-tuned pump and the set of parameters chosen in Fig. 2.

to Beer's law. Figure 4 illustrates the three-wave reflection coefficient for this case with the range of parameters given in Fig. 2 and assuming equal signal-conjugate amplitudes (Amplitude Modulation). The reflection bands are sharper than in Fig. 2b. This is due to the fact that the signal and conjugate both induce population pulsations that scatter pump energy into both signal and conjugate frequencies, thereby doubling the amount of scattering found in the four-wave case.

An interesting case of the three-wave modulation spectroscopy is that for Frequency-Modulation. A square-law detector sees no beat frequency. Hence the production of a beat frequency is totally due to the interaction with the nonlinear medium. This approach has been used, for example, to read out information stored by spectral hole burning,[7] and to stabilize dye lasers.[8] For pump-waves detuned from line center, one would expect that an incident FM wave would acquire AM components representative of the medium. Studies of this and various AM cases are in progress.

Given the amplitude transmission and reflection coefficients as functions of frequency, we can determine the transmission and reflection of signal pulses using Fourier analysis, provided the coupled-mode equations remain linear in the signal and conjugate amplitudes. As such we study linear deviations about a nonlinear operating point, and Fourier analysis applies. As Refs. 9 and 10 discuss, the reflection case requires an interchange between the signal and conjugate frequencies. With this in mind, one can carry out a new kind of transient laser spectroscopy using two, three, or four waves. The two and three-wave transmissions can often be solved in closed form. The four-wave case requires numerical Fourier transforms for all but the simplest choices of parameters. The technique provides us with one more way of studying relaxation rates in media.

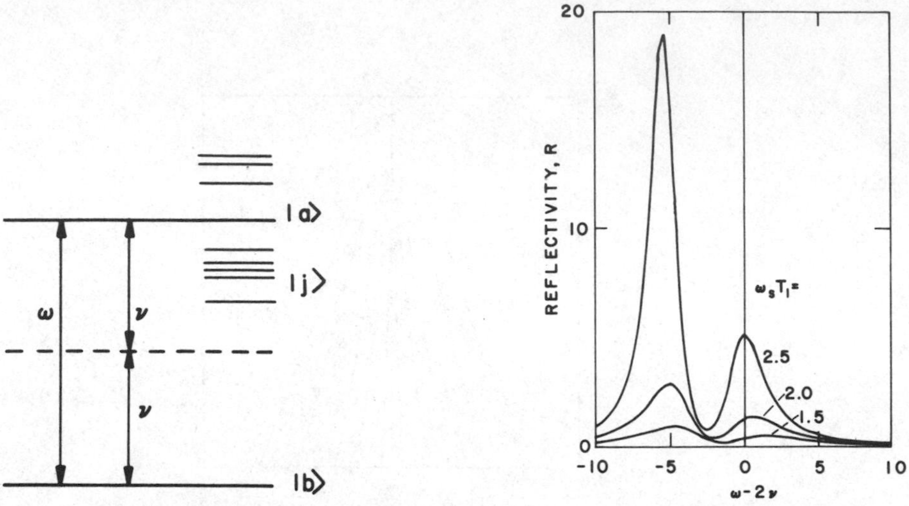

Fig. 5. (a) Two-photon transition level diagram. Electric-dipole transitions between levels a and b are forbidden, $\omega_{ab} \approx 2\nu_n$, and no allowed transition is resonant with the incident field frequencies ν_n. (b) Reflection vs pump detuning for various values of the Stark shift.

As a final example of phase conjugation for spectroscopy, we consider the two-photon transition diagrammed in Fig. 5a. The induced polarization can be determined from the steady-state solution of a "two-photon vector" model.[11,12] Specifically the polarization has the form

$$\mathscr{P}(\mathbf{r}) = \mathscr{E}(\mathbf{r})[k_{aa}\rho_{aa} + k_{bb}\rho_{bb}] + \mathscr{E}^*(\mathbf{r})k_{ab}{}^*\rho_{ab}e^{2i\nu t}. \tag{3}$$

The first two terms are just those one obtains from first-order perturbation theory, except that the populations ρ_{aa} and ρ_{bb} are functions of the incident electric fields. These terms both yield gratings oscillating at the frequency ω, and the pump reads these gratings out, as for the single-photon case. The third term involves a spatially isotropic term oscillating at ω and read out by the signal conjugated. The first two terms are pure index (Stark shift) terms, while the third has a complex Lorentzian dependence on frequency. The corresponding four-wave reflection spectrum vs pump detuning[13] reveals a double-peaked spectrum as illustrated in Fig. 5b. This is due to the interference of the third term in (3) with the first two. The process is reminiscent of the Fano spectrum discussed by Feneuille elsewhere in this volume. Recent results[14] show that pump-signal detuning decreases the reflection coefficient and reduces the depth of the valley between the peaks. These pronounced effects provide a method to measure T_1 and T_2 for this two-photon case. The theory also shows that the two-photon transition is unbleachable, unlike the single-photon case, for the Stark shift induced by an increasing pump intensity eventually exactly cancels the intensity factor in the saturation denominator.

This paper has shown some of the ways that phase conjugation can and has been used to study relaxation processes in matter. The reader is encouraged to read the references for details and further results of these approaches. A detailed review of the theory of resonant phase conjugation is in preparation.[15] It is a pleasure to acknowledge helpful discussions with Gary Bjorklund, Christian Bordé, Barry Feldman, Robert Fisher, Hyatt Gibbs, Sami Hendow, Fred Hopf, Juan Lam, Paul Liao, Pierre Meystre, and Rick Shoemaker.

References

1. For general reviews of phase conjugation, see A. Yariv, IEEE J. Quantum Electronics **QE-14**, 650 (1978); B. Ya. Zel'dovich, N. F. Pilipetskii, V. V. Reful'skii, and V. V. Shkunov, Kvantovaya Electron. Moscow **5**, 1800 (1978) [Sov. J. Quantum Electron. **8**, 8 (1978)]; and R. W. Hellwarth, IEEE J. Quantum Electron. **QE-15**, 101 (1979).
2. M. Sargent III, Appl. Phys. **9**, 127 (1976).
3. For a review, see M. Sargent III, Phys. Rep. **43**, 223 (1978).
4. T. Fu and M. Sargent III, Opt. Lett. **4**, 366 (1979).
5. D. M. Pepper and R. L. Abrams, Opt. Lett. **3**, 212 (1978).
6. Ch. Bordé, Compt. Rend. Acad. Sci. **271**, B371 (1970); J. J. Snyder, R. K. Raj, D. Bloch, and M. DuCloy, Opt. Lett. **5**, 163 (1980).
7. G. Bjorklund, Opt. Lett. **5**, 15 (1980).
8. J. Hall, see article in this volume.
9. A. Yariv, D. Fekete, and D. M. Pepper, Opt. Lett. **4**, 52 (1979).
10. R. A. Fisher, B. R. Suydam, and B. J. Feldman, Phys. Rev. A **23**, 3071 (1981).
11. D. Grishkowsky, M. M. T. Loy, and P. F. Liao, Phys. Rev. A**12**, 2514 (1976).
12. L. M. Narducci, W. W. Eidson, P. Furcinitti, and P. C. Eteson, Phys. Rev. A **16**, 1665 (1977).
13. T. Fu and M. Sargent III, Opt. Lett. **5**, 433 (1980).
14. M. Lu and M. Sargent III, to be published.
15. J. F. Lam and M. Sargent III, to be published.

CORSCAT: THE RELATIVISTIC ELECTRON SCATTERER FOR COHERENT UV AND X-RAY GENERATION

Francesco De Martini

Quantum Optics Group
Istituto di Fisica "G. Marconi", Università di Roma
Piazzale Aldo Moro, 2 - 00185 Roma, Italy

1. INTRODUCTION

The present paper deals with the process of coherent Doppler frequency upconversion of an electromagnetic field by a periodic array of relativistic charged particles. In particular, we shall focus our attention on electrons which are accelerated in conventional machines (linear accelerators, storage rings, etc.) to an energy such that a "free electron" analysis is adequate, i.e., on particles in the "ballistic" regime. In this regime collective (plasma) effects are negligible. The scattering process we are dealing with is a linear one, to first approximation, and it does not suffer from the limitations which affect the behavior of the free electron laser (FEL) at short wavelengths. It is expected that the frequency dependence of the gain of the free electron laser may prevent in the future the operation of this device at wavelengths belonging to the UV or X-ray region of the spectrum unless a new generation of accelerators with very advanced performances is developed. The process we are considering is essentially composed of two steps: the formation of a "coherent" array of electrons (bunching) and the coherent scattering by this array of a static periodic magnetic field. The two steps are considered separately emphasizing on the most critical process of bunching.

The physical principles on which the operation of the coherent relativistic scatterer (CORSCAT) is based have been considered in previous work[1,2,3,4]. They will be briefly summarized in this section as introduction to a more thorough discussion of the theory that will be presented in later sections of the paper. A coherent electromagnetic (e.m.) wave travelling in the same direction with a relativistic electron beam in an undulator ("buncher") will produce an energy modulation of the beam which is periodic with the same wavelength λ of the wave. The energy modulation is then transformed in charge density modulation after drifting of the beam through an opposite drift space. This Klystron-type bunching process[4] can be thought to consist of the mutual interference of charge density harmonic waves with wavelengths $n^{-1}\lambda$; n=1,2...; which travel with the average speed of the particles $\vec{v} = \vec{\beta}c$. We shall see later in the paper that each harmonics reaches its absolute maximum intensity at a distance from the exit of the buncher undulator depending linearly on the amplitude of coherent energy modulation $\Delta\mathcal{E} = m_0c^2\Delta\gamma$ which has been impressed on the beam. Suppose now that in the region of maximum intensity of one particular harmonics n, the beam is made to interact with another undulator ("scatterer") with a suitable intensity, polarization and wavelength λ_q of the corresponding periodic B-field structure[5,6]. An intense highly monochromatic coherent synchrotron radiation field is backscattered in forward direction with wavelength $\lambda_{sc} = \lambda/n$. We can see that if the condition of the momentum conservation (phase matching) in the coherent backscattering kinematics is satisfied by a large order harmonics, the wavelength of the scattered radiation can be very short, e.g. belonging to the UV or even X domain of the spectrum if also the excitation wavelength λ is short enough. In idealized conditions n can be very large and λ_{sc} very short. However the beam characteristics of the today accelerators impose drastic limitations on the maximum value of n which corresponds to a sizeable efficiency of the devices. This will be seen in the following sections of the paper where a detailed analysis of the electron bunching and coherent scattering processes will be presented in the single-particle approximation[7]. The results of the theory will be applied to the discussion of a numerical example corresponding to a detailed planning of a CORSCAT experiment to be performed with the Frascati Storage Ring "Adone".

2. BUNCHING IN THE RELATIVISTIC KLYSTRON

In the present section we apply a single-electron theory of klystron bunching in the relativistic regime to the wiggler structure of the conventional Optical Klystron. This one includes, as an essential feature, a dispersive magnetic drift space (DS). The single particle trajectory in DS will be determined analytically for the magnetic configuration of the transverse undulator which has been built for the Frascati free-electron laser facility LELA and will operate with the high energy ($\gamma = 1200$) electron beam of the "Adone" Storage Ring[5,6].

In figure 1 the sinusoidal evolution of the transverse magnetic field component B_y as function of the longitudinal coordinate $\hat{z}$ in the undulator is shown for the conventional - and for the Optical Klystron - configuration of the B-field polarization.

In the present case the change from one to the other configuration can be easily obtained by changing the direction of the electric current in the excitation windings of some of the electromagnetic poles in the middle of the undulator.

We note that in the optical klystron configuration the maximum intensity B_0 of the transverse field is the same for the three sections in which the overall undulator is subdivided: buncher, drift space, scatterer. In our case the "buncher" and the "scatterer" undulators have the same amount of magnetic periods λ_q.

In this configuration the lengths of the "buncher" and of the "scatterer" undulators in the optical klystron are the same $L_1=L_3=8\lambda_q$ while the one of the DS section is: $L = 3\lambda_q$. Further details on the undulator are reported by R. Barbini and G. Vignola in papers[5,6]. Apart from second-order corrections, the evolution of B_y along the coordinate $z' \equiv z \cdot 2\pi/\lambda_q = zk_q (k_q \equiv 2\pi/\lambda_q)$ in the "buncher" and "scatterer" wigglers may be represented by $B_y=B_o \cos z'$, while in the DS section it is represented by $B_y = + B_0$ in two sections of "lengths" $\pi/2$ and $B_y = - B_0$ in one central sections of length π. These three sections of constant B_y in DS are connected by 2 half-period cos z' functions (see Fig. 2). Since in the present paper we are mainly interested in the evolution of the harmonic content of the bunched beam in DS and in the scatterer wiggler, let us assume that the coordinate $z' = 0$ corresponds to the entrance transverse section of DS (Fig. 2). In order to include in the calculations the effect of angular spread, let us assume that, at $z' = 0$, the velocity $\vec{v} \equiv \vec{\beta}c$ makes an angle $\varphi \ll 1$ with $\hat{z}$!

The above B-field distribution determines the trajectory of the electron in the plane $(\hat{x}-\hat{z})$ as a solution of the dynamical equation $d\vec{p}/dt=e/c(\vec{v}\wedge\vec{B})$ (Ref. 7). This can be solved analytically for the given B-field distribution in DS. The lengths of the sections of the electron trajectory corresponding to zones of different functional z'-dependence of B_y in DS are:

a) $l_1 = (\sin^{-1}(h\pi - \sin\varphi) + \varphi))(hk_q)^{-1} \sim (\pi - (h\pi - \varphi)^3/(6h))k_q^{-1}$

(in zones with $|B_y| = B_o$, length $\pi/2$)

1)

b) $l_2 = \pi(1 + h^2((a^2/2) + (1/4) + (2a/\pi))(k_q^{-1})$

(in zones with $B_y = B_o \cos z'$, length $\pi/2$)

where $h = (K_o/\gamma) << 1$, K_o (strength parameter) $= \dfrac{eB_o\lambda_q}{2\pi m_o c^2}$, $a = (\pi - \dfrac{\sin\varphi}{h})$.

The total trajectory length in DS results to be:

$$l = 4l_1 + 2l_2 = 3\lambda_q(1 + 3,5\,h^2 + 0,5\,\varphi^2)$$

for $L = 3\lambda_q$ [8].

The dispersion properties of the magnetic drift space which lead to a large acceleration of the "bunching" process in DS are due to the $B_o^2 \cdot \gamma^{-2}$ dependence of the second term in the parenthesis. On the other hand, this and the third term in the parenthesis make the device very efficient in dephasing the electron wavefunctions when stochastic energy spread, $(\delta\gamma/\gamma) \neq 0$, and angular spread effects, $\delta\varphi \neq 0$, are present. These effects result in a reduction of the harmonic content of the beam, as we shall see later in the paper.

The evolution of the bunching process along $z' \equiv k_q z$ will be analyzed in the single-particle approximation. In the actual case we are considering (LELA experiment[5,6]) this is justified by the "ballistic regime" of the ultrarelativistic particles of the "Adone" beam (γ = 1200) (Ref. 3). Furthermore, the effects of the energy - and angular-spreads will be taken into account in the evaluation of the corresponding reducing factors of the final harmonic content of an idealized beam (i.e. with $\delta\varphi = 0$ and $(\delta\gamma/\gamma) = 0$ for all particles). If s(t) is the length of the one electron trajectory starting at the entrance section of DS, x = 0, z' = 0, at the initial time $t = t_o$, $s(t_o) = 0$, the subsequent time is $t = t_o + (s/\beta c)$, where βc is the modulus of the particle velocity which is not perturbed by the B-field static distribution. After the particle has passed through the "drift space" (i.e. $z' \gtrsim k_q L$) we can write $s/\beta c = 1/\beta c + k_q^{-1} \int_{k_q L}^{z'} (v_z)^{-1}\, dz$ where $\tau = (4\tau_1 + 2\tau_2)$ is given by Eqs. 1 in which we set $\varphi = 0$ at $z' = 0$.

For sake for simplicity we may consider the trajectory length of the particle which has passed through N periods of the "scatterer" undulator. Accordingly, the upper limit of the integral can be written as $z' = k_q L + N$. Assuming an oscillatory motion of the electron in the transverse (x, z') plane in the "scatterer" wiggler, the integral can be written in the form:

$$K_q^{-1} N \int_0^{2\pi} (1 - h^2 \sin^2 z')^{-1/2} dz' \cong N\lambda_q \left(1 + \frac{h^2}{4} + \frac{9h^4}{8^2} + \dots\right)$$

Consider only the first order approximation term of the power expansion of the elliptic integral[9]. We can write the expression of the time t for $\varphi = 0$ and $L = 3\lambda_q$, in the form[10]:

$$t = t_0 + (\lambda_q/\beta c)\{(3+N) + (K_0/\gamma)^2 (10{,}50 + N/4)\} \qquad 2)$$

Assume that a coherent harmonic modulation with frequency $\omega = 2\pi c/\lambda$ of the particle energy, $\mathcal{E} = m_0 c^2 \gamma$, is induced in the "buncher" section of the Optical Klystron by the interaction of the beam with a laser field[1,11] i.e. $\gamma = \gamma_0 + \Delta\gamma \cos \omega t_0$ for $\gamma_0 = (1-\beta^2)^{-\frac{1}{2}}$ and βc = average velocity of the particles.

The steady state electric current associated with the electron stream at time t is related to the one at time t_0 by the equations:

$$I = e\,\frac{dn}{dt} = e\,\frac{dn}{dt_0}\,\frac{dt_0}{dt} = I_0 \{1 + \eta \sin \omega t_0\}^{-1} \qquad 3)$$

where $I_0 = e\, dn/dt_0$ is the unperturbed beam current at s=0, $t = t_0$ and (dt_0/dt) is derived from 2).

The "bunching parameter" $\eta \equiv -\frac{2\pi}{\lambda}\, \mathcal{E} \left(\frac{\partial s}{\partial \mathcal{E}}\right)\left(\frac{\Delta\gamma}{\gamma}\right)$

can be written in our case in the form: $\eta = 4\pi\,(42+N)(\Delta\gamma/\gamma)$

having assumed λ_q = 11,6 cm, γ = 1200, K_0 = 4,93 (Ref. 5,6) and a value of the bunching wavelength corresponding to the II[d] harmonic of the Nd-YAG laser output radiation: $\lambda = 0{,}53\mu$.

The Fourier expansion of expression 3) yields the corresponding charge density in the form:

$$\rho(z',t) = \rho_0\left\{1 + 2\sum_{n=1}^{\infty} J_n(n\eta)\cdot\cos\left[n(\omega\tau + \pi/2)\right]\right\} \qquad 4)$$

where $\tau = t - (z'/k_q v_z)$. The effect of bunching results in the development along z' of harmonic charge density waves with wave-vectors $q_n = n\omega/v_z$. The intensity of the nth harmonic is the Bessel function $J_n(n\eta)$ which reaches its first and absolute maximum at a value of η that monotonically decreases from 1,89 (n = 1) to the asymptotic value 1 ($n \gtrsim 20$) (Ref. 9 page 126 ff.). We note the very interesting properties of the dispersive magnetic drift space. Eq. 2 and the expression of η show that in our case (i.e. $L=3\lambda_q$, maximum value of the B-field, $B_y = B_0$, equal for the three sections of the overall undulator) each period of the B-field in DS is about 15 times more efficient in determining the value of η than a corresponding period in the scatterer. This has two important physical consequences: a) the maximum intensity of the harmonics can be reached in a rather short path ($\sim$L) with a reasonable value of $\Delta\gamma/\gamma$ i.e. of the bunching laser intensity. b) once a certain value of J_n is established at the entrance of the scatterer wiggler, it keeps nearly constant through this wiggler if this one is not too long. This is very useful situation which provides the largest scattering efficiency in the operation of a FEL or of a CORSCAT. Equation 4) is valid for an idealized beam. If we want to take into account the energy and angular spreads of the beam we may evaluate the corresponding reduction factors δ_ε and δ_φ

$$\delta_\varepsilon \equiv (\sqrt{2\pi}\,\sigma_\varepsilon)^{-1}\cdot\int_{-\infty}^{+\infty} \exp(-(\Delta\varepsilon)^2/(2\sigma_\varepsilon^2)).\cos(2\pi n\Delta\varepsilon\left|\frac{\partial s}{\partial\varepsilon}\right|/\lambda).d\Delta\varepsilon =$$

$$= \exp(-(n^2/2).(2\pi\delta_\varepsilon\left|\frac{\partial s}{\partial\varepsilon}\right| / \lambda)^2).$$

$$\delta_\varphi \equiv (\sqrt{2\pi}\,\sigma_\varphi)^{-1}\cdot\int_{-\infty}^{+\infty} \exp(-\varphi^2/(2\delta_\varphi^2)).\cos(\pi nL\varphi^2/\lambda).d\varphi =$$

$$=((1+b)/(2b^2))^{1/2}.$$

$$\Delta\varepsilon \equiv \Delta\gamma\, m_0 c^2.$$

where δ_E, δ_φ are the r.m.s. deviations of the energy and angle spreads of the beam and the arguments of the cos functions are the corresponding phase deviations due to trajectory length spreads in DS, $b \equiv [1+(2\pi n L \delta_\varphi^2/\lambda)^2]^{1/2}$. In actual cases involving the typical parameters of a storage ring we find $\delta_\varphi \ll \delta_E$. Note that an efficient generation of higher order harmonics requires an increasingly lower value of $\sigma_E \propto n^{-1}$.

Let us give a numerical example corresponding to the CORSCAT that is designed for the LELA facility in Frascati[7,12]. Assume that the LELA undulator (Fig. 1) is divided symmetrically in 3 sections. The beam of the "Adone" storage ring with $\gamma = 1200$ enters in the first undulator section of length $8\lambda_q = 8 \times (11{,}6$ cm) superimposed with a high power laser beam with wavelength $\lambda = 0{,}53\mu$ (II^d harmonic of Nd-YAG laser radiation). The bunching develops in a drift section $L = 3\lambda_q$ and in a third section of length $8\lambda_q = L_W$: the "scatterer" wiggler. The wiggler parameter that establishes the resonance condition in the "buncher" is $K_o = 4{,}8$ corresponding to approximately a value of field $B_o = 5$ KG. This value of B_o is common to all sections of the undulator. Suppose that we are interested in the coherent scattering of the third harmonics of the fundamental bunching wavelength: $\lambda_{sc} = \lambda/3 = 1766$ Å.

Suppose also that we want to reach the maximum efficiency for scattering from the harmonics n = 3 by placing the absolute maximum of $J_3(3\eta)$ in the middle of the "scatterer" wiggler i.e. N=4. Since the value of η for that condition is 1,4 (Ref. 8), the amount of coherent modulation of the beam is $\Delta\gamma/\gamma = 1{,}4 \cdot [2\pi(42+4)]^{-1}$ In the "buncher" undulator of length $8\lambda_q$ the linear interaction of the beam with the laser field $E = E_o \sin\omega(t - z/c)$ gives rise to an harmonic modulation with relative amplitude[1,2,3]: $\Delta\gamma/\gamma = eE \alpha\, 8\lambda_q K_o/(2m_o c^2\gamma^2)$ The previous spatial condition for maximum bunching at N = 4 corresponds to a laser intensity I = 331 Mw/mm^2 over the transverse section of the electron beam[5]: $\sigma \equiv \pi d_e^2/4 = 2{,}8 \cdot 10^{-1}$ mm^2 The harmonics reduction factors δ_E and δ_φ are evaluated on the basis of Eqs. 5 in which the charateristic parameters of the "Adone" machine are inserted: $\sigma_E = 2{,}3 \cdot 10^{-4} E_o; \sigma_\varphi = 1{,}5 \cdot 10^{-4}$ rad We find: $\delta_E = 0{,}57$ and $\delta_\varphi = 0{,}98$ for n = 3.

3. COHERENT SCATTERING

The peak current and the number of electrons per burst that are effective for coherent scattering at the 3^d harmonics are respectively: $I_{(3)} \equiv \mu_3 I_o = 2I_o \cdot J_3(3\eta_{max}) \cdot \delta_E \delta_\varphi$ and $N_{eff} =$

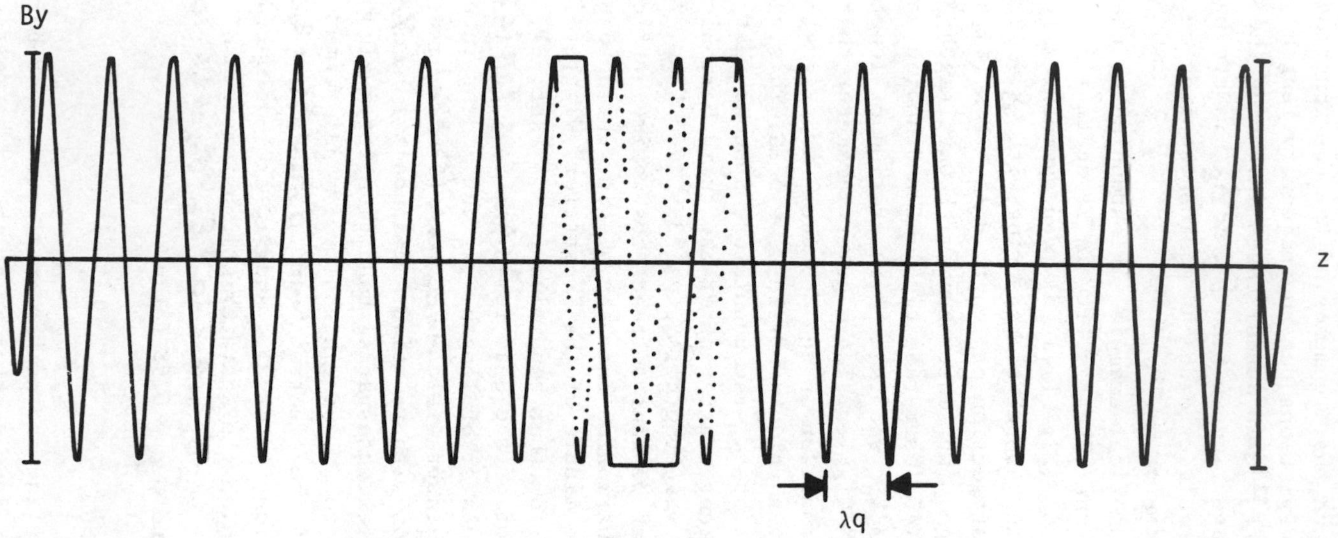

Fig. 1 - Evolution of the transverse component of the static magnetic field in the LELA undulator (Ref. 5 and 6). The "normal" configuration of $B_y(z)$ (dotted line) can be changed into the Optical Klystron configuration (full line). In the latter case the three parts of the undulator (buncher, drift space (DS), scatterer) are clearly identified.

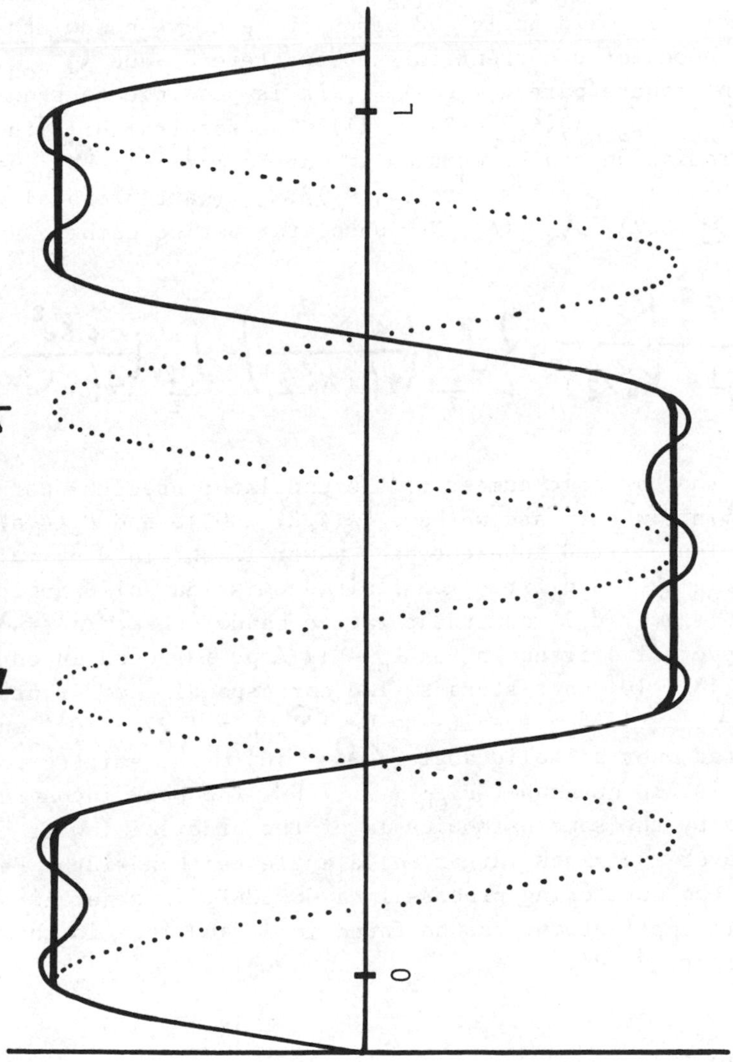

Fig. 2 - Evolution of B_y in the drift space (DS) corresponding to the plot of Fig. 1. The single-particle trajectory has been determined analytically in the paper by smoothing the ringing parts of B_y in the "flat" portions of the diagram.

$= \mu_3 N_{tot}/\sqrt{2}$ where I_o and N_{tot} are the unmodulated peak current and total number of electrons associated with each electron burst circulating in the Storage Ring. Assuming a gaussian shape for the bursts with HWHM duration $\Delta\tau$ = 1 nsec, we find: I_o = 17 A and $I_{(3)}$ = 8,2 A [9]. If $\mathcal{P}_{inc} \equiv (dP_{inc}/d\Omega \cdot d\nu/\nu)$ is the peak incoherent power per solid angle and per unit relative bandwidth emitted by an unmodulated current in the "scatterer", and $\mathcal{P}_{coh}$ is its coherent counterpart due to $I_{(3)}$, it is possible to show that: $\mathcal{P}_{coh} = \mathcal{P}_{inc} \cdot (N_{tot} \cdot \mu_3^2/2)$ (Ref. 11)). The spectral brightness of undulator radiation can be written in the form[11,12]: $\mathcal{P}_{inc}$ = $= 3{,}44 \cdot 10^8 \cdot I_o \cdot \mathcal{E}^2 \cdot N^2 \cdot F_3(K_o)/\lambda_{sc}$ (watt/sterads) where I_o (Amps), $\mathcal{E}$ (GeV), λ_{sc} (Å), N = undulator period number, and:

$$F_i(K_o) = \frac{i^2 K_o^2}{(1+K_o^2/2)^2} \cdot \left\{ J_{\frac{i-1}{2}}\left(\frac{i K_o^2}{4(1+K_o^2/2)}\right) - J_{\frac{i+1}{2}}\left(\frac{i K_o^2}{4(1+K_o^2/2)}\right)\right\}$$

where i is the harmonic number of the undulator spectrum corresponding to λ/n. In our case we have $F_3(4,8) = 0{,}15$ and $F_3(4,8) =$ $= 0{,}20$[12]. The emitted coherent peak power is obtained by multiplying $\mathcal{P}_{coh}$ by the diffraction limited emission solid angle $\Delta\Omega = 1{,}16\ (\lambda_{sc}/d_e)^2$ and the relative bandwidth $\Delta\nu/\nu = \lambda_{sc}/c\Delta\tau$. The peak spectral brigthness for I_o = 17 Amp. is found in our case: $\mathcal{P}_{inc} = 1{,}13 \cdot 10^7$ watt/sterads. The corresponding coherent peak power at λ_{sc} = 1766 Å results to be $\mathcal{P}_{coh} = 1{,}27 \cdot 10^{17}$ watt/sterads, emitted over a solid angle $\Delta\Omega = 10^{-7}$. The emitted total peak power is, in our case, P_{coh} = 7, 7 KW. The peak incoherent power emitted by the same harmonics is of the order of 1 watt and distributed over a much larger solid angle and bandwidth. Further details on the scattering process in a CORSCAT, in general and in our specific application, can be found in[11] and in a forthcoming "ad hoc" papers[13,14].

4. CONCLUSIONS

Although it is clear that very high harmonics will be more difficult to obtain, it seems feasible on the existing apparatus in Frascati to observe the third harmonic generation of light by an electron beam subject to a strong laser pulse.

If one considers reversing 5 poles instead or 3 in the Adone undulator, the power necessary for bunching is much lower ($\sim$50 MW/ /mm^2) and still considerable 3rd harmonic generation at λ_{sc}=1766Å can be obtained: $P_{coh} \simeq$ 1 KW (in this case $\delta_\varepsilon^2 \simeq 0.2$).

The necessary laser power is even lower if we produce the 3rd harmonic from a beam that has been modulated at the Nd-YAG laser wavelength: λ= 1,06 μ. In this case, if we wish to keep the resonant condition with the same undulator at the same value of K_o, we should work with a beam energy corresponding to γ = 850.

We acknowledge useful discussions with R. Coisson, R.H. Pantell, J. Edinghoffer and D. Deacon. This work has been supported in part by NATO Res. Contract RG 213-80 and Centro Ricerche FIAT S.p.A.

5. REFERENCES

1. N.A. Vinokurov and A.N. Skrinsky, Preprints INP 77.59 and 77.67 Novosibirsk 1977.
2. G.A. Kornyukhin, G.N. Kulibanov, V.N. Litvinenko, N.A.Mezentsev, A.N. Skrinsky, N.A. Vinokurov and P.D. Voblyi, Nucl. Instr. and Meth. 177, 247 (1980).
3. F. De Martini, in Physics of Quantum Electronics, Vol. 7, Edited by Jacobs, Pilloff, Scully, Sargent, Spitzer, Addison Wesley, Publ. Co. Inc. Readings, Mass. 1979.
4. F. De Martini and J.A. Edinghoffer in Proc. of the Course "Physics and Technology of free electron Lasers", Erice, Italy, 17-29 August 1980. In course of publication. Similar effects have been considered recently by: V. Stagno, I. Boscolo, G. Brautti and T. Clauser, Nuovo Cimento 56B, 219 and 58B, 267 (1980) and by: R. Coisson, Part. Accelerators 11,4 1981.
5. R. Barbini and G. Vignola, Report LNF-80/12 (R), March 10 (1980)
6. R. Barbini, A. Cattoni, B. Dulach, C. Sanelli, M. Serio and G. Vignola, Report LNF-80/62 (P), December 9 (1980)
7. J.D. Jackson, Classical Electrodynamics, Wiley, New York, 1975 page 547.
8. In the case $L = 5\lambda_q$ the total length of the trajectory in DS results to be: $l = 5\lambda_q \{1+10(K_o/\gamma)^2 + 0,5\,\varphi^2\}$.
9. E. Jahnke and F. Emde, Tables of functions, Dover N.Y. (1945) p. 52.
10. The present theory is valid for a length L_1 of the "buncher" section which is short compared to the coordinate z of maxi-

mum bunching when this one is established by spontaneous drifting of the particles along $\hat{z}$ in a non dispersive drift space (i.e. $B_y = 0$ in DS). In our case we can verify "a posteriori", that this is indeed verified for the LELA undulator and for values of ($\Delta\gamma/\gamma$) corresponding to a sizeable bunching obtained in the actual magnetic undulator. A non relativistic single-particle theory of bunching in a non dispersive drift space is given in: J.C. Slater, Microwave Electronics, Van Nostrand, Princeton (1950). Page 222. In the present paper the amplitude of the coherent harmonic energy modulation ($\Delta\gamma/\gamma$) should not be confused with the incoherent energy spread ($\delta\gamma/\gamma$).

11. F. De Martini and R. Coisson, Proc. of ONR workshop on free--electron lasers, Sun Valley, Idaho, June 1981 (to be published).
12. European Synchrotron Radiation Facility Suppl. II; ESF Strasbourg 1979, pages 52-65.
13. F. De Martini and M. Foresti - Submitted for publication.
14. S. Baccaro, F. De Martini and A. Ghigo, to be published in Optics Letters.

POSSIBLE NEW APPROACH TO FREE-ELECTRON LASERS

(Note after the Lecture by F. DeMartini)

V.S. Letokhov

There is one more possibility of light amplification due to the kinetic energy of free electrons. It is based not on amplifying the amplitude of a light wave as it interacts with electrons but on increasing its frequency on account of the Doppler effect caused by reflection of the laser pulse from a relativistic pulsed beam of high-density electrons. This effect may be called as amplification of the energy of a light pulse during its reflection from a relativistic mirror. For example, with the electron energy

$$E_{el} = m_0 c^2 / (1-\gamma^2)^{1/2} \simeq 1 \mathrm{MeV} \; (\gamma = v/c \simeq 0.8)$$

the reflected pulse wavelength of CO_2 laser $\lambda_0 = 10$ µm can be shifted to a region with $\lambda \cong 1$ µm. The critical density of electron cloud for $\lambda = 1$ µm $N_{cr} \cong 10^{21}$ cm^{-3}.

This technique may prove useful for applications in powerful multistage amplifiers of CO_2 laser pulses as the last most powerful amplification step when, for example, the energy of a nanosecond laser pulse should be increased from 10^4 J to 10^5 J and over with a corresponding shift of laser light wavelength to the short-wave region. The total energy of electron beam, of course, must be an order higher, i.e. it must lie in the region of 10^6 J and over to prevent electron beam decay during reflection of a powerful laser pulse.

It is very difficult to obtain good divergence as light is reflected from an electron cloud with a poor mirror surface but, nevertheless, this problem can be solved by using distributed reflection from the electron cloud under the conditions of four-wave nonlinear interaction (in the presence of an additional well-collimated laser beams) with phase conjugation. From this point of view, of particular significance are nonlinear processes of interaction of powerful lasers radiation with dense electron beams or also with relativistic plasma bunches.

INDEX